The Atlas of Mars

Mapping Its Geography and Geology

Planetary scientist and educator Ken Coles has teamed up with Ken Tanaka from the United States Geological Survey's Astrogeology team and Phil Christensen, Principal Investigator of the Mars Odyssey orbiter's THEMIS science team, to produce this all-purpose reference atlas, *The Atlas of Mars*. Each of the 30 standard charts includes: a full-page color topographic map at 1:10,000,000 scale, a THEMIS daytime infrared map at the same scale with features labeled, a simplified geologic map of the corresponding area, and a section describing prominent features of interest. The *Atlas* is rounded out with extensive material on Mars' global characteristics, regional geography and geology, a Glossary of Terms, and an indexed Gazetteer of up-to-date Martian feature names and nomenclature. This is an essential guide for a broad readership of academics, students, amateur astronomers, and space enthusiasts, replacing the NASA atlas from the 1970s.

Kenneth S. Coles is Associate Professor and Planetarium Director at Indiana University of Pennsylvania. An award winning teacher, he has dedicated his career to sharing planetary science and geology discoveries with university students, schoolchildren, and the public.

Kenneth L. Tanaka is a Geologist retired from the US Geological Survey Astrogeology Science Center in Flagstaff, Arizona. He has 35 years of experience in the geologic mapping of Mars, informing NASA's exploration missions, and has received the US Department of Interior's Distinguished Service Award.

Philip R. Christensen is Regents' Professor of Planetary Geoscience at Arizona State University. He is the Principal Investigator for the Mars Odyssey THEMIS instrument and has received the Geological Society of America's G. K. Gilbert Award, NASA's Exceptional Scientific Achievement Medal, and NASA's Public Service Medal.

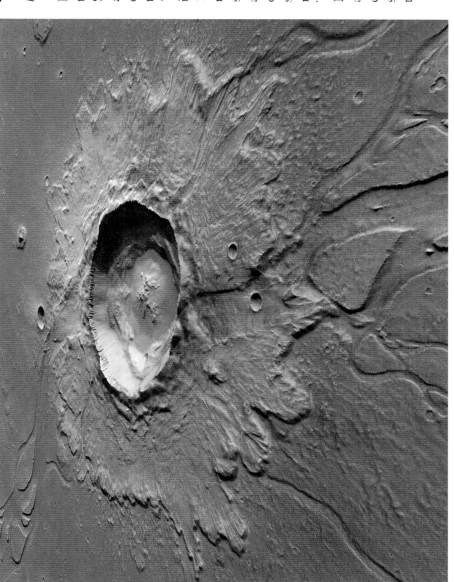

Half title image caption: An impact crater, about 15 km across, shows ejecta emplaced by fluidized flow during the impact. The crater overlies channels of Hebrus Valles. While Hebrus may appear to have been caused by the impact, streamlined islands within channels on both sides of the crater record flow in the same direction, consistent with channel origin prior to the impact. The view is toward the south in Utopia Planitia at 20° N, 126° E (see map MC-14 in this atlas; HRSC orbit 5122, 16 m/pixel, ESA/DLR/FU-Berlin).

Title image caption: The western hemisphere of Mars shown in two datasets. The Mars Orbiter Laser Altimeter (MOLA) representation in the lower left and upper right portions of the globe shows elevation (by color) and topographic relief (by artificial illumination from upper left; see Figure 3.2b in Chapter 3 for a more detailed explanation). Colored units from the Geologic Map of Mars are merged with the MOLA relief view and displayed in a swath from upper left to lower right (further explanation accompanies Figures 5.1a–d in Chapter 5). The enormous Olympus Mons and the three aligned Tharsis Montes are immense shields that dominate the Tharsis volcanic region to the west (largely red). Heavily cratered terrain lies to the east, including the Argyre impact basin (the blue-green area at bottom). The extensive Valles Marineris canyon system, just below and left of center, forms part of the Thaumasia plateau's northern edge. Emanating to the north and east of Valles Marineris are broad outflow channels that cut through lava plains and cratered terrain. The channels extend into the northern lowlands (blue region), wherein the Viking 1 and Mars Pathfinder spacecraft landed. At top, the north polar ice cap, Planum Boreum, rises above the surrounding lowlands. (View centered at 20° N, 300° E; grid spacing 30 degrees.)

The Atlas of Mars

Mapping Its Geography and Geology

KENNETH S. COLES
Indiana University of Pennsylvania

KENNETH L. TANAKA
United States Geological Survey

PHILIP R. CHRISTENSEN
Arizona State University

With contributions from

James M. Dohm
University Museum, University of Tokyo

Corey M. Fortezzo
United States Geological Survey

Trent M. Hare
United States Geological Survey

Jonathon R. Hill
Arizona State University

James A. Skinner, Jr.
United States Geological Survey

CAMBRIDGE
UNIVERSITY PRESS

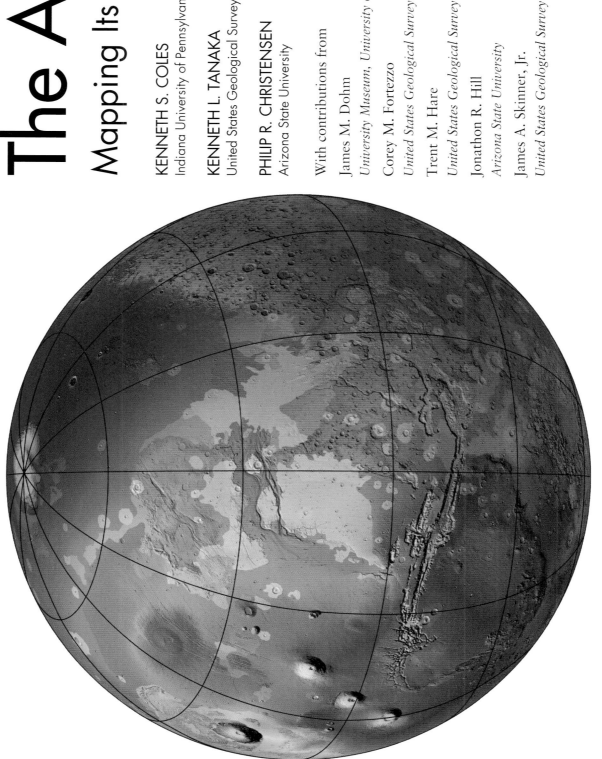

CAMBRIDGE
UNIVERSITY PRESS

University Printing House, Cambridge CB2 8BS, United Kingdom

One Liberty Plaza, 20th Floor, New York, NY 10006, USA

477 Williamstown Road, Port Melbourne, VIC 3207, Australia

314–321, 3rd Floor, Plot 3, Splendor Forum, Jasola District Centre, New Delhi – 110025, India

103 Penang Road, #05–06/07, Visioncrest Commercial, Singapore 238467

Cambridge University Press is part of the University of Cambridge.

It furthers the University's mission by disseminating knowledge in the pursuit of education, learning, and research at the highest international levels of excellence.

www.cambridge.org
Information on this title: www.cambridge.org/9781107036291
DOI: 10.1017/9781139567428

First published 2019
Reprinted 2021

Printed in Singapore by Markono Print Media Pte Ltd

A catalogue record for this publication is available from the British Library.

Library of Congress Cataloging-in-Publication Data
Names: Coles, Kenneth S., 1957– creator. | Tanaka, Kenneth L., author. |
   Christensen, Philip R. (Philip Russel), author.
Title: The Atlas of Mars : mapping its geography and geology / Kenneth S. Coles,
   Kenneth L. Tanaka, Philip R. Christensen.
Other titles: Mapping its geography and geology
Description: Edition 1. | New York : Cambridge University Press, [2019] | "Kenneth S. Coles,
   Kenneth L. Tanaka and Philip R. Christensen 2019."
Identifiers: LCCN 2018058698 | ISBN 9781107036291 (hardback)
Subjects: LCSH: Mars (Planet)–Maps. | Mars (Planet)–Geology–Maps. | Mars (Planet)–Geography. |
   Mars (Planet)–Research–History–20th century. | LCGFT: Atlases.
Classification: LCC G1000.5.M3 C6 2018 | DDC 559.9/230223–dc23
LC record available at https://lccn.loc.gov/2018058698

ISBN 978-1-107-03629-1 Hardback

Additional resources for this publication at www.cambridge.org/atlasofmars

Table of Contents

Preface

The decade that followed the arrival of the Viking landers and orbiters on Mars in 1976 saw the release of a global geologic map (Batson *et al.*, 1979),[1] a new global geologic map (Scott and Tanaka, 1986; Greeley and Guest, 1987; Tanaka and Scott, 1987), and even a digital image of the entire surface (the Mars Digital Image Model, or MDIM). Kieffer *et al.* (1992b) summarized the major conclusions of this work. The stage was set for a series of spacecraft missions to Mars that began in the 1990s and continues today. Every one of the realizations about Mars in the pages that follow is a discovery in progress. They continue even as we write. Since the information we present will undoubtedly continue to grow and change rapidly, why this atlas now?

This book comes about as a result of a confluence of new information and understanding about Mars. The global imagery is more detailed than in the past, and a small percentage of the surface can be studied at the level of individual boulders and gullies. Mapping from new data has greatly improved our knowledge of the geology of the entire planet compared to the last synthesis, a generation ago. The relative dating of units and events, particularly by crater counting, is also more refined. Sensors that gather infrared and radar data now tell us the composition and mineralogy of the surface and the structure beneath the surface, not seen in visible light. Details of subsurface water ice have emerged, not only in polar regions but all over Mars. Observations on the surface by increasingly sophisticated landers and rovers have confirmed some ideas and discounted others.

Some aspects of Mars exploration have been covered in detail elsewhere:

- Collections of great pictures (many published, with more issued each year);
- Histories (in detail) of pre-space age ideas about Mars (e.g. Sheehan, 1996);
- Histories of the robotic exploration of Mars (e.g. Morton, 2002; Hubbard, 2011; Stooke, 2012, 2016);
- Detailed analyses of findings by individual landers (e.g. Mishkin, 2003; Squyres, 2005) or orbiting instruments (e.g. Beyer *et al.*, 2012);
- Arguments for (or against) Mars exploration (e.g. Zubrin, 2011);
- The geology of Mars, presented in the framework of geologic time (in the book by Hartmann, 2003, which we commend to the reader); and
- The atmosphere and climate of Mars (Haberle *et al.*, 2017).

Our focus here is a synoptic view of geology: how geology controls or influences some typical landforms, how the environment has changed over the history of Mars, and current debates and outstanding questions in Mars research. In sum, this overview of what we know and what we don't know will suggest what is important to study next on the red planet.

While some of this will be of use to active Mars researchers, we particularly hope to reach other readers: scientists from other fields, interested non-scientists, and persons who wonder what all the missions to Mars have told us. It is an astonishing assortment of facts, ideas, and most of all entire new questions and mysteries to motivate further work. But then, why should the study of the fourth planet differ from any other field of human inquiry?

Acknowledgments

Staff at the justly renowned US Geological Survey (USGS) Astrogeology Science Center gave input and suggestions from the first ideas for this atlas through to the final drafts. While we couldn't incorporate every idea they gave us, we appreciate their willingness to share ideas and point us to useful images and other resources: Jen Blue, Colin Dundas, Ken Herkenhof, Chris Isbell, Randy Kirk, Baerbel Lucchita, Moses Milazzo, David Portree, Larry Soderbloom, and Tim Titus. Also at the USGS, Rose Hayward checked the nomenclature maps and Gazetteer for completeness and Chris Okubo reviewed the entire manuscript.

Images were generously created or provided for our use by Vic Baker, Donald Barker, Serina Diniega, Henrik Hargitai, Rose Hayward, Jack Holt, Eric Peterson, Than Putzig, and Alexis Rodriguez. Additional valuable ideas came from Nick Deardorff, Robert Jacobsen, Gregory Michael, and Thomas Platz. The many researchers who have been studying Mars for over half a century and building and operating robotic missions have both inspired this work and made much of it possible. Many of their ideas remain useful and have held up well under prolonged study. We would like to acknowledge the staff (past and present) who have collected and processed images and data of Mars, as well as those who have made them very accessible and useful, at the USGS, NSSDC (NASA), JPL-Caltech, THEMIS (Arizona State University), HiRISE (University of Arizona), HRSC (Freie Universität Berlin), and MSSS, along with others listed in the credits for each image. The collections, staff, and facilities of the US Geological Survey Library, the Indiana University of Pennsylvania Libraries, and the Logansport-Cass County Public Library were essential to the research for this atlas.

A key figure throughout this effort has been Vince Higgs at Cambridge University Press. He first proposed that this work be based on the new geologic map of Mars, and he and his staff patiently and professionally shepherded us through the entire process, in spite of how lengthy it became. We are sure we speak for all the contributing authors as we offer gratitude to the teachers who trained us, the colleagues and students who work with us, and the loved ones who support us as we worked in a focused way on a time-consuming but gratifying project.

We dedicate this book to Ron Greeley (1939–2011), who pioneered the application of geologic and cartographic mapping in planetary science. Ron helped bring rigorous mapping standards to planetary geology, and he inspired a generation of scientists to apply geologic principles and techniques to study the planets and moons in our solar system.

K. S. C.
K. L. T.
P. R. C.

---

[1] A revision of the 1979 atlas is listed in one online bibliography as having been assigned the designation NASA Special Publication SP-506. Neither the authors nor the archivist at the US Geological Survey Astrogeology office can find any evidence that such a revision was ever completed or published.

Abbreviations

CNES (*Centre National d'Études Spatiales*) National Centre for Space Studies, space agency of France

CNRS (*Centre National de la Recherche Scientifique*) National Centre for Scientific Research, France

CRISM Compact Reconnaissance Imaging Spectrometer for Mars, on MRO

CTX Context Camera, on MRO

DEM Digital elevation model

DLR (*Deutsches Zentrum für Luft- und Raumfahrt*) German Aerospace Center, space agency of Germany

ESA European Space Agency

ExoMars Exobiology on Mars, series of ESA missions (as of this writing, the first mission is the Trace Gas Orbiter [TGO] mission)

GSFC Goddard Space Flight Center of NASA

GRS Gamma Ray Spectrometer, on MO.

HiRISE High Resolution Imaging Science Experiment, on MRO

HRSC High Resolution Stereo Camera, on ESA Mars Express orbiter

IAS (*Institut d'Astrophysique Spatiale*), Institute of Space Astrophysics, France

ISRO Indian Space Research Organisation, the national space agency of India

JHUAPL Johns Hopkins University Applied Physics Laboratory

JPL Jet Propulsion Laboratory, operated for NASA by the California Institute of Technology (Caltech)

MAVEN Mars Atmosphere and Volatile Evolution, NASA orbiter

MC Mars Chart, system of 30 maps, originally at 1 to 5 million scale, created for Mars in the early 1970s

MDIM Mars Digital Image Model, an image mosaic of the Mars surface derived from Viking orbiter imagery

MER Mars Exploration Rover, NASA twin missions A and B, also known as Spirit and Opportunity, respectively

MGS Mars Global Surveyor, NASA orbiter

MO Mars Odyssey, NASA orbiter

MOC Mars Orbiter Camera, on MGS

MOLA Mars Orbiter Laser Altimeter, on MGS

MRO Mars Reconnaissance Orbiter, NASA mission

MSL Mars Science Laboratory, also known as Curiosity, a NASA surface rover

MSSS Malin Space Science Systems, involved in the engineering and/or operation of MOC, CTX, and THEMIS-visible, among others

NASA National Aeronautics and Space Administration, space agency of the USA

NSSDC National Space Science Data Center, data archive for NASA

OMEGA (*Observatoire pour la Minéralogie, l'Eau, les Glaces et l'Activité*) Visible and Infrared Mineralogical Mapping Spectrometer, on ESA Mars Express orbiter

SHARAD Shallow Radar, on MRO

TES Thermal Emission Spectrometer, on MGS

THEMIS Thermal Emission Imaging System, on MO

USGS United States Geological Survey

How to Use this Atlas

Global maps of Mars are found in Chapter 2 (maps predating exploration by spacecraft), Chapter 3 (showing various properties or abundances, see also Figures 15.E and 23.B), Chapter 4 (regional geographic features in Figures 4.1 and 4.2), and Chapter 5 (geology, Figures 5.1 through 5.6, 5.9, and 5.11). **Thirty map sheets** showing physiography, geology, and feature names follow Chapter 5, as shown in the accompanying index map below.

Elevation on many of the maps is depicted by the colors shown here.

The names and correlation of the **geologic map units** used on the geologic maps are summarized in

the Appendix, while Chapter 5 describes the units in more detail.

The **figures** in the first five chapters are denoted by chapter number, figure number in that chapter and, where needed, a lower-case letter indicating the figure part. For example, Figure 3.1 and Figures 3.2a and 3.2b are the first three figures in Chapter 3.

In the map sheets the figures are denoted by Map Chart (MC) number and an upper case letter; sometimes a number follows the upper case letter. Thus Figure 3.A accompanies MC-3, Arcadia, and Figures 8.A1–8.A5 accompany MC-8, Amazonis.

- By geologic period: use Figure 5.9 together with the index map of Mars Charts below to locate maps showing units of a particular age.
- By region: use the index map of Mars Charts below with Figures 4.1 and 4.2 to find the appropriate numbered map(s).

To find features

- By known name: turn to the Gazetteer, or for informal names see the Index.
- By feature type: use the Index to find examples (see the Glossary of Terms for definitions).

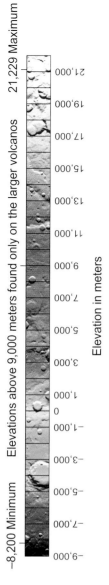

−8,200 Minimum   Elevations above 9,000 meters found only on the larger volcanos   21,229 Maximum

−9,000 −7,000 −5,000 −3,000 −1,000 0 1,000 3,000 5,000 7,000 9,000 11,000 13,000 15,000 17,000 19,000 21,000

Elevation in meters

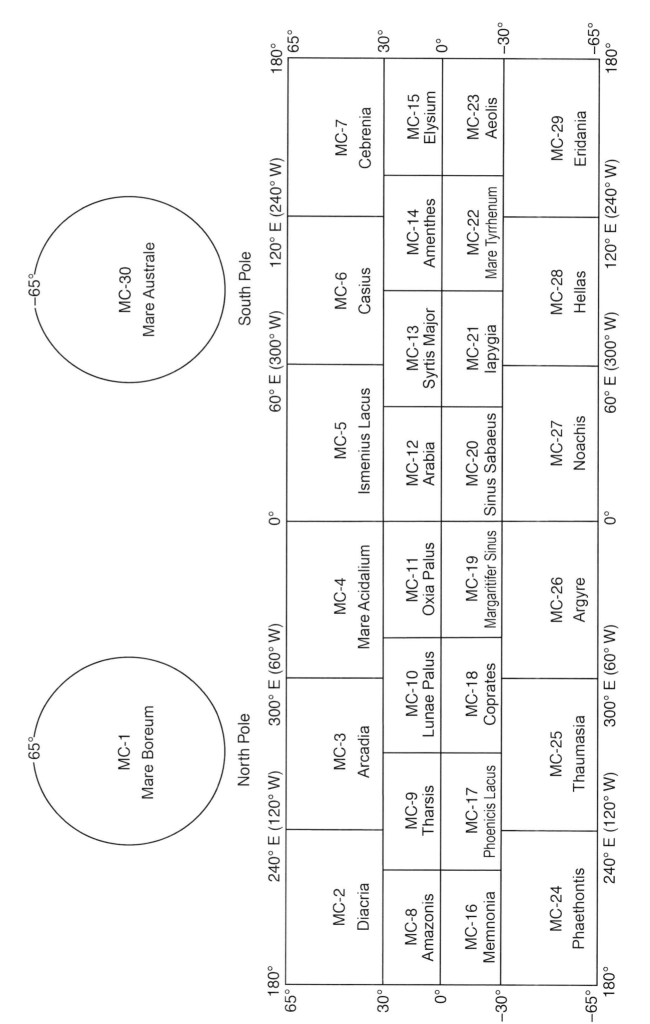

North Pole

MC-1
Mare Boreum

South Pole

MC-30
Mare Australe

Sources of Images

Individual images from particular instruments or provided by colleagues are credited in captions. Some images (particularly using the MOLA dataset) were created by the authors for this work. Details for all these sources are given below. The web page for this atlas, www.cambridge.org/atlasofmars, includes links to some of the most useful sites that give access to these and other images of Mars.

CRISM (Compact Reconnaissance Imaging Spectrometer for Mars): NASA/JPL-Caltech/Johns Hopkins University Applied Physics Laboratory

CTX (Context Camera): NASA/JPL-Caltech/Malin Space Science Systems (see Malin et al., 2007; Bell et al., 2013)

HiRISE (High Resolution Imaging Science Experiment): NASA/JPL-Caltech/University of Arizona (see McEwen et al., 2007a)

HRSC (High Resolution Stereo Camera): European Space Agency/Deutsches Zentrum für Luft- und Raumfahrt (German Aerospace Center)/Freie Universität Berlin (Free University of Berlin; see Jaumann et al., 2007; 2015; Gwinner et al., 2016)

MDIM (Mars Digital Image Model) derived from Voyager Orbiter imagery, Version 2.1 from NASA Ames Research Center/USGS

MOC (Mars Orbiter Camera): NASA/JPL-Caltech/Malin Space Science Systems (see Malin et al., 1991)

MOLA (Mars Orbiter Laser Altimeter): Except where otherwise noted, images were created using the MOLA color and/or MOLA hillshade datasets from NASA/JPL-Caltech/Goddard Space Flight Center (see Smith et al., 2001)

OMEGA (Observatoire pour la Minéralogie, l'Eau, les Glaces et l'Activité): ESA/Centre National d'Études Spatiales/Centre National de la Recherche Scientifique/Institut d'Astrophysique Spatiale/Université Paris-Sud, Orsay (see Ody et al., 2012)

SHARAD (Shallow Radar): NASA/JPL-Caltech/Sapienza University of Rome/Southwest Research Institute

TES (Thermal Emission Spectrometer): NASA/JPL-Caltech/Arizona State University (see Christensen et al., 2001)

THEMIS (Thermal Emission Imaging System): NASA/JPL-Caltech/Arizona State University (see Christensen et al., 2004)

CHAPTER 1
Introduction

Mars is the fourth planet from the Sun and the outermost of the rocky, terrestrial planets that make up the inner solar system. Mars is the second smallest planet; only Mercury is smaller. Surface gravity on Mars is 3.71 m s^{-2}, which is 37.6% that of the Earth. The present atmospheric pressure is low (~0.6 kPa) relative to Earth's (101 kPa), and the atmosphere is mostly carbon dioxide (95%). The obliquity of Mars (tilt of the axis of rotation relative to the plane of orbit) is presently 25 degrees and may have varied by tens of degrees over the past tens of millions of years and longer (Laskar *et al.*, 2004).

The rotational period (sidereal) of Mars is slightly longer than that of Earth, at 24 hours 37 minutes, while the Martian solar day (sol) is 24 hours 40 minutes. The orbit of Mars is more elliptical (eccentricity = 0.093) than that of Earth (0.017), so the seasons vary in length with northern spring being the longest. The sidereal period is 687 days (670 Martian sols), while the synodic period relative to Earth, and the interval between oppositions of Mars, averages 780 days. The progress of the seasons is commonly given by L_s, which is the orbital longitude, or angle in degrees at the Sun measured from the northern hemisphere spring equinox to the position of Mars. Thus, L_s = 90° is the summer solstice in the northern hemisphere and the winter solstice in the south (Carr, 2006 gives additional detail on the motion of Mars).

Organization of atlas; map scale and projections

The chapters following this one give the reader a short summary of the history of spacecraft exploration of Mars (Chapter 2), global views of datasets (Chapter 3), the geography of Mars (Chapter 4), and the geology of Mars (Chapter 5).

The heart of the atlas considers the geography and geology found in each of 30 maps. We follow the Mars Chart (MC) system of quadrangle[1] maps that were first created for the Mariner 9 images in the early 1970s. These maps cover Mars in 30 sheets, originally at a nominal scale of 1:5,000,000, though published maps vary from this scale when matching along map boundaries is a priority (Batson *et al.*, 1979). We have instead maintained a consistent numerical scale of 1:10,000,000, using Mercator projection scaled at the equator for sheets at 0–30° latitude, in Lambert conformal projection for the sheets covering 30–65° latitude scaled on standard parallels of 36.15° and 59.47°, and polar stereographic scaled at 90° latitude for the two polar sheets covering 65–90° latitude.

Each sheet is first presented in a pair of facing maps. The first is in color to show elevations in the MOLA (Mars Orbiter Laser Altimeter) dataset with a minimum of annotation beyond a latitude and longitude graticule every 10 degrees. The map margins indicate the numbers of adjacent map sheets. The second map shows the surface using the THEMIS (Thermal Emission Imaging System) daytime infrared dataset, with named

---

[1] We follow a common usage in geology, where the term *quadrangle* denotes any map sheet, even those not having four corners.

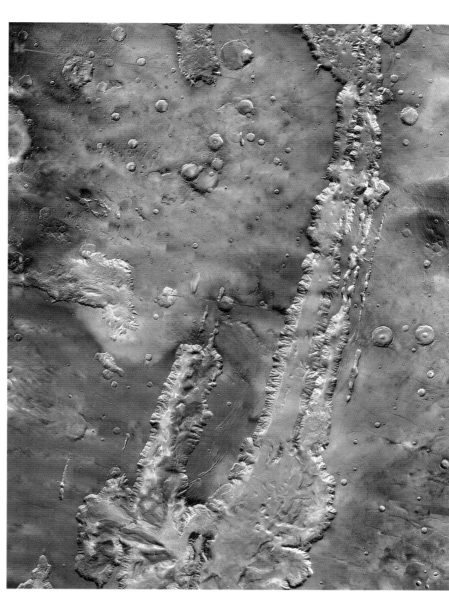

FIGURE 1.2 The same region as Figure 1.1 is shown in a mosaic of THEMIS daytime infrared images. A thermal infrared image is bright where the surface is warmest. The warm regions in an infrared image are those that either retain heat or are dark and absorb more sunlight. This correspondence between warm in THEMIS and dark in the visible is strikingly evident in Figure 1.1. While daytime infrared images from THEMIS, which we use extensively in this atlas, cover all of Mars and have great detail, this comparison is a reminder that visible and infrared images show different properties of the Martian surface (NASA/JPL-Caltech/Arizona State University).

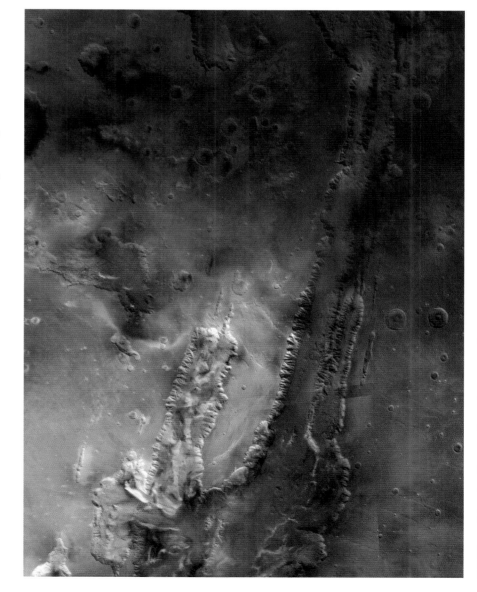

FIGURE 1.1 Mosaic of images from the Mars Orbiter Camera (MOC-WA) view taken through a red filter. The view shows the northeastern portion of Valles Marineris and adjacent plana (NASA/JPL-Caltech/MSSS).

features (Figures 1.1 and 1.2 show how infrared and visible images of Mars differ). For the sake of clarity, we generally omit names of minor features, where crowded, and of craters smaller than 20 km in diameter on the maps, though they are included in the Gazetteer in plain (non-bold) font. Albedo features (defined solely by amount of reflected light) are not included on the maps or in the Gazetteer. The location map and other images (except where noted) showing elevation derived from MOLA data in this atlas use the "rainbow" color scheme of the first MOLA release (Smith *et al.*, 1999).

The third page of each map sheet includes summaries of the geography and geology. Also shown is a color MOLA topography index map with boxes locating the images featured in one or more following pages. The geology (generalized from Tanaka *et al.*, 2014) is shown at 1:20,000,000 scale as well. The image location map uses the elevation color scheme of the first, overview, map, while a brief legend for the geologic map is shown (the full list of units is given in the Appendix). The accompanying images and text for each map describe prominent, unique, or enigmatic features with the intention of covering each important feature type and process at least once among the 30 map sheets. Some of the images are well-known classics while others were chosen by searching all those available that cover a given area.

Throughout this work we aimed to strike a balance between listing every possible reference to published scientific work and omitting references entirely. The goal is a small but representative list of sources to enable the reader to learn more. While we cannot include all of the thousands of Mars publications that come out each year or all the explanations put forth to explain what we see, or don't see, on Mars, we acknowledge this work and commend it to the reader who wishes to seek it out.

The atlas concludes with an Appendix, which lists the units of the various geologic maps, unit conversions, and Latin descriptors, a brief Glossary of Terms, as well as a Gazetteer, which gives basic data about named features and lists the map(s) where they are shown.

Creation of THEMIS base maps

Each quadrangle of a previous THEMIS daytime infrared global mosaic (Hill *et al.*, 2014), which was composed of THEMIS band 9 (12.57 μm) images, was visually reevaluated, image by image. Images that reduced the quality of the overall mosaic were removed and replaced by higher-quality images as follows (Hill and Christensen, 2016).

1. Images with poor geometry, where offsets with adjacent images were evident (due to extrapolations over gaps in spacecraft position and pointing telemetry, etc.), were removed and replaced by images with better geometry data. The previous THEMIS global mosaics, as well as this one, do not attempt to tie images to ground points, though offsets are usually small relative to the 100 m/pixel resolution of the mosaic.

2. Images with significant noise resulting from low surface temperatures, atmospheric effects, etc. have been replaced by higher-quality images where possible.

3. Small gaps due to line dropouts in various images were filled with the highest-quality images available.

4. Coverage was extended poleward of 87.3 degrees north and south latitude, the limit of the Mars Odyssey orbit, with images taken off-nadir.

Coordinates on Mars

Latitude and longitude

Coordinates for Mars exist in two different systems: planetographic latitude with longitude increasing to the west, and planetocentric latitude with longitude increasing to the east. The definition of latitude also differs, being either the angle between the equatorial plane and a vector directed at the point of interest measured at the center of mass of Mars (planetocentric), or the angle between a vector perpendicular to a (nonspherical) reference surface at the point of interest and the equatorial plane (planetographic; Planetary Data System, 2009). Both systems are

approved for use on Mars by the International Astronomical Union (IAU) (Seidelmann *et al.*, 2002).

Planetographic latitude with west longitude was the system used for most maps and publications prior to 2002. Planetocentric latitude with east longitude has been adopted for use by the USGS and other organizations for use in making more recent Mars maps and imagery (Duxbury *et al.*, 2002). The maps and descriptions in this atlas follow this recent practice and use planetocentric coordinates. Converting longitude between the systems requires only simple arithmetic, but the latitude conversion is more complex. The nomenclature maps at the USGS Gazetteer of Planetary Nomenclature website (https://planetarynames .wr.usgs.gov/) show both coordinate systems for conversion between them.

Elevation and datum definition

Elevations on Mars are derived from MOLA measurements of radius relative to the center of mass of the planet. An equipotential surface derived from these measurements is chosen to match the mean equatorial radius at the equator (3396 km) and extended to other latitudes (Smith *et al.*, 1999; 2001). This surface incorporates the effects of the local gravity field and the rotational flattening of the planet; the surface is commonly referred to as "datum" in this atlas. This serves as a reference for positive and negative elevations on Mars much as mean sea level (which is also an equipotential surface) does on Earth. Chapter 4 discusses the surface elevation of Mars in terms of regional geography.

Geographic nomenclature

Planetary nomenclature, like terrestrial nomenclature, is used to uniquely identify a feature on the surface of a planet or satellite so that the feature can be easily located, described, and discussed. Names are intended to be descriptive rather than imply a specific origin. The USGS Gazetteer of Planetary Nomenclature (the basis for names in this atlas, https://planetarynames .wr.usgs.gov/) contains detailed information about

all names of topographic and albedo features on planets and satellites (and some planetary ring and ring-gap systems) that the IAU has named and approved from its founding in 1919 through to the present time.

Note on Latin terms

Feature names on Mars are commonly given in Latin form. Terms for types of features (descriptors) are listed in the Appendix of this atlas. While the naming and spelling is systematized, in our experience the pronunciation of feature names (as for Latin in general) within the scientific community is not. We have not attempted to reconcile these variations; one attempt to provide guidance is that of Hargitai and Kereszturi (2010).

Image resolution – how much detail do we see?

Spacecraft images of Mars, which now number over two million, consist of picture elements, or pixels, recorded and returned to Earth in digital form. Each pixel represents an area (typically a square) with a uniform level of gray or color. When pixels are displayed as an image, features about four pixels across are the smallest that can be resolved. For images we report the size shown by one pixel. For example, an image with a pixel size of 6 m would reveal features down to about 24 m in size. Although we do not display the finest detail in printed versions of all images, the available resolution may be useful information for readers wishing to consult the original images. In general descriptions of spacecraft instruments, the term resolution in this atlas serves as a synonym for pixel size.

An example from Earth illustrates the effect of smaller pixel size on the resolution of features. Figure 1.3 shows a portion of the Earth's surface with 4-km pixels. While ridges and valleys are evident, little other detail is seen.

Compare this to Figure 1.4, where the increase in resolution is five-fold, to 800 m per pixel. It is easier to recognize that the view includes one of the most spectacular valleys on Earth. The Grand Canyon might not be recognized on the first, coarser image if it were the only information from another planetary body.

A number of cameras have produced the images of Mars used in this atlas; the names and common abbreviations for each are given in Chapter 2 (see Table 2.1, which gives typical pixel sizes and the amount of Mars covered). In image captions where applicable we give coordinates, dimensions of the area shown, and the north direction.

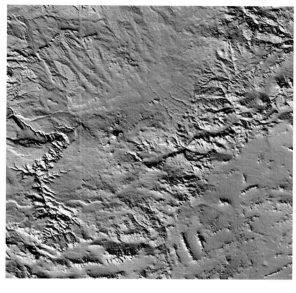

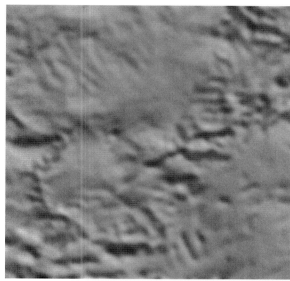

FIGURE 1.3 A portion of Earth's surface depicted using a digital elevation model (DEM), artificially illuminated from the west (blurred to the equivalent of 4-km pixels, view about 450 km by 450 km, north at top, DEM from US Geological Survey).

FIGURE 1.4 The same area and dataset as Figure 1.3, shown using 800-m pixels, i.e. five times better resolution. The view shows part of Arizona in the southwest USA, extending from the Grand Canyon in the north to Phoenix in the south.

CHAPTER 2

History of Exploration of Mars

Mars has attracted study ever since its motions were first apparent to ancient skywatchers. Summaries of early observations and ideas are listed in, e.g., Collins, 1971; Hartmann and Raper, 1974; Moore, 1977; Kieffer *et al.*, 1992a; Martin *et al.*, 1992; Sheehan, 1996; Morton, 2002. Hubbard (2011) gives an interesting example of the planning of Mars missions.

Pre-spacecraft studies

Although Mars is Earth's second closest planetary neighbor (after Venus), the detail in telescopic observations since Galileo's, in 1609, has been limited by the small size of Mars and the effects of Earth's atmosphere (Slipher, 1962; Kieffer *et al.*, 1992a; Martin *et al.*, 1992).

Schiaparelli and Lowell: The "canals"

Reports of "canali" (Italian for channels; commonly "canals" in English), especially by Schiaparelli in Italy (starting in 1877) and Lowell in the USA, starting in the 1890s, influenced ideas about Mars for nearly a century (Figures 2.1a and 2.1b; Hartmann and Raper, 1974; Kieffer *et al.*, 1992a). Despite a number of claims by these and other visual observers (Sheehan, 1996), no convincing photographic evidence was presented, and eventually spacecraft found no traces of canals. The canals or channels

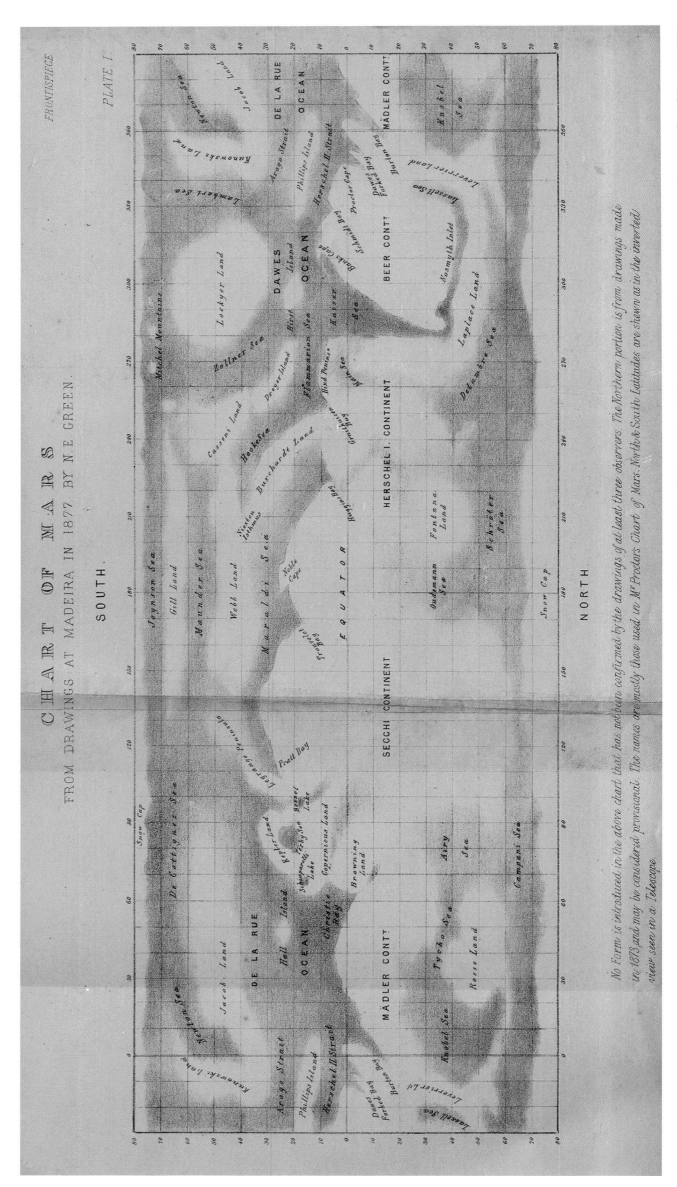

FIGURE 2.1 (a) Map of Mars markings ("albedo"), by English amateur astronomer and artist Nathaniel Green in 1877. While the names used by Green did not persist in the later work of others, he was rather careful about the limits of observation and doubted the so-called canals (map as reproduced in Ledger, E., 1882, *The Sun, Its Planets and Their Satellites*: London, courtesy H. Hargitai/T. Lindemann).

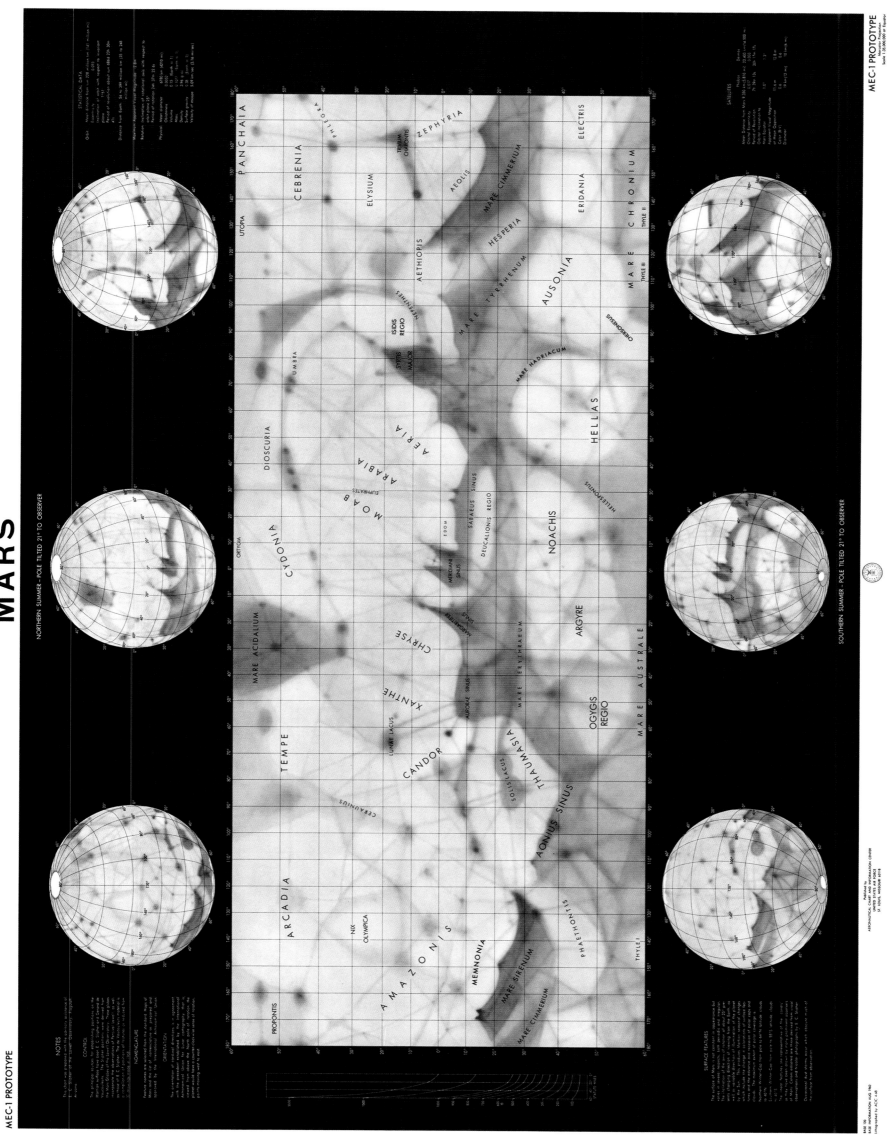

FIGURE 2.1 (b) Map of Mars albedo prepared by the US Air Force Aeronautical Chart and Information Center in 1962, primarily from the maps of E. C. Slipher at Lowell Observatory. Note the fine, canal-like features (courtesy of Lunar and Planetary Institute).

Pre-spacecraft studies

The Atlas of Mars

5

evidently were an imagined perception rather than a real feature.

Telescopic observation

Studies of Mars from telescopes on Earth are insufficient to discern the planet's geologic nature, but they do yield information about dust storms in the atmosphere and seasonal changes in the polar caps. Albedo markings (areas of light and dark; see Figure 3.1 in Chapter 3) are readily apparent and change over time, though the consistency of seasonal patterns in these changes is a matter of debate (Hartmann and Raper, 1974; Martin et al., 1992). By the time of the first spacecraft missions, maps of Mars emphasized these major markings (Figures 2.1a and 2.1b; Martin et al., 1992; Morton, 2002).

First Mars spacecraft

A large number of robotic spacecraft have been launched toward Mars. Orbiters and landers and results from them that are noteworthy in the mapping and geologic study of the planet are summarized here. Progress in the coverage and resolution of imaging is summarized in Table 2.1 and illustrated in Figures 2.2 and 2.3. For views from surface landers and rovers see Chapters 4 and 5.

Mariner 4 and Mariner 6/7

The first evidence of the nature of the Martian surface came from observations by the Mariner 4 spacecraft, in 1965. Images covering about 1 percent of Mars at pixel size near 1 km showed a heavily cratered surface similar to the ancient highland terrain on the Moon (Figure 2.2a; Leighton et al., 1965). Mariner 6 and 7, a pair of flyby spacecraft in 1969, imaged 10 percent of Mars at a resolution similar to Mariner 4 and gave a similar impression of the surface. A few images did show featureless plains, regions termed "chaotic" (imaged by a later mission in Figure 2.3a), and the south polar cap (Leighton et al., 1969; NASA, 1969; Collins, 1971; Snyder and Moroz, 1992).

Mariner 9

The greatest single advance in knowledge of Mars was due to the Mariner 9 mission of 1971–72. This

revolution in thinking about the geology of Mars (Hartmann and Raper, 1974; NASA, 1974; Hartmann, 2003) and set the stage for the planned landings of the ambitious Viking program.

Soviet spacecraft

Among a number of attempted Soviet missions to Mars are several noteworthy achievements. The lander vehicle of the Mars 3 mission apparently landed successfully on the surface in December 1971, and broadcast radio signals for less than a minute before falling silent. The companion

spacecraft, the first to orbit another planet, made images of the entire surface that resolved features of 2–3 km over the course of one Earth year (NASA, 1974; Jaumann et al., 2007). It also observed the gravity field and character of, and variations in, the atmosphere (Snyder and Moroz, 1992). Among the many discoveries were Moon-like cratered highlands, immense volcanoes, the enormous Valles Marineris canyon system, numerous channels and branching valley networks, suggesting erosion by a moving fluid such as water, evidence of dust transport by wind, and layered deposits near the polar caps. Mariner 9 spurred a

orbiter spacecraft returned a number of images, as did the Mars 4 flyby and Mars 5 orbiter (both in 1974; Perminov, 1999; Mitchell, 2004). Phobos 2 entered Mars orbit (in 1989) and returned a number of images and other observations of Mars and of Phobos, but contact was lost before a planned close approach to Phobos (Sagdeev and Zakharov, 1989).

Viking 1 and 2

The Viking 1 and 2 spacecraft, two orbiters and two landers, arrived at Mars in 1976. Both

TABLE 2.1 Spacecraft imaging of Mars used in this Atlas

Mission	Per cent of Mars covered	Brightness levels	Pixel size at close encounter/periapsis	Spectral bands	Lines × samples
Mariner 4 (1965)	1	64	1.25 km	One (two color filters)	200 × 200
Mariner 6/7 (1969)	100[a] 10[b,c] <1[b,d]	256	4–43 km[a] 1 km[b,c] 100 m[b,d]	One (three color filters)[c] One[d]	704 × 945
Mariner 9 (1971–72)	100[c] 1–2[d]	512	500 m[c] 50 m[d]	One (eight color or polarizing filters)[c] One[d]	700 × 832
Viking 1/2 Orbiters (1976–80)	100 <1	128	150–300 m 8 m	One (six color filters)	1,056 × 1,182
Mars Global Surveyor – MOC (1997–2006)	100[c] 5[d]	256	240 m[c] 1.5–12 m[d]	Two[c] One[d]	var. × 3,456[a] var. × 2,048[b]
Mars Odyssey – THEMIS (2002–)	65 100	256	18 m (visible) 100 m (infrared)	Five visible Ten infrared	1,024 × 1,024 (visible) 240 × 320 (infrared)
Mars Express – HRSC (2004–)	95 (at mean pixel size of 18 m)	256	10–20 m	Four spectral bands, three bands for three dimensional images	var. × 5,184
Mars Reconnaissance Orbiter – HiRISE (2006–)	~3	16,000	0.3 m	Three	var. × 20,264 (red) var. × 4,048 (green and infrared)
Mars Reconnaissance Orbiter – CTX (2006–)	>99	4096	6 m	One	var. × 5,064

For Mariner 6/7, Mariner 9, Mars Global Surveyor, MOC: a = far encounter, b = near encounter, c = wide-angle camera, d = narrow-angle camera . "Brightness levels" gives the number of shades of gray coded in the image. "Lines" gives the number of rows of (generally square) pixels in an image, while "samples" is the number of pixels in a line. The number of lines varies for some instruments. Sources: Bell et al., 2013; Carr and Evans, 1980; Christensen et al., 2004; Collins, 1971; Gwinner et al., 2015; Jaumann et al., 2007; Leighton et al., 1965; 1969; Levinthal et al., 1973; Malin et al., 1991; Masursky et al., 1970; Masursky, 1973; NASA, 1967; 1969; 1974; 2017; Snyder and Moroz, 1992; Spitzer, 1980; Tanaka et al., 1992; also mission information from HiRISE, HRSC, MSSS, NSSDC.

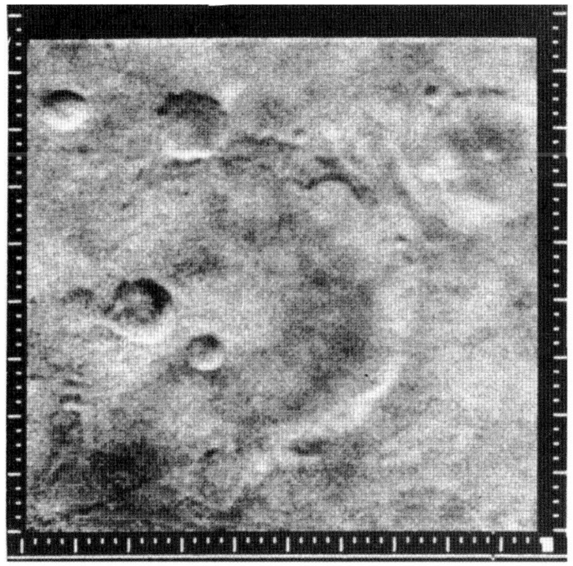

(a)

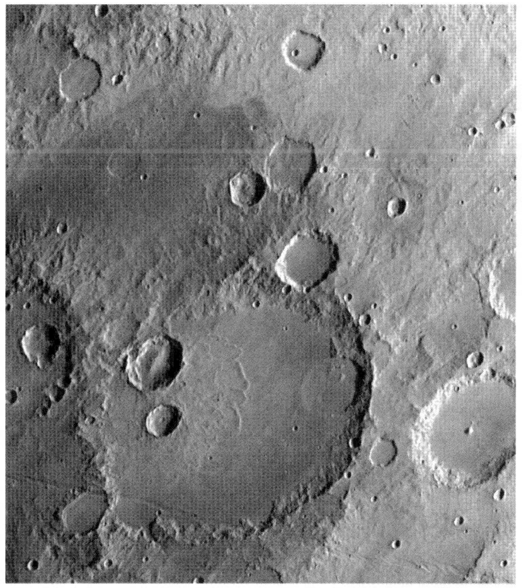

(b)

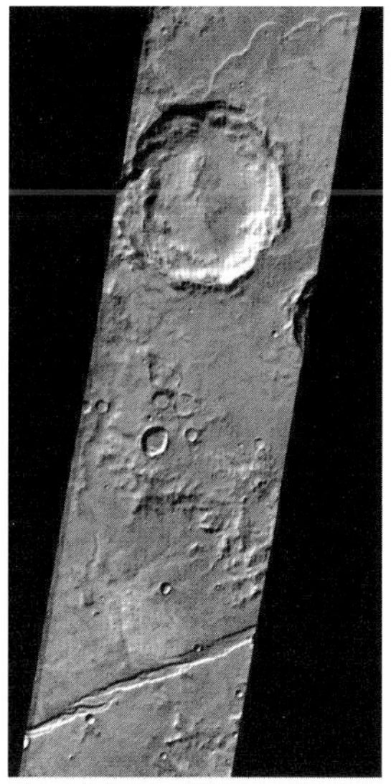

(c)

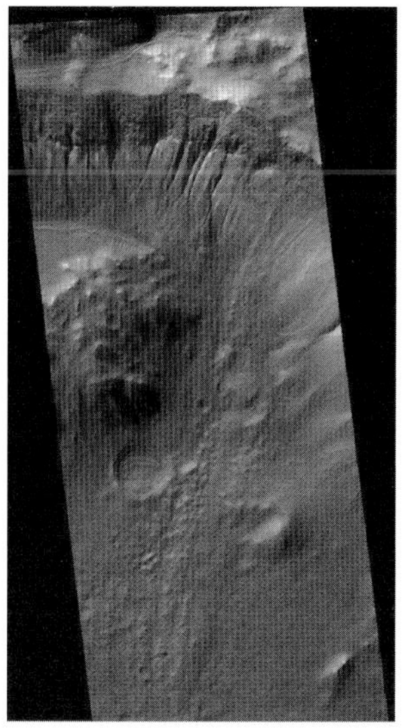

(d)

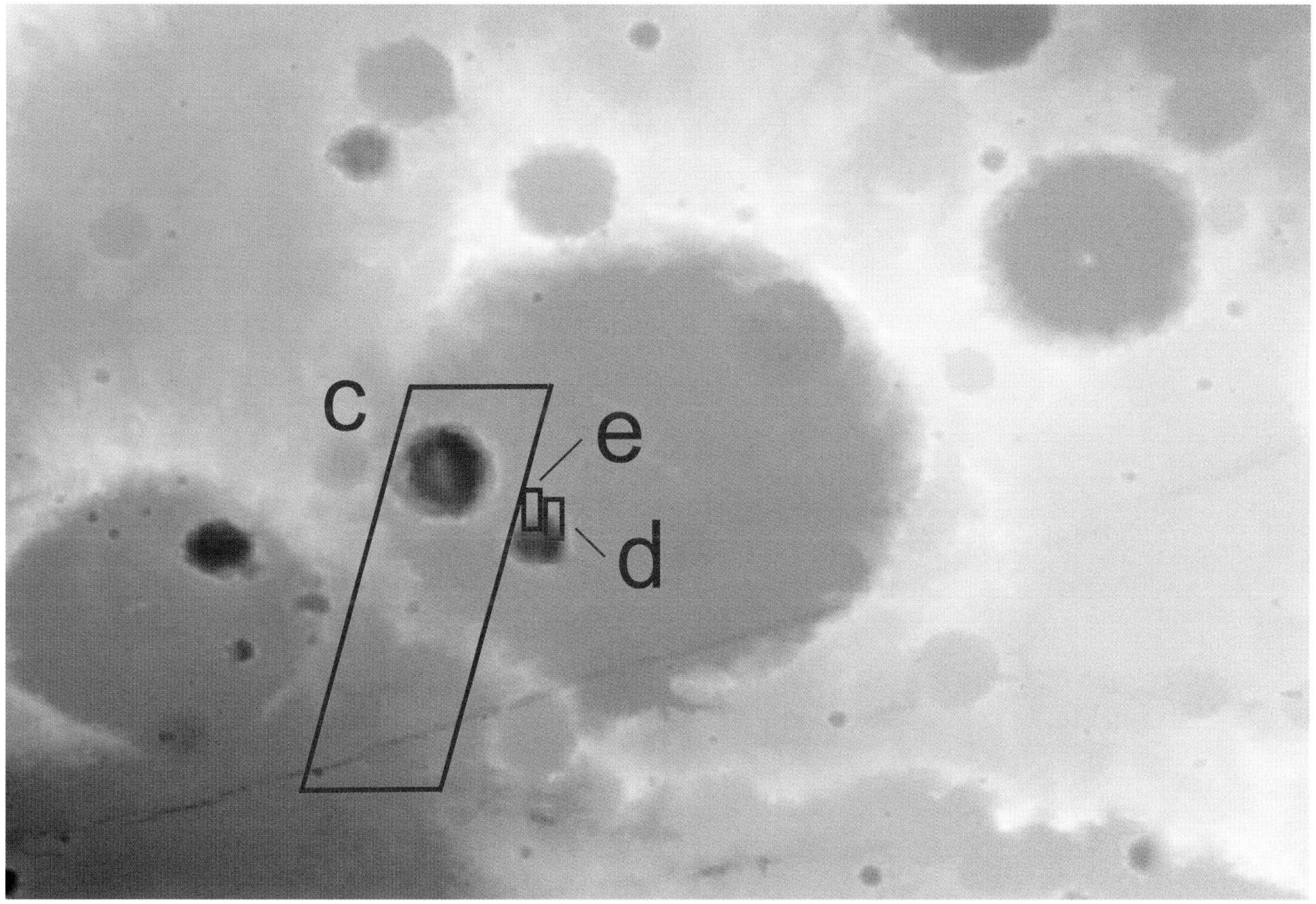

(e)

FIGURE 2.2 Mariner crater (151 km in diameter, at 34.68° S, 195.76° E) named for the Mariner 4 spacecraft that first imaged it, in 1965. (a) Mariner 4 frame 11 (pixel size about 1.2 km, NASA/JPL-Caltech). (b) Viking orbiter 1 (image 635A72, red filter, pixel size about 300 m) showing detail in craters and the ejecta around two younger craters inside Mariner crater (NASA/JPL-Caltech). (c) THEMIS infrared daytime image of the larger interior crater and a fracture of the Sirenum Fossae (image I34775002, pixel size about 100 m, NASA/JPL-Caltech/Arizona State University). (d) and (e) Two views of the north rim of the smaller of the two interior craters (d: MOC-NA image E02/00757, pixel size 6 m, view about 3 km across, NASA/JPL-Caltech/MSSS; e: HiRISE color image PSP_002317_1445, 25 cm/pixel, view about 1 km across, NASA/JPL-Caltech/University of Arizona). Locations of images in Figures 2.2c–e shown over MOLA grayscale elevation at lower left.

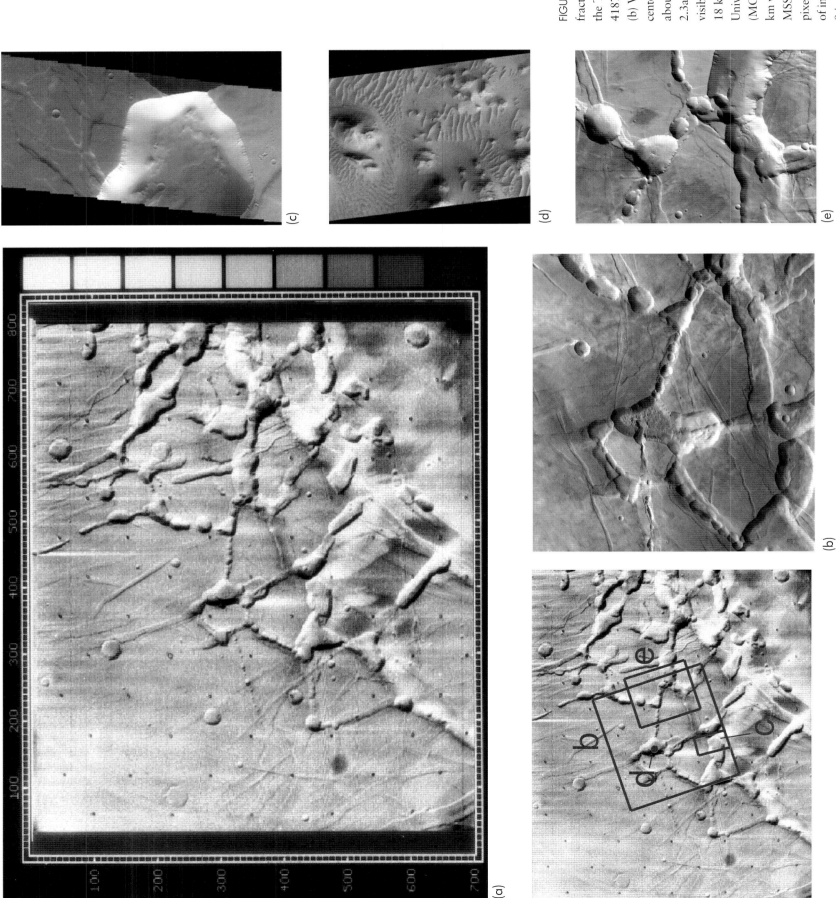

FIGURE 2.3 Noctis Labyrinthus, a system of intersecting fracture valleys (at 5.25° S, 255° E) near the center of the Tharsis uplift. (a) Mariner 9 image (MRVS 4187–45, pixel size about 0.5 km, view 400 km across). (b) Viking 1 Orbiter image, showing the area in the center of the Mariner 9 image (image 049A26, pixel size about 120 m, view 160 km across, images in Figures 2.3a and 2.3b both NASA/JPL-Caltech). (c) THEMIS visible image (V06633001, pixel size about 35 m, view 18 km wide, NASA/JPL-Caltech/Arizona State University). (d) Detail in floor of collapsed region (MOC-NA image M09/06504, pixel size 3 m, view 1.5 km wide), showing hills and dunes (NASA/JPL-Caltech/MSSS). (e) HRSC image H3210_0000_ND4 (12.5 m/pixel, 55 km by 70 km, ESA/DLR/FU-Berlin). Locations of images in Figures 2.3b–e are shown over the Mariner 9 image at lower left.

landings were successful and gave the first detailed information from the surface (see Chapter 4). Experiments that were designed for the purpose of detecting living organisms on Mars returned negative results. The orbiters returned images and other measurements for 4 years and mapped the entire planet over all Martian seasons at a pixel size of several hundred meters, and selected areas at tens of meters, ten to one hundred times better resolution than Mariner 9 (Carr and Evans, 1980; Spitzer, 1980; Snyder and Moroz, 1992). Two decades passed after Viking before another spacecraft successfully reached Mars; during this time, maps of the entire planet and numerous geologic studies came out of the Viking data.

Missions since 1996

The most recent phase of Mars exploration, since the late 1990s, has led to a significant increase in the types, spatial resolution, and amount of data returned both from Mars orbit and on the surface (for summaries of results see e.g., Carr, 2006; Barlow, 2008; Bell, 2008; Jaumann et al., 2015; Haberle et al., 2017).

Mars Pathfinder landed in the vast, ancient flood plain known as Chryse Planitia in 1997, hosting the first arrival of a rover, named Sojourner, to the planet. The landing included the use of a heat shield for aerobraking through the atmosphere, followed by deployment of a parachute and then airbags to cushion the lander's impact with the surface. Once safely landed, the rover was guided off a landing platform. Cameras and other instruments on both the platform and rover were able to survey rocks and soil surrounding the platform, as well as the landscape farther afield and the surrounding atmosphere. The spacecraft operated successfully for almost 3 months, returning thousands of images, chemical measurements of rocks and soils, and data on the weather and atmospheric dust.

The **Mars Global Surveyor** was active in Mars orbit between 1997 and 2006. It measured the magnetic and gravity fields, mapped surface mineral composition, monitored the atmosphere, and made a detailed map of elevation ("Mars Orbiter Laser Altimeter [MOLA] topography," used as base for index and overview maps in this atlas). The Mars Orbiter Camera (MOC) returned over 240,000 images. These included daily global coverage in the visible range at 7.5 km pixel size, and coverage in visible/near infrared wavelengths of 5.45 percent of Mars at 12 m, and 0.5 percent showing details as small as a few meters (Malin et al., 2010). The MOC included a narrow-angle (MOC-NA) and two wide-angle (MOC-WA) cameras, one with a red filter and the other with a blue filter.

Mars Odyssey has orbited Mars since 2001, while sensing gamma rays and neutrons to study the composition of the surface and search for water, measuring space radiation, and operating a visible/infrared camera, the Thermal Emission Imaging System (THEMIS). THEMIS yields 18-m pixels in five visible bands and 100-m pixels in ten infrared bands (Christensen et al., 2004).

The **Mars Express** orbiter mission (ESA, arrived at Mars in 2003 and began imaging in 2004) carries a host of experiments to study minerals on the surface, to search for water ice beneath the surface using radar, to study the atmosphere and interactions with the solar wind, and to image the surface in color and near-infrared with pixels as fine as 2 m using a super resolution channel (High Resolution Stereo Camera, HRSC; Jaumann et al., 2007).

NASA's ambitious, second-generation **Mars Exploration Rover** (MER) mission landed two rovers on opposite sides of the planet in 2004. These rovers, which were better equipped for scientific investigations and were larger than Sojourner, landed with the same techniques, and were able to operate independently from their landing platforms. The Spirit rover (MER-A) investigated the floor of and ancient hills within Gusev crater, before failing to respond in 2011, after traversing more than 7.7 km. Meanwhile, the Opportunity rover (MER-B) landed on the highland plain known as Meridiani Planum and was able to document rock strata in a series of impact crater walls, as well as other geologic and atmospheric features and dynamics. At the time that contact was lost (2018), Opportunity had logged more than 45 km of travel, and had explored the rim of Endeavour crater.

The **Mars Reconnaissance Orbiter**, since 2006, has looked for the history of water in the planet, including in the subsurface, the atmosphere, and in minerals, and has studied layered deposits and possible ancient shorelines. The instruments include shallow radar (SHARAD) and a 50-cm telescope with a visible/near-infrared camera (High Resolution Imaging Science Experiment; HiRISE), which has a resolution near 1 m (30-cm pixels; McEwen et al., 2007a). The Context Camera (CTX, 6-m pixels) provides a wider view of the region imaged by HiRISE).

The **Phoenix** mission was the first successful lander in a polar region, arriving in 2008. For 5 months it investigated the presence of water ice and minerals in the arctic soil, using a robotic arm to scoop samples for analysis. Instruments included a heating unit to drive off volatiles for analysis by a mass spectrometer and a chemical analyzer that tested soil pH and minerals. Cameras characterized the landscape at the landing site, and a weather station recorded conditions in the northern spring and summer (Smith et al., 2009).

The latest rover mission, **Mars Science Laboratory** (MSL), Curiosity, arrived on Mars in 2012 in the floor of Gale crater, which hosts a 5-km-high mound known as Aeolis Mons (or, informally, Mount Sharp). The Curiosity rover can be characterized as a mobile laboratory given its sophisticated capabilities to detect and measure compositional information, both remotely and with on-board equipment. As of the time of this writing (2018), Curiosity had investigated water- and wind-formed strata and landforms near its landing site and had begun ascending Mount Sharp. Here, a host of experiments are planned with the objective of reconstructing the geologic history and potential habitability of the paleo-environments in which the strata were lain down.

The **Mars Atmosphere and Volatile Evolution** (MAVEN) mission arrived in 2014 to study the history of Mars climate through the interactions of the Mars atmosphere with space, including how the solar wind may have stripped away much of the gas and water in the once-thick atmosphere. MAVEN's orbit takes it through the uppermost atmosphere, where the instruments study gas composition, behavior, and solar-wind interactions, while imaging of the atmosphere in the ultraviolet occurs in higher parts of the elliptical orbit.

The **Mars Orbiter Mission or Mangalyaan** (Indian Space Research Organisation, ISRO; it arrived at Mars in 2014) is a technology demonstration mission with instruments to measure hydrogen/deuterium, methane, and other gases, a thermal infrared imaging spectrometer, and a visible camera.

The first **ExoMars Trace Gas Orbiter and Schiaparelli Mission** (ESA) arrived at Mars in late 2016. The orbiter has several instruments to measure methane and other trace gases of possible biologic significance, as well as the Colour and Stereo Surface Imaging System (CaSSIS), to identify and study the location of any trace gas sources on the Mars surface. The unsuccessful Schiaparelli lander was intended to test entry, descent, and landing technology in preparation for a future rover mission, for which the orbiter will serve as the communications relay.

The **Interior Exploration using Seismic Investigations, Geodesy and Heat Transport (InSight)** mission will employ geophysical methods to study the deep interior of Mars. InSight landed on Elysium Planitia (MC-15) in December 2018. Instruments will measure seismic waves traveling through Mars, heat flow from the subsurface, and careful tracking of the site from Earth to establish the effect of interior structure on the rotation of Mars. Among a number of other robotic missions to Mars planned by several nations is the NASA **Mars 2020** rover mission, with a planned landing in Jezero crater (see MC-13). The rover is based on the Curiosity design and will cache drill-core samples for possible return to Earth by a future mission.

CHAPTER 3

Global Character of Mars

The global views presented here in Figures 3.1 to 3.14 convey properties of the surface (albedo, elevation, dust cover, relative abundance of various minerals), near-surface region (ice, thermal inertia), and interior (local gravity, crustal thickness, magnetic field). The maps are in simple cylindrical projection, using planetocentric coordinates where possible. Some datasets are only available in other views and are reproduced as they were made available.

Albedo

FIGURE 3.1 Albedo denotes the light or dark appearance of different parts of the surface, originally as seen by telescope. The albedo of Mars shows seasonal variations, possibly due to movement of windblown dust and changes in ice deposits. Albedo features do not appear exactly the same after each seasonal cycle, giving rise to longer-term, permanent changes. Data, here plotted over MOLA shaded relief, are from the Thermal Emission Spectrometer (TES, on Mars Global Surveyor; Christensen *et al.*, 2001; 7.5 km/pixel, simple cylindrical projection, planetocentric).

Elevation (shaded relief)

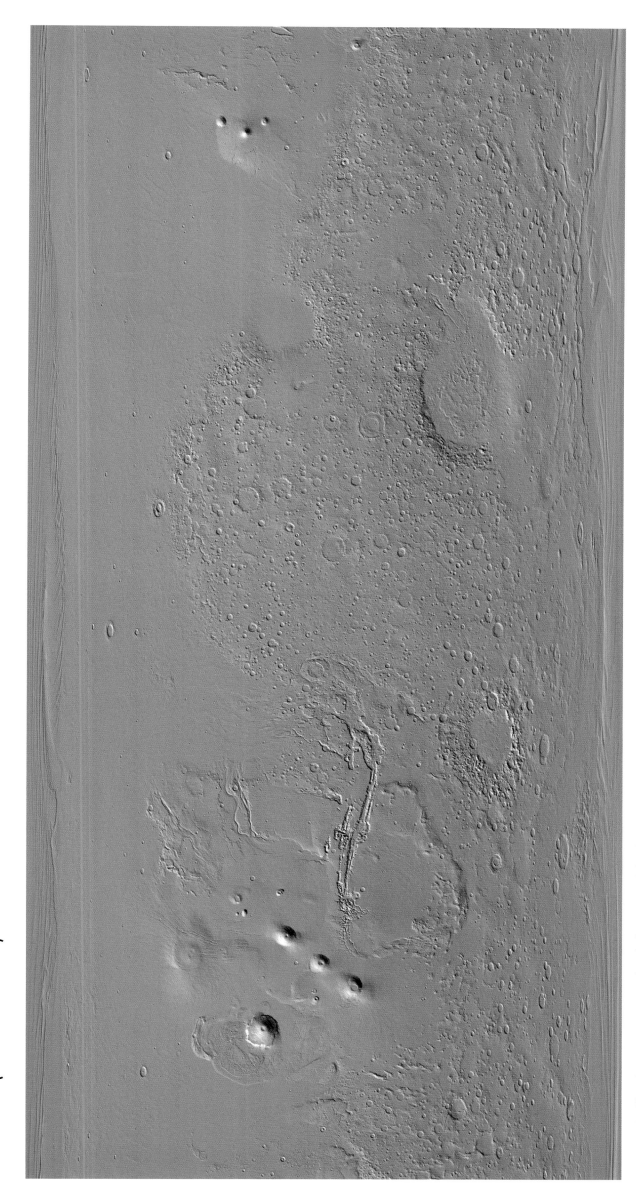

FIGURE 3.2 (a) Elevation measurements (altitude for Mars is defined in Chapter 1) taken by the Mars Orbiter Laser Altimeter (MOLA, on Mars Global Surveyor; Smith *et al.*, 2001) may be shown in several different ways.

Depicting the topography (shape of the landscape) by illumination from a uniform direction gives a shaded-relief image (460 m/pixel, illumination from the northwest, simple cylindrical projection, planetocentric).

Elevation (color)

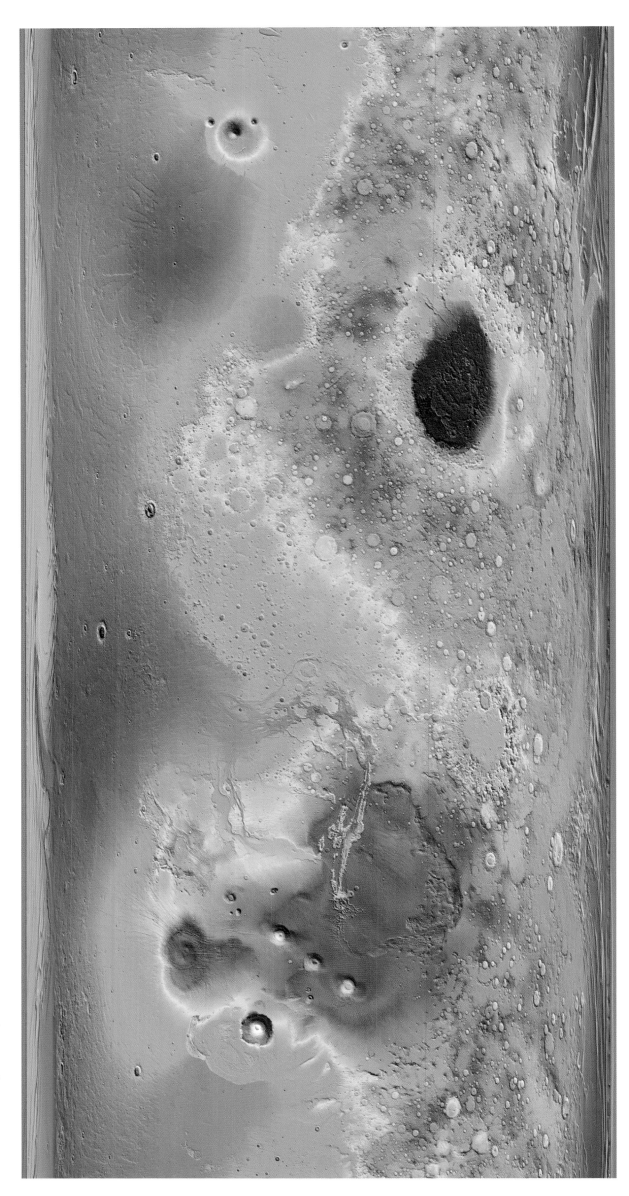

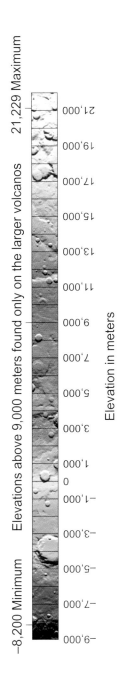

−8,200 Minimum   Elevations above 9,000 meters found only on the larger volcanos   21,229 Maximum

Elevation in meters

FIGURE 3.2 (b) An alternative way to view of the MOLA dataset is to use shades of gray or a range of colors to indicate various elevations. The most commonly used current color scheme, overlain upon shaded relief, is shown here and used for the overview maps and other views in this atlas (Smith *et al.*, 2001; 460 m/pixel, simple cylindrical projection, planetocentric).

Bouguer gravity

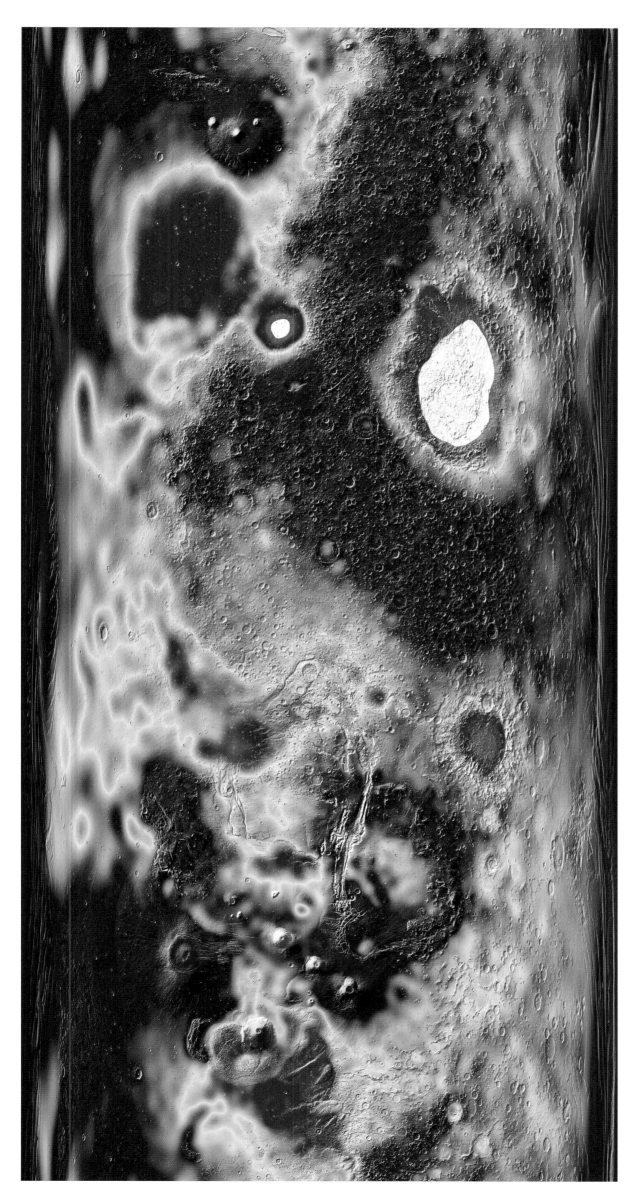

−600   −300   0   300   600   900   mGal

FIGURE 3.3 Tracking of spacecraft in orbit around Mars reveals small changes in the path due to variations in gravity. A large number of measurements yields the regional gravity. Here (Goddard Mars Model 3 by Genova *et al.*, 2016) the variations in gravity due to topography (relative to datum; see Chapter 1) have been removed by modeling the density (mass per volume) of the rock (this gives the Bouguer gravity anomaly). If crustal rocks are relatively constant in density, the resulting gravity map may be interpreted as showing the depth to the crust–mantle boundary. Where the anomaly is more positive (for example, in Hellas basin in the lower right of the map), the denser mantle is likely closer to the surface than average. Where the anomaly is more negative (for example, under the major volcanoes in Tharsis in the left center part of the map), the crust–mantle boundary is deeper than usual for Mars (Neumann *et al.*, 2004; Genova *et al.*, 2016; 1 Gal = 1 cm s^{-2}, gravity plotted upon MOLA shaded relief, simple cylindrical projection; image credit NASA Scientific Visualization Studio). This dataset allows resolution of features larger than ∼150 km wavelength.

Crustal thickness

Crustal thickness

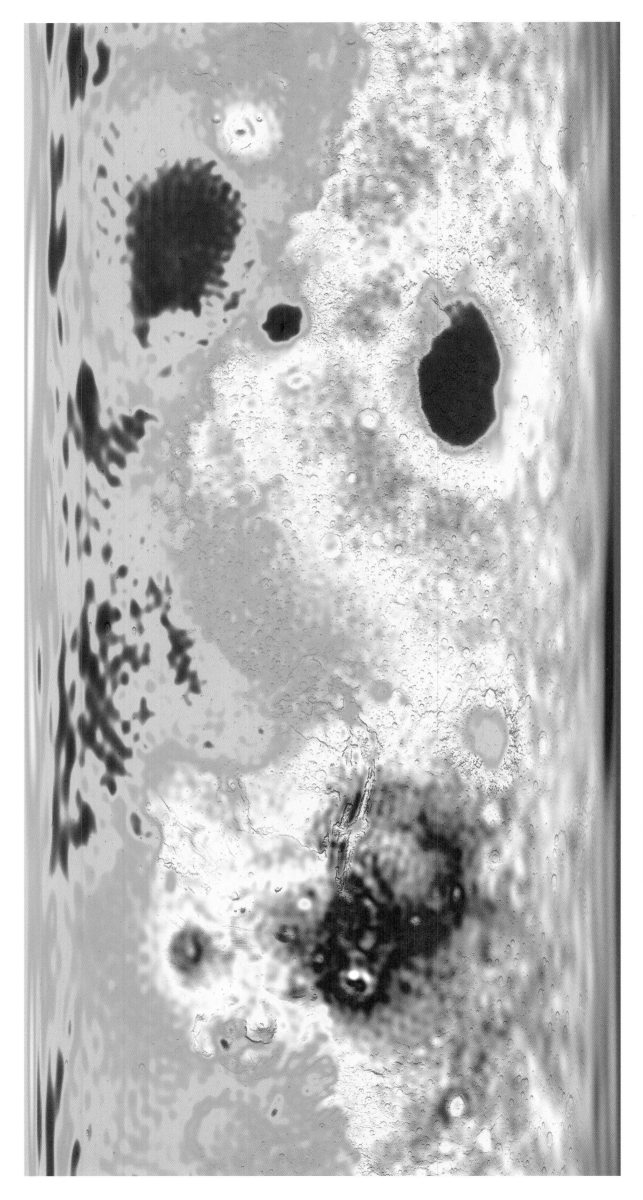

FIGURE 3.4 Here the topography and gravity are combined with estimates of mantle density to model the crustal thickness on Mars (from Genova *et al.*, 2016). The global dichotomy seen at the surface also characterizes the entire crust. The northern half of Mars has crust less than 40 km thick in this model, while the southern highlands' crust is close to 60 km thick. The Utopia (thin crust at upper right) and Hellas (thin region at lower right) impacts postdate the formation of the dichotomy, as does the thickening of the crust in the Tharsis region (left central part of map; Neumann *et al.*, 2004; Genova *et al.*, 2016; thickness plotted upon MOLA shaded relief, simple cylindrical projection; image credit NASA Scientific Visualization Studio).

km

0   20   40   60   80   100

Magnetization

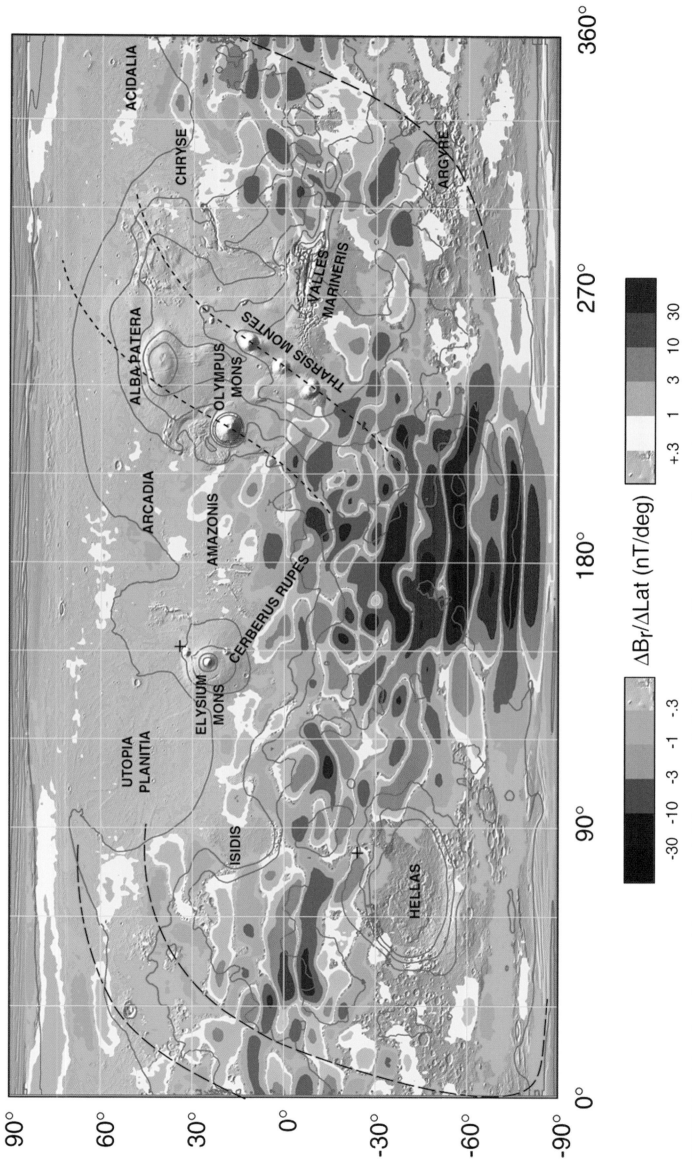

FIGURE 3.5 Mars lacks an internally generated planetary magnetic field, in contrast to the Earth. Measurements of the magnetic field from Mars orbit, however, have revealed areas of crustal rocks having remanent magnetization (image from Connerney et al., 2005). This probably dates from the earliest history of Mars, more than 4 billion years ago, when Mars did have a planetary magnetic field. Because the crust of Mars dates from this early era, regions lacking magnetization may have lost it through chemical alteration or by heating and erasure of the signal by younger igneous rocks or by large impacts. Several regions of stripes in the magnetization are reminiscent of magnetic stripes created by sea-floor spreading in the presence of a reversing magnetic field on Earth. This has led to suggestions that plate tectonics could have operated on early Mars, although the stripes could also reflect intrusion of the crust by linear bodies (dikes; Connerney et al., 2005; Nimmo and Tanaka, 2005). Data presented at 1-degree (60-km) resolution. Note that this map is centered on longitude 180° E, rather than 0°, (filtered radial magnetic field variation with respect to latitude $\Delta B_r/\Delta Lat$, plotted upon MOLA shaded relief, simple cylindrical projection; annotations are described in Connerney et al., 2005; courtesy of and copyright 2005 National Academy of Sciences, USA).

Water Ice (in shallow soil)

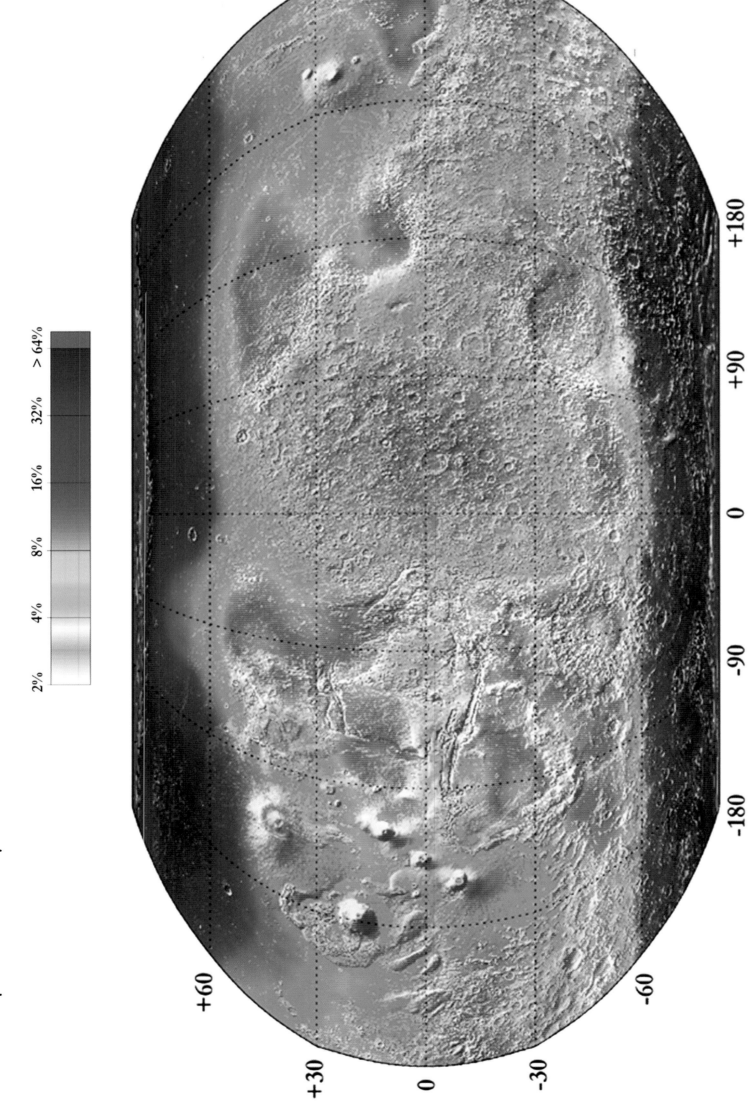

FIGURE 3.6 While liquid water is not stable at the surface of present-day Mars, water ice is present. Cosmic rays striking the surface of Mars cause the scattering of neutrons that can be measured to estimate the abundance of hydrogen, and by inference water, in the shallow subsurface (uppermost meter) of Mars (Feldman *et al.*, 2004). Radar, which penetrates more deeply, applied to several regions of Mars, also indicates substantial water present as permafrost in the subsurface (plot of percentage of uppermost meter that is water ice plotted upon MOLA shaded relief; data are from the neutron spectrometer aboard Mars Odyssey, Robinson projection, image credit NASA/JPL/Los Alamos National Laboratory).

Thermal inertia

J m^{-2} K^{-1} s$^{-1/2}$

25   600

FIGURE 3.7 Comparison of daytime and nighttime temperatures at the surface yields thermal inertia, or the tendency of the surface material to resist changes in temperature. High thermal inertia (areas that cool less at night) are interpreted as bare rock, while low thermal inertia indicates dust-covered terrain (Christensen *et al.*, 2003).

Data are thermal inertia from TES measurements (Mellon *et al.*, 2000; 3 km/pixel, highest in red, lowest in violet, plotted upon MOLA shaded relief, simple cylindrical projection, planetocentric).

Dust cover

FIGURE 3.8 Dust-covered terrain (interpreted as silicate particles finer than 100 μm) indicated by thermal–infrared spectral behavior (Ruff and Christensen, 2002; Bandfield, 2002; Christensen *et al.*, 2003). Shown is relative dust cover from TES measurements (Ruff and Christensen, 2002; 7.5 km/pixel, highest in red, lowest in blue, plotted upon MOLA shaded relief, simple cylindrical projection, planetocentric).

Low

High

Plagioclase

FIGURE 3.9 Plagioclase, a compound of silicon, aluminum, and oxygen with varying proportions of calcium, potassium, and sodium, is one of the most common rock-forming minerals, particularly in igneous rocks. Data are from TES measurements (Bandfield, 2002; 7.5 km/pixel, rock fraction [0.1 means 10%] shown highest in red, lowest in blue, no color plotted where abundance below detection limit, plotted upon MOLA shaded relief, simple cylindrical projection, planetocentric).

0   0.1   0.2

High-calcium pyroxene

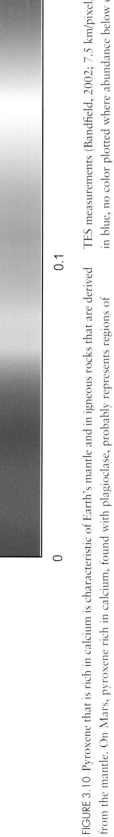

0   0.1   0.2

FIGURE 3.10 Pyroxene that is rich in calcium is characteristic of Earth's mantle and in igneous rocks that are derived from the mantle. On Mars, pyroxene rich in calcium, found with plagioclase, probably represents regions of basalt. Both Syrtis Major (MC-13) and eastern Solis Planum (MC-18) near the equator are examples. Data are from TES measurements (Bandfield, 2002; 7.5 km/pixel, rock fraction [0.1 means 10%] shown highest in red, lowest in blue, no color plotted where abundance below detection limit, plotted upon MOLA shaded relief, simple cylindrical projection, planetocentric).

Olivine

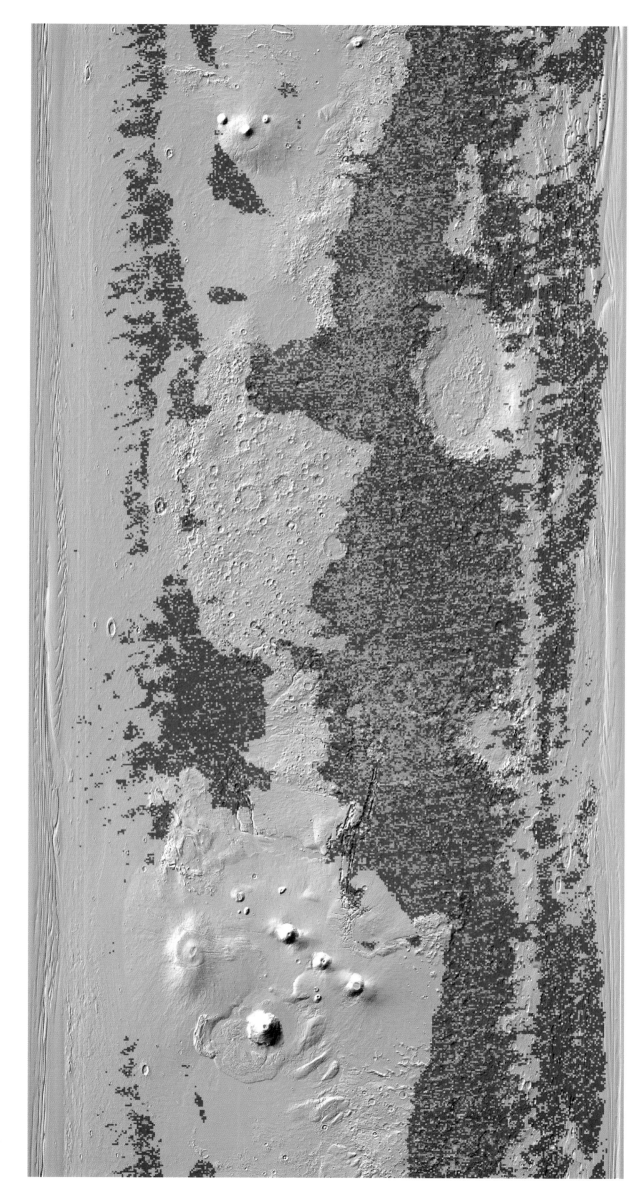

FIGURE 3.11 Olivine is an important mineral in Earth's mantle and in igneous rocks formed by melting of the mantle. This appears to be the case on Mars as well (Koeppen and Hamilton, 2008; Ehlman and Edwards, 2014). Data are from TES measurements (Koeppen and Hamilton, 2008; 7.5 km/pixel, rock fraction [0.1 means 10%] shown highest in red, lowest in blue, no color plotted where abundance below detection limit, plotted upon MOLA shaded relief, simple cylindrical projection, planetocentric).

OMEGA/CRISM hydrated
minerals

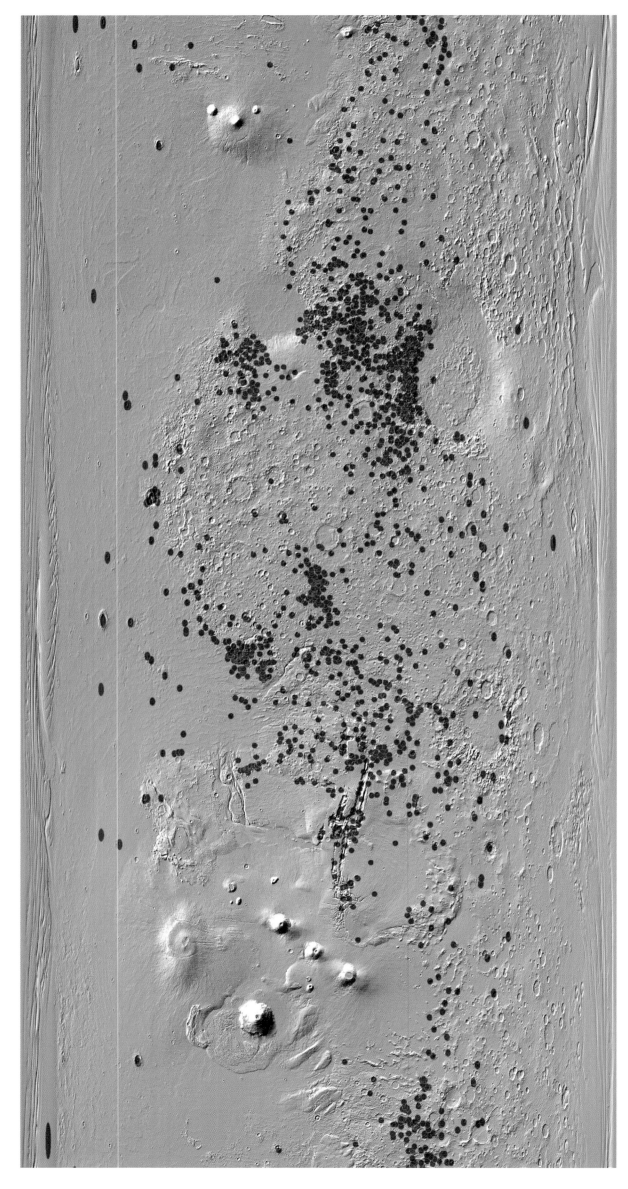

FIGURE 3.12 This global map show sites at the surface of Mars where hydrated, or water-bearing, minerals were detected, using the reflectance data acquired by the Mars Express OMEGA (Observatoire pour la Minéralogie, l'Eau, les Glaces et l'Activité) hyperspectral camera from January 2004 to September 2010, and from the Mars Reconnaissance Orbiter (MRO) Compact Reconnaissance Imaging Spectrometer for Mars (CRISM) (Carter et al., 2013; plotted upon MOLA shaded relief, simple cylindrical projection).

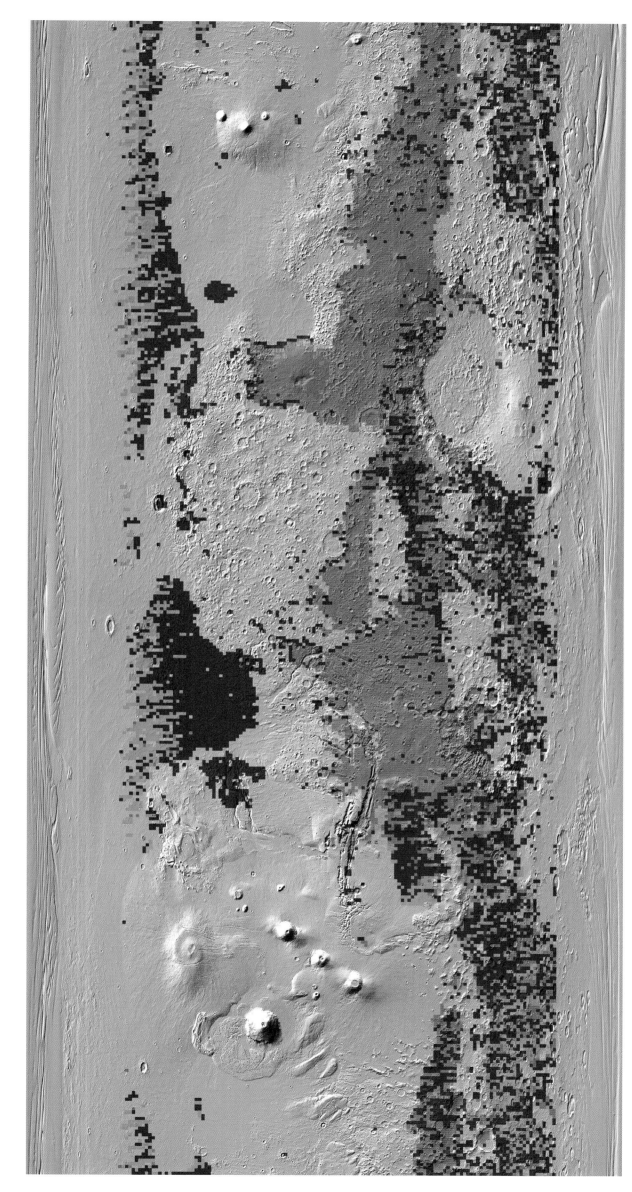

FIGURE 3.13 Volcanic rock units derived from TES mineral mapping. The dust-free surface of Mars can be divided into five distinct units, based on the varying abundances of the major volcanic minerals (plagioclase, pyroxene, olivine) and glass components. These five units are shown here in different colors. Data are from TES measurements (Rogers and Christensen, 2007, which describes the rock units; 7.5 km/pixel, plotted upon MOLA shaded relief, simple cylindrical projection, planetocentric).

Ferric oxide

Low
Abundance

High
Abundance

FIGURE 3.14 Ferric oxides (minerals of iron and oxygen, and lacking water) are common on the surface, both on rock surfaces and as dust. They appear to represent the part of Mars history after which water became scarce at the surface and the formation of hydrated (water-bearing) minerals ended (see the section in Chapter 5 on the Chemical evolution of Mars environment). This map, by the OMEGA instrument on ESA's Mars Express, plots the distribution of ferric oxides, with highest concentration in red, lowest in blue (Ody *et al.*, 2012; ESA/CNES/CNRS/IAS/Université Paris-Sud, Orsay; plotted upon MOLA shaded relief; cylindrical projection, 32 pixels per degree, or 2 km/pixel).

CHAPTER 4

Regional Geographic Features and Surface Views of Mars

The global landscape of Mars is diverse, and this diversity tells a story of its surface history.

Geographic regions and zones, each unique in character, are dominated by specific landforms and materials. These in turn express processes and histories, including geologic and climate-driven activity. Understanding the major geographic regions of Mars, as on Earth, then, is essential to unraveling the factors involved in global and regional geologic activity. It also provides context for study of local areas and individual features or sets of features. In addition, some of the major regions of Mars are indicative of heterogeneities in the crust and mantle (see Chapter 5).

In this chapter, we describe features of global significance and classes of features of regional importance. We also highlight observations made by surface landers at several locations across the planet. Formal names for geographic features on Mars are overseen by the International Astronomical Union (see Chapter 1) and are shown on individual map sheets. Figure 4.1 shows the largest named features.

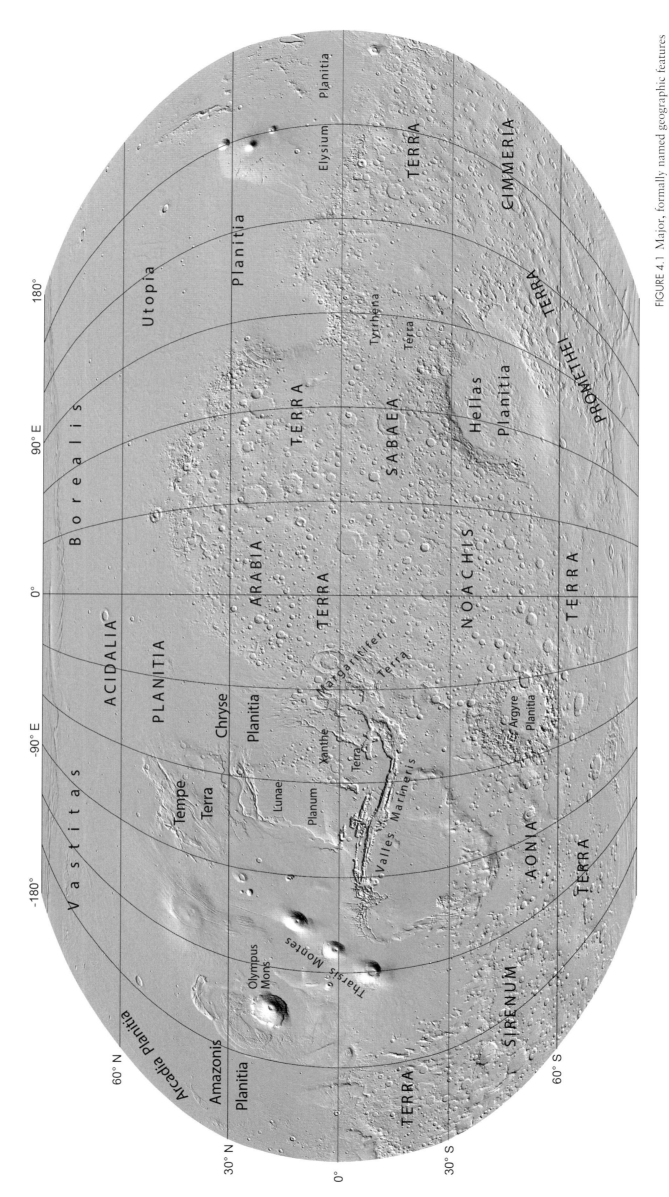

FIGURE 4.1 Major, formally named geographic features on Mars (Robinson projection, MOLA hillshade).

Global dichotomy boundary

Viewing Mars in any number of representations (image maps of elevation, Chapter 3; geologic maps, Chapter 5) reveals the **most important** global feature, the so-called **global dichotomy**. The global dichotomy is the division of the surface of Mars between the southern highlands and the northern plains. It is expressed along a topographic boundary (Figure 4.2), characterized by gradual to abrupt changes in elevation and relief of the surface, density of cratering, and crustal thickness (Nimmo and Tanaka, 2005; Carr, 2006; Carr and Head, 2010). The southern hemisphere of Mars is above the global datum (by up to 5,000 m; datum is defined in Chapter 1) and

appears heavily cratered. Much of the northern hemisphere of Mars is well below the datum elevation (−1,000 to −5,000 m) and displays few ancient impact depressions. These include the Isidis and Utopia basins and Chryse Planitia, which may also be the site of an early impact (e.g. Schultz et al., 1982). In Tharsis, the boundary is concealed by younger volcanic deposits

The overall form of the dichotomy boundary is nearly circular. In some regions, it appears defined and modified by the extents of large, ancient impact depressions. These include the Isidis and Utopia basins and Chryse Planitia, as on Earth, though the reason may be very different (Figure 4.3; Smith et al., 2001).

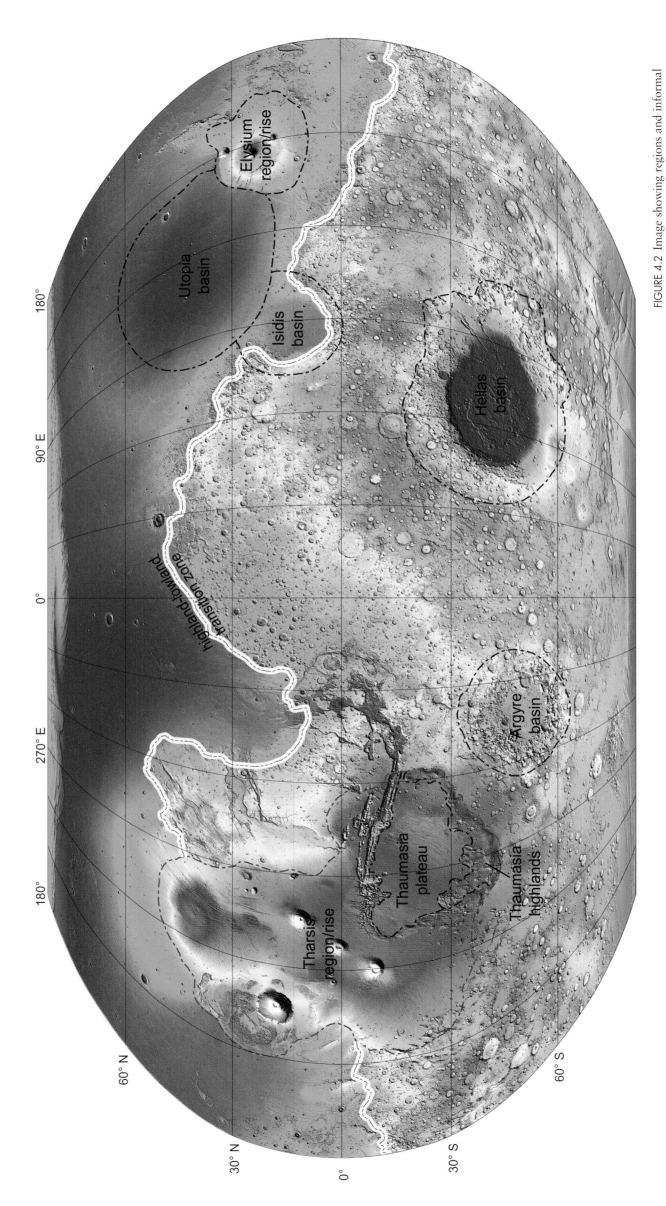

FIGURE 4.2 Image showing regions and informal feature names (adapted from Tanaka et al., 2014; Robinson projection, MOLA color hillshade).

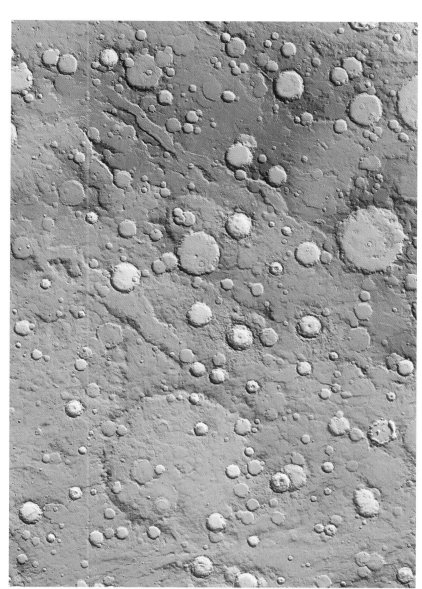

FIGURE 4.4 Highlands terrain, view spans about 355° E to 30° E, 30° S to 50° S (1650 by 1150 km, MOLA color hillshade).

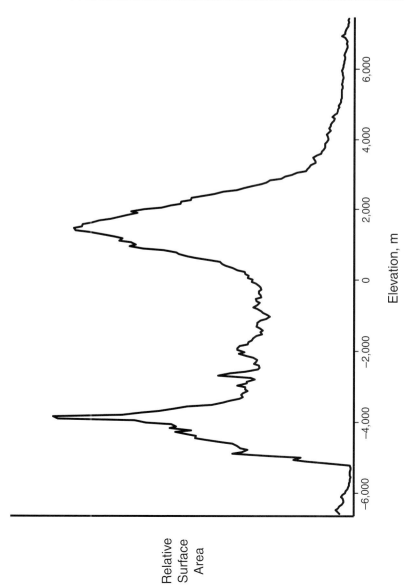

FIGURE 4.3 Hypsometric curve (distribution of elevation on Mars relative to the center of mass), plotted from a MOLA elevation dataset (Smith *et al.*, 2001). In this digital elevation model (DEM) peaks occur at about −3,900 m and 1,400 m. Elevations on Mars range from −8,200 m to 21,229 m; these extremes are clipped off as a result of the sorting method used (generated from an equal-area map of DEM at 8-bit resolution sorted into 256 bins).

(Figure 4.2). The zone of transition between the southern highlands to the south and the lowlands to the north is marked by diverse landforms. These include scarps, knobs, mesas, and plateaus that are hundreds of meters to a few kilometers in relief. They make up Deuteronilus, Nilosyrtis, Protonilus, and Cydonia Mensae. These landforms extend eastward from the southern margins of Acidalia Planitia to the southern margin of Amazonis Planitia.

The much higher crater density of the southern highlands attests to its relative antiquity, whereas the northern lowlands are less cratered and younger. However, subdued "quasi-circular depressions," tens to hundreds of kilometers in diameter, in the northern plains suggest a potentially dense pre- to Early Noachian population of partly and fully buried impact structures (Frey *et al.*, 2002; see Figure 5.11 and the discussion below).

Southern highlands

The southern highlands consist of densely cratered, rugged terrain (Figure 4.4), which covers much of the equatorial to southern latitudes. Included are Noachis Terra, Tyrrhena Terra, Terra Cimmeria, Terra Sirenum, Xanthe Terra, Margaritifer Terra, Aonia Terra, and Promethei Terra, and some parts of the northern hemisphere of Mars, mostly in Arabia Terra, Terra Sabaea, and Tempe Terra. These terrae (plural of "terra") divide the highlands into subregions that extend for 2,000 to nearly 6,000 km. Most highlands are between 0 and 5,000 m above datum (Smith *et al.*, 1999; 2001). However, Arabia Terra and parts of other highland areas north of the equator lie at 0 to −3,000 m elevations.

Highland crater forms include a broad mix of morphologies, reflecting their preservation state (e.g. Barlow and Bradley, 1990). Impacts onto rugged highland surfaces that are made up of regolith commonly result in irregular ejecta blankets, which are extended on radiating spokes, marked by elongate clusters of secondary craters. Most of the larger highland impacts appear to be moderately to intensely degraded and/or buried. Parts of the highlands were resurfaced by expanses of less-cratered, high-standing plains, 800 to 1,600 km across, including Hesperia Planum, Meridiani Planum, and Syrtis Major Planum. Craters on the highland plains commonly have well-preserved, single- to multi-lobed ejecta forms.

Much of Xanthe and Margaritifer Terrae are dissected by a series of broad valley systems, tens to hundreds of kilometers wide and hundreds to thousands of kilometers long. These connect canyons of eastern Valles Marineris and depressions containing chaotic assemblages of knobs and mesas to the northern lowlands. These valley systems are commonly called "outflow channels" and include Kasei, Simud, Tiu, and Ares Valles. Uzboi and Ladon Valles are part of a system that may have at one time provided drainage from Argyre basin into the northern plains via Ares Valles. Similar but smaller valleys, such as Mangala and Ma'adim Valles, drain across the dichotomy boundary into the northern plains. Others drain into Hellas Planitia, including Dao, Harmakhis, and Niger Valles. Reull Vallis winds toward, but terminates prior to, reaching Hellas basin.

Less dramatic but more pervasive in the highlands are systems of smaller valleys, less than a few kilometers wide and tens to hundreds of kilometers long, many of which form dendritic networks (e.g. Paraná and Warrego Valles). Most of these eventually drain into local basins or crater floors, which are generally smooth. Some, however, include polygonal to rounded mesas and knob forms (e.g. Atlantis and Gorgonum Chaoses). Sinuous, commonly discontinuous broad ridge forms in many cases are coupled with crenulations, especially across highland planar surfaces. Those with crenulations are called "wrinkle ridges," whereas those without crenulations are called "simple ridges" or

27

familiar, such as tongue-shaped lava flows, abandoned river channels, and fault grabens and ridges, to exotic and strange, including fields of small ridges and cone features, some forming whorled patterns that resemble thumbprints. In the Scandia region north of Alba Mons, fields of knobs and mesas make up Scandia Colles. Farther north, near Planum Boreum, the irregular mounds of Scandia Tholi and the depressions of Scandia Cavi are tens to hundreds of kilometers across.

Impact basins

The largest, well-preserved impact basins (Table 4.1) in the Martian cratered highlands consist of Hellas (~2,400 km in diameter, which includes the deepest surface point on the planet in Badwater crater at –8,200 m elevation; Figure 4.6), Isidis (~1,500 km in diameter) along the highland–lowland margin, and Argyre (~900 km in diameter). Hellas is one of the largest recognizable impact basins in the solar system, comparable in size to the South Pole–Aitken basin on the far side of the Moon.

These basins include rugged rims, typically a few hundred kilometers wide, made up of jumbled, degraded massifs (e.g. Nereidum and Charitum Montes, surrounding Argyre basin, and Libya Montes, south of Isidis). In places, there are wide troughs with radial and concentric orientations separating the Argyre rim massifs. The eastern rim of Hellas basin is unusually wide, extending for >1,000 km. The inner basins are covered by thick sequences of plains- and plateau-forming deposits. Surrounding the Hellas and Isidis basins are discontinuous zones of circumferential ridges and troughs that make up Hellespontus Montes; Chalcoporos Rupēs; Nia and Amenthes Fossae; and Scylla, Charybdis, Eridania, and Oenotria Scopuli.

In addition to well-preserved basins, other buried or inferred large impact basins on Mars include Utopia, Acidalia, and Chryse (Table 4.1). Many other buried basins and inferred quasi-circular depressions have been proposed (Frey, 2006; Figure 5.11).

Dozens of smaller basins, a few hundreds of kilometers across, occur in the cratered highlands. Many of these have two nested circular rims that are relatively narrow and continuous, in contrast to the massif-dominated rims of the largest basins.

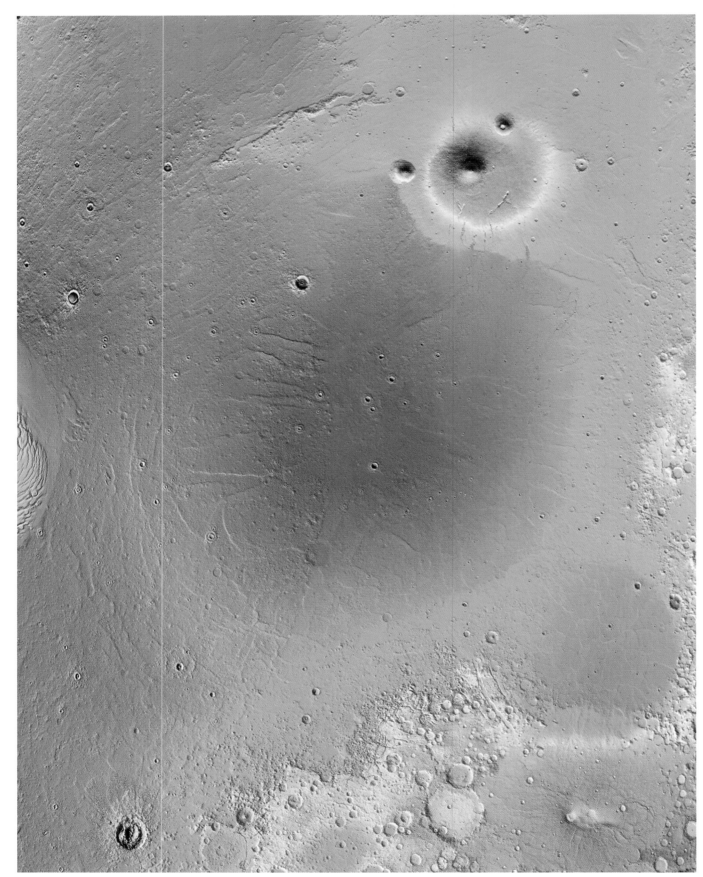

FIGURE 4.5 Lowlands terrain of Utopia Planitia, view spans about 60° E to 160° E, 0° N to 80° N (northern half of eastern hemisphere of Mars, MOLA color hillshade, cylindrical projection).

"lobate scarps," where they have a predominant flank (Watters, 1993).

Northern lowlands

The northern lowlands include the circumpolar plain of Vastitas Borealis and the middle-latitude lowlands of Utopia, Isidis, Chryse, Acidalia, and Arcadia Planitiae, among others. These lie lower in elevation than the highlands or transition zone, mostly in the range of –4,000 to –5,000 m. The plains constitute the northern third of the planet and generally have low regional surface gradients of <1 degree. Amazonis Planitia is one of the flattest regions known on any planetary surface. Broadly, the northern lowlands include three basins: the north polar, Utopia, and Isidis basins.

The lowland surfaces are made up of relatively smooth, moderately cratered plains (Figure 4.5). They are marked by a variety of landforms of local to regional extent. The landforms range from the

TABLE 4.1 Impact basins

Buried or inferred large impact basins referred to in this atlas

Name	Diameter (km)	Latitude (°N)	Longitude (°E)	Age[a]
Utopia	~3,400	45	115	pre-N
Acidalia	~2,800	58	341	pre-N
Chryse	~1,700	25	318	pre-N

Readily visible[b]/preserved impact craters and basins (and possible impact-like features of other origin), >200 km in diameter

Name[c]	Diameter (km)	Latitude (°N)	Longitude (°E)	Age
Hellas	~2,400	-42	71	eN
Isidis	~1,500	13	89	mN
Argyre	~900	-50	318	mN
Huygens	467	-13.9	55.6	eN
Schiaparelli	446	-2.5	16.8	mN
Greeley	427	-36.8	2.8	eN
Cassini	408	23.4	32.1	eN
Antoniadi	401	21.4	60.8	eN
—[d]	376	36.7	192.4	~N
Dollfus	359	-21.6	356.2	eN
Tikhonravov	344	13.3	35.9	eN
—[d]	341	-58.8	283.2	eN/mN
—[d]	340	23.5	53.2	eN
—[d]	327	-52.7	250.5	eN
—	326	-0.4	28.9	eN
Newton	312	-40.4	201.9	eN
de Vaucouleurs	312	-13.3	171.1	mN
—	302	-59.9	275.9	eN
Copernicus	302	-48.8	191.2	eN
Herschel	298	-14.5	129.9	mN
Schroeter	292	-1.9	56.0	mN
Koval'sky	285	-29.6	218.6	eN
—[d]	285	12.9	41.9	eN
—	279	54.4	82.0	H/N
— (Aram Chaos)	276	2.8	338.8	mN
Orcus Patera	263	14.2	178.6	IN
—[d]	261	-4.3	333.1	N
— (Atlantis Chaos)	261	-34.4	182.4	eN
—	257	-58.5	265.5	eN
Newcomb	256	-24.2	1.1	eN
—[d] (Gorgonum Chaos)	250	-37.7	189.3	eN
Flaugergues	236	-16.8	19.2	mN
—	229	-50.1	243.1	eN
Galle	223	-50.6	329.1	eA
Kepler	222	-46.7	141.2	eN
Lyot	220	50.5	29.3	eA
—[d]	218	43.7	18.4	N
Secchi	217	-57.8	102.0	mN
—[d]	215	11.7	0.8	eN
—[d]	211	-41.8	130.9	eN
Vinogradov	210	-19.8	322.3	eN
—[d]	204	-42.8	215.0	eN
—[d]	202	47.9	89.8	N/H
Kaiser	202	-46.2	19.1	eN
Schmidt	201	-72.0	281.9	eN

[a] Chronostratigraphic period and epoch designations include Noachian (N), Hesperian (H), Amazonian (A), Early (e), Middle (m), and Late (l), as dated by Tanaka et al. (2014). [b] Visible basins as identified by Robbins et al. (2013). [c] As given by the International Astronomical Union, where available. [d] The impact origin of these features is considered to be uncertain due to poorly preserved or inferred, buried rims.

The more prominent of these smaller basins are named after the early astronomers who observed Mars and studied its motions: Schiaparelli, Antoniadi, Huygens, Cassini, Herschel, Kepler, Lyot, Copernicus, Newton, Lowell, and others.

Tharsis and other volcanic provinces

The Tharsis rise (Figure 4.7) consists of an immense flow field that extends 5,000 km east-west by 6,000 km north-south, overlapping the highland–lowland dichotomy boundary. The rise is capped by some of the largest volcanic features in the solar system, including Olympus Mons, Alba Mons, and the three aligned Tharsis Montes (Ascraeus Mons, Pavonis Mons, and Arsia Mons). In addition, the Tharsis rise includes intermediate-sized volcanoes, tens of kilometers across, and fields of kilometer-sized volcanoes and vent structures. Also present are tectonically deformed, elevated terrains (Syria Planum, Tempe Terra, Noctis Fossae, Ceraunius Fossae, and the southern and eastern margins of the Thaumasia plateau (informal name), which extends about 2,000 km south and east of Syria Planum). The summit of Olympus Mons stands at 21,229 m – the highest elevation on Mars – and the mountain is surrounded by several blocky, lobate aureoles that extend for hundreds of kilometers.

All of these structures are surrounded and interconnected by low to high plains, most of which preserve lobate flow structures. These include Daedalia, Syria, Solis, Sinai, Thaumasia, Lunae, and Icaria Plana, as well as the low- and flat-lying Amazonis Planitia.

The Elysium province (Figure 4.8) occurs in the northern lowlands between Utopia and Arcadia Planitiae. It consists of the broad Elysium rise, surmounted by the shields of Elysium Mons, Hecates Tholus, and Albor Tholus. The rise includes lightly cratered lobate flow fields, radiating from summit areas. South of the rise, extensive flows also emanate from Cerberus Fossae into Elysium Planitia, including Cerberus Palus, and eastward into western Amazonis Planitia via Marte Vallis.

Northeast and south of Hellas Planitia, a variety of volcanic landforms and channel systems interrupt the rugged ancient rim features of Hellas basin. Prominent forms northeast of Hellas include Hadriacus Mons and Tyrrhenus Mons and the channel systems of Dao, Harmakhis, Niger, and Reull Valles. Flow features and smooth plains of Hesperia Planum are marked by a crisscrossing network of wrinkle ridges. South and southwest of Hellas, circular and oval forms make up Amphitrites, Peneus, Malea, and Pityusa Paterae. These features are surrounded by ridged and dissected plains, mostly within Malea Planum.

FIGURE 4.6 Image of Hellas: the view spans about 40° E to 100° E, 20° S to 60° S (about 3,300 by 2,400 km, MOLA color hillshade).

rise, beginning at the cross-hatched Noctis Labyrinthus trough system (Figure 4.7; see also Figure 18.A). The main, interconnected troughs include, from west to east: Ius, Tithonium, Candor, Ophir, Melas, and Coprates Chasmata. Hebes, Juventae, and Ganges Chasmata form discrete depressions, north and east of the main system. In addition, substantial linear chains of depressions and the northeastern and eastern troughs of Valles Marineris connect with additional irregular and complex depressions. These include the chaotic fields of knobs and mesas in Xanthe and Margaritifer Terrae and broad outflow channels that dissect highland surfaces to depths of several kilometers over courses that stretch a few thousand kilometers in length, prior to their entrance into Chryse Planitia. The most prominent channels – Kasei, Tiu, Simud, Ares, and Shalbatana Valles – have astounding widths of tens to a few hundred kilometers and lengths of one to two thousand kilometers.

Polar plateaus

Both of the Martian poles are capped by plateaus that are elevated a few kilometers above the surrounding plains. The circular, north polar plateau – Planum Boreum – occurs near the topographic center and low point (–5,000 m elevation) of the north polar basin, within the northern lowlands (Figure 4.9). Planum Boreum is ~1,000 km in diameter, contains numerous swirled troughs, and is partly surrounded by vast dune fields, including Olympia Undae. The south polar plateau – Planum Australe – is within the southern cratered highlands at 1,000–1,500-m elevation and also is troughed, having curved, elongate, and irregular forms (see Figures 30.A, 30.B). Planum Australe has an ellipsoidal planimetric shape (~1,100 × 1,400 km²). Both poles contain characteristic "chasma" troughs, which expose the depths of the polar plateau, with one large reentrant in the north (Chasma Boreale) and three reentrants in the south (Chasma Australe, Promethei Chasma, and Ultimum Chasma). In addition, Planum Australe is surrounded by plains (Argentea Planum, Promethei Planum, and others), marked by bifurcating ridge systems (including Dorsa Argentea), as well as regions of broad, deep depressions (Cavi Angusti and Sisiphi Cavi) and local mountain systems (Sisyphi

Valles Marineris

A gigantic system of depressions makes up the linear Valles Marineris troughs that extend for >2,000 km from the eastern flank of the Tharsis

Montes, Sisyphi Tholus, and Australe Montes, see MC-30). Planum Boreum, on the other hand, has nearby pancake-shaped mounds, tens to a few hundreds of kilometers across (Scandia Tholi), the largest of which include irregular depressions that are hundreds of meters deep (Scandia Cavi, see MC-1).

Views of the surface from landers

Table 4.2 gives specific mission dates and coordinates. See Stooke (2012; 2016) for extensive details on lander missions.

Viking 1 and 2 landers

In the first successful mission to the surface of Mars, Viking 1 lander touched down on July 20, 1976 in the plain of southwestern Chryse Planitia (see map MC-10). This lies in the main area of discharge of the extensive Maja Valles outflow channel system. Panoramic views showed a rocky landscape, made of angular basaltic cobbles intermingled with a crusted soil of finer-grained material (Figure 4.10). Several weeks later, Viking 2 lander landed in eastern Utopia Planitia (MC-7). This was also a rocky landscape that was interspersed with soil (Figure 4.11). The site is about 200 km west of Mie crater and may include remnants of the crater's ejecta. Both landing sites occur in the northern plains of Mars and were chosen from Viking orbiter images showing what appeared to be the safest areas to land.

Mars Pathfinder

On July 4, 1997, this mission landed both a fixed station and the first rover ever to be remotely driven on the surface of Mars. The site is in eastern Chryse Planitia, downstream from the confluence of the Tiu and Ares Valles outflow channel systems (MC-11). Like the Viking landers, Pathfinder's site was chosen on the basis of Viking orbiter images. Mars Pathfinder's Sojourner rover visited rocks within a few meters of the lander (Figure 4.12), named the Carl Sagan Memorial Station.

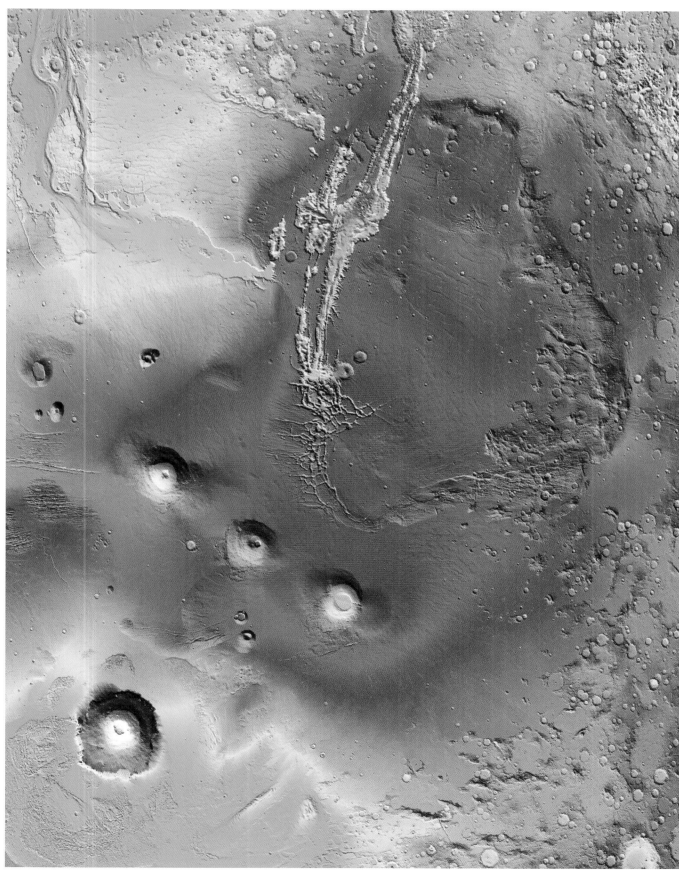

FIGURE 4.7 Image of Tharsis, including the prominent Tharsis Montes and Olympus Mons circular volcanic shields with Valles Marineris at the center right and the Thaumasia plateau at the lower right; the view spans about 210° E to 310° E, 30° N to 45° S (view about 6000 km E–W, cylindrical projection, MOLA color hillshade; northernmost extension of Tharsis includes Alba Mons, not shown).

Planum (Figure 4.14; map MC-19). Orbital data from the Mars Global Surveyor for this site (see MC-19) showed the presence of hematite, an iron oxide that is commonly formed in the presence of water. Remarkably, the landed spacecraft, rolling along in its protective airbag cocoon, prior to deployment, ended up in a small crater named Eagle. The rover was able to visit a variety of layered rock outcrops that were exposed within crater walls (see Chapter 5). Opportunity concluded a traverse of over 45 km by exploring the west rim of the 22-km-diameter crater Endeavour where the mission ended.

These rovers collectively have viewed a variety of landscapes and terrains, encountering rocky soils, layered rock outcrops, minerals and weathered spherules indicative of deposition in water, small sand dunes, meteorites, and more (see Chapter 5 for discussion).

Mars Phoenix Lander

At 68° N latitude, the Mars Phoenix Lander is the highest-latitude surface investigation on the planet. It resides in the Vastitas Borealis plains (MC-1), which contain dense, near-surface ice that the spacecraft was able to dig into and investigate (Chapter 5). The surrounding landscape is flat with scattered pebbles but marked by interconnected troughs, forming a network of polygons, 2–3 m across, which demonstrate cracking and buckling of the permafrost soil (Figure 4.15).

Mars Science Laboratory rover – Curiosity

This most recent (and most capable) rover put down in the northwest floor of 154-km-diameter Gale crater (see MC-23), atop an alluvial fan. The rover is visiting a variety of science targets, with the emphasis on parts of the rock record that may reveal relatively wet climate conditions (Chapter 5). The rover's main objective is the lower slopes of Mount Sharp (officially named Aeolis Mons), where orbital observations show layering with sulfate and phyllosilicate-rich sequences (Figure 4.16).

FIGURE 4.8 Image of Elysium; the view spans about 125° E to 175° E, 5° N to 45° N (view about 3000 km E–W, cylindrical projection, MOLA color hillshade).

Mars Exploration Rovers – Spirit and Opportunity

These rovers successfully landed on Mars in January 2004 – Spirit first, on January 4, and Opportunity on January 25. Spirit landed on the spacious floor of 158-km-diameter Gusev crater (Figure 4.13;

MC-23). The crater floor occurs along the Martian highland–lowland dichotomy boundary and below the mouth of the ancient Ma'adim Vallis channel system, which crosses several hundred kilometers of the Martian highlands. However, the rover discovered that, instead of sediment, the plain was covered up of lava that had been gardened by

millennia of impacts. In search of a potential signature of water activity on early Mars, the rover traveled nearly 5 km to reach ancient outcrops, named the Columbia Hills (Figure 4.13), where the mission eventually came to an end.

The Opportunity rover landed near the equator on Mars, in a highland plain known as Meridiani

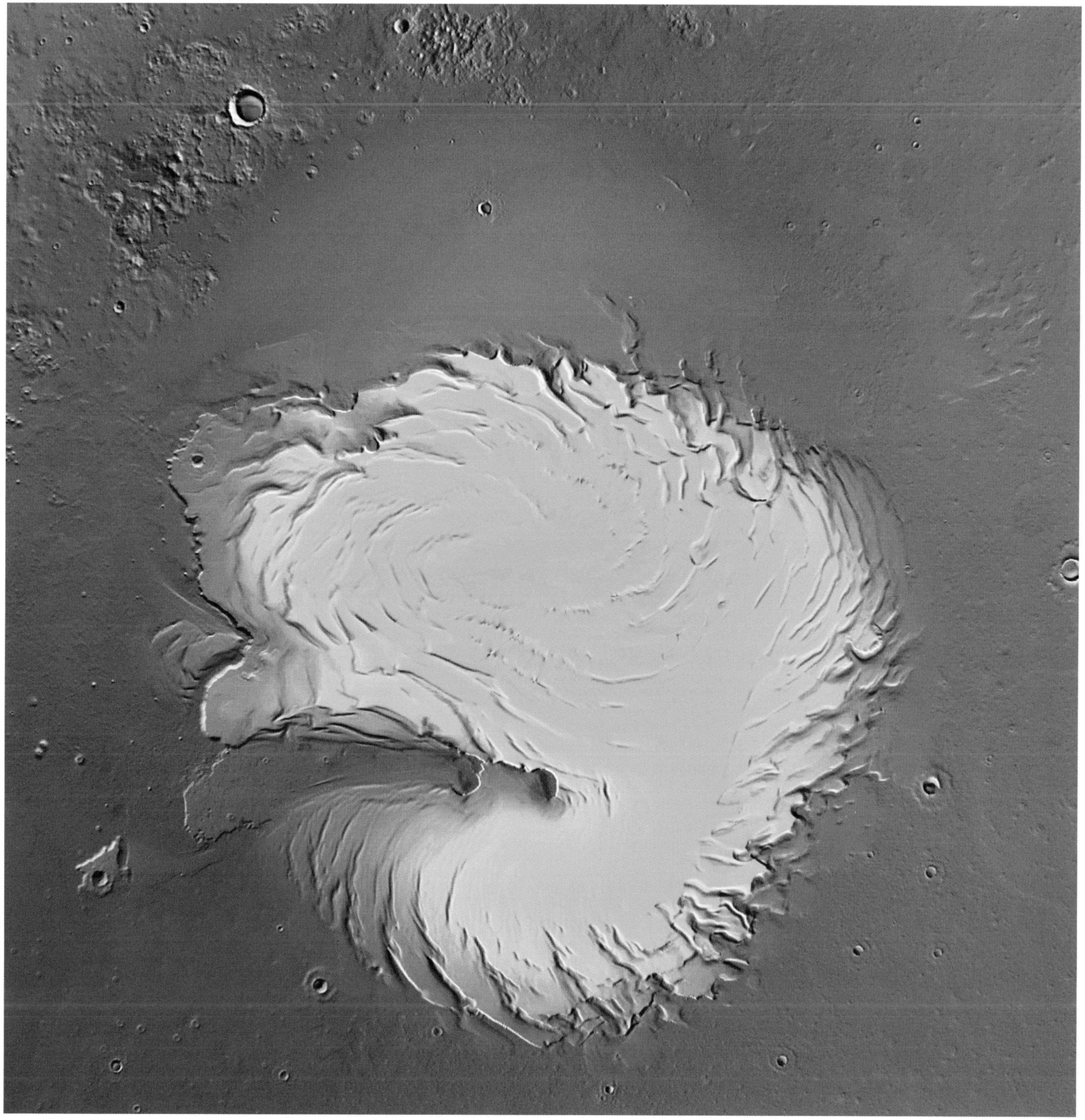

FIGURE 4.9 North polar plateau, north of latitude 77° N (view 1,500 km across, color MOLA hillshade image).

Views of the surface from landers

TABLE 4.2 Mars surface missions

Mission (rover name, where applicable)	Date(s) active on surface[a]	Entry, descent, landing	Landing site	Lander/ rover	Instruments	Major results
Mars 3	December 2, 1971	Aerobraking, parachute, retrorocket	Near 45° S 202° E	Lander/ rover	Soil penetrator and spectrometers (gamma and X-ray), measure temperature and gases in atmosphere, two panoramic television cameras	First to transmit from surface (only a few seconds)
Viking 1	July 30, 1976 to Nov 13, 1982	Aerobraking, parachute, retrorocket	22° N 310° E	Lander	Camera, surface sampler to find composition and test for chemical signs of life, meteorology measurements, seismometer	First successful science on surface: atmospheric composition, several years of weather data, soil composition resembles weathered basalt, signs of life lacking or ambiguous
Viking 2	September 3, 1976 to April 11, 1980	Aerobraking, parachute, retrorocket	48° N 134° E	Lander		
Pathfinder (Sojourner)	July 4, to September 27, 1997	Aerobraking, parachute and rockets, airbags	19° N 326° E	Lander/ rover	Alpha–proton–X-ray spectrophotometer, visible-near-infrared cameras, meteorology instruments	First successful rover on Mars, geometry of clasts in outflow channel, possible andesite or basaltic andesite and conglomerate rocks
Mars Exploration Rover-A (Spirit)	January 4, 2004 to March 22, 2010	Aerobraking, parachute and rockets, airbags	15° S 175° E	Rover	Cameras (panoramic, microscopic), Thermal Emission Spectrometer, alpha-particle X-ray spectrometer, Mossbauer spectrometer, rock abrasion tool	Basalt makes up floor of Gusev crater, water present during mineral alteration
Mars Exploration Rover-B (Opportunity)	January 25, 2004 to June 10, 2018	Aerobraking, parachute and rockets, airbags	2° S 355° E	Rover		Possible concretions and evaporite minerals formed by water, sedimentary rocks
Phoenix	May 25 to November 2, 2008	Aerobraking, parachute, retrorocket	68° N 234° E	Lander	Microscopy, electrochemistry and conductivity analyzer, stereo and arm cameras, gas analyzer, meteorology instruments	Ice found just below surface, perchlorate found in soil
Mars Science Laboratory rover (Curiosity)	6 Aug 2012[b]	Aerobraking, parachute and rockets, tether	5° S 137° E	Rover	Stereo camera, alpha particle X-ray spectrometer, laser-induced breakdown spectroscopy, X-ray diffraction/ fluorescence	Organic carbon in rocks, conglomerate indicating running water in past, methane in atmosphere

Chronology: http://nssdc.gsfc.nasa.gov/planetary/chronology_mars.html. [a] Dates are from landing to end of operations. [b] Still operating at the time of writing.

FIGURE 4.10 Viking 1 lander view to the east, 2 hours after sunrise, showing sharp-crested sand dunes. Large boulder at left lies 8 m from Viking 1 and is 1 × 3 m, while the meteorology boom crosses the center foreground (view is 100° across, Viking 1 image P17430, NASA/JPL-Caltech).

FIGURE 4.11 Viking 2 lander view to the northeast, in the afternoon, shows the variation in rocks, from porous (vesicular) to dense. The largest rocks are about 1 m wide. Material between the rocks has been redistributed by wind. The apparent tilt of the horizon is due to tilt of the spacecraft by 8°; the local horizon has no tilt (view is 85° across, L_s = 118°, Viking 2 image P17688, NASA/JPL-Caltech).

FIGURE 4.12 Pathfinder lander 360-degree color panorama. Three triangular lander petals and deflated airbags are visible across the bottom, while the "Twin Peaks" (informal name) are on the horizon to the west-southwest. The rover Sojourner is analyzing the large rock ("Yogi") at the center of the image, northwest of the lander (image MRPS94103, NASA/JPL-Caltech).

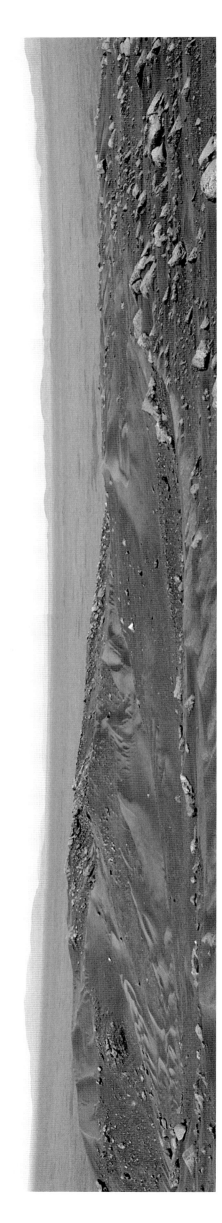

FIGURE 4.13 Spirit panorama of Gusev crater, seen looking east from "Columbia Hills" (informal name). The rim of Thira crater lies on the floor of Gusev, 15 km away. Sand has blown across the foreground hills (Spirit pan camera, blue filter, image taken Sol 581, view about 60° wide, NASA/JPL-Caltech/Cornell University).

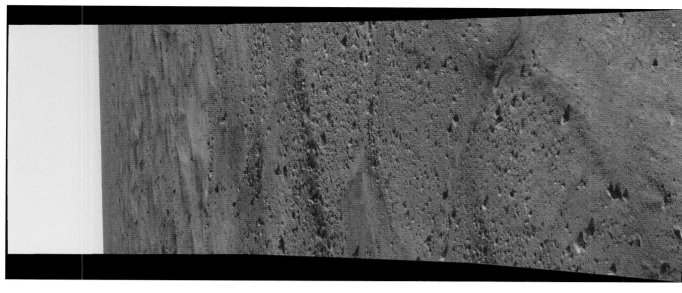

FIGURE 4.15 Polygonal cracks dominate this view of the northern plains that was taken by the Phoenix Lander. The cracks indicate seasonal expansion and contraction of subsurface ice (NASA/JPL-Caltech/University of Arizona).

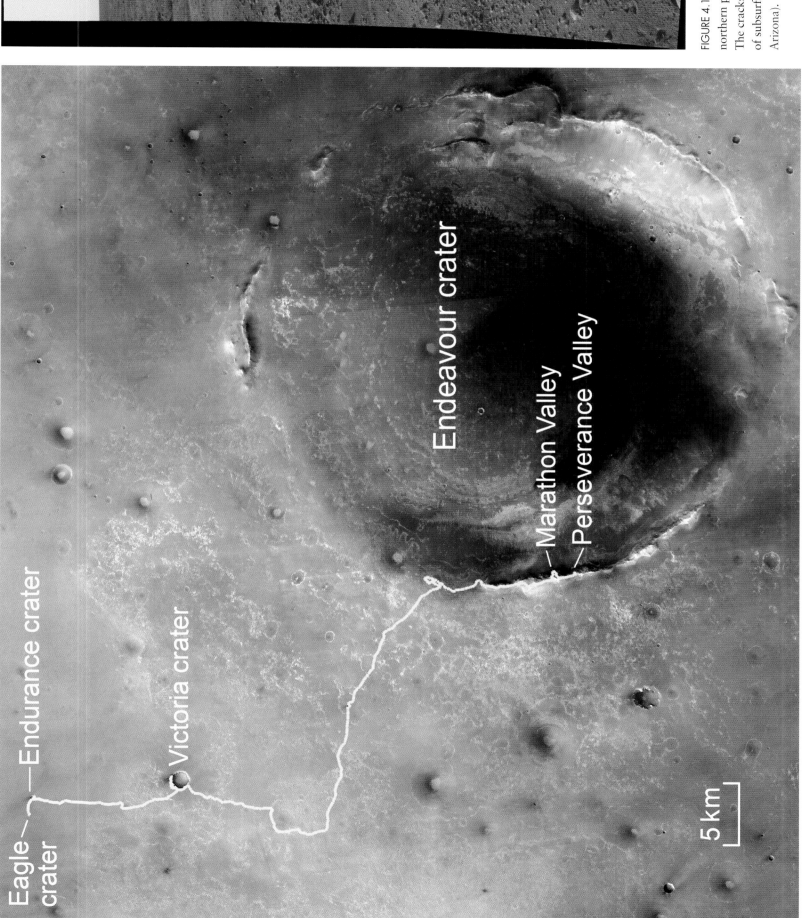

FIGURE 4.14 The traverse of Opportunity rover over 5,111 sols (more than 7 Martian years) covered 45 km on Meridiani Planum. Endeavour crater is 22 km in diameter (CTX mosaic, north at top, NASA/JPL-Caltech/MSSS).

Eagle crater

Endurance crater

Victoria crater

Endeavour crater

Marathon Valley

Perseverance Valley

5 km

Views of the surface from landers

The Atlas of Mars

The Atlas of Mars

FIGURE 4.16 The lower slopes of Mount Sharp (Aeolis Mons) consist of layered rocks that record the history of filling of Gale crater. The pointed mound at the center of the image is about 100 m high (Mast camera, NASA/JPL-Caltech/MSSS).

CHAPTER 5

Geology of Mars

A principal scientific objective for sending space-craft to Mars has been to explore its geology as interpreted from data returned to Earth (see Chapter 2 for a review of the history of Mars exploration). This exploration was spurred on when early spacecraft revealed a surface reflecting a long-lived, dynamic interplay of geologic forces, in many ways similar to those evident in our own Earth–Moon system. Thus, scientific investigation has focused on topics such as: What are Martian rocks made of, and how old are they? What do volcanic and tectonic features reveal about the interior of the planet? What is the history of impact bombardment? How much water has there been at or near the surface, and were there times of abundant surface water on the planet? Have there been gradual or sudden changes in the planet's climate? Is Mars geologically active today? Could life survive in environmental niches on the surface or below the surface, now or in times past?

These and other related questions have been researched extensively with a variety of imaging, topographic, spectral, radar, and other investiga-tions, from orbiters and flybys of Mars as well as landers and rovers. Many of the orbital results are summarized and/or find context in the geologic map of Mars (Tanaka *et al.*, 2014), which we present in a generalized form (the fully detailed map is available through the web page for this atlas: www.cambridge.org/atlasofmars). In add-ition, we highlight in additional sections the major geologic components and processes that are evi-dent on the surface and interior of Mars.

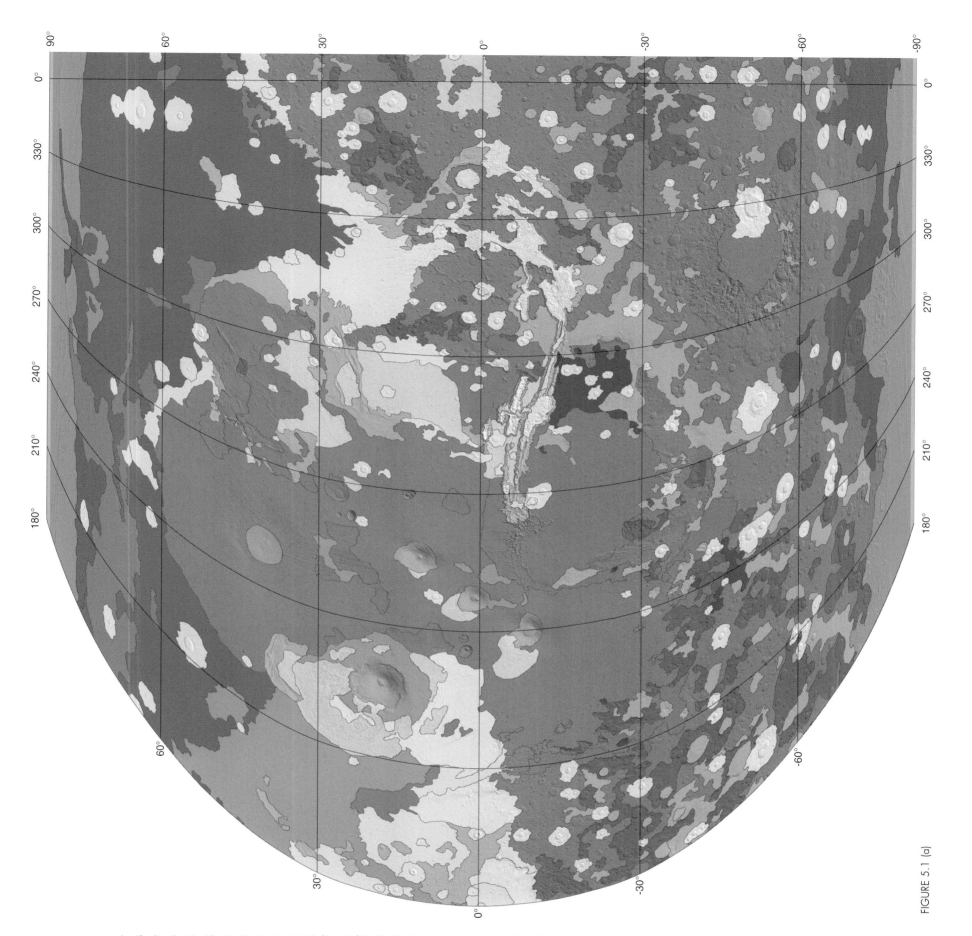

FIGURE 5.1 (a)

FIGURE 5.1 Simplified geologic map of Mars (see text for explanation; adapted from Tanaka *et al.*, 2014). (a) Western and (b) eastern hemispheres at 1:40,000,000 scale in Robinson projection over shaded relief base. (c) North (left) and south (right) polar regions at 1:25,000,000 scale in polar stereographic projection over topographic shaded relief base. (d) Correlation of map units, time-stratigraphic units (periods and epochs), and cumulative crater densities. Some colors appear darker on the maps than in this chart owing to the shaded relief base.

Geologic map overview

The geologic map of Mars (Tanaka *et al.*, 2014) was constructed primarily from landforms observable in Mars Global Surveyor Mars Orbital Laser Altimeter (MOLA) topographic reconstructions and in Mars Odyssey Thermal Emission Imaging System (THEMIS) infrared image mosaics, acquired during daylight hours. The surface is divided into 44 map units, separated according to geographic occurrence, geologic origin, and age. Each quadrangle in this atlas has a geologic map using these 44 units. Figures 5.1a–d shows a simplified version of the global map, where the units are combined, resulting in only 20 map units (the Appendix lists which units were combined). This is done according to unit group (ignoring further geomorphologic subdivisions) and age (Amazonian, Hesperian, and/or Noachian Period, ignoring epochal subdivisions; see explanation in the Appendix). The eight group types are: polar, impact, volcanic, apron, basin, lowland, highland, and transition. Table 5.1 describes each of these unit groups.

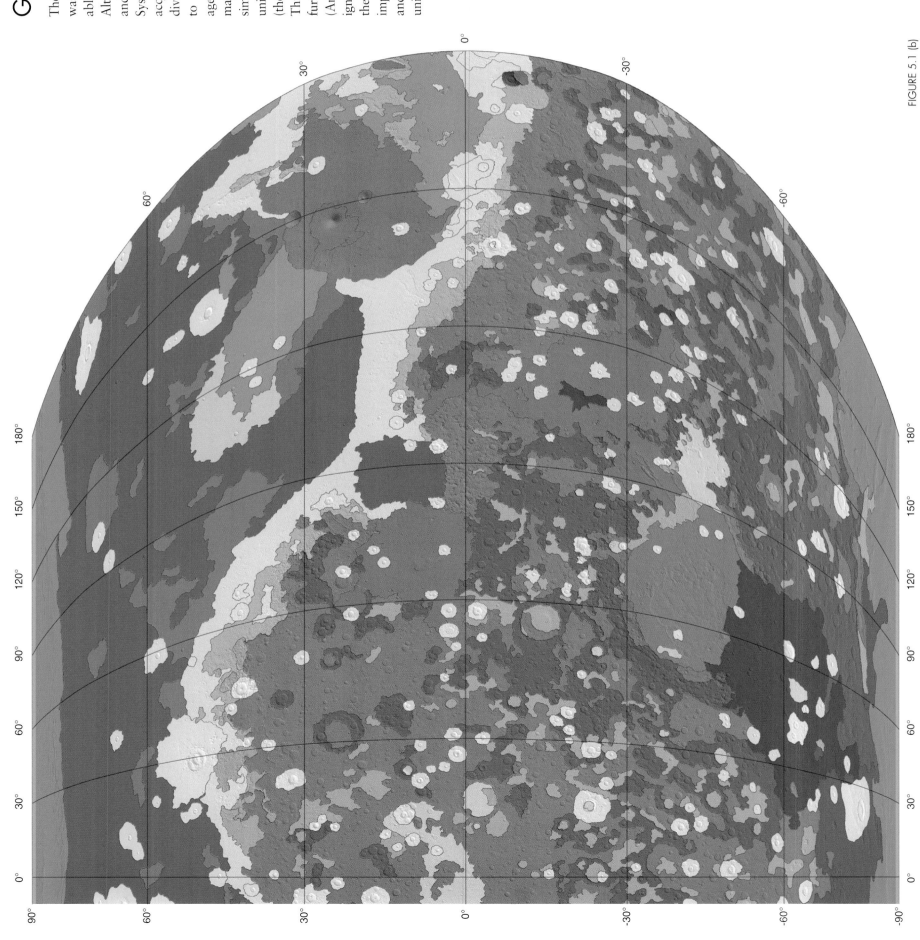

FIGURE 5.1 (b)

40

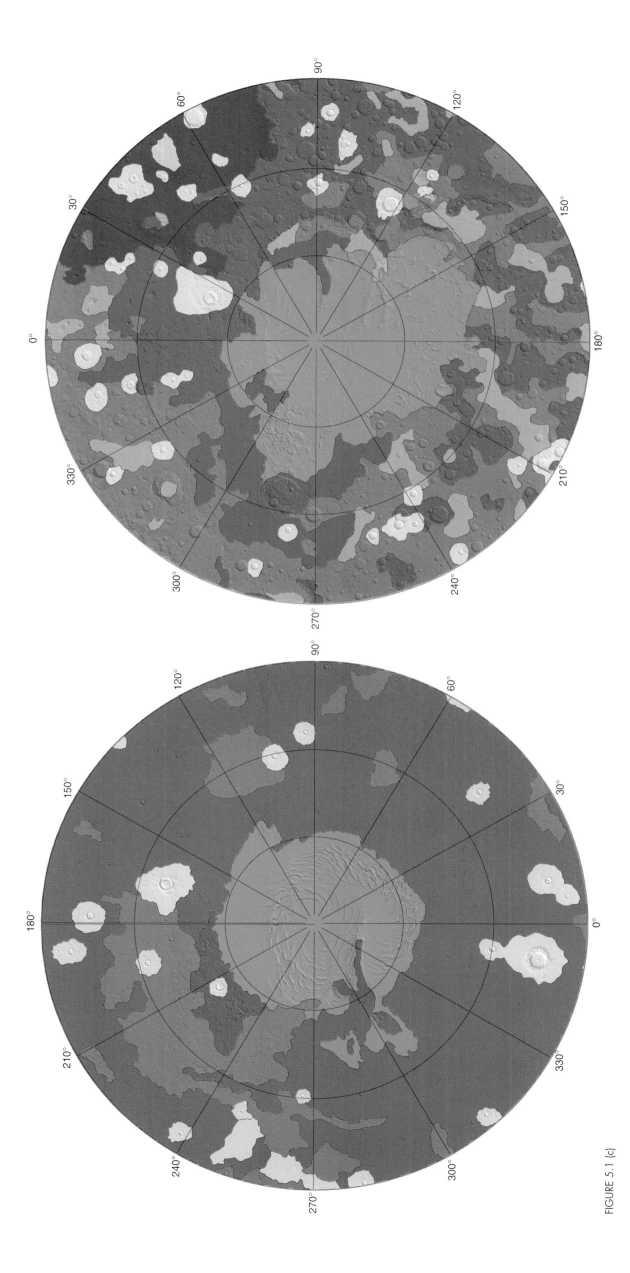

FIGURE 5.1 (c)

Geologic map overview

The Atlas of Mars

41

CORRELATION OF MAP UNITS (generalized to 20 units)

PERIOD	EPOCH	NO. CRATERS LARGER THAN 1, 5, AND 16 KM DIAMETER PER 1,000,000 KM²	LOW-LAND UNITS	IMPACT UNIT	POLAR UNITS	BASIN UNITS	VOLCANIC UNITS	APRON UNITS	TRANSITION UNITS	HIGHLAND UNITS	
AMAZONIAN	Late	1 5 16		AHi	Ap		Av	AHv	Aa	ANa	AHtu		
	Middle	---160---	mAl									
	Early	---600---										
HESPERIAN	Late	---2100---	lHl		Hp	eAb	Hv		ANa	Ht	HNt	Hh
	Early	---125---				HNb						
NOACHIAN	Late	---200---					Nv					Nh
	Middle	---100---										mNh
	Early	---200---										eNh

FIGURE 5.1 (d)

TABLE 5.1 Geologic unit groups

Unit group type	Description	Comments
Polar	Amazonian and Hesperian layered and other ice-rich deposits at both poles making up polar plateaus and other nearby deposits, as well as dune fields in north polar region	Although dunes occur all over Mars, large dune seas occur only in Vastitas Borealis plains encircling Planum Boreum
Impact	Largest, relatively undegraded Amazonian and Hesperian impact crater rims, interiors, and ejecta deposits	Noachian craters are generally highly degraded and incorporated into Noachian units
Volcanic	Broad flow fields and edifices that range from Noachian to Amazonian in age; local accumulations are kilometers in thickness	Mostly in Tharsis, Elysium, and circum-Hellas regions; many Noachian features possibly unrecognized
Apron	Deposits formed by ice-lubricated flow or avalanching of rock debris along relatively steep slopes of tallest Tharsis volcanic shields, canyon walls of Valles Marineris, and more abrupt parts of highland–lowland boundary (HLB)	Units formed during Amazonian, but include Noachian remnant materials in places along HLB
Basin	Floor deposits made up of sediments and volcanic rocks in broad, deep Hellas, Argyre, and Utopia basins	Local landforms indicate presence of water or ice during and after accumulation
Lowland	Relatively flat and smooth-surfaced deposits north of the HLB. Fairly ubiquitous Hesperian lowland unit possibly hundreds to thousands of meters thick; overlying Amazonian unit is patchy and mostly <100 m thick	Hesperian unit likely deposited by catastrophic floods; its margins possibly reworked by tsunami waves. Amazonian unit formed by ice and dust accumulation and largely eroded away
Highland	Noachian highland units dominated by impact materials, including rims of Hellas, Isidis, and Argyre basins; much more extensive and rugged than Hesperian highland unit, consisting of plains-forming, largely undifferentiated sedimentary and volcanic materials	Noachian surfaces largely degraded, including extensive valley incision, indicative of warmer, wetter climate
Transition	Materials crossing the geographic divide between highland and lowland units; ranging from Noachian to Amazonian in age	

The map also includes several types of line features. The most common types include: tectonic wrinkle ridges (Figure 5.2); grabens and pit chains (Figure 5.3); volcanic caldera rims and lobate flow directions (Figure 5.4); erosional valleys, outflow channel directions, spiral troughs, and yardangs (Figure 5.5); and impact craters (Figure 5.6).

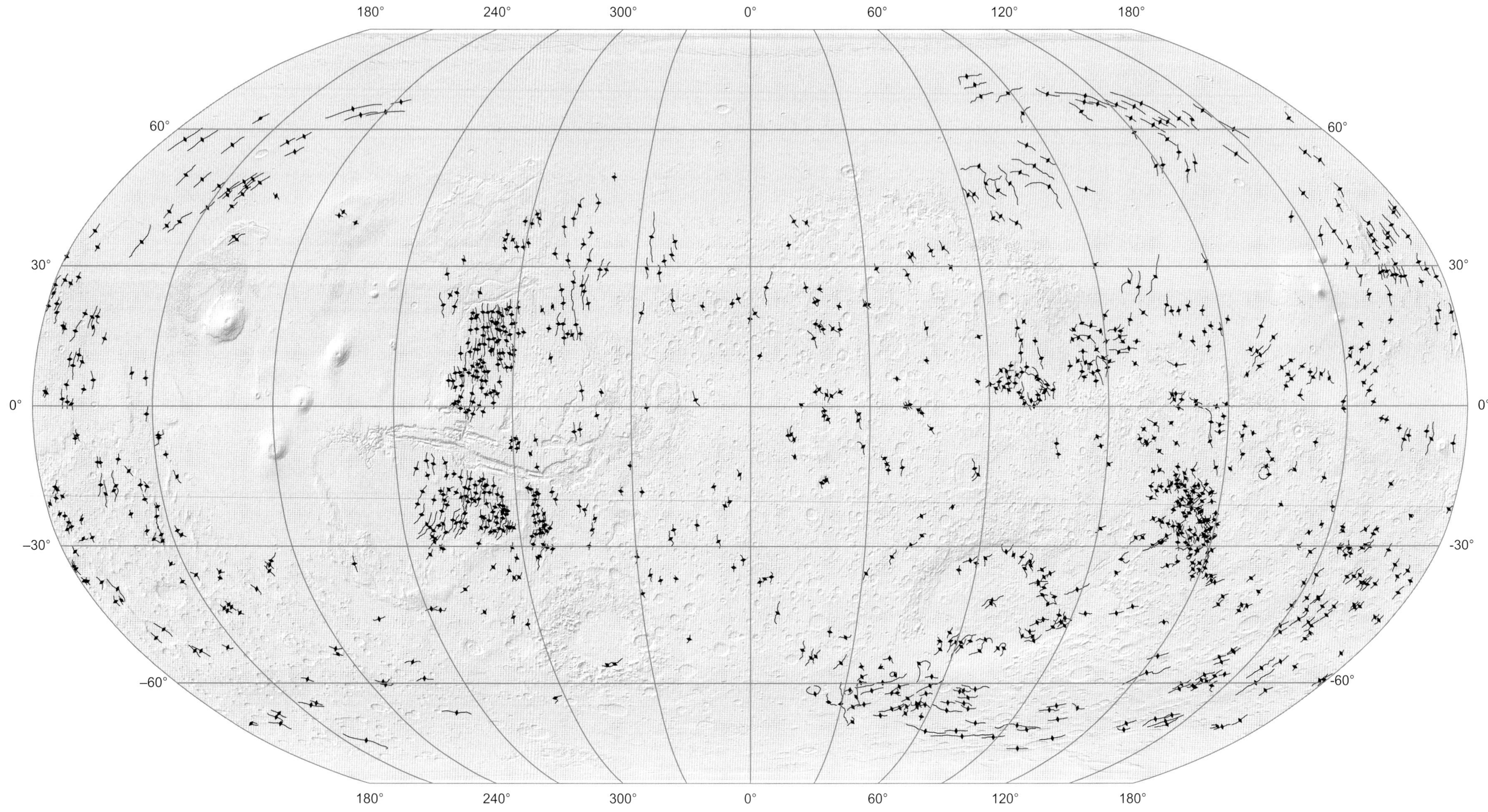

FIGURE 5.2 Wrinkle ridges on Mars (1:60,000,000 scale, Robinson projection over shaded relief base; adapted from Tanaka *et al.*, 2014).

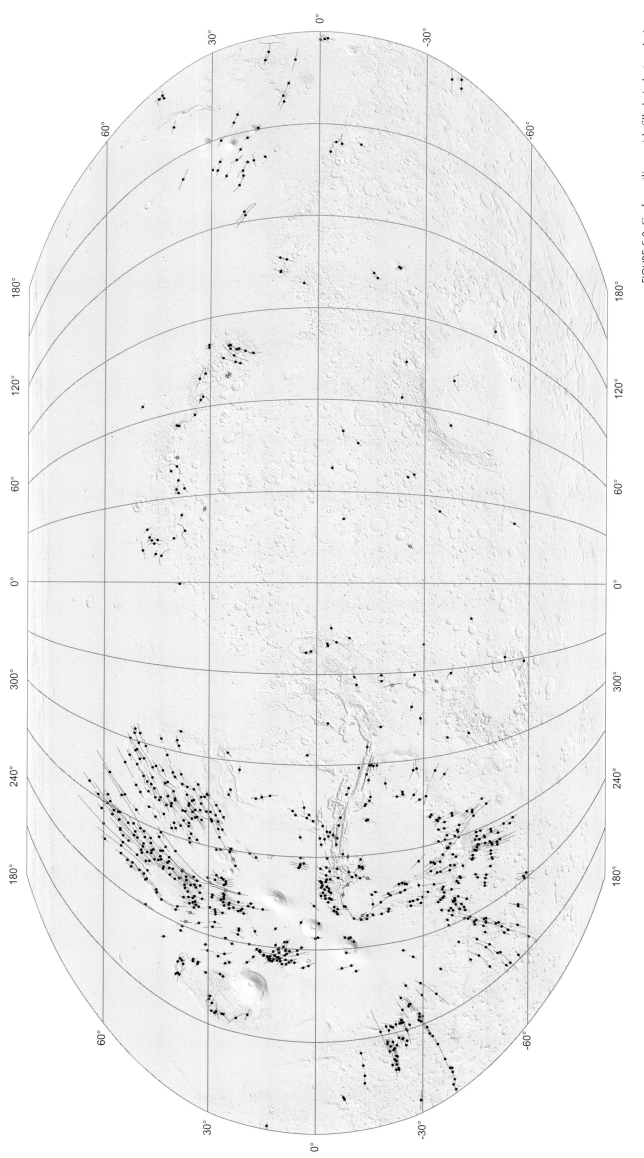

FIGURE 5.3 Grabens (lines with filled circles) and pit chains (lines with open red circles) on Mars (1:60,000,000 scale, Robinson projection over shaded relief base; adapted from Tanaka *et al.*, 2014).

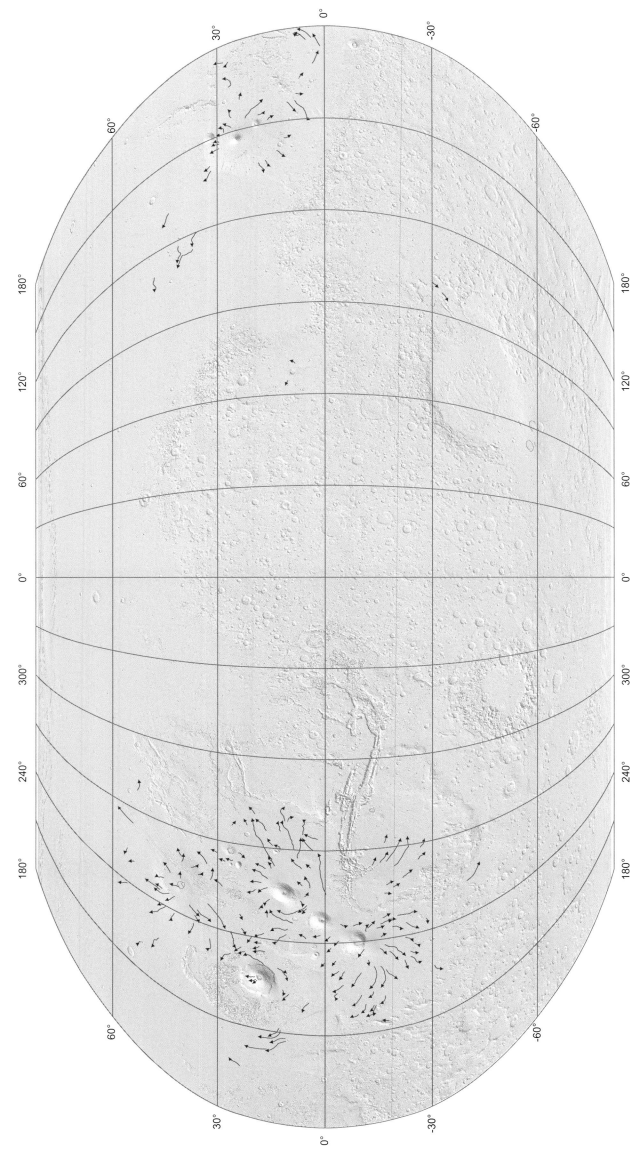

FIGURE 5.4 Volcanic caldera rims (arcs) and lobate flow directions (arrows) on Mars (1:60,000,000 scale, Robinson projection over shaded relief base; adapted from Tanaka *et al.*, 2014).

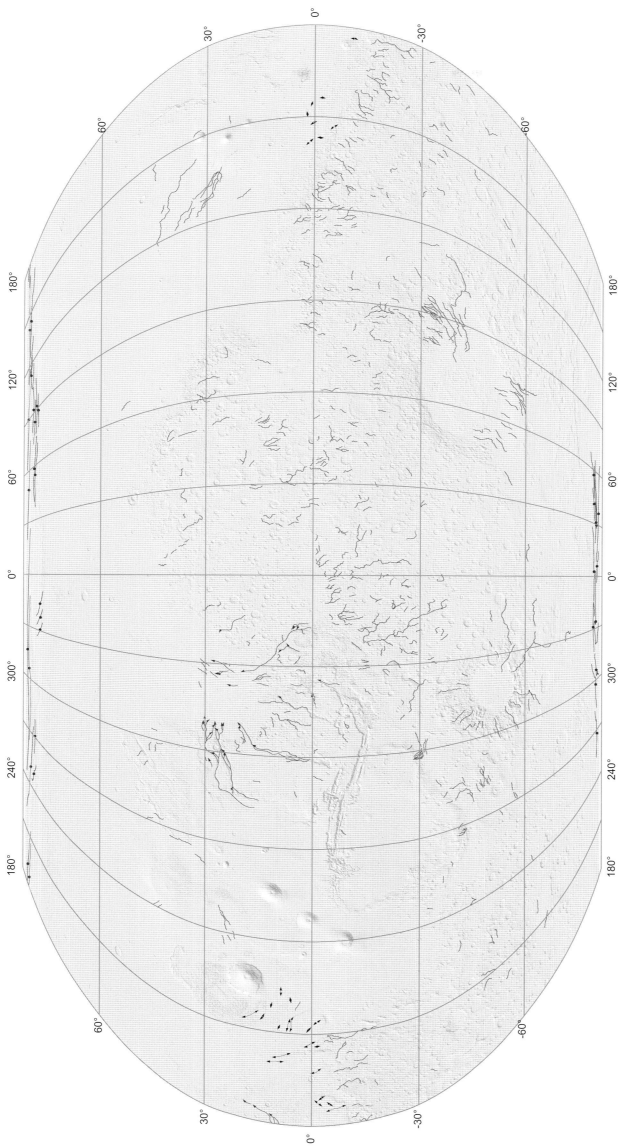

FIGURE 5.5 Erosional valleys (blue lines), outflow channel directions (blue arrows), spiral troughs (lines with solid circles in polar regions), and yardangs (black double-ended arrows) on Mars (1:60,000,000 scale, Robinson projection over shaded relief base; adapted from Tanaka *et al.*, 2014).

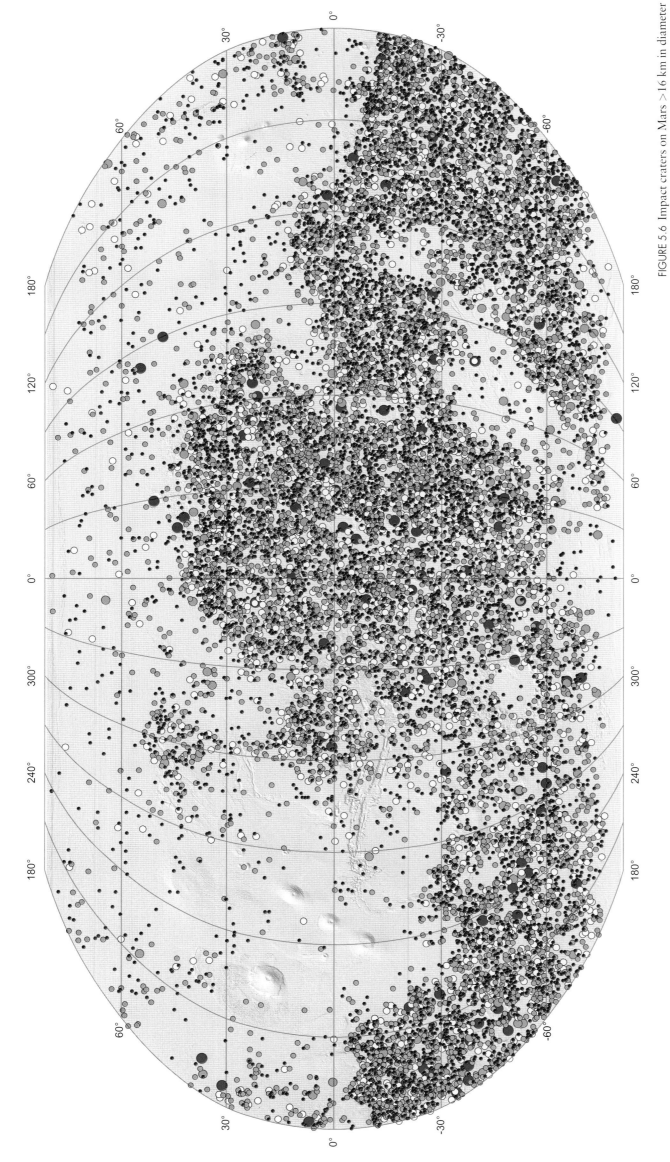

FIGURE 5.6 Impact craters on Mars >16 km in diameter (1:60,000,000 scale, Robinson projection over shaded relief base; Robbins and Hynek, 2012; map adapted from Tanaka *et al.*, 2014). Crater diameters within ranges indicated by colored circles of graduated size: red, 16–25 km; orange, 25–50 km; yellow, 50–100 km; green, 100–150 km; blue, >150 km.

Geologic timescale for Mars

The surface history of Mars is defined by major geologic eras, as revealed by geologic mapping (Figures 5.7, 5.8; see also the Geologic map overview above). The resulting chronologic units are the Noachian, Hesperian, and Amazonian Periods, which are subdivided into the Early Noachian, Middle Noachian, Late Noachian, Early Hesperian, Late Hesperian, Early Amazonian, Middle Amazonian, and Late Amazonian Epochs. The time boundaries of these units are dated in a relative manner by impact crater size–frequency distributions.

Assuming spatially random cratering, older surfaces collect higher densities of craters at given diameters. For each epoch boundary, a given diameter is chosen to optimize the statistical dataset, in which the cumulative number of craters larger than the given diameter per million square kilometers is determined (see Figure 5.1d and the Appendix). The older boundaries require a larger minimum diameter because smaller craters tend to get destroyed, thus yielding less reliable crater densities. However, for the younger boundaries, smaller diameters are chosen to increase the sample set. Alternatively, crater size–frequency distributions, based on two diameters, can be fitted to a standardized crater-production function for the planet. These methods have statistical uncertainties that affect the accuracy and precision of the results.

In turn, crater size–frequency distributions can be converted to estimated absolute ages by use of model impact cratering rates. Such models are based on the cratering rate history determined for the Earth's Moon and an estimation of the relative cratering rates of the Moon and Mars. Another approach is to use an estimated cratering rate based on the populations of observed Mars-crossing asteroids. The timescale for Mars given in Figures 5.7 and 5.8 shows epoch age boundaries, based on two equally regarded, current models of crater production for Mars (Hartmann and Neukum, 2001), using the approach involving the cratering rate for Mars relative to the Moon.

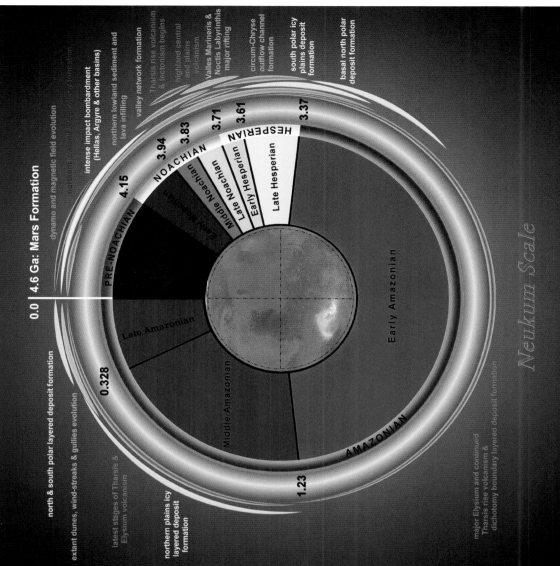

FIGURE 5.8   Martian geologic time wheel using the Neukum timescale (courtesy of Donald C. Barker).

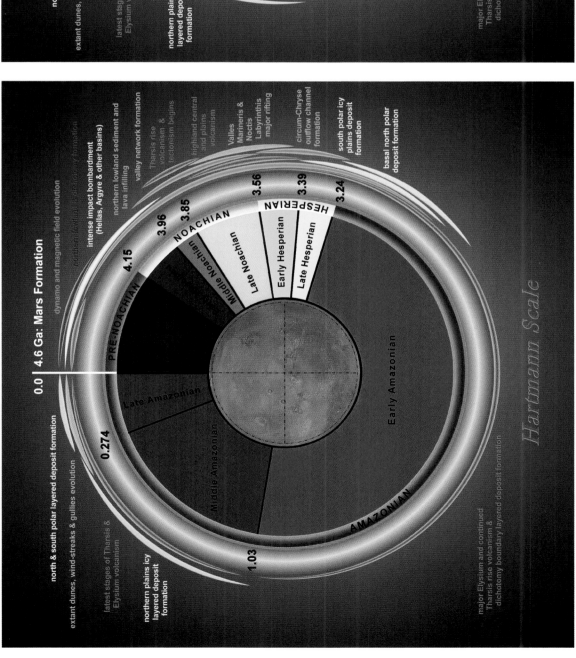

FIGURE 5.7   Martian geologic time wheel using the Hartmann timescale (courtesy of Donald C. Barker).

Geologic history

The geologic map and feature maps (Figures 5.1 to 5.6) portray the surface of Mars through time at regional to global scale (some of the specific geologic processes and resulting features are described in more detail in following sections). Major episodes of geologic processes are shown on the timescale (Figures 5.7, 5.8), while Figure 5.9 shows units on each appropriate time slice. The oldest unit, the Early Noachian highlands unit (eNh, >3.94 Ga; ages in this discussion are from the Neukum timescale[1]), comprises the most densely cratered and rugged parts of the southern highlands, particularly surrounding Hellas basin. Presumably, Early Noachian materials form the upper crust of Mars, which shows no clear evidence of having been recycled by plate tectonics. However, the northern plains, volcanic flow fields, basin and crater floors, and intracrater lows are largely covered by sufficient thicknesses of younger materials to bury the Early Noachian landscapes.

The Middle Noachian highland unit (mNh, 3.94–3.83 Ga) spans the entire cratered highlands and is the most abundant highland unit. This unit, like the unit eNh, is buried by younger materials in the lower areas, but less so. The Noachian highland (Nh), Hesperian and Noachian transition (HNt) and Amazonian and Noachian apron (ANa) units are made up of substantial amounts of Early to Late Noachian materials. In many cases these have been modified by younger degradational processes. The latter units (HNt and ANa) include younger deposits reworked from Noachian rocks, whereas the unit Nh includes Noachian materials that in some cases were eroded at various times during the Late Noachian (3.83–3.71 Ga), and later. Noachian highland erosion included widespread valley incision as well as degradation of steeper parts of the highland–lowland boundary. This was perhaps the result of a thicker, warmer atmosphere with precipitation that resulted in surface runoff and erosion. The Hesperian and Noachian basin unit (HNb) represents sedimentary fill deposited in Hellas and Argyre basins. Meanwhile, in parts of the Tharsis region, on the southwestern flank of Hellas basin, and in various highland locations, Noachian

volcanic flow fields and vents arose and are mapped as the Noachian volcanic unit (Nv). Throughout the highlands, Noachian surfaces are marked by wrinkle ridges (Figure 5.2), indicative of global contraction as a consequence of planetary cooling. Surrounding the Tharsis volcanic rise, radial fracture systems (grabens in Figure 5.3) developed, apparently as a result of loading of the lithosphere by a massive accumulation of volcanic rocks. The most dramatic tectonic feature – the Valles Marineris canyon system – began opening by the Late Noachian Epoch. Meanwhile, the impact cratering declined dramatically over the course of the Noachian Period, so that Late Noachian surfaces are moderately cratered in comparison to older surfaces.

The beginning of the Hesperian Period (3.71–3.37 Ga) is marked by a near-cessation of widespread highland valley formation. As a consequence, highland infilling of low areas lessened and resulted in an accumulation of the Hesperian highland unit (Hh) and younger sequences of the Hesperian and Noachian basin unit (HNb). In addition, greatly decreased surface erosional rates allowed for better preservation of larger impact craters and their ejecta, mapped as the Amazonian

and Hesperian impact unit (AHi). Various erosional and depositional processes along the highland–lowland boundary contributed to the degradation of Noachian terrains and nearby sedimentation, resulting in the Hesperian and Noachian transition unit (HNt). More continuous deposits, particularly along the upper plains lying north of and below the highland–lowland boundary, are mapped as the Hesperian transition unit (Ht). In addition, this unit includes layered sediments that accumulated within Valles Marineris as well as deposits filling the outflow channel floors that connect some of the Valles Marineris canyons to Chryse Planitia – also covered by unit Ht. These deposits transition northward into the Late Hesperian lowland unit (lHl), covering much of the northern plains – some consider the unit an enormous sedimentary unit resulting from catastrophic floods and mass wasting of the highlands. Also, layered deposits along the highland–lowland boundary west of the Tharsis rise, mapped as the Amazonian and Hesperian transition undivided unit (AHtu), appear to be primarily deposited by the wind. Volcanic flows that are mapped as the Hesperian and Amazonian and Hesperian volcanic units (respectively, Hv and AHv) were emplaced in parts of the

Tharsis and Elysium rises and across Hesperia Planum and Syria Planum. Many of these volcanic rocks and parts of unit lHl are crossed by wrinkle ridges and grabens, indicating that planetary contraction and volcanic loading continued to drive tectonic deformation globally and regionally. The first clear indication of polar geologic activity produced icy plains deposits and mound-like structures, mapped as the Hesperian polar unit (Hp), perhaps resulting from ice-related deformation and/or cryovolcanism.

In the Amazonian Period (3.37 Ga to present), geologic activity continued but at lower rates, resulting in further development of the units AHi, AHv, and AHtu. In addition, patches of the Amazonian volcanic unit (Av) form the upper parts of large shield volcanoes in Tharsis and lava plains scattered around Tharsis rise, southeast of Elysium rise, and in Utopia basin. The Early Amazonian basin unit (eAb) includes sediments accumulated in Hellas and Utopia basins. The Amazonian and Noachian apron unit (ANa) resulted from mass wasting of, and ice accumulation on, mid-latitude, Noachian slopes of the highland–lowland boundary and massifs along the northeastern rim of Hellas basin. The

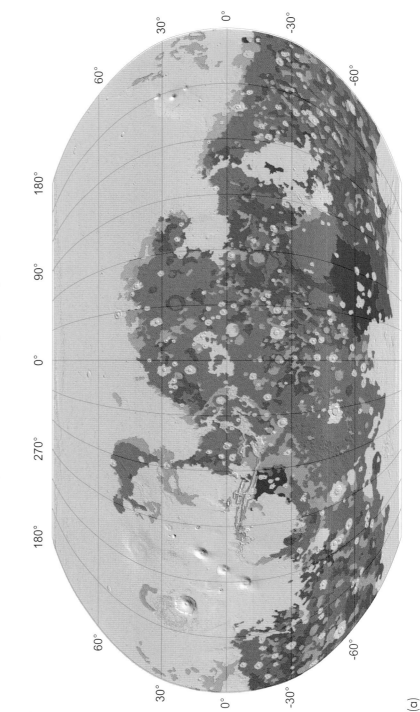

(a)

[1] Ga denotes giga-annum, for ages in billions (10[9]) of years before present.

Amazonian apron unit (Aa) consists of landslide deposits within Valles Marineris and what appear to be glacial deposits on the northwestern flanks of Tharsis Montes and Olympus Mons. Patches of the Middle Amazonian lowland unit (mAl) are scattered throughout the northern lowlands, superposed on the unit lHl. The patches testify to an episode of broad, lowland accumulation of dust and ice, followed by extensive erosion of the unit. The polar regions display near-pole centered plateaus made of layered sequences of the Amazonian polar unit (Ap). In the north polar region, frozen sand dunes form part of the lower sequence of the unit, and recently active dune seas surround the polar plateau.

FIGURE 5.9 Units of the geologic map of Mars (Figure 5.1) plotted by geologic period, showing how the location and type of unit varied through time. (a) Noachian; (b) Hesperian; (c) Amazonian. Units that span more than one period are shown on both appropriate time slices.

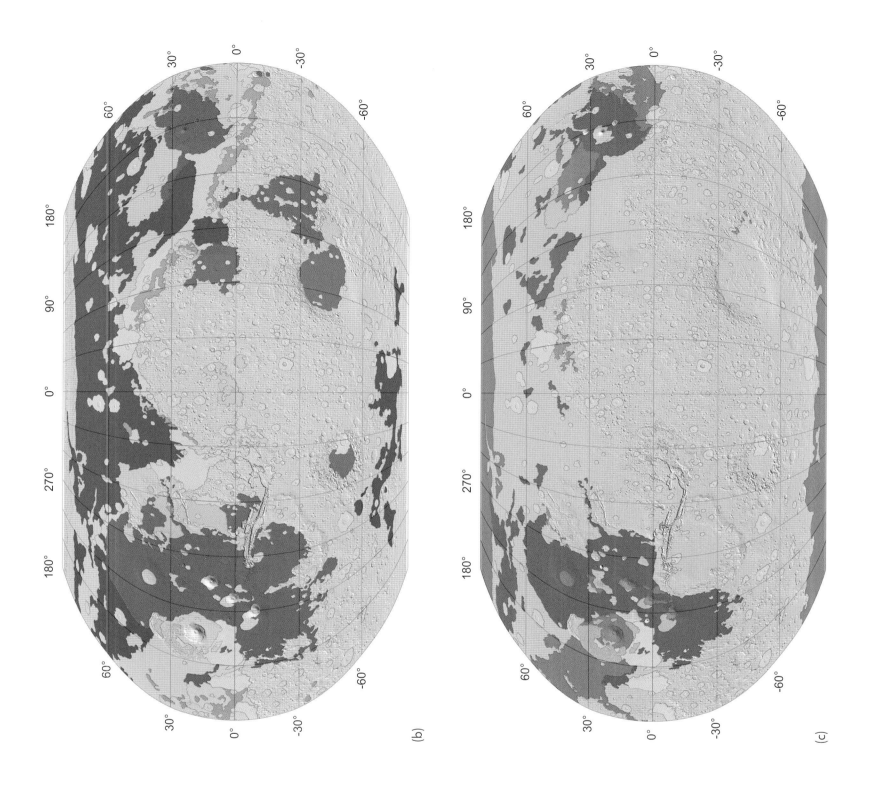

(b)

(c)

Interior of Mars

Like all the terrestrial planets, Mars is differentiated, or layered, inside. The migration of material into layers began very early, less than 20 million years after accretion, as indicated by the abundances of short-lived isotopes that were incorporated into Mars and separated from their decay products by formation of the core (Kleine *et al.*, 2002). Similar evidence shows that the extraction of much of the crust from the mantle happened less than 100 million years after Mars formed (Nimmo and Tanaka, 2005).

Magnetic anomalies discovered by Mars Global Surveyor in the ancient crust of the southern highlands (Figure 3.5 in Chapter 3) indicate that early Mars had a magnetic dynamo in its core that created a global magnetic field, a feature lacking at present. The strength of the anomalies suggests that a substantial layer (tens of kilometers thick) carries the magnetization. The anomalies are not found in the large Noachian impact basins, which are thus thought to post-date the demise of the magnetic dynamo, sometime before the start of the Noachian Period (Nimmo and Tanaka, 2005).

Crust

Gravity and topographic data from Mars together with estimates for the density of the mantle indicate that the crust thins from south (50–60 km) to north (less than 40 km) across the dichotomy boundary (Figure 3.4 in Chapter 3; thickness estimates of Neumann *et al.*, 2004; Carr, 2006, p. 277; Genova *et al.*, 2016). The dichotomy itself and the subsurface structure of the crust were present by the start of the Noachian Period (Nimmo and Tanaka, 2005).

What is the origin of the crustal dichotomy? This is one of the most significant questions about the geology of Mars. One possibility is that one huge, or several overlapping, large impact basins account for the low elevation of the northern hemisphere of Mars (see the next section; Carr, 2006; Andrews-Hanna *et al.*, 2008; Carr and Head, 2010). Conversely, an early impact into the south polar region may have resulted in relative crustal thickening of the southern hemisphere (Leone *et al.*, 2014). Another explanation involves internal convection processes (Nimmo and Tanaka, 2005; Carr and Head, 2010). A key question about the northern half of Mars is why the

surface is so smooth and free of craters. Volcanic rocks and a sediment layer, from lake or ocean water and/or from mud or debris flows, probably buried the older crust. This gives the present appearance in which only outliers of highland material and ghost outlines of some older craters are visible.

Impacts

Impact cratering is ubiquitous on solid surfaces in the solar system. Craters occur on every scale, from microscopic to a substantial fraction of the diameter of the target body. The form of craters varies with their size (Figure 5.10), but all display a circular, raised rim and a depressed interior sometimes with later fill. Craters that are well-preserved commonly show *ejecta*, material deposited outside the rim by the impact process. As a surface ages, crater density increases; counting craters is the primary method of relative dating of planetary surfaces. Table 4.1 (Chapter 4) lists identified large impact basins as well as several inferred or buried basins.

Even larger buried or inferred basins on Mars are proposed, based on geologic or geophysical

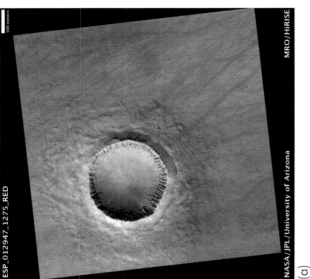

ESP_012947_1275_RED

MRO/HiRISE

NASA/JPL/University of Arizona

(a)

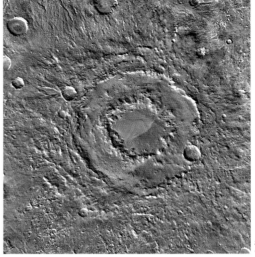

(c)

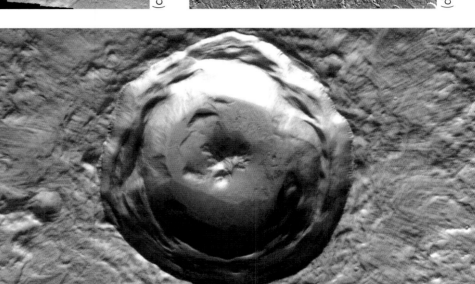

(b)

(d)

FIGURE 5.10 (a) Simple crater about 1.5 km in diameter (HiRISE image ESP_012868_1275, 25 cm/pixel, north at top, NASA/JPL-Caltech/University of Arizona). (b) Crater in Isidis basin, about 18 km across, having a central peak and slump terraces on the inner crater wall (THEMIS visible image V05914015, north slightly to left of top, 35 m/pixel). (c) Crater about 19 km across showing central peak (THEMIS visible image V05391006, 17 m/pixel, north at top). (d) Lowell, a multi-ring basin formed by impact, is 200 km in diameter (THEMIS daytime infrared mosaic, 100 m/pixel, north at top, all THEMIS images NASA/JPL-Caltech/Arizona State University).

cornerstone of the Tharsis rise, peaks out at an amazing elevation of 21,229 m and is covered by Amazonian lava flows. Surrounding this edifice is a boundary scarp rising as high as 8 km (MC-8). It may be the composite head scarp of huge, lobate avalanches, known as the aureole deposits. Amazonis Planitia includes the flattest plain on Mars and displays Amazonian lava flows that originate from southern and eastern parts of the Planitia. The Tharsis rise is capped by the broad ridge formed by the three aligned Tharsis Montes – Ascraeus Mons, Pavonis Mons, and Arsia Mons (MC-9, MC-17). These volcanoes are surrounded by broad fans of Late Noachian to Amazonian lava flows, intermediate-sized shields including Ceraunius, Biblis, Jovis, Tharsis, Ulysses, and Uranius Tholi, and local small shield fields. South of Arsia Mons, the flows of Daedalia Planum onlap Noachian highland terrain, which is dotted by dozens of isolated massifs, suspected to be Noachian volcanic edifices.

evidence (see Table 4.1). Utopia Planitia forms the floor of a giant circular basin northeast of Isidis Planitia. Utopia approaches 3,000 km in diameter and is probably the vestige of an ancient impact basin, as are other circular but generally less well-defined broad depressions across Mars (Schultz et al., 1982; Frey, 2006; Figure 5.11).

Volcanism

Volcanic activity is recorded from the formation of the crust early in the history of Mars up to the Late Amazonian Epoch. Volcanic deposits and landforms are varied and include volcanoes with summit calderas, flood lavas that erupted from vents or fissures, ash deposits, and related intrusive activity.

Estimates for relief (base to summit) and volume of some of the largest volcanoes on Mars (Table 5.2) show striking contrasts with Earth. On Earth, plate-tectonic motion ensures that individual volcanic centers are active for at most a few million years; on Mars, their lifetime is generally much longer. A more representative comparison is the entire Hawaii–Emperor seamount chain (eruption duration less than 100 million years) with the largest Tharsis volcanoes, which took longer (perhaps more than 3 billion years) to reach their present size.

Volcanism in Tharsis was regional and long-lived, creating a complex, nearly hemispherical rise. One of the oldest parts is the Thaumasia plateau (informal name, Figure 4.2), bordered by contractional scarps and ridges on most of its flanks, whereas the northern margin includes the troughs and grabens of Noctis Labyrinthus and Valles Marineris. The plateau is locally dissected by rifts and grabens along which irregular volcanic edifices developed during the Noachian Period. Volcanism continued during the Hesperian Period in the interior of the plateau, including large lava flows as well as local small shields and vents (see MC-17). North of Valles Marineris, local rilles and fractures are vents for Early Hesperian flows that cover Lunae Planum. In Tempe Terra, low shields and small vents source lava-flow fields (MC-3). Alba Mons and Ceraunius Fossae form the locus of broad flow fields that radiate as far as 1,200 km from fissures and partial caldera structures at the summit of Alba Mons. Olympus Mons, the northwestern

The Elysium volcanic rise (MC-7, MC-15) is similar to, but smaller than, the central and western parts of the Tharsis rise. It is capped by the Elysium Mons shield, from which Amazonian and Hesperian flows radiate and onlap the Hecates Tholus and Albor Tholus shields. The western flank of the rise is dissected by complex troughs and rilles (MC-15) that form the sources of an extensive field of lava flows and possible lahars (MC-7), which cover much of the broad Utopia basin floor. Southeast of the rise, Late Amazonian flows originated from Cerberus Fossae and flooded the Athabasca and Marte Valles channel systems (MC-15), producing the broad lava plains of Cerberus Palus and southwestern Amazonis Planitia.

Volcanic plains originating from broad depressions and low shields occur along the southern and northeastern rims of Hellas basin. The Hellas volcanic materials are mainly friable, likely consisting of pyroclastic deposits emplaced during the Late Noachian Epoch. These include the furrowed and ridged plains of Malea Planum (MC-27), marked by the circular Amphitrites and Peneus Paterae, as well as Malea and Pityusa Paterae. Northeast of Hellas basin, Hadriacus Mons is furrowed and the basal materials of Tyrrhenus Mons (MC-22) are deeply etched and thus are likely made up of Late Noachian pyroclastic materials. However, Tyrrhenus includes a set of

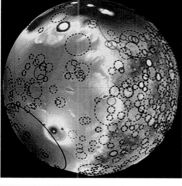
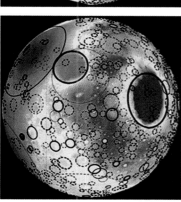
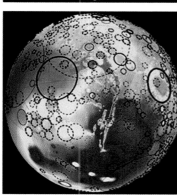

FIGURE 5.11 "Quasi-circular depressions," identified in MOLA data, are combined with visible basins, >200 km in diameter, over colored MOLA topography (blues are lowlands, and reds are highlands; from Frey, 2006). The number of buried basins inferred in the northern lowlands suggests the ancient nature of the buried crust there (solid circles are visible basins, dashed circles are buried or inferred basins; from left to right are equatorial views centered at longitudes 300° E, 60° E, and 180° E).

TABLE 5.2 Major volcanoes on Mars and Earth

Volcano	Map (MC number)	Relief (km)	Dimensions of base (km × km)	Volume (10³ km³)
Olympus Mons	8/9	21.9	840 × 640	2,400
Alba Mons	3	5.8	1015 × 1150	1,800
Ascraeus Mons	9	14.9	375 × 870	1,100
Arsia Mons	17	11.7	461 × 326	920
Pavonis Mons	9/17	8.4	380 × 535	390
Syrtis Major	13	4	1000 × 1400	160–320
Elysium Mons	15	12.6	375 (diameter)	200
Apollinaris Mons	23	5.4	189 × 278	73
Hecates Tholus	7	6.6	177 × 187	67
Uranius Mons	9	3.0	242 × 280	35
Tharsis Tholus	9	7.4	131 × 158	31
Albor Tholus	15	4.2	157 × 164	29
Ceraunius Tholus	9	6.6	98 × 130	24
Tyrrhenus Mons	22	1.5	215 × 350	21
Biblis Tholus	9	3.6	128 × 176	18
Hadriacus Mons	22/28	1.1	330 × 550	16
Mauna Loa	(Earth)	9	120 (diameter)	42.5
Entire Hawaii–Emperor Chain	(Earth)	NA	NA	1,081

Sources: Bargar and Jackson, 1974; Hiesinger and Head, 2004; Plescia, 2004; Carr, 2006.

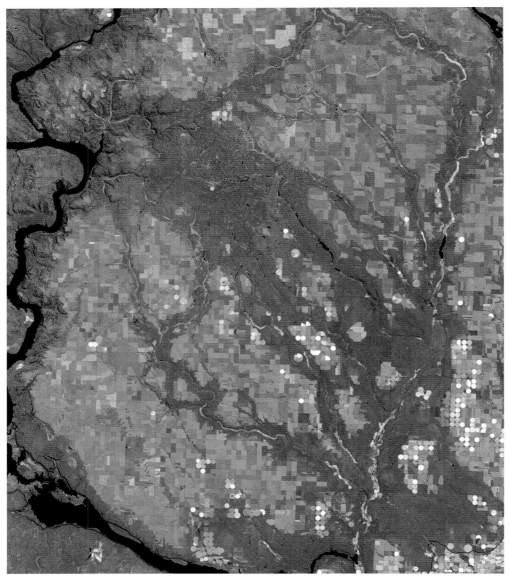

FIGURE 5.12 View from Earth orbit of the Channeled Scablands in Washington State, USA. Catastrophic flooding from northeast to southwest (upper right to lower left) left the dark regions of bare rock. Square and circular farm fields appear where surface soils escaped the erosive flow (stretched monochrome of Landsat-8 image LC80440272013202LGN00 taken July 21, 2013; 30 m/pixel, view 80 km by 95 km, north at top, courtesy of the US Geological Survey).

Hesperian lava flows, and the surrounding plains of Hesperia Planum may also comprise volcanic materials. A variety of other features on Mars have been proposed to be of volcanic origin, including several large irregular depressions in northwestern Arabia Terra (Michalski and Bleacher, 2013), the large outflow channel systems (Leverington, 2011), and the degraded, scattered mountains across the highlands (e.g. MC-24) that follow a dozen possible alignments (in association with other identified volcanoes) planetwide. The alignments are proposed to be the result of migrating mantle plumes caused by a primordial south polar giant impact (Leone, 2016).

The compositions of volcanic rocks on Mars have been analyzed from Mars orbit, by surface measurements, and in Martian meteorites. Most common are basalt and basaltic andesite (analyses reveal less than 52 weight percent [wt%] SiO_2), indicative of sources in the mantle without extensive evolution of magma observed in the continental crust settings on Earth. Analyses of data from the Curiosity rover, however, indicate the presence of igneous rocks, richer in silica (SiO_2, up to 67 wt %), some of which are also rich in alkali metals (sodium and potassium). The diversity and coarse textures of these rocks are consistent with those expected during the formation of continental crust (Sautter et al., 2015).

Tectonics

On Earth, plate tectonics, or the motion of segments of the lithosphere, accounts for many of the largest physiographic features, including the contrast between continental and oceanic lithosphere. Whether or not plate tectonics ever operated on early (pre-Noachian) Mars and which features it may have created is unresolved. Rollback of a subducting slab has been suggested as controlling the location and timing of Tharsis volcanism (Yin, 2012), while sea-floor-type spreading (Sleep, 1994) or global-scale convection (Zhong and Zuber, 2001) have been proposed as the cause of the global dichotomy (see the discussion in Nimmo and Tanaka, 2005) rather than the oft-cited giant impacts.

The geologic map of Mars (described above) records tectonic development by way of resulting landforms beginning in the Noachian Period until recent times. Wrinkle ridges (Figure 5.2) indicate tectonic contraction resulting from deep-seated (several kilometers or more) thrust faulting (e.g. Golombek et al., 2001). For the most part, wrinkle ridges deform mainly Noachian and Hesperian surfaces and were produced by global compression that resulted from planetary cooling, Tharsis-induced regional stresses arising from volcanic weighting of the lithosphere, and local contraction such as within volcanic calderas. Broad grabens (>10-km wide), largely in the Thaumasia region (MC-17, MC-25), indicate deep-seated rifting (e.g. Hauber et al., 2010). Narrow grabens (<10-km wide) within and radiating from the Tharsis and Elysium regions (e.g. MC-16) are indicative of either hourglass-configured normal faults (e.g. Schultz et al., 2007) or shallow detachment along horizontal mechanical interfaces that may be underlain by dikes or hydrofractures (e.g. Tanaka et al., 1991; Grosfils and Head, 1994).

Valles Marineris is one of the largest tectonic features on Mars and in the entire solar system, yet its mechanism of formation is still debated. Among the mechanisms that have been proposed for the opening and subsidence of the great valley system are pull-apart rifting (e.g. Carr, 2006, and earlier workers), downdropping of blocks bounded by vertical dikes (Andrews-Hanna, 2011), and transform faulting that connects pull-apart zones in Noctis Labyrinthus and the chaos regions east of Valles Marineris (Yin, 2012). Geologic mapping indicates that the canyons began forming in the Late Noachian Epoch and that structural development largely ceased by the Early or Late Hesperian Epochs.

Water and ice

Liquid water is not stable on Mars at present owing to low atmospheric pressure. One of the most surprising discoveries by Mariner 9, however, was widespread evidence of flowing liquid in the past. Water (in liquid and frozen forms) is now commonly accepted as an important, if not the only, agent in creating these features.

The global geologic map (Figure 5.1) shows the largest stream-carved erosional valleys (see also Figure 5.5). Outflow channels likely formed as the result of catastrophically erupted, pressurized groundwater mainly during the Hesperian Period (e.g. Baker et al., 1992), whereas narrow valleys likely formed through outburst floods and/or precipitation-driven surface runoff (e.g. Howard et al., 2005). Features similar to the outflow channels of Mars appear in the Channeled Scablands on Earth that formed by breakout floods during glacial retreat (Figures 5.12, 5.13). The narrow valleys are widespread on Noachian surfaces, but more restricted on younger surfaces. These relationships indicate a thicker atmosphere during the Noachian Period, whereas later activity may have been governed by weaker climate excursions and temporary, localized, wet micro-climates. A good example of Late Noachian hydrologic activity is Ma'adim Vallis (MC-23), which drained a vast system of interconnected highland basins (Figure 29.B, MC-29) and discharged via Gusev crater into the northern plains. (For more detailed mapping of valleys, see Hynek et al., 2010). In some cases mass flows, glaciers, and lava flows may also be responsible for channel formation. Rilles form a specific channel type that generally narrows downslope and is attributed mainly to thermal erosion (e.g., Nummedal and Prior, 1981; Lucchitta, 1982; Leverington, 2011).

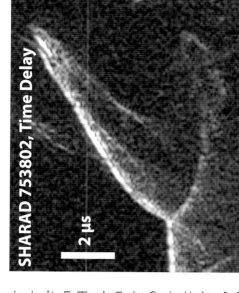

SHARAD 753802, Time Delay

2 µs

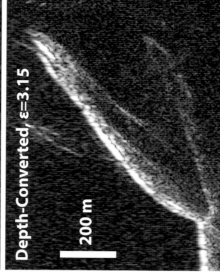

Depth-Converted, ε=3.15

200 m

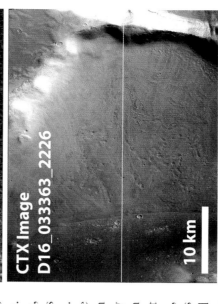

CTX Image D16_033363_2226

10 km

The northern plains and Hellas basin form vast topographic sinks into which many of the largest outflow channel systems on Mars drained. Possible paleoshorelines, tsunami margins (MC-5), and sedimentary deposits within these basins indicate the presence of temporary seas or oceans, perhaps ice-covered, during the Hesperian Period and possibly earlier. Other enclosed highland basins likely collected water, mostly during the Noachian Period, as indicated by valley network drainage systems and related landforms, the formation of hydrated minerals, and the suspected deposition of salts (which can be indirectly inferred from spectral mapping results).

Another surprise was the suggestion that the Mars Phoenix Lander developed droplets on its lander legs, which could have been water that had its freezing point greatly depressed by dissolved perchlorate salts. The Mars Reconnaissance Orbiter's (MRO's) HiRISE camera has also detected patches in Valles Marineris where seasonal, dark slope streaks (termed "recurring slope lineae") form that may be the result of transient surface water (see the Mass wasting and slope processes section below). Spectra taken by the Compact Reconnaissance Imaging Spectrometer for Mars (CRISM) instrument indicate hydrated salts in proximity to recurring slope lineae, consistent with current water activity on Mars (Ojha et al., 2015).

Frozen water, originally suspected and since confirmed at the poles of Mars, turns out to be widespread and possibly substantial in the subsurface over a range of low to high latitudes. The subsurface ice, inferred by landforms and confirmed in places by radar sounding (Holt et al., 2008), includes debris aprons on the Tharsis Montes and Olympus Mons (MC-9, MC-10), along parts of the highland-lowland boundary, particularly the northern margin of Arabia Terra (MC-5; Figure 5.14), and surrounding massifs east of Hellas basin (MC-28). Radar sounding also suggests buried ice in the northern plains, including Arcadia Planitia and Utopia Planitia. The Mars Phoenix Lander also revealed shallow ice (see the Surface science section below), as have recent

impacts in many northern plains locations (Figures 7.G, 7.H). The variation in the paleo-obliquity of Mars to values much higher than the current 25° could have made glaciation possible in equatorial regions since insolation and associated solar heating would be concentrated at the poles. Features at low latitudes suggest glaciation happened in the past (MC-9). Lobate flows may be debris-covered glaciers (for example, see map sheets MC-5, MC-28). At present, neutron spectrometer results from the Mars Odyssey spacecraft indicate that water ice occurs within the upper meter of the surface at latitudes generally >45°, and in more discontinuous fashion within the upper 5 m above 30° latitude (see Figure 3.6). This is consistent with the widespread occurrence of polygonal terrains at meter scales that may have resulted from seasonal expansion and contraction of ice-rich frozen ground (Figure 4.15).

The Martian polar regions also display seasonal ice caps made up of carbon dioxide (CO_2), which freezes out of the atmosphere (MC-1, MC-30). Parts of the south polar region also apparently accumulate translucent CO_2 ice, which results in geyser-like eruptions during the spring. This happens as the ice sublimates at the base of the CO_2 ice layer and becomes pressurized, leading to fracturing of the ice and escape of the CO_2 gas. Dust is incorporated in these CO_2 gas eruptions, and spider-like dark forms result where dust gets channeled to eruption sites (Figures 30.K–30.N). Carbon dioxide also has accumulated on the uppermost part of the south polar plateau known as Australe Mensa, forming a deposit generally <10 m thick. The deposit is actively receding in places, enlarging enclosed depressions that result in a Swiss-cheese-like texture (Figures 30.H, 30.I, 30.J). A more substantial deposit, as much as 300 m thick of what may be CO_2 ice, is buried within Australe Mensa, as revealed by radar sounding and geologic mapping (Figure 30.G; Phillips et al., 2011). There would be sufficient CO_2 in this deposit, if reintroduced into the atmosphere, to increase atmospheric mass by as much as 80 percent.

FIGURE 5.14 Radargram in western Deuteronilus Mensae (MC-5) showing time section of an apron deposit next to a raised hill (upper panel) and conversion to a depth section using a water-ice composition (middle panel). CTX image in bottom panel shows surface view along radar line (SHARAD observation 753802 centered near 18.47° E, 41.94° N, NASA/JPL-Caltech/Sapienza University of Rome/Southwest Research Institute; CTX image D16_033363_2226, NASA/JPL-Caltech/MSSS; courtesy J. W. Holt and E. I. Peterson).

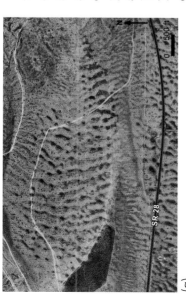

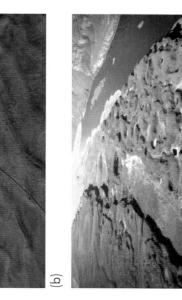

FIGURE 5.13 (a) and (b) Two examples of giant current ripples found in the deep channels cut by breakout floods in the Channeled Scablands of the northwestern USA (Baker et al., 2016). The ripples in image a average 2.5 m in height and 65 m in spacing (this image is US Soil Survey photo provided by V. R. Baker; image b is an aerial photo courtesy of Bruce Bjornstad). (c) Example of butte-and-basin scabland produced by scouring of basalt during a breakout flood; roadways show the scale of features (aerial photo courtesy of V. R. Baker).

Wind

Wind is the most active surface agent on Mars today, and it has been an important long-term cause of surface erosion, sedimentation, and redistribution of loose granular material. Dust storms can obscure the atmosphere locally (Figures 5.15, 5.16), regionally, or can encircle the entire planet (Figure 5.17). Redistribution of dust and sand may be responsible for observed changes in albedo markings. Prevailing winds can be shown by albedo contrast downwind of craters and other landforms (e.g. MC-12, MC-22). Ripples and dunes, in scales from centimeters to kilometers and varying in form, are common on Mars and are similar to features on Earth (Figure 5.18; Hayward et al., 2007a).

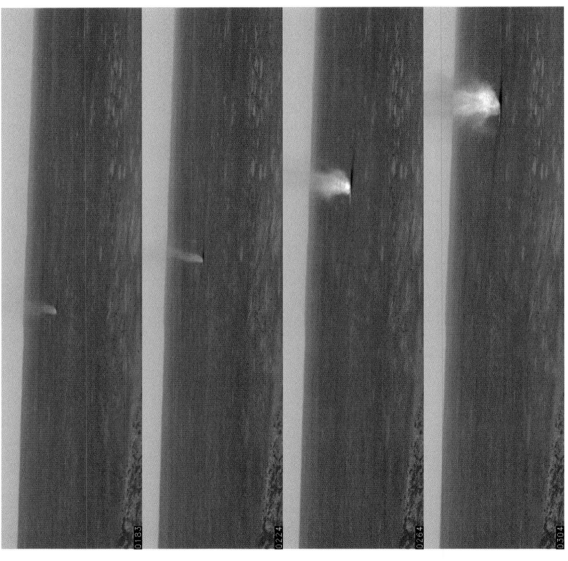

FIGURE 5.15 (right) A wind vortex, or "dust devil," moves across the field of view in these frames taken by the Navigation Camera on Spirit rover. Taken in northern autumn ($L_s = 218°$), frames 40 seconds apart (NASA PIA 07139, image credit NASA/JPL–Caltech/ Texas A&M University).

FIGURE 5.16 (above) A Mars sunset view in color from the Mast Camera of Curiosity rover shows the dusty atmosphere (NASA PIA 19400, image credit NASA/JPL–Caltech/MSSS/Texas A&M University).

FIGURE 5.17 (below) A dust storm develops and dissipates in this Mars Orbiter Camera wide angle (MOC-WA) view of the Hellas/Syrtis Major hemisphere of Mars. Observations suggest that the global expanses of storm clouds seen in southern spring and summer occur by a coincidence in time of multiple regional storms (left to right: sol 365 at $L_s = 175°$; sol 414 at $L_s = 205°$, sol 468 at $L_s = 239°$; MOC release 2–290, image credit NASA/JPL–Caltech/MSSS).

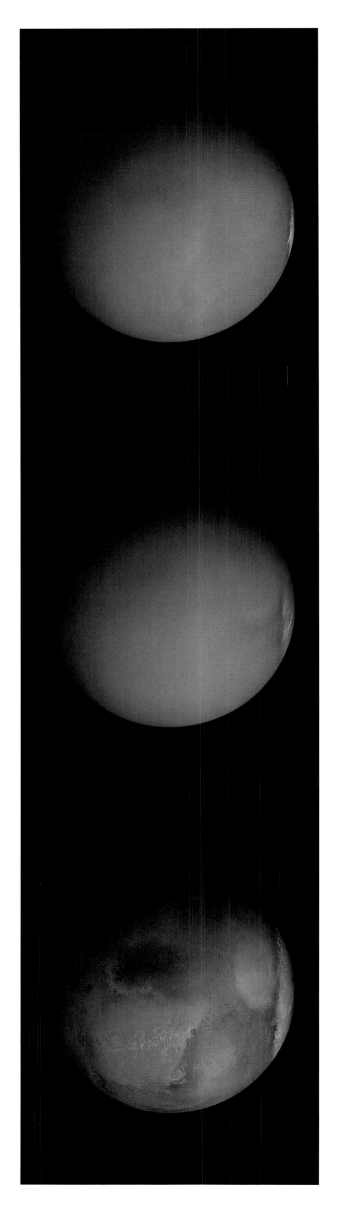

FIGURE 5.18 Examples of dune types used to classify dune fields. Panels (a), (b), and (d) through (i) are based on McKee's (1979) Earth-based classification system. Panel (c) is another type also found on Mars. Panel (j) is a dune type first recognized on Mars. Each image description includes dune type, image ID number, dune field ID number (Hayward et al., 2007a; 2007b), longitude and latitude of dune field centroid and discussion, if needed: (a.1) barchan, E0902707, 0304–475, 30.4° E, 47.5° S; (a.2) barchan, R0300926, 1370–050, 137.0° E, 5.0° S; (a.3) barchan, M0204432, 0194–468, 19.4° E, 46.8° S, some joining of dunes; (a.4) barchan, R0400598, 1283–141, 128.3° E, 14.1° S, elongated horns on one side indicate influence of secondary wind direction; (b.1) barchanoid, M2300263, 1283–141, 128.3° E, 14.1° S, dune field occurs within large (~300-km-diameter) crater, detail here shows barchanoid dunes nearly obscuring small (~1-km-diameter) crater; (b.2) barchanoid, E0302016, 0671+088, 67.1° E, 8.8° N, edge of dune field in Nili Patera, shows transition from barchans with elongated horns to a barchanoid; (b.3) barchanoid, E0302016, 0671+088, 67.1° E, 8.8° N, same image and dune field as in image b.2, but shows interior of dune field where barchanoid form is more uniform; (c.1) reticulate, V01048003, 0347–437, 34.7° E, 43.7° S; (d.1) transverse, M0806802, 2938–497, 293.8° E, 49.7° S; (d.2) transverse, R1001964, 1586–633; 158.6° E, 63.3° S; (e.1) barchan, barchanoid and transverse, M0806802, 2938–497, 293.8° E, 49.7° S, dune types shown as they occur together in a dune field; (f.1) dome, R1901441, 0380–447, 38.0° E, 44.7° S; (g.1) linear, M2001808, 0168–589, 16.8° E, 58.9° S; (g.2) linear, formed in lee of obstruction, PSP_007676_1385, no dune database number, 76.29° E, 41.47° S; (h.1) star, R0300863, 2975–411, 297.5° E, 41.1° S; (h.2) star, M0702777, 0304–475, 30.4° E, 47.5° S; (i.1) sand sheet, E1302032, 2086–603, 208.6° E, 60.3° S; (j.1) transverse eolian (aeolian) ridges (TARs), these light-colored, ridge-like eolian features are not dunes, PSP_007676_1385, no dune database number, 76.31° E, 41.45° S; (j.2) TARs, these light-colored, ridge-like eolian features are not dunes, PSP_007676_1385, no dune database number, 76.30° E, 41.41° S (figures and descriptions courtesy of R. Hayward, US Geological Survey; image c.1 from THEMIS, NASA/JPL-Caltech/Arizona State University; g.2, j.1, j.2 are HiRISE images from NASA/JPL-Caltech/University of Arizona; others are MOC images from NASA/JPL-Caltech/MSSS).

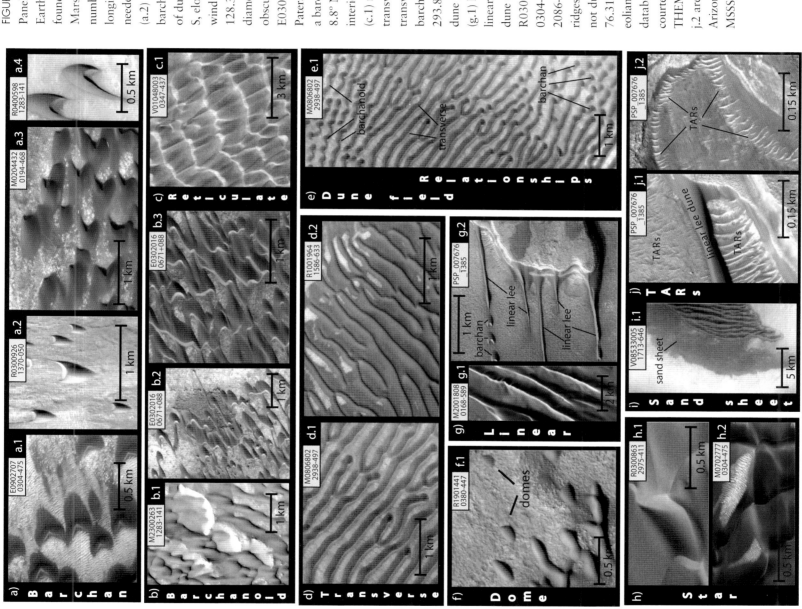

	Classic (alcove-channel-apron) gullies [1-5,9,10,17,19,22-27,35]	Slope streaks [28-34]	Linear gullies [6-8,11,17,18,21]	Recurring slope lineae (RSL) [12-14]	Dark (frost/flow) streaks [15,16,20]	Boulder tracks [22]
Typical OoM W x L [m]	100 x 1,000	1-10 x 10-1,000	1-5 x 100 (-1,000)	2 x 100	2 x 50	1-5 x 100
Depth	Alcoves can cut deeply into surface, channels 1-few m	<1 m	<2 m	Follow bottom of small gullies; the features are superficial	<1 m	<1 m
Surface material	Rocky crater wall, sand dune	Dusty, bright	Sandy (crater walls or dunes)	Rocky, dark	Frost-covered sand dune	Varied
Slope orientation	Pole-facing in midlatitudes, w/ transition to equator-facing poleward of 45° S, and varied poleward of 50° N	Varies (perhaps related to regional wind flow)	Pole-facing only	Slopes that are warm (>250 K)	Various (down the slipfaces)	Varied
Slope steepness	Steep	Steep	Originate within steep slopes, extend over low (<15°) slopes	Steep	Steep	Originate in steep slopes, can extend over low slopes
Typical spacing/abundance	Closely spaced on crater walls, up to 10 s along slope	Up to 10 s on a slope, spacing varies?	Few – 100 s on the slope, spacing varies	Up to 1,000 s on a slope, closely spaced	1-10 within a cluster on a slope; Several clusters on a slope	Isolated to cluster of ~10
Global distribution	Poleward of ±30°, mostly 30-40° [2]; some authors have identified equatorial examples [27]	Equatorial dusty areas [28]	30-70° S; Largest in Russell Crater	40° N-60° S	Poleward of 40° S	Anywhere
When active (present day)?	Late winter-early spring [16-18]	All seasons [29]	Early spring [11,17]	Only season with warmest temperature [12]	Winter-early spring	No seasonal, etc. control suspected
Formation event(s)	Likely one large event per season; not necessarily every year	One event per streak	Incremental [11,17]	Simultaneous incremental growth of many flows	Incremental growth each winter	One event per track
Degradation timescale	Centuries (infilling of dune gullies) or millennia+ (crater gullies) [17]	Decades [30,31]	<1 year for new ~1 m wide troughs, longer for large features [17]	Weeks to months	Disappear when frost sublimes	Years
Repeat activity within feature?	Yearly [16,17,19]	None	Yearly [11,17]	Yearly, or twice per year in equatorial regions [12]	Yearly	None
If water is involved, estimate(s) of volume to form a feature:	10^2-10^4 m^3 [5,9,35]	Brine of ~3 wt% of H_2O in the soil + ~1 wt% of $CaCl_2$ = ~8 %vol [32]	10^2-10^4 m^3 for largest (Russell) [6,8]	2-10 m^3 [14]	Only interfacial water is available [15]; no estimates found for how much would be needed	N/A
Other (dry) theories	Seasonal frost-driven [16,17]; aeolian controls [19] for north polar erg	Dry dust avalanches triggered by dust devils, marsquakes, or falling boulders [33,34]	Sublimating "hovercraft" blocks of dry ice [18]	N/A	Sand and dust avalanches/deposits [16,20]	Falling boulder
If recent activity is representative, timescale to form a typical feature:	Few million years, in southern hemisphere [27]	Seconds-minutes	>Centuries-millennia (a very loose extrapolation with yet unverified assumptions from [11,17,18])	Weeks	Weeks	Seconds-minutes
Evidence of water in present day?	Unlikely	Unlikely	Unlikely	Likely	Unlikely	No
Evidence of water in a past climate?	Debated	N/A	TBD	N/A	N/A	No

FIGURE 5.19 A summary of Martian slope features (reproduced by permission of Diniega et al., 2014; courtesy NASA/JPL-Caltech). Ojha et al. (2015) observed water-bearing minerals present during the formation of recurring slope lineae. References: [1] McEwen et al., 2007b. [2] Harrison et al., 2015. [3] Head et al., 2008. [4] Johnsson et al., 2014. [5] Kereszturi, 2012. [6] Mangold et al., 2003. [7] Mangold et al., 2010. [8] Jouannic et al., 2012. [9] Malin and Edgett, 2000a. [10] Malin et al., 2006. [11] Reiss et al., 2010. [12] McEwen et al., 2013. [13] Grimm et al., 2014. [14] Stillman et al., 2014. [15] Kereszturi et al., 2011. [16] Hansen et al., 2011. [17] Dundas et al., 2012.

[18] Diniega et al., 2013. [19] Horgan and Bell, 2012. [20] Gardin et al., 2010. [21] Pasquon et al., 2016. [22] Hansen et al., 2015. [23] Pelletier et al., 2008. [24] Dundas et al., 2017. [25] Treiman, 2003. [26] Diniega et al., 2017. [27] Dundas et al., 2015. [28] Aharonson et al., 2003. [29] Schorghofer and King, 2011. [30] Schorghofer et al., 2007. [31] Bergonio et al., 2013. [32] Kreslavsky and Head, 2009. [33] Sullivan et al., 2001. [34] Chuang et al., 2007. [35] Heldmann et al., 2005.

FIGURE 5.20 Martian meteorite NWA 7034, a breccia that includes fragments over 4 billion years old (image courtesy of NASA).

FIGURE 5.21 Timescale of chemical evolution of Mars environment proposed by Bibring et al., 2006.

phyllosian	theiikian	siderikian
clays	sulfates	anhydrous ferric oxides
Noachian	Hesperian	Amazonian

Meteorites from Mars

Several dozen meteorites are identified as samples of the surface of Mars, based on crystallization age, chemical properties, and trapped gases (McSween, 2008). Nearly all of these have Amazonian crystallization ages, ranging from 170 Ma to 1.3 Ga. They originate in extrusive or intrusive igneous rocks of basaltic to ultramafic composition. Two meteorites sample Noachian crust, ALH 84001 and NWA 7034/7533, with ages of 4.4–4.5 Ga. ALH 84001 is an ultramafic cumulate rock with alteration minerals that include carbonate and magnetite. NWA 7034 and NWA 7533 (which Agee et al., 2013 and Humayun et al., 2013, respectively, argue are also from Mars) are two fragments of the same breccia containing fine mafic igneous and impact melt grains (Figure 5.20), with ten times as much water as younger meteorites from Mars. Formation of the crust of Mars very soon after accretion of the planet would account for the age of these two meteorites.

Mass wasting and slope processes

Processes driven by gravity are evident wherever slopes have been steepened by wind erosion, impact deformation, tectonic movements, or frost action. A variety of talus deposits, landslides, rock and ice avalanches, glaciers, debris flows, and fans have resulted, depending on the latitude, presence of water or ice (including CO_2 ice near the poles), local relief and slope gradients, and bedrock type. While a slope is clearly requisite to the formation of gullies that are common on many slopes, their origination due to the presence and activity of water, water or CO_2 ice, or other factors is largely unclear. Some of the most active slopes, where seasonal modification has been observed, include polar dunes, where sublimating CO_2 ice deposits may be the trigger, and steep walls of ice on the margins of Planum Boreum, which are undergoing calving and avalanching.

Four decades of images show that streaks are common on Mars, and that they appear and disappear over time (see also recurring slope lineae under the Water and ice section above). Categories of streaks are shown in Figure 5.19 (Diniega et al., 2014). Factors influencing the type, timing, and location of streak formation include the type of terrain, the season, the presence of frost, sun illumination, and gravity (e.g. Bergonio et al., 2013 on slope streaks; Diniega et al., 2013 on linear gullies).

Polar deposits contain units and individual layers that have a significant wind-blown component, including what appear to be dark sand dunes entrapped by ice layers (MC-1). At the outcrop scale, cross-bedding in rocks of Meridiani Planum observed by the Opportunity Mars Exploration Rover (Figure 5.26 in the Surface science section below) indicate transfer of surface particles during the Late Noachian Epoch. Spiral troughs (Figures 1.C and 30.B) are particular erosional features related to erosion and transport of the polar plateaus by insolation and katabatic winds during the Late Amazonian Epoch (e.g. Howard, 1978; Smith and Holt, 2010). Cyclones have also been observed moving east to west around the north polar plateau, Planum Boreum.

Chemical evolution of Mars environment

Visible and spectral mapping from Mars orbit suggests global changes through time in the chemical environment on the Mars surface (Bibring et al., 2006; Figure 5.21). The era of clay-mineral formation (the "phyllosian" era) indicates wet conditions at the surface or hydrothermal activity. A change to acidic conditions, possibly caused by volcanic outgassing, led to formation of sulfates (the "theiikian" era). Finally, before the end of the Hesperian Period, dry conditions saw formation of iron oxides (the "siderikian" era). Detailed surface studies may well show a more complex pattern of cycles of mineral types in time, local or regional variations, or the presence of minerals that imply different conditions (Bell, 2008; Grotzinger et al., 2011).

Habitability

Orbiting at a distance on average about 50 percent farther from the Sun than the Earth, Mars is the only other planet besides the Earth that is within the solar system's potentially habitable zone (Kopparapu et al., 2013). Furthermore, the Martian polar ice caps, observable from Earth, are made up mostly of water ice. Thus, the possibilities of

liquid water and habitable environments on or near the surface of Mars have driven interest in the planet. Water may occasionally occur at present on the surface during summertime afternoons, particularly on Sun-facing slopes or where water can be precipitated via deliquescence. The latter process involves the absorption of atmospheric water by substances on the surface, sufficient to produce a solution. However, the resulting droplets, examples of which may have formed on the Mars Phoenix Lander's supports, would be short-lived and concentrated with salts that, on Earth, would make the water toxic to most life. A host of paleo-environments may have been present on Mars during which life could have

survived. These include bodies of water, likely ice-covered, in local- to regional-scale basins, including the northern plains. Many of the volcanoes and rims of larger impacts on Mars are dissected by valley networks, some of which may be the result of groundwater discharge. Thus, long-lived groundwater systems may have existed that would have provided potentially favorable environmental conditions for microbial life. However, as Mars became generally colder after the Noachian Period, due both to a possibly thinning atmosphere and decreased heat flow, the upper crust became frozen to depths of kilometers. Climate excursions, however, may have led to more temperate conditions during later epochs.

Surface science

Many of the questions regarding the makeup and history of Mars can only be addressed close-up by observations and experiments performed on the surface at strategic locations. The surface exploration of Mars is a story of repeated attempts and, eventually, great successes. Landers have arrived safely and have accomplished – and sometimes greatly exceeded – their intended missions.

Two fundamental prongs of strategic research, development, and implementation have enabled this highly productive surface exploration of Mars. First, engineering stipulates what is needed in terms of where and when a spacecraft should land to ensure functional mission success. Among the key engineering parameters are the dimensions of the landing ellipse, the amount of solar exposure (critical for solar power needs and daytime experiments), surface hazards and obstacles (including high slopes, rocks, sand dunes, and dust that may either ruin a landing or reduce rover mobility), and anticipated winds (which might interfere with landing). Each lander is unique in terms of engineering requirements, given its design, makeup, and intended application. Second, the mission's science objectives require strategic placement of the lander/rover to optimize the likelihood that the objectives will be met. For example, is the objective to search for evidence of life, or water, or to understand the geology of a particular type of landscape or material? Furthermore, the science community has been increasingly concerned about the possibility of "infecting" Mars with terrestrial microbes that are hitchhiking on landed spacecraft, and thus measures have been taken to sterilize spacecraft prior to launch from Earth as well to avoid possible surface contamination upon landing (even in crash scenarios; Rummel *et al.*, 2014). When all aspects have been mutually optimized for a given mission, the chosen landing sites have yielded amazing results (see also images in Chapter 4).

In 1976, Viking 1 and 2 landers both arrived safely in northern plains locations – Chryse (Figure 5.22; MC-10) and Utopia Planitiae (MC-7), respectively. Both were equipped as surface laboratories, with the primary goal of determining if life has been present in Martian soils, via chemical experiments on scooped samples. The result was largely negative (Klein *et al.*, 1992), although some aspects of the data had ambiguous

FIGURE 5.22 Viking 1 lander view to the southeast in color shows the red-orange surface veneer that covers darker rocks (Viking 1 image P17165; NASA/JPL-Caltech).

implications regarding the possible presence of organic material. Other detectable aspects of the surface environment (atmosphere, soil, rocks, landforms, magnetic field, frost; Figure 5.23) were observed and recorded for the first time.

Mars Pathfinder brought the first rover, named Sojourner, to a "return visit" to Chryse Planitia (specifically, a part of the plain littered with the debris from the long-extinct Ares Vallis flood channel, see MC-11) in 1997. This was a low-cost mission designed to test engineering capability to perform rover-hosted experiments, which would require communication relay to Earth via

the lander platform. The tests were fully successful, beginning with its hard landing in a cocoon of airbags. Sojourner navigated to rocks and terrains of interest, and both the rover and its lander base returned a variety of science data on the Martian environment. Some boulders at this site lean against each other in "trains" that may be relicts of their emplacement at the tops of debris flows (Figure 5.24).

The Mars Exploration Rovers *A* and *B* (Spirit and Opportunity) landed on Mars in 2004. These rovers were designed as hefty cousins to the Sojourner rover (being 17 times heavier), with a similar landing system and rover chassis design. Spirit and Opportunity had ambitious science

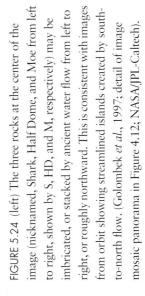

FIGURE 5.23 (above) Viking 2 lander view of the left half of the panorama shown in Figure 4.11 shows thin water-ice frost on rocks and the surface. This was the second Martian year that frost formed at the site during this season (Stooke, 2012; view to northeast, $L_s = 289°$, Viking 2 image P21841, NASA/JPL-Caltech).

FIGURE 5.24 (left) The three rocks at the center of the image (nicknamed, Shark, Half Dome, and Moe from left to right, shown by S, HD, and M, respectively) may be imbricated, or stacked by ancient water flow from left to right, or roughly northward. This is consistent with images from orbit showing streamlined islands created by south-to-north flow. (Golombek *et al.*, 1997; detail of image mosaic panorama in Figure 4.12; NASA/JPL-Caltech).

objectives, including the search for evidence of past life on Mars, understanding the surface geology and climate history, and preparing for future human exploration of the planet's surface. At the nexus of all these objectives is water, and some of the goals have been addressed in concert with orbit-based investigations, as part of the broader Mars exploration program being conducted by NASA and ESA. For 6 years, Spirit investigated the floor of the ancient Gusev crater, an impact feature with inlet and outlet channels indicative of the former presence, albeit transitory, of a crater-filling lake (MC-23). However, instead of lake-beds, Spirit found soil-covered basaltic lava plains, scattered meteorites, and ancient hills

FIGURE 5.29 Layering visited by Opportunity and studied with the alpha particle X-ray spectrometer (APXS) reveals likely clay minerals, rich in aluminum and silica and probably altered by water (false-color Pancam image, NASA/JPL-Caltech/Cornell University/ Arizona State University).

released as hard nodules, which locally collect abundantly on the surface. Like Spirit, Opportunity also found meteorites resting on the surface (Figure 5.28). Opportunity has characterized an unprecedented amount of the Martian surface. Opportunity operated on Mars for more than 14 years and traversed more than 45 kilometers, characterizing many surface features and rock outcrops along the way. During the last 6 years of its mission Opportunity explored the immense, 22-km-diameter Endeavour crater. This crater shows evidence of clay minerals that formed in the presence of water (Figure 5.29).

Meanwhile, in 2008, the Mars Phoenix Lander came to rest within the planet's arctic circle, at a latitude of 68° N (MC-1). This mission investigated the polar soil, including imaging of dust particles only 1 μm in diameter, during the Martian north polar summer. The soil is rich in water ice a few centimeters below the surface (Figure 5.30), where it is somewhat protected from sublimating into the hyperarid atmosphere. The landscape surrounding the lander was found to be flat, barren, somewhat rubbly, and marked by polygonal troughs that are typical of permafrost

FIGURE 5.27 Opportunity color image of spherules made of hematite, originally formed in the presence of water. The spherules are weathering out of the rock, which was analyzed after cleaning of a circular area with the rock abrasion tool (NASA/JPL-Caltech/Cornell University).

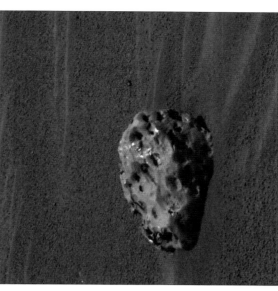

FIGURE 5.28 Iron meteorite discovered on Meridiani Planum by Opportunity. Spectrometer readings confirmed the iron–nickel composition that is suggested by the rock's appearance (NASA/JPL-Caltech/Cornell University).

subsurface. The plains-forming material is layered in forms reflecting accumulation of small rock particles in standing water, which have been rearranged into rippled and cross-bedded deposits by wind action (Figure 5.26). The deposits include the mineral jarosite, which indicates formation by the evaporation of salt-rich waters. Also, small spherules (Figure 5.27) of hematite within the deposits show that iron oxide precipitated to form concretions within the deposits, later to be

FIGURE 5.25 Spirit view downslope in the Columbia Hills showing layering in rock outcrops across the middle of the mosaic (color view taken Sol 454, NASA/JPL-Caltech/Cornell University).

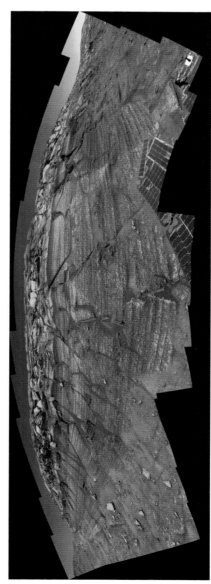

FIGURE 5.26 Rock layers in "Burns Cliff" in the southeastern wall of Endurance crater. The lower layers show lamination patterns typical of wind-transported deposits, while minerals in layers like those in the upper part of the cliff formed in association with water (composite of 46 color images, 180° across, NASA/JPL-Caltech/Cornell University).

(Figure 5.25). The latter are composed of several rock types, which in turn include minerals formed in the presence, or by the alteration, of water. Possibly, Gusev did enclose an ancient lake whose Noachian sediments and impacted host materials were later largely buried by Hesperian basaltic lavas. Impact gardening and airfall dust then modified the otherwise dormant surface in the area over the past 3 billion years.

The Opportunity rover meanwhile landed on the opposite side of Mars and in a very different geologic setting – the highland plain known as Meridiani Planum. Nestled within the ancient cratered Arabia Terra highland region, the surface of Meridiani had been noted from orbit to be rich in the mineral hematite, which generally is formed in the presence of water (MC-19). On this plain, largely worn flat by eons of exposure to the wind and elements, telltale outcrops that may date near the end of the Noachian Period have been found in the walls of impact craters that have served to excavate and expose the rich geology in the

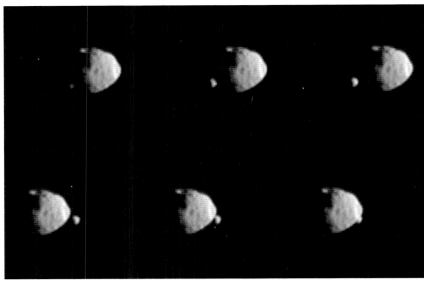

FIGURE 5.33 Phobos occults, or passes in front of, Deimos in this view from the Curiosity Mast Cam. The start of the occultation steps through the three images on the left from top to bottom; the images on the right show three steps in the reappearance of Deimos (NASA/ JPL-Caltech/MSSS/Texas A&M University).

rovers, Curiosity soft-landed with the help of tethered lowering from an overhead skycrane system, and it is powered by the heat generated by the radioactive decay of plutonium. Its unprecedented suite of scientific instruments and navigational capabilities has been applied to examine the rock record of current and past environmental conditions within Gale. Curiosity landed on an alluvial fan and has detected a variety of sedimentary rock structures (Figure 5.31) on its way toward the base of Mount Sharp. The structures include even, rippled, and cross-bedded layering and multiple generations of deltas, formed on the crater floor, indicative of a long-lived or recurring lake system. These observations indicate past environments that included precipitation of snow and perhaps rain on the crater rim, water runoff, rock weathering and erosion of the crater rim, precipitation of minerals from water (Figure 5.32), and transport of granular material in flowing water into standing bodies of water. Further discoveries await Curiosity's ascent of Mount Sharp (Figure 4.16).

Moons of Mars

Mars has two natural satellites, Phobos and Deimos (Figure 5.33; see also the nomenclature and images at the end of the map sheets). They are small and irregular in shape. They appear similar in composition, resembling carbonaceous asteroids, while contrasts include the morphology of craters and the presence or absence of grooves (Thomas *et al.*, 1992; Basilevsky *et al.*, 2014). Although the composition suggests these could be captured objects that originated in the asteroid belt, the orbital dynamics (nearly circular orbits close to the equatorial plane of Mars) are instead consistent with a local origin (Burns, 1992). Accretion from a debris disk in the equatorial plane can account for the present positions and motions of Phobos and Deimos. One scenario to provide a disk would be debris thrown into Mars orbit by a basin-forming impact, which could also account for the current rotation period of Mars (Craddock, 2011; Citron *et al.*, 2015).

FIGURE 5.31 The Curiosity rover found this outcrop several weeks after landing. The rounded gravel suggests the host rock is a conglomerate. This association and the rounding of pebbles are typical of water-transported material on Earth, implying a former stream or river (Mast camera, sol 27, NASA/JPL-Caltech/MSSS).

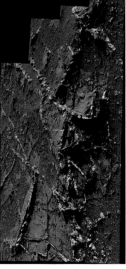

FIGURE 5.32 Veins appear in this Curiosity image from the floor of Gale crater. They stand as much as 6 cm above the surface and are eroding more slowly than the enclosing rock. Veins form when fluids move through fractures in rock and precipitate minerals (Mast camera, sol 929, NASA/JPL-Caltech/MSSS).

Curiosity, the Mars Science Laboratory (MSL) rover, is the latest, largest, and most sophisticated rover mission to the Red Planet. It landed in 2012 in Gale crater (MC-23) to further assess the planet's potential habitability for microbes. Gale, located along the planet's highland–lowland dichotomy boundary, formed at the beginning of the Hesperian Period. It contains an impressive 5,000-m-high mountain called Aeolis Mons (or, informally, Mount Sharp). In contrast to previous

FIGURE 5.30 Trench dug by the Phoenix Lander showing water ice beneath soil. The three small lumps in the lower left part of the trench vanished by turning to vapor over 4 sols (NASA/JPL-Caltech/ University of Arizona/Texas A&M University).

ground on Earth and common in high latitudes on Mars (Figure 4.15). The lander's robotic arm was used to scoop soil trenches and feed samples to the on-board chemical laboratory. In addition to confirming the presence of water ice, the experiments also discovered minor concentrations of perchlorate (Smith *et al.*, 2009), which can both inhibit and promote microbial activity, depending on circumstances, as well as depress the freezing point of water.

MAP SHEETS

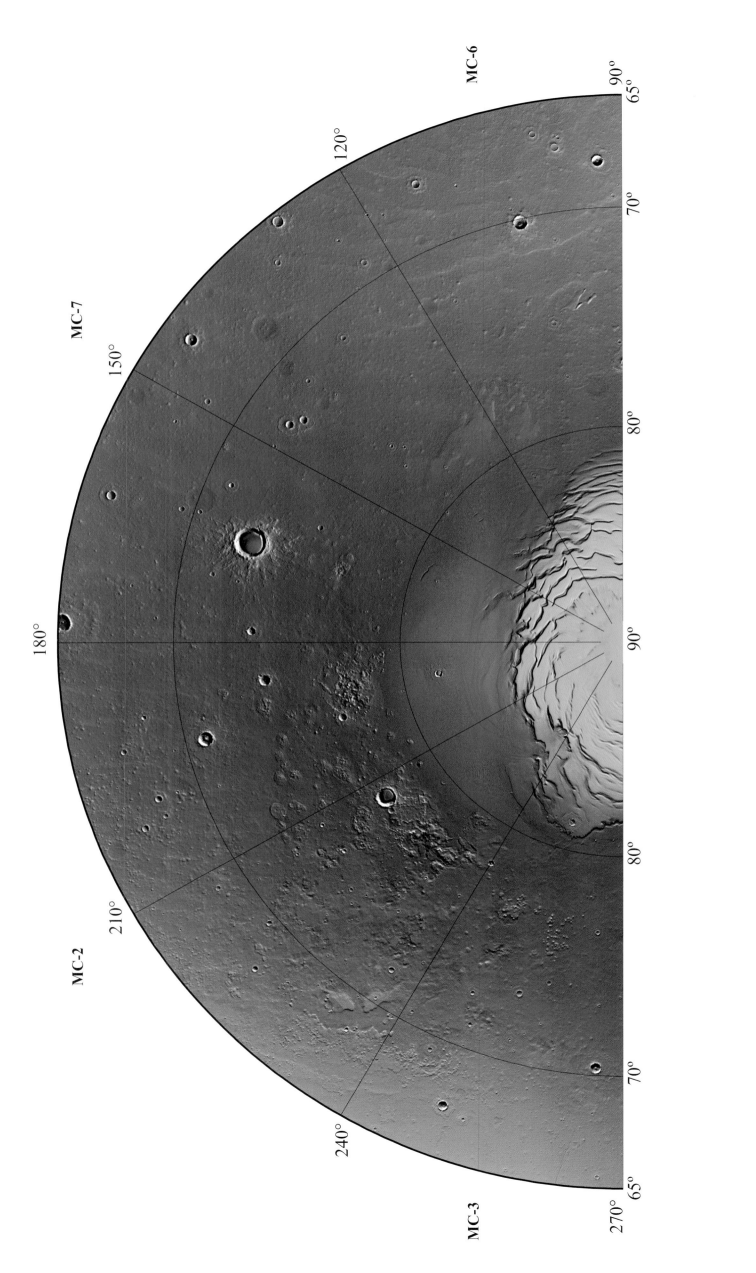

Mars Chart 1: Mare Boreum

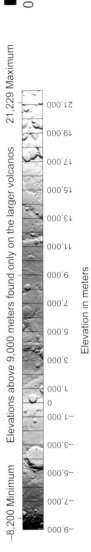

−8,200 Minimum   Elevations above 9,000 meters found only on the larger volcanos   21,229 Maximum

Elevation in meters

km

1:10,000,000

scale at standard parallel

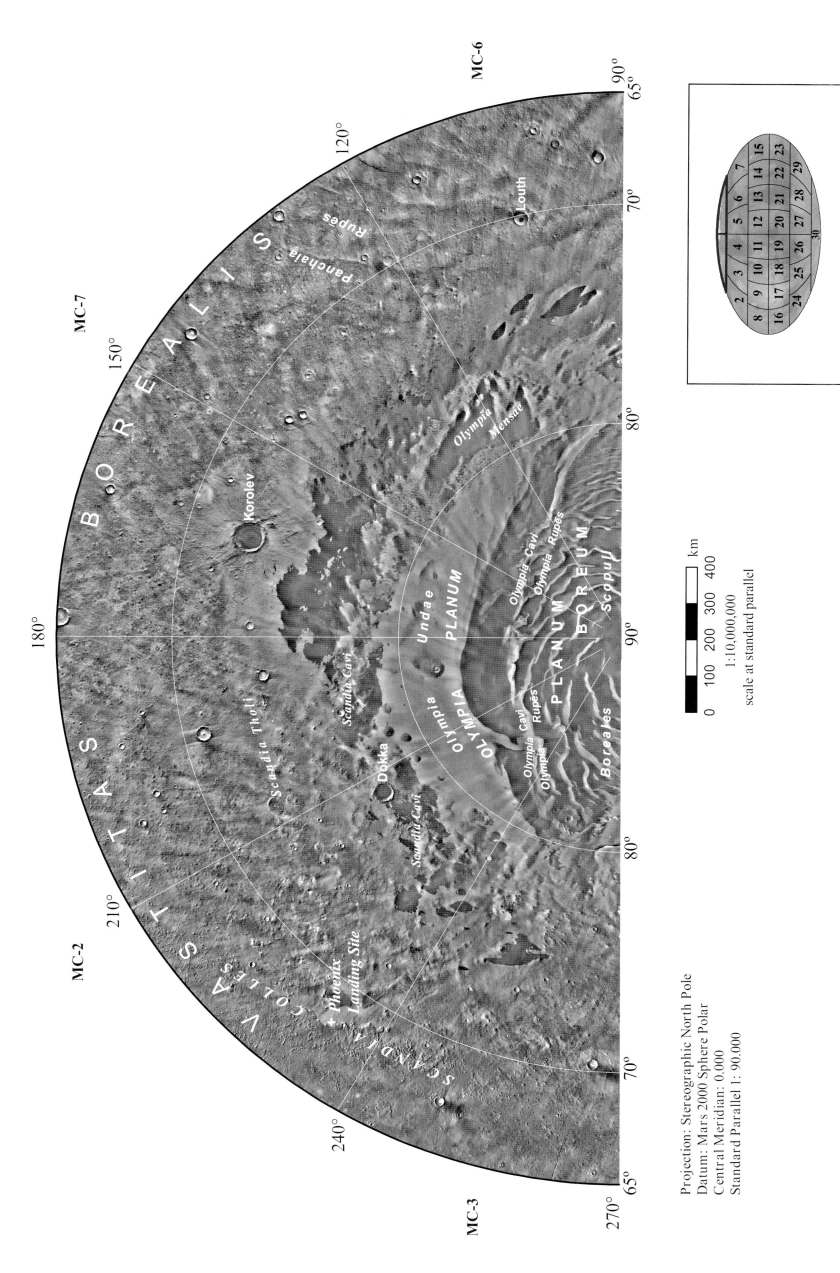

MC-6

90°

65°

120°

MC-7

70°

Louth

150°

B O R E A L I S

Panchaia Rupes

80°

180°

Olympia Mensae

Korolev

Olympia Cavi

Olympia Rupes

Undae PLANUM

OLYMPIA PLANUM OLYMPIA

Scandia Cavi

P L A N U M   B O R E U M

MC-2

210°

Scandia Tholi

Dokka

Olympia Cavi

Olympia Rupes

Boreales

Scopuli

T I T A S

+ Phoenix Landing Site

Scandia Cavi

80°

S C A N D I A   C O L L E S   V A S T I T A S

70°

240°

MC-3

270°

65°

km

0   100   200   300   400

1:10,000,000

scale at standard parallel

Projection: Stereographic North Pole
Datum: Mars 2000 Sphere Polar
Central Meridian: 0.000
Standard Parallel 1: 90.000

	2	3	4	5	6	7
1	8	9	10	11	12	13	14	15
	16	17	18	19	20	21	22	23
	24	25	26	27	28	29
			30		

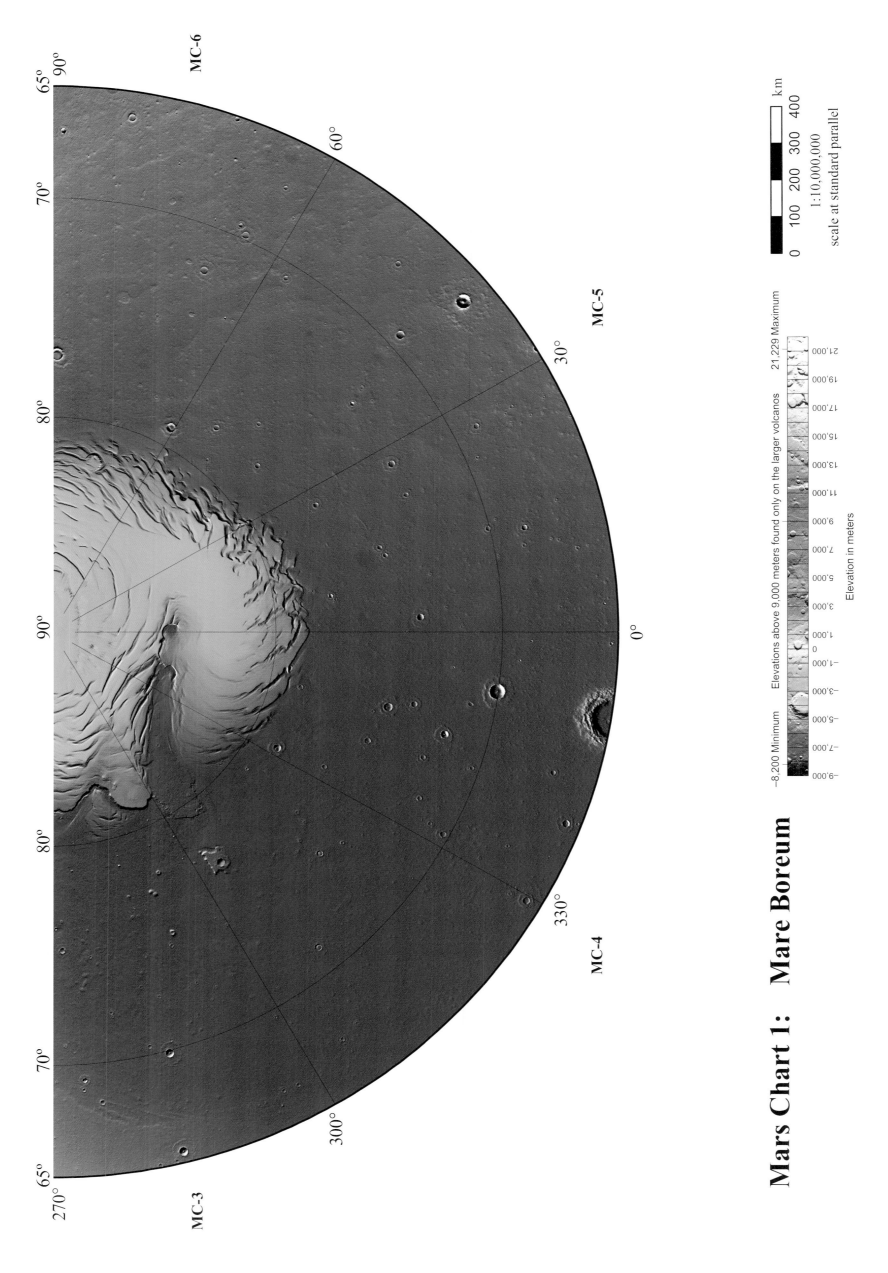

Mars Chart 1: Mare Boreum

MC-6

MC-5

MC-4

MC-3

1:10,000,000
scale at standard parallel

km

0   100 200 300 400

−8,200 Minimum   Elevations above 9,000 meters found only on the larger volcanos   21,229 Maximum

Elevation in meters

−9,000 −7,000 −5,000 −3,000 −1,000 0 1,000 3,000 5,000 7,000 9,000 11,000 13,000 15,000 17,000 19,000 21,000

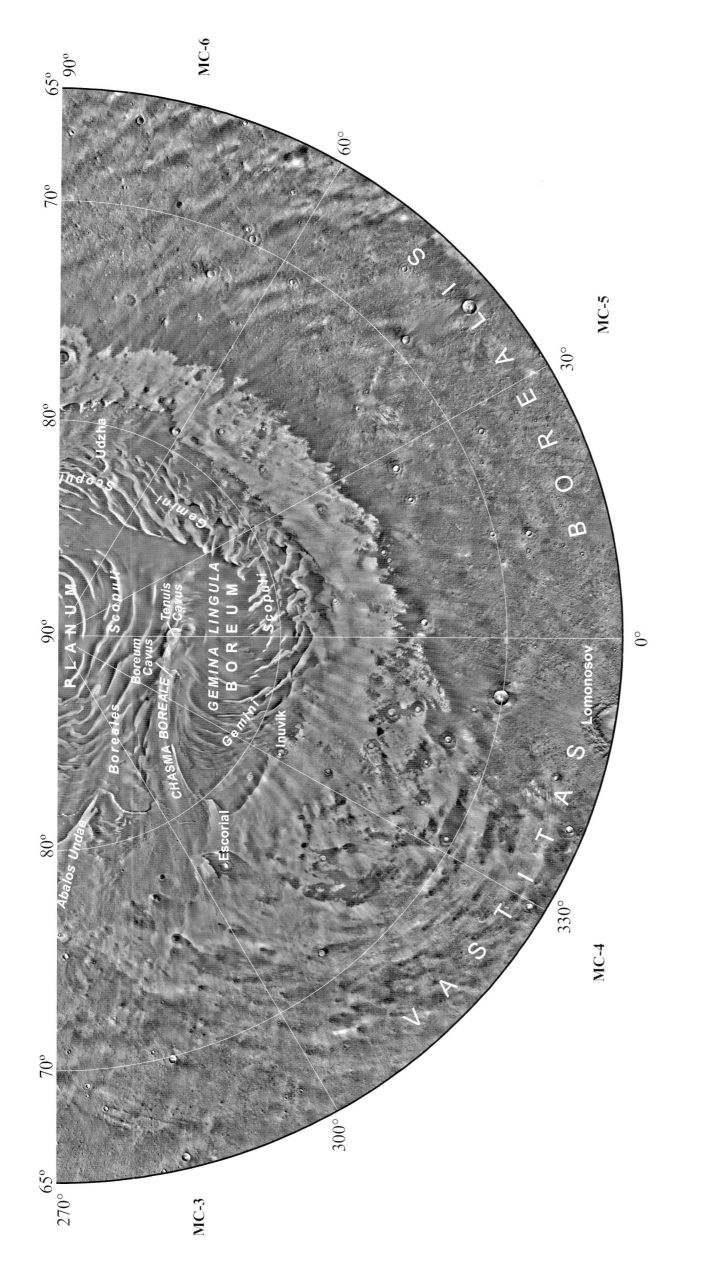

90°

65°

65°

70°

60°

70°

80°

30°

80°

90°

90°

B O R E A L I S

P L A N U M

Udzha

Scopuli

Gemini

Scopuli

Boreum
Cavus

Tenuis
Cavus

GEMINA LINGULA
BOREUM

Boreales

CHASMA BOREALE

Scopuli

Abalos Undae

Gemini

Escorial

Inuvik

Lomonosov

V A S T I T A S

0°

330°

300°

70°

80°

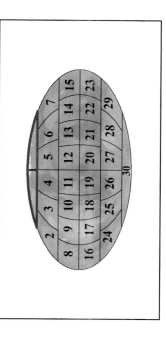

Projection: Stereographic North Pole
Datum: Mars 2000 Sphere Polar
Central Meridian: 0.000
Standard Parallel 1: 90.000

km

0   100 200 300 400

1:10,000,000

scale at standard parallel

Mare Boreum (MC-1)

Geography

The MC-1 quadrangle includes the north pole of Mars, which is covered by a 1,000-km-diameter plateau, Planum Boreum. This plateau rises 2–3 km above the surrounding northern lowland plains of Vastitas Borealis, which lie 3–5 km below datum. Planum Boreum is dissected by a large canyon, Chasma Boreale, which sets apart a secondary plateau, Gemina Lingula. These plateaus are dissected by the swirling, spiral troughs of Borealis and Gemini Scopuli. On the opposite side from Chasma Boreale, Olympia Planum

forms a kidney-shaped rise that is partly buried by the Olympia Undae dune sea. These dunes also overlap the pancake-shaped rises of Scandia Tholi and the irregular depressions that form Scandia Cavi. The lowest parts of the northern plains occur around the tholi and at the mouth of Chasma Boreale. Dense fields of low knobs make up Scandia Colles. The Phoenix Lander site (see Chapters 4 and 5) lies in Vastitas Borealis.

Geology

This region tells how the climate of Mars has varied over the past 3.5 billion years, beginning with voluminous sedimentation, brought forth by

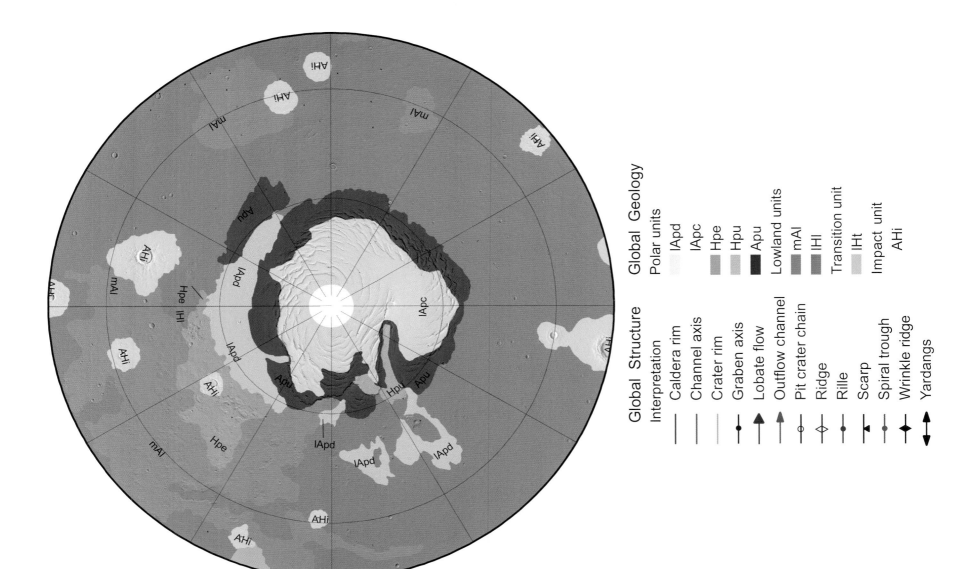

Global Geology

Polar units
- IApd
- IApc
- Hpe
- Hpu
- Apu

Lowland units
- mAl
- lHl

Transition unit
- lHt

Impact unit
- AHi

Global Structure

Interpretation
- Caldera rim
- Channel axis
- Crater rim
- Graben axis
- Lobate flow
- Outflow channel
- Pit crater chain
- Ridge
- Rille
- Scarp
- Spiral trough
- Wrinkle ridge
- Yardangs

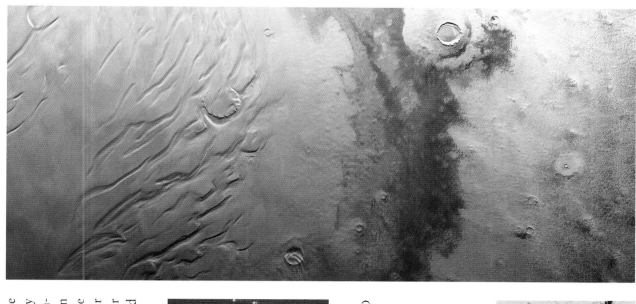

outflow channels that discharged groundwater from the equatorial highland regions at the end of the Hesperian Period. The waters brought along eroded debris that filled the northern lowlands, and some evidence suggests that initially the waters may have ponded into a vast sea, which would have frozen rapidly. Scandia Tholi and Cavi may be large features that result from buoyant instabilities in the sediments, leading to discharges of slurry onto the surface.

The basal part of Planum Boreum is thickly layered and has electrical properties, consistent with a deposit that is about half ice and half rock grains. This "basal unit" is eroded down more than 1,000 m in places, and the displaced material may account for a discontinuous, subtle, ice-rich Middle Amazonian lowland unit, tens of meters thick, covering parts of Vastitas Borealis. Later, the basal and lowland units eroded and may be the source of patches of fine sandy layers that transition laterally and vertically into icy, finely layered deposits, forming the upper part of Planum Boreum. These fine layers may be a few million to a few hundred million years in age. They may mark individual climate episodes, controlled by variations in the rotational tilt (obliquity) and the degree to which the orbit of Mars differs from a circle (eccentricity).

The dune seas surrounding Planum Boreum are the largest on Mars and include hydrated minerals, with gypsum particularly concentrated in eastern Olympia Undae. Dunes on Mars, for the most part, move extremely slowly compared to their counterparts on Earth, likely because they are largely ice-cemented (near the poles) and because wind forces on Mars are relatively weak – most of the time. Sand and gas emerge from beneath the seasonal carbon dioxide (CO_2) ice layer during northern spring as the seasonal ice sublimates from its base, a process analogous to the formation of "spiders" at the south pole (MC-30; Hansen et al., 2013).

Mars layer-cake geology in three dimensions

How does Mars look underneath its surface? Remarkably, two orbiting subsurface radar mapping systems onboard the Mars Express and Mars

Reconnaissance Orbiter spacecraft have been effective at penetrating the entire sedimentary sequence making up Planum Boreum (Figure 1.A; Figure 1.B). This sequence includes a cratered, thickly layered basal unit that may date back to the Late Hesperian Epoch. The radar mapping reveals that it has a box-like shape, but layers are not evident in the radar data (Brothers et al.,

2015). The younger polar layered deposits are draped over the basal unit and include laterally extensive fine layering. Local and regional unconformities are seen in the radar data as well as in eroded deposit surfaces. The spiral troughs expose details of the layering across much of the polar plateaus (Figure 1.C). It is a mystery how the radar layers, spaced tens of meters apart, correspond

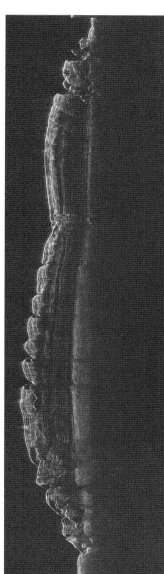

FIGURE 1.A Radargram across Planum Boreum showing the basal unit (strong, broad reflector at bottom) overlain by the internally layered deposits that constitute the bulk of Planum Boreum (Putzig et al., 2009; SHARAD profile about 1100 km long, converted to show depth, vertical range of depth is 3 km, vertical exaggeration about 100:1, view towards longitude 90° E with Gemina Lingula on right, courtesy of N. Putzig, SHARAD data courtesy NASA/JPL-Caltech/Sapienza University of Rome/Southwest Research Institute).

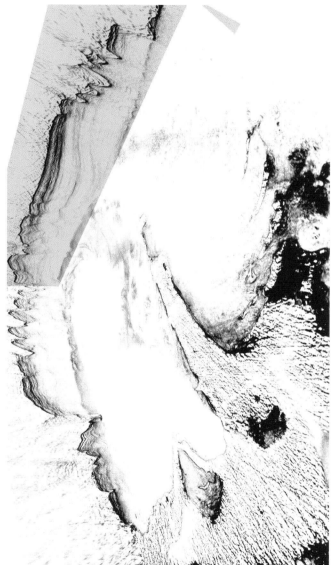

FIGURE 1.B Perspective view looking toward the north pole (which is in the white circular region not covered by radar imaging) from a longitude of 330° E. Two vertical sections (top panels) and a horizontal slice (bottom panel) show the geometry of layering in Planum Boreum (about 900 km in horizontal extent; Putzig et al., 2014, courtesy of N. Putzig, SHARAD data courtesy NASA/JPL-Caltech/Sapienza University of Rome/Southwest Research Institute).

FIGURE 1.C A variety of terrain: The upper, or northern, half of this color HRSC image shows the spiral troughs of Gemini Scopuli cutting Planum Boreum, which lies at an elevation of –3,000 to –4,000 m. The crater Udzha in the midst of the spiral troughs is partly buried by the polar ice and sediment layers. The dark band in the middle of the image includes numerous dunes. Pedestal craters and ground patterns typical of periglacial processes appear in the lowland plain (elevation –4,500 m) at the bottom of the image (HRSC color image, orbit 10247, 94 m/pixel, spanning about 900 km north to south by 400 km, north at top, ESA/DLR/FU-Berlin).

with exposed sequences of meter-thick layers (Lasskar *et al.*, 2002; Putzig *et al.*, 2009; Fortezzo and Tanaka, 2010; Tanaka and Fortezzo, 2012).

Scandia Tholi and Cavi: giant mud volcanoes?

Dozens of pancake- and irregularly shaped bodies and associated depressions, tens to hundreds of kilometers across, occur in a lower part of Vastitas Borealis, near Planum Boreum (Figure 1.D). These might result from the extrusion of large volumes of buoyant slurry – perhaps slushy mud. These masses may have arisen from the deposition of water-saturated sediments from outflow channels that originate in the Martian highlands. Their low elevations could have led to high, near-surface groundwater pressure. Also, the Alba Mons igneous center to the south (MC-3) may have contributed geothermal heat and seismic activity. Combined, such conditions may explain these unusual and unique Martian features. Scandia Tholi and Cavi are surrounded by large networks of polygonal troughs. Similar networks occur in association with submarine mud volcanoes on Earth (Tanaka *et al.*, 2008; 2010; 2011).

Dunes beget dunes

Some of the most active surfaces on Mars form steep scarps and cliffs along some margins of Planum Boreum (Figure 1.E; Figure 1.F), where clouds of dust and debris, caused by landsliding, have been captured in high-resolution images (Figure 1.G). The bases of these cliffs reveal sequences of steeply dipping stacks of dark rock, intermittently covered by ice lenses. These are frozen sand dunes, individually preserved under thin ice sheets, which are now being eroded away. The sands are re-accumulating into fields of sand ripples and dunes in the downwind direction (Figure 1.H; see Tanaka *et al.*, 2008). The winds are dominantly downslope, or katabatic, resulting from the topographic form of Planum Boreum, and swing westward in the adjacent plain where Coriolis forces guide the wind.

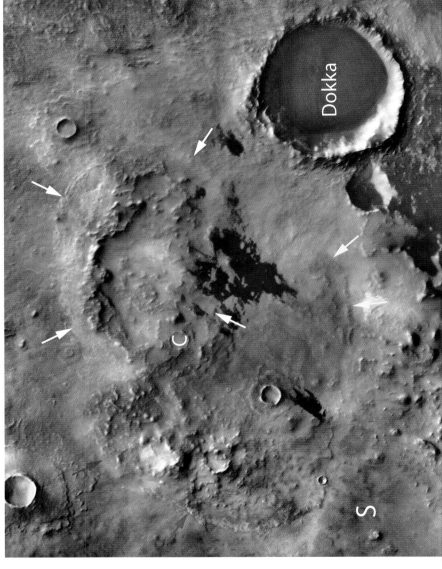

FIGURE 1.D Tholus: View of area south of Dokka crater (north is toward lower right), showing an oblong tholus feature, partly surrounded by a moat (white arrows), which includes a rugged outline and central lows. The tholus cuts the Dokka ejecta and includes a crater with a partly removed rim (C). A second tholus (yellow arrows) appears to be partly buried by Dokka ejecta. Two coalesced tholi occur to the left (red arrows). Smooth deposit (S) at lower left is pitted and overlaps knobs. (Tanaka *et al.*, 2011; THEMIS daytime infrared mosaic, view about 150 km by 200 km, centered at 76.1° N, 217.1° E, NASA/JPL-Caltech/Arizona State University)

FIGURE 1.E Layer cake: Perspective view (see also Figure 1.F) showing the layering in Planum Boreum. The darker, basal unit contains water ice and sediment; active erosion and disaggregation of this unit may be a source for regional sand dunes. The upper, finely layered unit shows an unconformity – a break in the depositional record followed by a change in the angle of beds (this view of Olympia Cavi, and that in Figure 1.F, courtesy of J. Skinner; HiRISE image ESP_018943_2655, centered at 85.38° N, 186.10° E, view to northwest; draped over topography from HiRISE stereo pair with no vertical exaggeration).

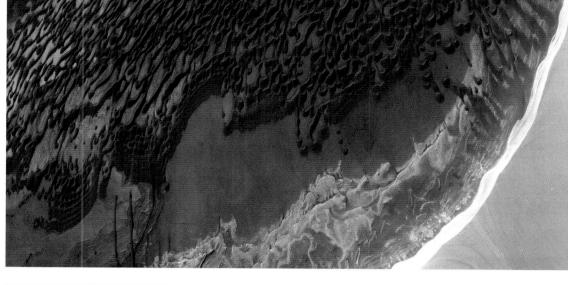

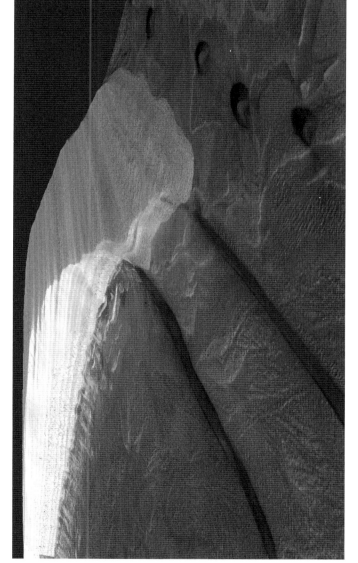

FIGURE 1.F Another perspective "layer-cake" view of Planum Boreum; details as for Figure 1.E. Here the regional sand dunes, formed from material possibly eroded from the basal unit, are seen as dark streaks and mounds (HiRISE image ESP_019298_2640, centered at 83.75° N, 235.70° E, view to north-northeast, NASA/JPL-Caltech/University of Arizona).

FIGURE 1.G Avalanche!: Several avalanches have been imaged as they happen along the steep cliffs of Planum Boreum during northern spring. This HiRISE image shows the cloud of dust and vapor from an event in progress. The cliff is 700 m high; CO_2 frost covers the steep, upper part, while the red-brown basal layers are ice and dust. Sunlight is from the upper right (image PSP_007338_2640, $L_s = 34.0°$, 32 cm/pixel, view about 250 m by 400 m, north to lower left, 83.8° N, 235.5° E; this is close to, but does not overlap, the location of Figure 1.F, NASA/JPL-Caltech/University of Arizona)

FIGURE 1.H On edge: Layering of Planum Boreum is evident in the south-facing margin of one of the Olympia Cavi, along Olympia Rupēs. Finely layered deposits of Planum Boreum make up the light-colored, upper part of the cliff. Beneath it, a darker basal unit of sand (that gives the unit its color) and ice includes cross-bedding, typical of transport by dunes. Light-colored landslides are evident on this unit. Barchan-type dunes that are made of the dark sand originate from the darker unit and continue onto the lowland plain at the base of the cliff (Tanaka et al., 2008; HiRISE image PSP_001341_2650, 32 cm/pixel, 6 km by 12 km, north at lower left, 85.0° N, 150.4° E, NASA/JPL-Caltech/University of Arizona).

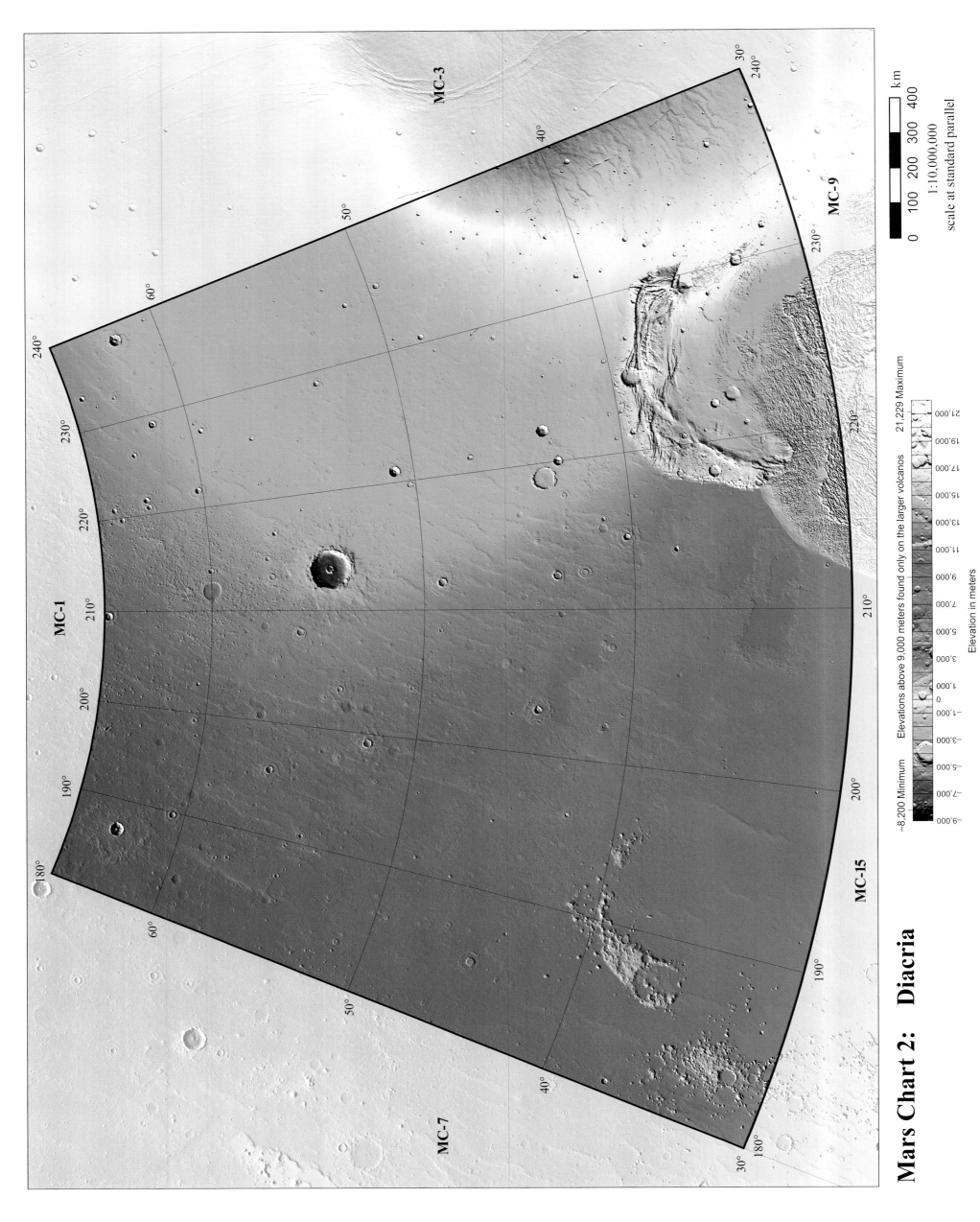

Mars Chart 2:   Diacria

MC-1

MC-3

MC-9

MC-15

MC-7

1:10,000,000
scale at standard parallel

0   100   200   300   400
km

−8,200 Minimum   Elevations above 9,000 meters found only on the larger volcanos   21,229 Maximum

−9,000 −7,000 −5,000 −3,000 −1,000 0 1,000 3,000 5,000 7,000 9,000 11,000 13,000 15,000 17,000 19,000 21,000

Elevation in meters

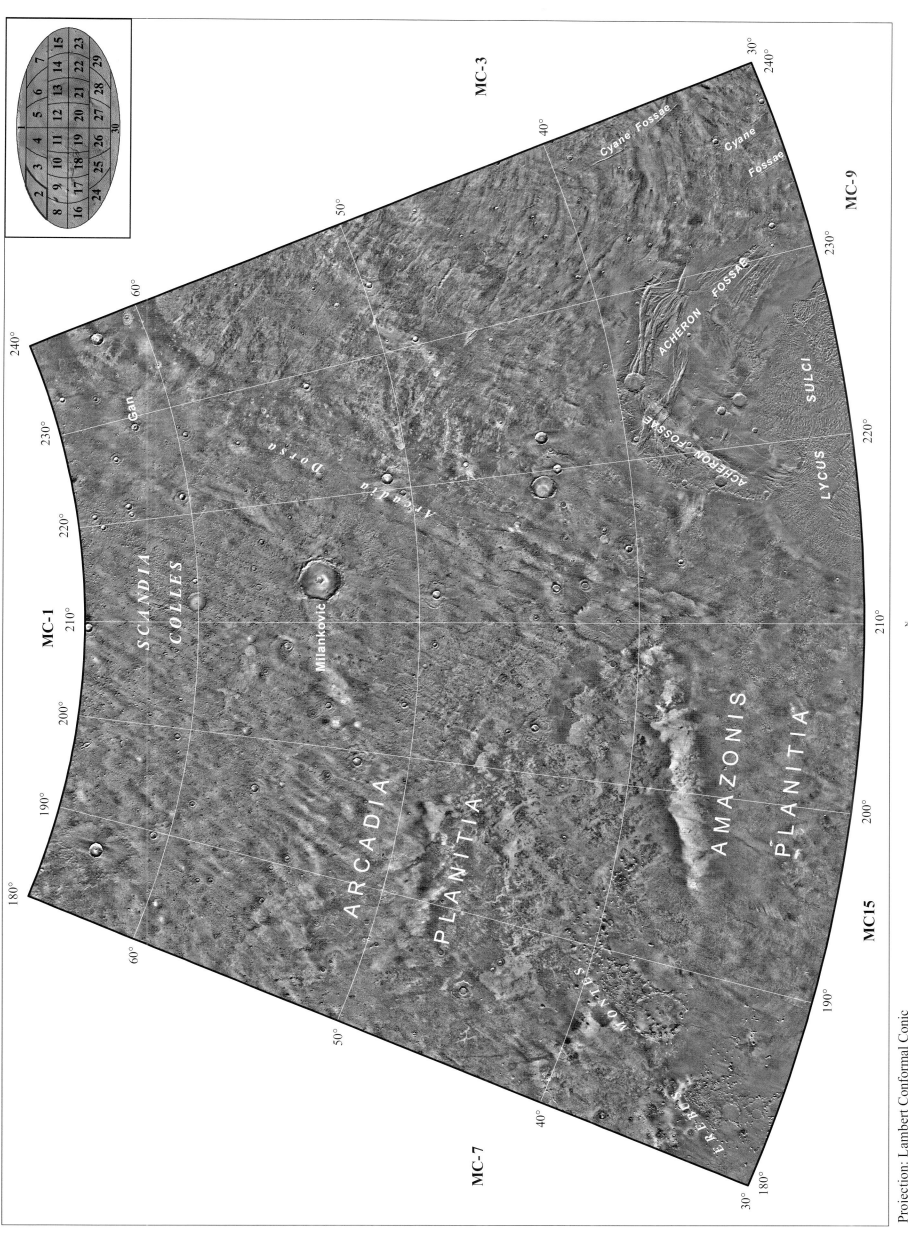

MC-3

MC-9

MC-1

MC-7

MC15

Cyane Fossae

Cyane
Fossae

ACHERON FOSSAE

ACHERON FOSSAE

LYCUS SULCI

Gan

Arcadia Dorsa

Milankovic

SCANDIA
COLLES

ARCADIA

PLANITIA

AMAZONIS

PLANITIA

MONTES

EREBUS

30°

40°

50°

60°

60°

50°

40°

30°

240°

230°

220°

210°

200°

190°

180°

240°

230°

220°

210°

200°

190°

180°

N
W   E
S

Projection: Lambert Conformal Conic
Datum: Mars 2000 Sphere
Central Meridian: 210.0000
Standard Parallel 1: 36.1516
Standard Parallel 2: 59.4669

km

0   100 200 300 400
1:10,000,000
scale at standard parallel

Diacria (MC-2)

Geography

The Diacria quadrangle includes parts of Arcadia and Amazonis Planitiae on the planet's northern plains (around −4000 m elevation), the northwestern flank of Alba Mons (reaching >3000 m elevation), and part of Vastitas Borealis to the north. Other prominent features include the northern part of the rugged Lycus Sulci terrain, the Acheron Fossae that cut a broad, arcuate ridge, and knobby and cratered plateaus of Erebus Montes.

Geology

Erebus Montes and the ridge cut by Acheron Fossae form degraded inliers of Noachian materials (Tanaka *et al.*, 2005). The Acheron Fossae

structures may have resulted from a local mantle plume. Most of the Noachian substrate has been buried by Hesperian sediments infilling Vastitas Borealis. The flank of Alba Mons was built up by effusion of lava flows during Hesperian and Early Amazonian time (Tanaka, 1990). Loading of the crust by these flows contributed to thrust faulting and folding to form the circumferential Arcadia Dorsa ridge system (Tanaka *et al.*, 1991). The Arcadia and Amazonis Planitiae are dominated by Amazonian volcanic flows that originated from buried vents beneath the Lycus Sulci region and fissures in Amazonis south of the quadrangle (Tanaka *et al.*, 2005). Some areas north of Milanković crater are covered by the scattered knobs and low mesas of Scandia Colles, which may be remnants of a Middle Amazonian ice-rich mantle deposit (Skinner *et al.*, 2012).

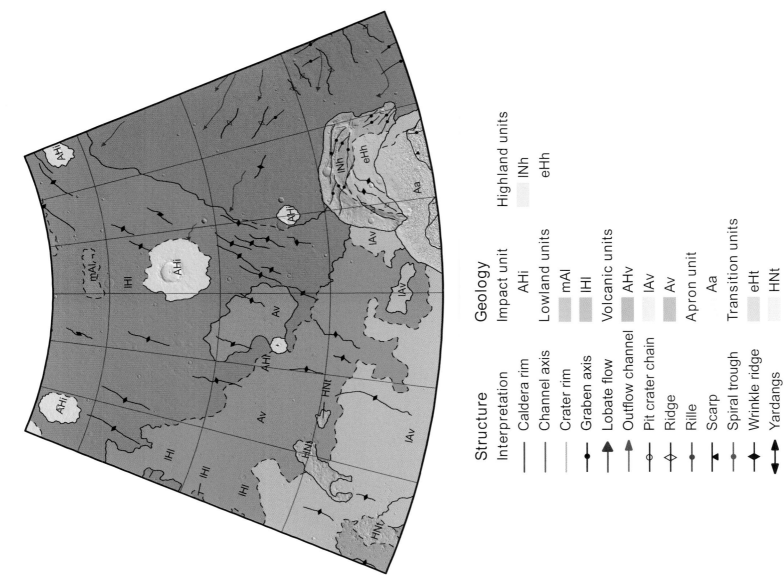

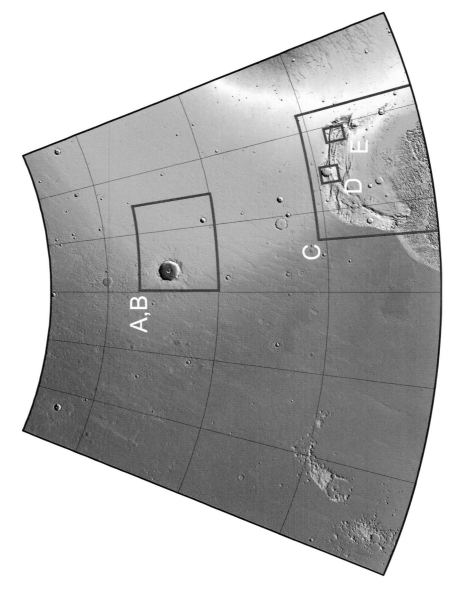

Shake, break, and flow

Milankovič crater forms a huge, 114-km-diameter hole (Figure 2.A). The impact appears to have led to fracturing and faulting along a possibly pre-existing system of cracks, oriented radially to Alba Mons, from which emanated three large flows of material. The flows are several hundred kilometers long, tens of meters thick, and include well-developed channels with relatively horizontal floors (Figure 2.B). These well-developed channels are uncommon in other flows of Alba Mons, many of which are volcanic. It may be that the cratering shook the ground so hard that loose, water-saturated debris in the subsurface liquefied and flowed out onto the surface. (Tanaka *et al.*, 2011).

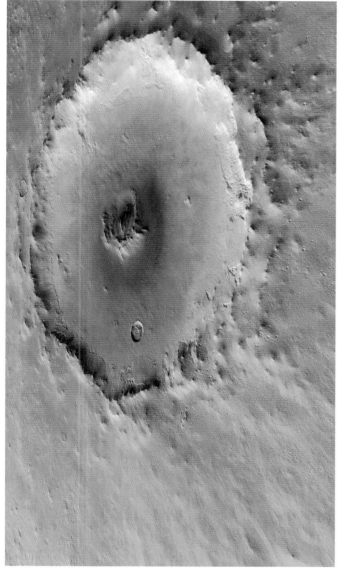

FIGURE 2.A Milankovič crater (HRSC image, orbit 8259, 50 m/pixel, view toward the north, ESA/DLR/FU-Berlin).

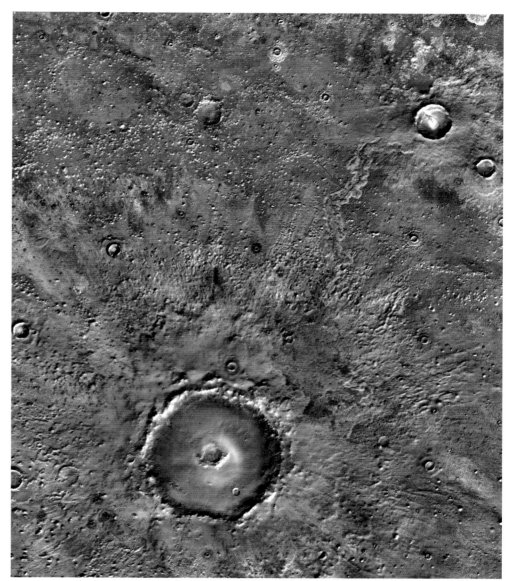

FIGURE 2.B Flow features (red arrow) extend northwest toward Milankovič, possibly along a prior fracture that was reactivated by the impact (THEMIS daytime infrared mosaic, 100 m/pixel, view about 400 km by 500 km, north at top, 50° N to 57° N, 210° E to 222° E, NASA/JPL-Caltech/Arizona State University).

Acheron Fossae: Cooling like custard

An unusual set of fractures, Acheron Fossae, crosses Noachian terrain that stands higher than the surrounding plains. The arc shape of this region suggests it is part of a circular structure, some 700 km in diameter, most of which is covered by probable volcanic deposits associated with the Tharsis province to the south (Figure 2.C). While the overall circular feature could be due to impact or some sort of collapse, its resemblance to features on Venus that are known as "coronae" suggests a different story. Magma that rose from deep in the planet collected below the surface, creating a circular uplift. Escape of magma to surface volcanism or elsewhere underground led to collapse and fracturing in radial and circumferential directions, much like the cooling surface of a custard or pudding. The fractures clearly cut an older crater (Figure 2.D). Later, moving fluid, probably water, formed channels and dendritic drainage on both sides of the crater rim, and possibly in the fault valleys (Figure 2.E) elsewhere in Acheron Fossae (Farmer and Landheim, 1995; Tanaka *et al.*, 1996).

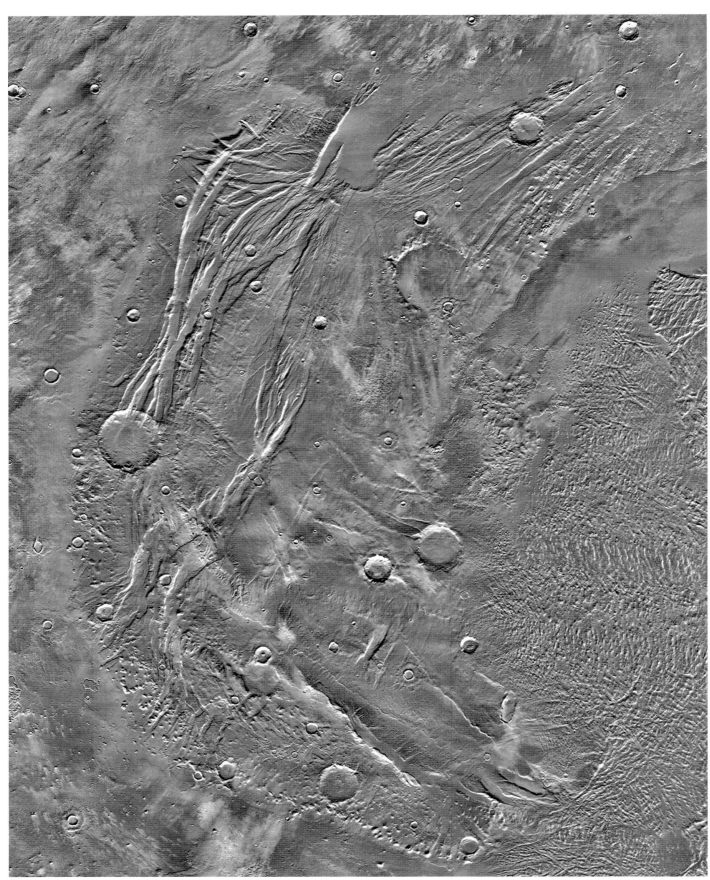

FIGURE 2.C THEMIS infrared scene of Acheron Fossae (THEMIS daytime infrared mosaic, 100 m/pixel, view about 600 km by 800 km, north at top, 31° N to 41° N, 216.5° E to 232.5° E, NASA/JPL-Caltech/Arizona State University).

FIGURE 2.E HRSC view of Acheron Fossae horst-and-graben structure. Vertical relief across the faults is between 1 km and 2 km (HRSC orbit 37, 30 m/pixel, view about 65 km by 90 km, north to right, 38° N, 228.7° E, ESA/DLR/FU-Berlin).

FIGURE 2.D Crater disrupted by fractures and subsequent channel activity in Acheron Fossae (HRSC image H0143_0009, 50 m/pixel, 80 km by 100 km, north at top, 39.2° N, 224.5° E, ESA/DLR/FU-Berlin).

77

The Atlas of Mars

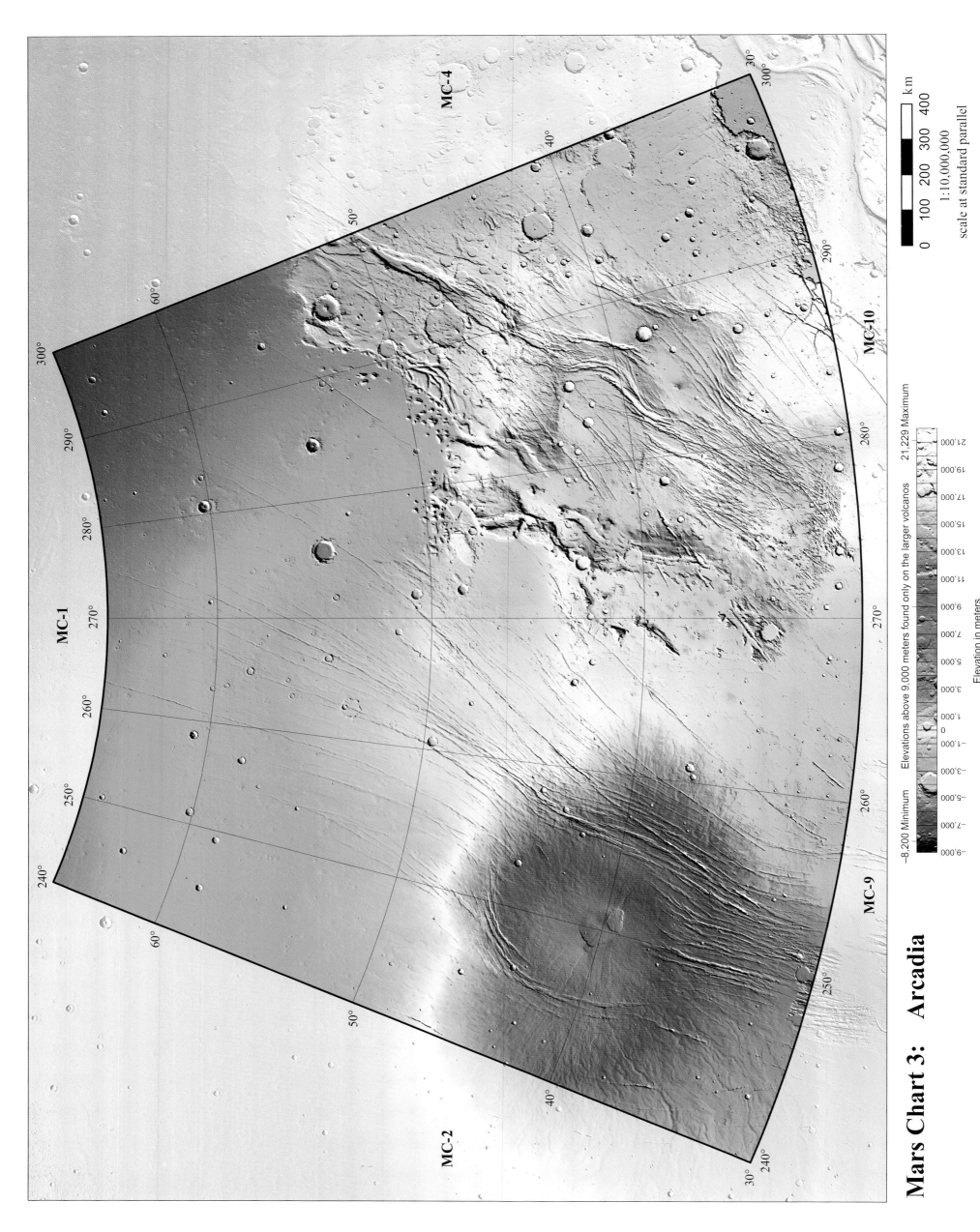

Mars Chart 3: Arcadia

MC-4

MC-1

MC-2

MC-9

MC-10

1:10,000,000
scale at standard parallel

km

0   100 200 300 400

−8,200 Minimum   Elevations above 9,000 meters found only on the larger volcanos   21,229 Maximum

Elevation in meters

−9,000 −7,000 −5,000 −3,000 −1,000 0 1,000 3,000 5,000 7,000 9,000 11,000 13,000 15,000 17,000 19,000 21,000

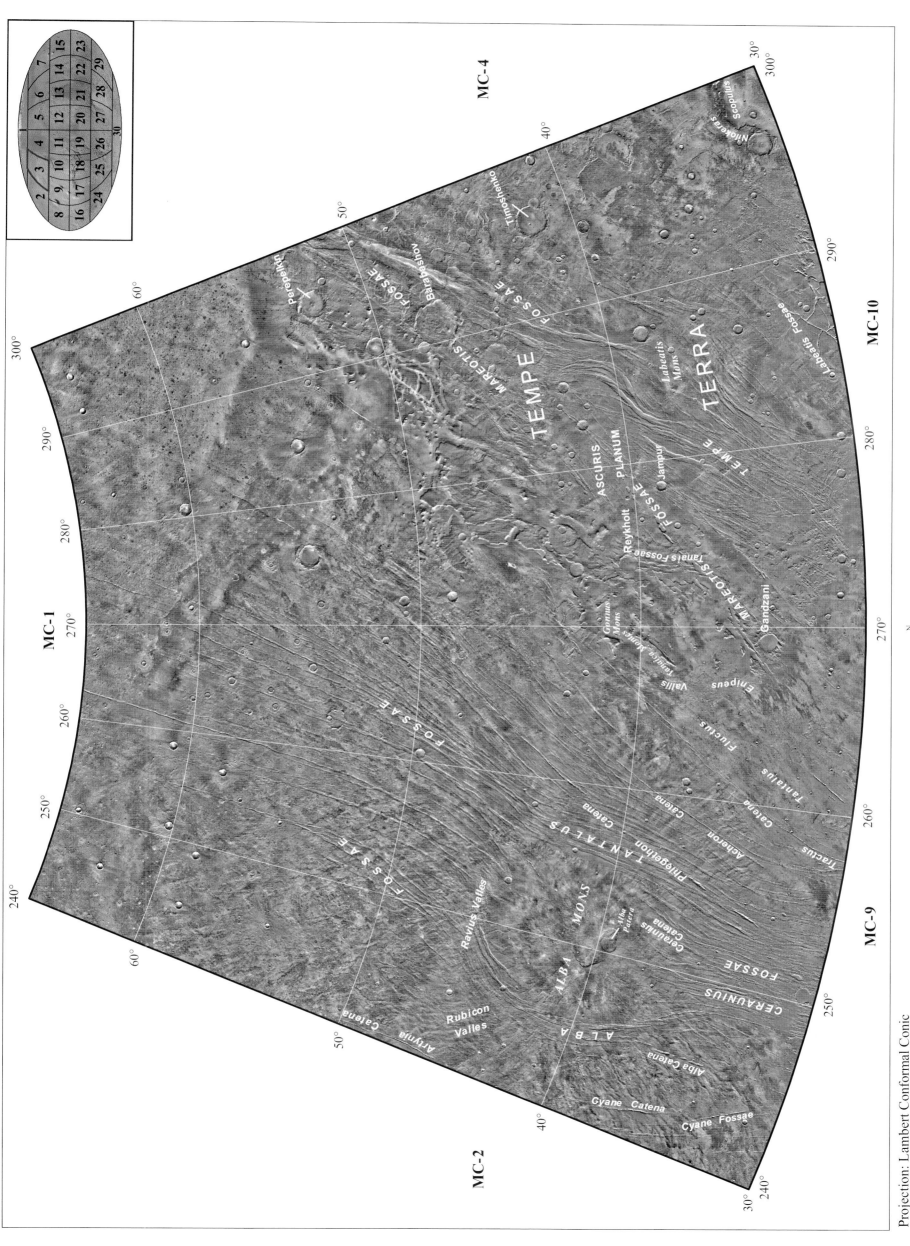

MC-1

MC-2

MC-4

MC-9

MC-10

300°
290°
280°
270°
260°
250°
240°

60°
50°
40°
30°

TEMPE

TERRA

ASCURIS PLANUM

MAREOTIS FOSSAE

TEMPE FOSSAE

FOSSAE

TANTALUS FOSSAE

CERAUNIUS FOSSAE

ALBA FOSSAE

ALBA MONS

Timoshenko

Perepelkin

Barabashov

Labeatis Mons

Jampur

Reykholt

Tanais Fossae

Gandzani

Gonnus Mons

Tenjin Mons

Vallis Enipeus

Flucius

Tantalus

Acheron Catena

Tractus

Phlegethon Catena

Ceraunius Catena

Alba Patera

Rubicon Valles

Ravius Valles

Arrynia Catena

Alba Catena

Cyane Catena

Cyane Fossae

Nilokeras Scopulus

Tabaeus Fossae

N
S
E
W

Projection: Lambert Conformal Conic
Datum: Mars 2000 Sphere
Central Meridian: 270.0000
Standard Parallel 1: 36.1516
Standard Parallel 2: 59.4669

1:10,000,000
scale at standard parallel

km

0   100   200   300   400

Arcadia (MC-3)

Geography

The Arcadia quadrangle is dominated in the western part by fractured Alba Mons, the broadest volcanic shield on Mars, extending more than 2,000 km across and attaining nearly 10 km in relief above the northern plains. The eastern half of the quadrangle mostly consists of the rugged and troughed Tempe Terra. North of Tempe, the Vastitas Borealis plains slope down northeastward to the lowest point in the quadrangle, reaching 4,700 m below the Martian datum.

Geology

Arcadia includes the northernmost outcrops of Noachian highland materials on Mars, making up northern Tempe Terra, which is faulted by northeast-trending grabens and irregular trough systems. Volcanism from small shields and fissure vents resurfaced southwestern Tempe in the Early Hesperian Epoch. Alba Mons consists of exposed Hesperian and Amazonian flows, forming a largely older, broad, heavily faulted basal rise, superposed by a ~400-km-diameter central shield that contains caldera structures and contractional ridges. This huge volcanotectonic structure may result from the development of a large mantle plume beneath it. Flows of Alba overlie possibly ice-rich sediments of the northern plains that mostly display Late Hesperian surface ages, along with some patches of Middle Amazonian icy dust deposits.

Service for one (plate)

Alba Mons forms the broadest volcanic shield within the immense Tharsis region (Figure 3.A). Unlike the other large shields of Tharsis, Alba is cut by large systems of concentric and radial graben systems known as Alba and Tantalus Fossae. The radial systems extend both from Alba Mons and from the center of the Tharsis rise (Tanaka, 1990). The concentric systems define a broad ring, about 500 km in diameter, within which a central, younger, inner shield has been built. The crest of the inner shield is cut by two nested calderas, forming Alba Patera, and several contractional ridges radiate across the Hesperian and inner shield. The overall structure of Alba mimics "corona" structures on Venus, where concentric and radial rifting and folding are common. Such structures are indicative of

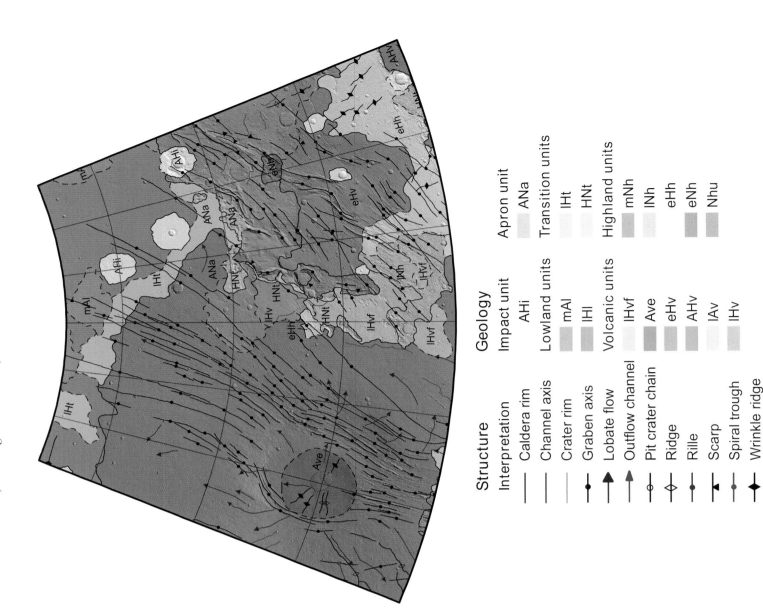

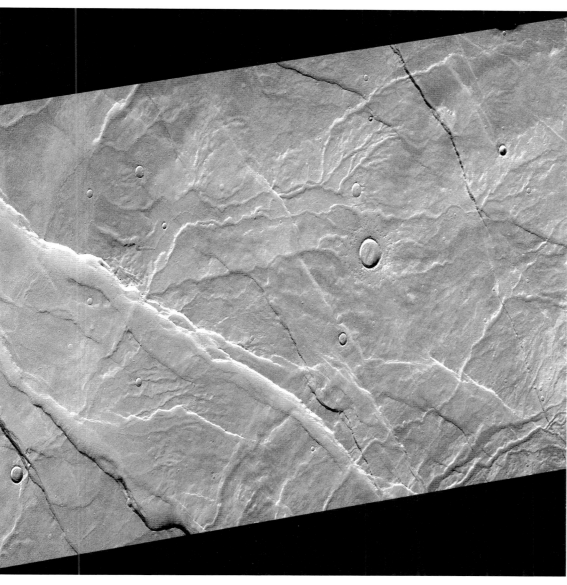

FIGURE 3.B A part of Ravius Valles, formed during the Amazonian Period and cut by grabens on the north flank of Alba Mons (CTX image P20_008960_2263, 6 m/pixel, view 40 km by 50 km, north toward top, 46.4° N, 249.5° E, NASA/JPL-Caltech/MSSS).

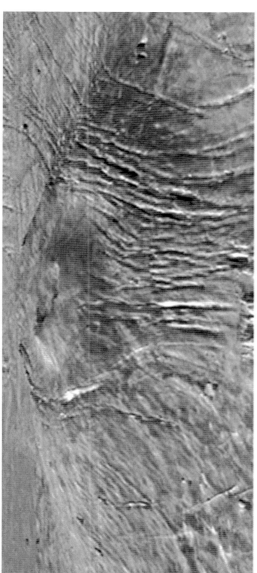

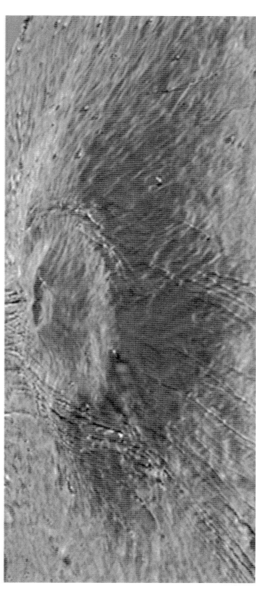

FIGURE 3.A Two views (vertically exaggerated by a factor of 10) of Alba Mons: top view looking north, and bottom view toward the south. The Viking image mosaic is draped over MOLA-derived topography (NASA/MOLA science team).

stationary crust, as on a one-plate planet, unlike that of Earth, whose crust is broken up into slowly moving plates. The one-plate structure results in long-lived centers of magmatism over evolving hot mantle plumes. The northern flank of Alba Mons is dissected by dendritic valley systems, Rubicon and Ravius Valles (Figure 3. B). These developed during the Amazonian Period, when Mars was generally arid and freezing, possibly indicating local precipitation during climate excursions, or intense episodes of volcanism.

Cyane Catena

Some of the youngest Alba Mons tectonic features consist of pit chains, called catenae, along trends that are oblique to many of the grabens. These indicate the formation of subsurface tension cracks into which surface material collapses. The example of Cyane Catena in Figure 3.C shows dozens of rimless pits, along with mostly smaller rimmed impact craters, within a graben trough. Arcuate scarps and other linear forms within the trough indicate complex overlapping styles of deformation.

81

The Atlas of Mars

Tempe Terra

Tempe Terra lies in the southeastern part of the quadrangle. It displays numerous fractures and fault valleys (grabens; Figure 3.D), ridges probably underlain by igneous dikes (Manfredi and Greeley, 2012), and small volcanoes (Moore, 2001). These features reflect mostly Hesperian igneous and tectonic activity, which cut across and altered the underlying Noachian highland terrain. The margin of Tempe Terra drops steeply in places, to lowlands to the north, volcanic plains to the west, and the floor of Kasei Valles to the south (MC-10). Along northern parts of this margin are lobate debris flows (Figure 3.E), likely caused by downslope creep in which near-surface ice played a role, perhaps like glaciation on Earth. More recently, the lobate debris flows have been modified by wind-transported cover materials and loss of ice by sublimation (van Gasselt *et al.*, 2010; Chuang *et al.*, 2011).

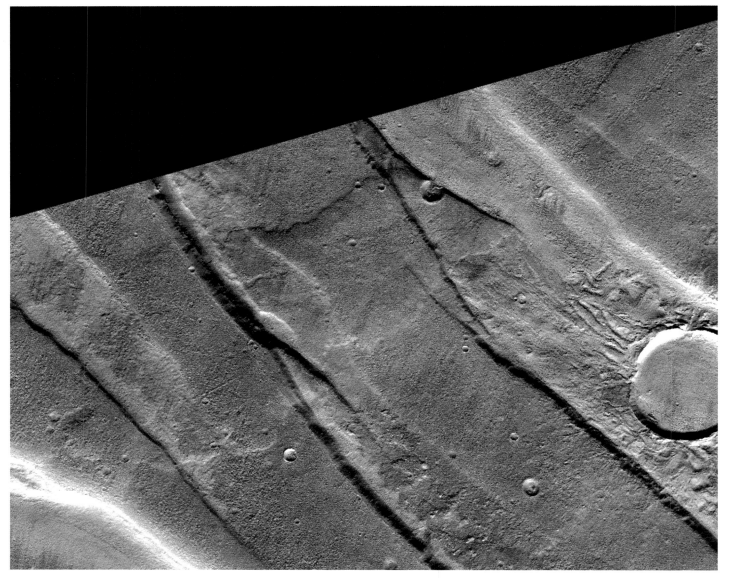

FIGURE 3.D CTX image of lava flow, flowing north (towards top) from Labeatis Mons, which lies to the south. Fractures of Tempe Fossae cut the flow and thus postdate it (Neesemann *et al.*, 2014; image P20_008972_2175, 6 m/pixel, view about 15 km across, 38.8° N, 283.2° E, NASA/JPL-Caltech/MSSS).

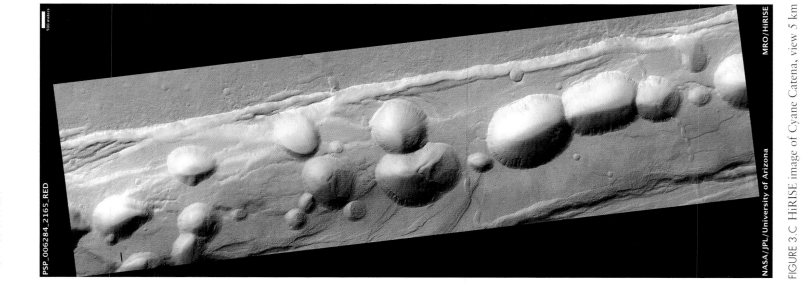

FIGURE 3.C HiRISE image of Cyane Catena, view 5 km by 17 km with north toward top (illumination from lower left, image PSP_006284_2165 red band, 25 cm/pixel, centered at 36.1° N 241.8° E, NASA/JPL-Caltech/University of Arizona).

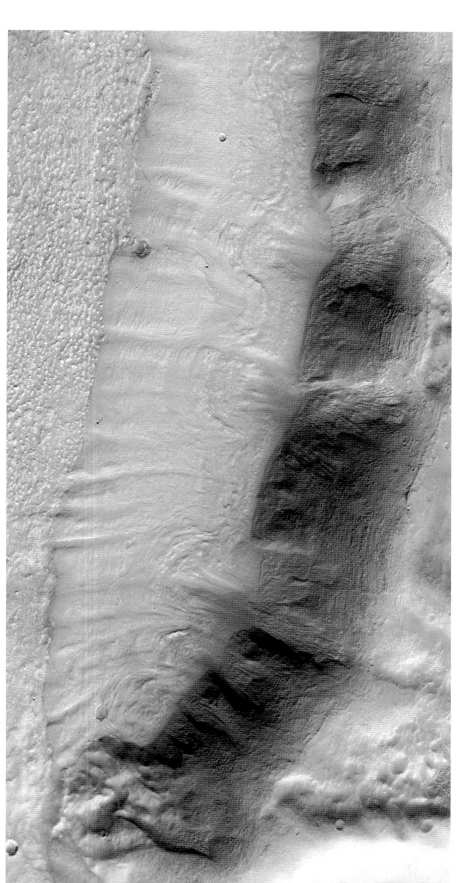

FIGURE 3.E Lobate aprons along Mareotis Fossae at the western edge of Tempe Terra. Downslope is toward the top of image; view 16 km by 8 km (CTX image P17_007852_2154_XN, 6 m/pixel, north at top, 35° N, 268° E, NASA/JPL-Caltech/MSS).

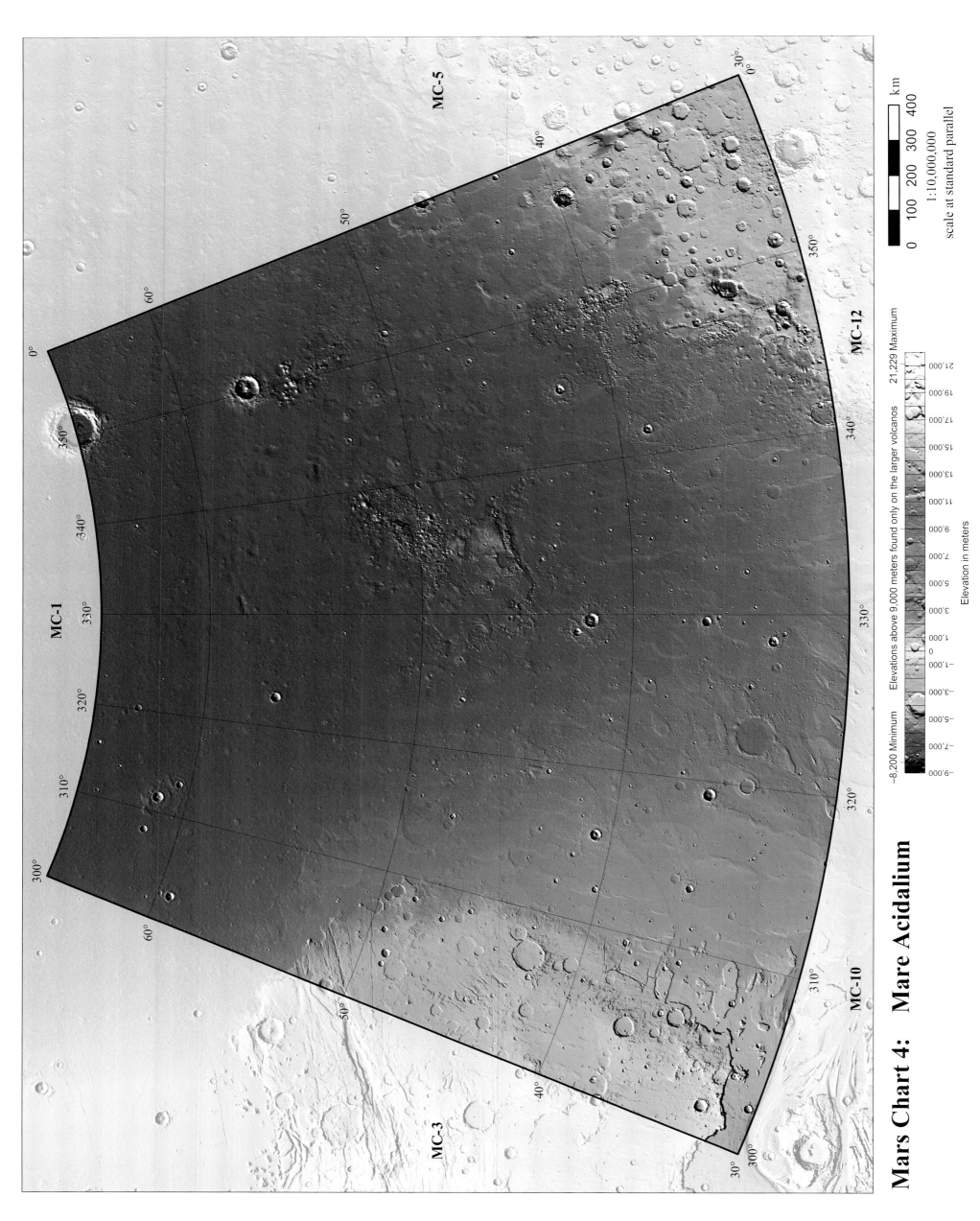

Mars Chart 4:   **Mare Acidalium**

1:10,000,000
scale at standard parallel

km
0 100 200 300 400

−8,200 Minimum   Elevations above 9,000 meters found only on the larger volcanos   21,229 Maximum

−9,000 −7,000 −5,000 −3,000 −1,000 0 1,000 3,000 5,000 7,000 9,000 11,000 13,000 15,000 17,000 19,000 21,000

Elevation in meters

MC-5

MC-1

MC-12

MC-3

MC-10

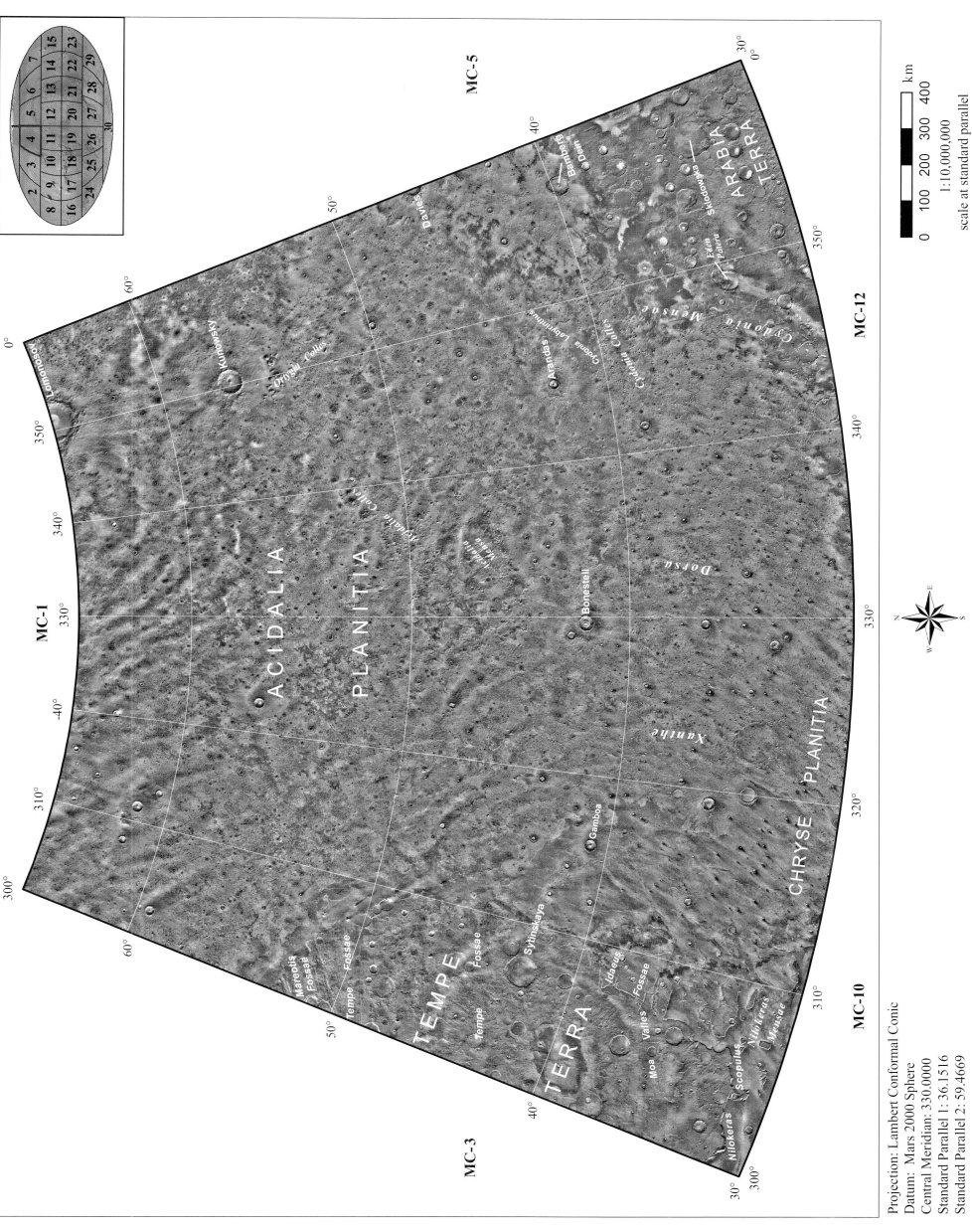

MC-5
MC-12

ARABIA TERRA

Bamberg
Dein
Sklodowska
Eden Patera

Cydonia Mensae

Davies

Cydonia Colles

Arandas
Cydonia Labyrinthus

Kunowsky
Lomonosov

Orygia Colles

ACIDALIA

PLANITIA

Acidalia Colles

Acidalia Mensa

Dorsa

Bonestell

Xanthe

Gamboa

MC-1

Sytinskaya

Tempe
Fossae

Mareotis
Fossae

Tempe
Fossae

TEMPE

Tempe

TERRA

Idaeus
Fossae

CHRYSE PLANITIA

Moa
Valles

Scopulus

Niliheras
Mensae

Niliheras

MC-3
MC-10

MC-1

N
E
W
S

Projection: Lambert Conformal Conic
Datum: Mars 2000 Sphere
Central Meridian: 330.0000
Standard Parallel 1: 36.1516
Standard Parallel 2: 59.4669

1:10,000,000
scale at standard parallel

0 100 200 300 400   km

Mare Acidalium (MC-4)

Geography

The Mare Acidalium quadrangle is dominated by the gently northward-dipping lowland plain of Acidalia Planitia and its contiguous southern neighbor, Chryse Planitia, both lying between 4,000 and 5,000 m below datum. These plains are partly bordered by cratered highlands that rise as much as 3,000 m above the plains – Arabia Terra to the southeast and Tempe Terra to the southwest. The outer margins of these terrae define the mouths of the largest fluvial-type channels anywhere on Mars – the circum-Chryse outflow channels (named for Chryse Planitia, into which the channels collectively emerge, see also MC-10 and MC-11). The regional planitiae contain the generally north-south trending ridge systems of Xanthe Dorsa. Acidalia Mensa is located in the center of the quadrangle and forms an irregularly shaped plateau that rises from 100 to >500 m above the surrounding plains.

Geology

In this quadrangle, the circum-Chryse outflow channels are overlapped by low-lying plains materials, commonly thought to be sediments deposited within a Late Hesperian ocean, possibly reworked along their margins by tsunami waves (Rodriguez *et al.*, 2016; see MC-5). Because material ages generally reflect the cessation of activity, it is likely that outflow occurred throughout the Hesperian Period. Temporal and some spatial correlations indicate that outflow activity was likely genetically linked to pulses in the geologic activity of Tharsis volcanoes and the Valles Marineris rift system. Younger units exist locally within the region, including a tens-of-meters-thick unit, potentially composed of fine-grained material, emplaced during the Middle Amazonian Epoch. This unit may be an older cousin to more recent, meters-thick mantles, locally observed throughout the Martian northern plains.

Geology

Although this plateau is low-lying relative to the closest cratered highlands, it contains impact craters, fractures, and knobs that are more similar to highland terrains than the lowland plains in which they reside. Noachian and Hesperian

A plateau in the plains

Acidalia Mensa, which means "flat-topped plateau with cliff-like edges in Acidalia," rises up in the midst of Acidalia Planitia (Figure 4.A).

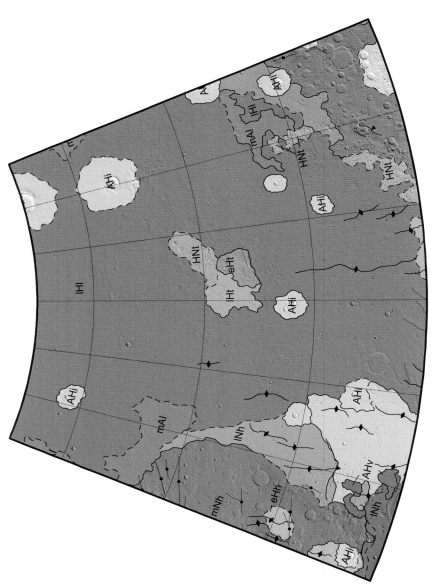

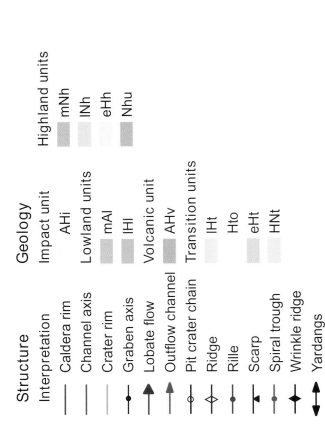

Structure

Interpretation

—	Caldera rim
—	Channel axis
—	Crater rim
●	Graben axis
▲	Lobate flow
▲	Outflow channel
⊖	Pit crater chain
⬦	Ridge
●	Rille
◀	Scarp
◆	Spiral trough
◆	Wrinkle ridge
↕	Yardangs

Geology

Impact unit

AHi

Lowland units

mAl
lHl

Volcanic unit

AHv

Transition units

lHt
Hto
eHt
HNt

Highland units

mNh
lNh
eHh
Nhu

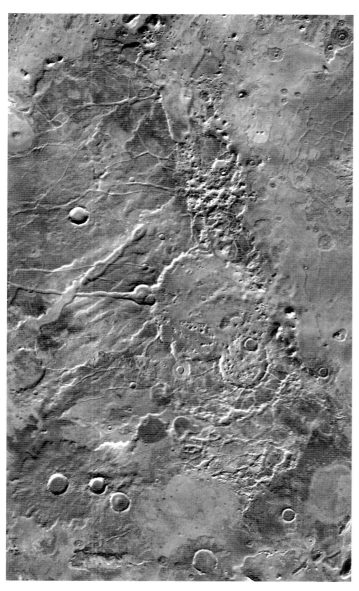

FIGURE 4.A Southern margin of Acidalia Mensa, located between Acidalia and Chryse Planitiae. Eroded craters and linear fractures give the appearance of cratered highland materials (THEMIS daytime infrared mosaic, view about 150 km by 250 km, north at top, 45° N to 47.5° N, 331° E to 337° E, NASA/JPL-Caltech/Arizona State University).

surface ages for Acidalia Mensa add to this comparison and underscore the interpretation that this feature and the nearby hills of Acidalia Colles are outcrops of older, now spatially partitioned, terrains. Though questions remain regarding the origin of Acidalia Mensa, one possibility is that it may be an erosional remnant of the central uplift of a now-buried, Acidalia-centered giant impact basin, formed early in the history of Mars.

Channel scours and evidence of absence

As noted, the circum-Chryse outflow channels are the largest fluvial-type channels on Mars, representing the apparent flow of thousands of cubic kilometers of water across the planet's surface. We naturally expect that such immense volumes of water would result in erosional and depositional features, just as large rivers on Earth carve valleys along their route and emplace deltas at their mouths. Teardrop-shaped bars located in the quadrangle (Figure 4.B) formed as water flowed

across the low-lying plains. Therein, topographic promontories (such as impact craters) acted as obstacles around which water was forced to flow, resulting in the teardrop shape. In some cases, these bars are >100 m tall, providing a gauge of how deep the water might have been. Despite this evidence of erosion, there is an absence of clear depositional features in Acidalia Planitia, which might indicate the termination of floods or their entrance into a standing body of water. This suggests a major question: Where is the evidence for deposition of the vast amounts of eroded material? The answer may be literally buried in the northern plains. Here HiRISE images reveal discontinuous bright, fine-grained materials that bury dark materials, possibly lavas (Figure 4.C; Salvatore and Christensen, 2014). Closer investigation of these features helps show that though evidence for deposition is subtle, it is not entirely absent. And while water-charged floods have been the most commonly advocated origin for the immense channel systems, some have also proposed origins involving glaciers, debris or mud flows, erosive lava flows, and carbon dioxide-charged floods.

FIGURE 4.B Teardrop-shaped bars located in southwestern Chryse Planitia. Features typically form around topographic obstacles, in this case impact craters. The point of the bar indicates the direction of flow. While such bars are seen in many of the major outflow channels on Mars (e.g. MC-11), these examples are well beyond the channel mouths, out in the northern lowland (MOLA color hillshade overlain on THEMIS daytime infrared image, view about 325 km by 400 km, north at top, 35.5° N to 41° N, 321° E to 329° E, NASA/JPL-Caltech/Arizona State University/GSFC).

Muddy manifestations?

Thousands of pitted cones, of kilometer scale, dot areas of the northern plains of Mars (Figure 4.D). Although the cones commonly resemble cinder cones that are produced by hot lavas, their association with large lowland areas, where water could have pooled and where water-saturated sediments could have accumulated, strongly suggests that the cones resulted from eruptions of water-charged sedimentary particles (Oehler and Allen, 2012, see their Figure 2B). Some of the plains areas where the cones occur are cut by trough networks, which form polygons that are several kilometers across. If water-rich sediments

hundreds of meters thick quickly accumulated in the northern plains, some water-rich pockets may have become pressurized, leading to eruptions of mud. Some eruption sites occur along sinuous ridges, possibly developed by folding of the sediment due to horizontal contraction (Figure 4.E; Tanaka et al., 2005). Also, as the sediments compacted and settled due to escape of water, they may have broken up into gigantic mud-crack forms. On Earth, mud volcanoes and large crack systems occur in some ocean basins, such as offshore from Norway and Morocco, and in the South China Sea (Oehler and Allen, 2012, their figures 7A–C).

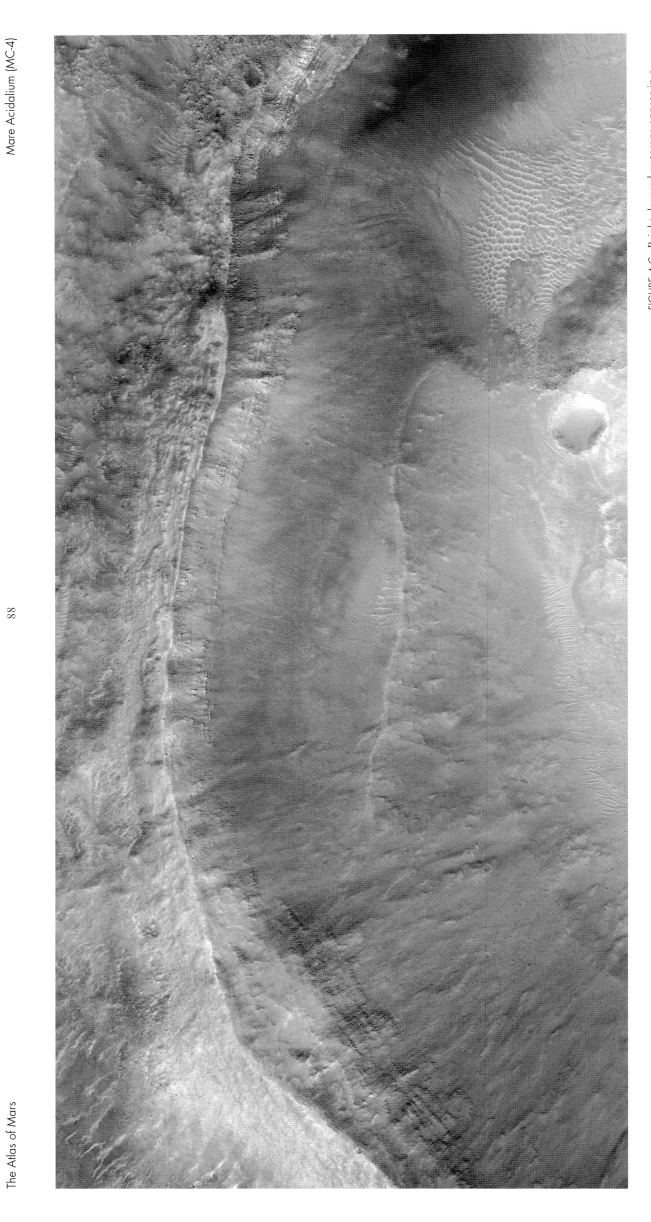

FIGURE 4.C Bright, layered outcrops appear in a northern plains impact crater wall, overlying dark, layered material. The bright material may be sedimentary deposits that resulted from outflow channel floods. The dark material may be lava (Salvatore and Christensen, 2014; HiRISE image ESP_026957_2200, 60 cm/pixel, view about 4 km across, north at top, 39.6° N, 310.3° E, NASA/JPL-Caltech/University of Arizona).

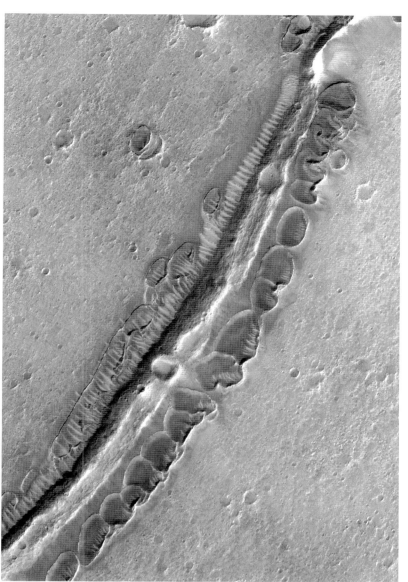

FIGURE 4.E HiRISE image of pits along a ridge in southeastern Acidalia Planitia (monochrome image ESP_031268_2115, 60 cm/pixel, 3 km across, north at top, 30.9° N, 339.4° E, NASA/JPL-Caltech/University of Arizona).

89

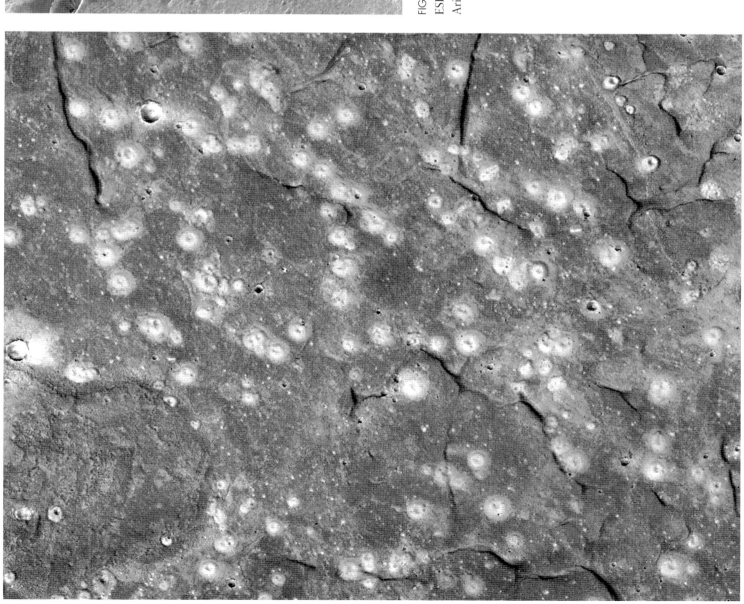

FIGURE 4.D Bright, pitted cones amidst polygonal troughs in Acidalia Planitia (Oehler and Allen, 2012; CTX image P19_008522_2211, 6 m/pixel, view about 25 km by 30 km, north at top, 40.8° N, 332.5° E, NASA/JPL-Caltech/MSSS).

The Atlas of Mars

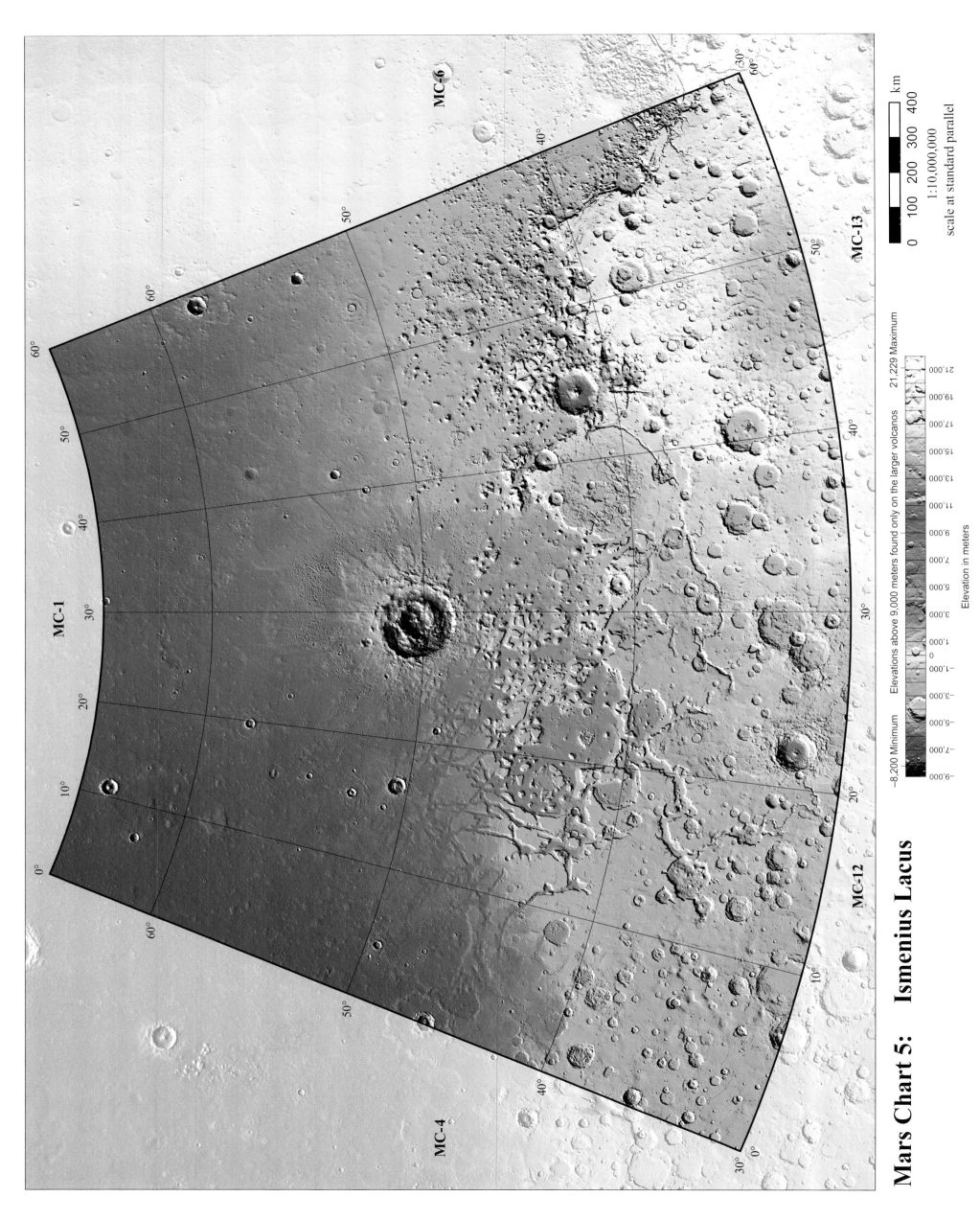

Mars Chart 5: Ismenius Lacus

MC-6

MC-13

MC-1

MC-12

MC-4

0   100   200   300   400
km

1:10,000,000
scale at standard parallel

−8,200 Minimum   Elevations above 9,000 meters found only on the larger volcanos   21,229 Maximum

−9,000 −7,000 −5,000 −3,000 −1,000 0 1,000 3,000 5,000 7,000 9,000 11,000 13,000 15,000 17,000 19,000 21,000

Elevation in meters

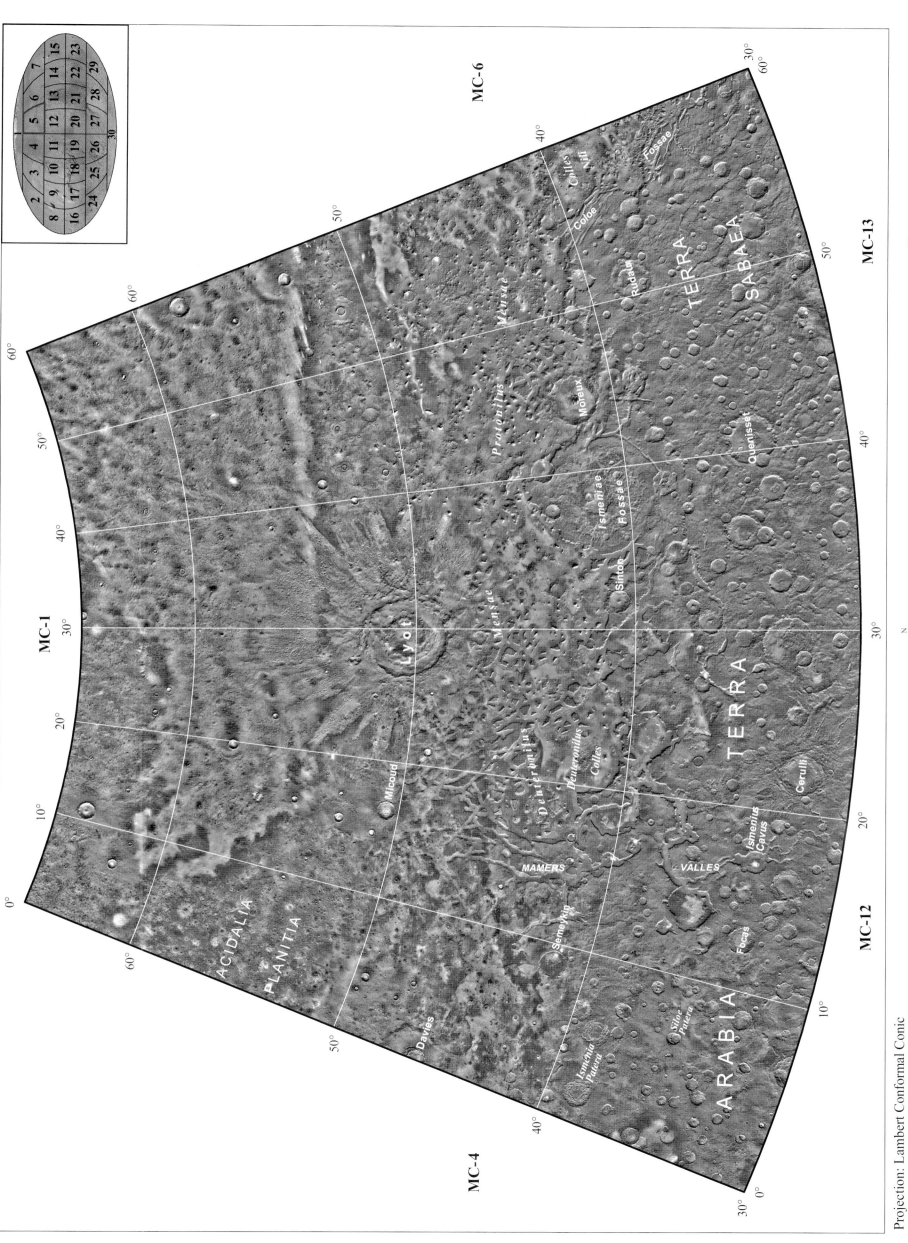

Projection: Lambert Conformal Conic
Datum: Mars 2000 Sphere
Central Meridian: 30.0000
Standard Parallel 1: 36.1516
Standard Parallel 2: 59.4669

1:10,000,000

scale at standard parallel

0   100   200   300   400   km

MC-1

MC-4

MC-6

MC-12

MC-13

ACIDALIA PLANITIA

Lyot

Micoud

Davies

Mamers

Deuteronilus Mensae

Deuteronilus Colles

Protonilus Mensae

Mensae

Moreux

Coloe Colles

Coloe Fossae

Nili Fossae

Ismeniae Fossae

Sinton

Rudaux

Querisset

TERRA SABAEA

TERRA

ARABIA

Cerulli

Focas

Siloe Patera

Ismenia Patera

Semeykin

VALLES

Ismenius Cavus

Ismenius Lacus (MC-5)

Geography

Ismenius Lacus is located in the northern mid-latitudes of the eastern hemisphere of Mars. It includes sections of both the southern highlands and northern plains. The topographic transition is defined by gently sloping surfaces, steep scarps, linear to sinuous channels, and isolated knobs and plateaus. The southern part of the quadrangle is defined by the northernmost extent of the high-standing, ancient cratered highlands of Arabia Terra and Terra Sabaea, at elevations near datum to –3,000 m. The regional highlands in Ismenius Lacus contain large, ancient channel systems – Okavango and Mamers Valles (Figure 5.A; Mangold and Howard, 2013) – as well as networks of linear depressions – Ismeniae and Coloe Fossae. These physiographic features all record complex geologic processes that are associated with the long-term break-up and marginal erosion of the cratered highlands. From the north to the south,

the highland–lowland transition in Ismenius Lacus is marked by the high-standing plateaus of Deuteronilus and Protonilus Mensae as well as the irregularly-shaped depressions of Deuteronilus Colles and Colles Nili. The lowlands in the north typically lie at –4,000 m or lower. The 236-km-diameter Lyot crater and its radial and lobate ejecta blanket dominate the center of the quadrangle.

Geology

The Ismenius Lacus quadrangle contains land-forms and terrains that typify the highland, transition, and lowland regions of Mars. The highlands contain a diversity of impact craters and basins, a testament to the long-lived bombardment of the planetary surface. The land surface between these ancient impact scars contains subtle undulations, particularly in Arabia Terra. This has been cited as evidence of a land surface that is slowly being lowered (subsiding), perhaps due to the compaction of weak rock layers in the subsurface by the

removal of water or ice (Skinner *et al.*, 2004). Evidence for water or ice is also observed in the transition regions of Ismenius Lacus, where many of the plateaus and knobs that jut out from the

highlands are surrounded by smooth aprons (Figure 5.B; Figure 5.C). These may be similar to glaciers on Earth and – based on a near total lack of impact craters on their surfaces – they may be

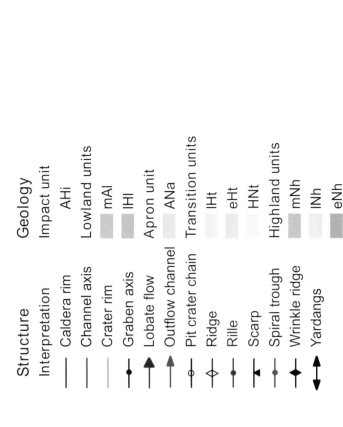

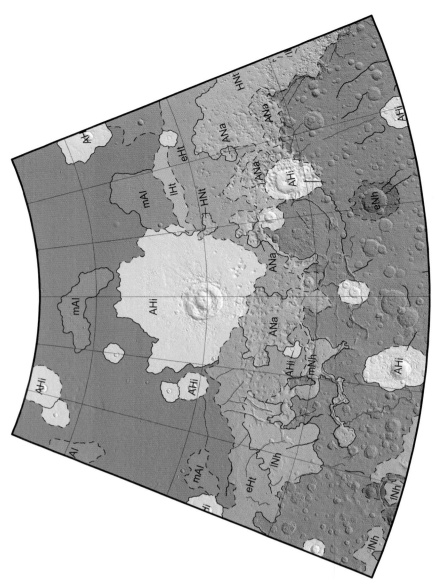

Structure

Interpretation

——	Caldera rim
——	Channel axis
——	Crater rim
●——	Graben axis
▲	Lobate flow
➤	Outflow channel
◇—○	Pit crater chain
◇—●	Ridge
●—●	Rille
◄—■	Scarp
◆—◆	Spiral trough
◆	Wrinkle ridge
↕	Yardangs

Geology

Impact unit

AHi

Lowland units

mAl

lHl

Apron unit

ANa

Transition units

lHt

eHt

HNt

Highland units

mNh

lNh

eNh

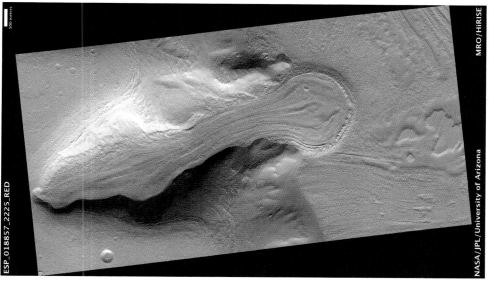

ESP_018857_2225_RED

NASA/JPL/University of Arizona · MRO/HiRISE

FIGURE 5.C View of a valley in an isolated plateau in Protonilus Mensae; parallel ridges on surface of valley fill indicate a possible valley glacier (Mellon and Byrne, 2010; HiRISE ESP_018857_2225_RED, 30 cm/pixel, view 6 km by 12 km, north toward top, 42.2° N, 50.5° E, NASA/JPL-Caltech/University of Arizona). On Earth, valley glaciers flow from regions of ice accumulation to regions of ice ablation, whereas on Mars they may occur where gravity causes motion of pre-existing ice, the distribution of which is a function of latitude and slope orientation rather than elevation or precipitation (Dickson et al., 2011; Souness et al., 2012; Souness and Hubbard, 2012).

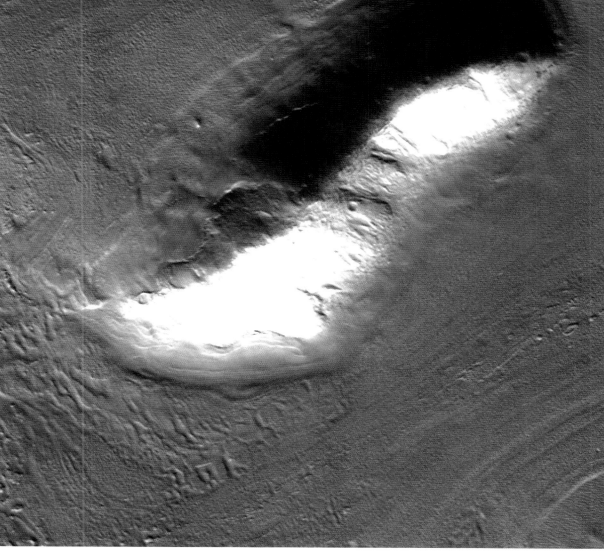

FIGURE 5.A This central portion of Mamers Valles is about 1,200 m deep and includes the remnant of a crater about 4 km across. Parallel lineations trend away from the channel walls before merging and turning downstream. These forms resemble the moraines, or ridges of rocky material, in valley glaciers on Earth. Flowing ice, possibly in recent Martian history, could account for the parallel ridges and the removal of part of the crater. Ice may still be present beneath the debris on the valley floor. The image is colored according to daytime surface temperatures; red is warmer, blue is colder. (THEMIS daytime infrared image, 18m/pixel, view 18 km by 31 km, north toward top, 37.4° N, 15.8° E, color by THEMIS team, NASA/JPL-Caltech/Arizona State University).

less than a few million years old. The northern part of the quadrangle is defined by the low-lying, relatively youthful plains of eastern Acidalia Planitia and western Utopia Planitia (Figure 4.1). These regions contain landforms that typify the

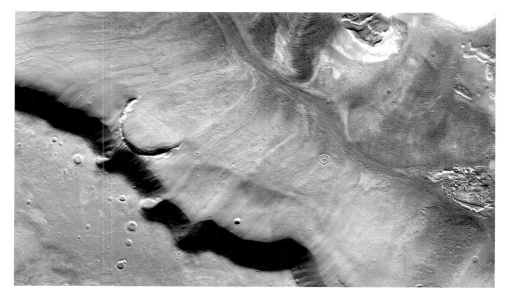

FIGURE 5.B Apron deposits surround an eroded piece of highland material in Ismenius Lacus. The striations are likely a result of slowly moving, ice-rich sediment, similar to terrestrial glaciers. Note the general lack of impact craters, which suggests the material is relatively young (part of THEMIS visible image V29216009, 19 m/pixel, view 15 km across, north at top, centered at 44.6° N, 30° E, NASA/JPL-Caltech/Arizona State University).

Martian lowlands: knobs, pedestal craters, subdued ridges, and circular to irregularly shaped shallow depressions.

There are always highs and lows

In many ways, Ismenius Lacus contains typical examples of the suite of landforms and geologic units that occur along the globe-encircling

highland–lowland transition. In addition, there is a notable variation in these transition terrains. To the east, in northern Terra Sabaea, the transition is defined by a rugged scarp, bordered by highstanding knobs and plateaus. This rugged terrain indicates a suite of geologic processes that essentially "break" the Martian highlands into pieces, which then fall apart due to gravity-assisted erosion. To the west, in northern Arabia Terra, the

transition is defined by a gently sloping (<1°) surface that gradually grades into low-standing northern plains. This sloping terrain implies a suite of geologic processes that essentially "bend" the Martian highlands, which then get buried by younger rock and sediment units from the north.

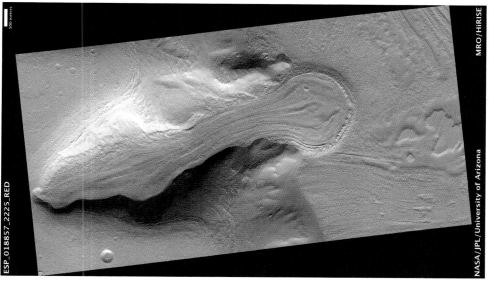

MC-5

The Atlas of Mars

93

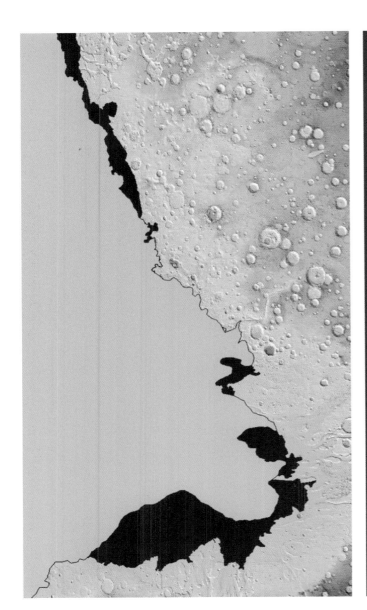

FIGURE 5.D Lyot crater dominates the center of the Ismenius Lacus quadrangle. It is a complex crater, composed of two rings that define a plate-shaped (flat-floored) crater. The ejecta blanket that surrounds Lyot crater is composed of both linear grooves and lobate masses, suggesting that the impact excavated and dispersed both dry and "wet" materials from the subsurface (MDIM 2.1 mosaic, view 900 km across; north at top, courtesy of USGS).

FIGURE 5.F Proposed older (top) and younger (bottom) tsunami deposits (in red) that originated from hypothetical bodies of water (in light and dark blue; Rodriguez *et al.*, 2016). The margins of the deposits form lobes where they were emplaced in up-slope directions. Presumably, large impacts within a northern plains-filling ocean generated the deposits, incorporating water, ice, soil, and rocks (both are MOLA color shaded relief views, spanning 10° N to 52° N, 305° E to 20° E, including parts of MC-4, MC-5, MC-10, and MC-11, elevation color mapping not the same as elsewhere in atlas; courtesy A. Rodriguez).

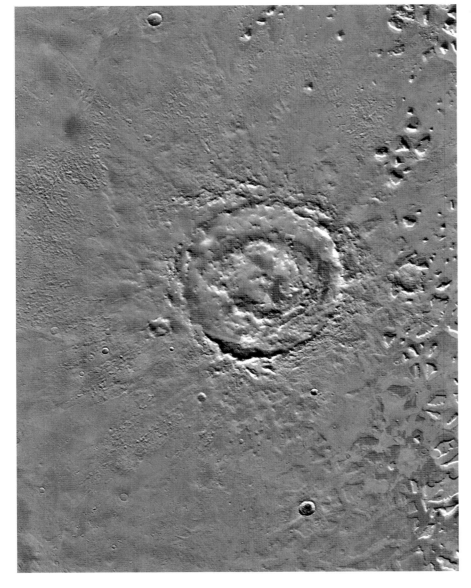

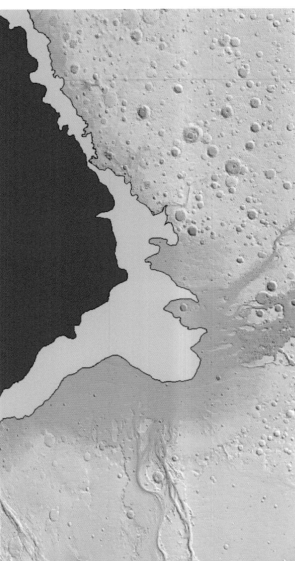

FIGURE 5.E (left) Proposed older (light blue) and younger (dark blue) shorelines inundating Chryse Planitia (Rodriguez *et al.*, 2016). These correspond to the tsunami deposits shown in Figure 5.F top and bottom, respectively (MOLA color shaded relief view 7° N to 50° N, 288° E to 10° E, including parts of MC-4, MC-5, MC-10, and MC-11, elevation color mapping not the same as elsewhere in atlas; courtesy A. Rodriguez).

Though the details of the "break" versus "bend" scenarios are still being investigated, the discrepancy between these two different – yet adjacent – terrains can be explained by an eastward thickening of the Martian crust in the Ismenius Lacus quadrangle (see the map of crustal thickness in Figure 3.4) as well as deeply seated faults and folds that are perhaps remnants of planetwide geologic processes that operated very early in the formation of Mars.

Making an impact

Lyot crater (Figure 5.D) dominates the center of the Ismenius Lacus quadrangle; it is 236 km in diameter, and is one of the largest of the younger craters on Mars. Craters on Mars that are larger than approximately 50 kilometers in diameter are named for deceased scientists, writers, and others who have made significant contributions to the study and "lore" of Mars. Lyot crater is named after Bernard Lyot, a French astronomer (1897–1952) who made important discoveries of Mars using telescopic observations. The crater itself is a "complex" crater, meaning that it consists of a "plate-shaped" depression defined by an outer and inner "ring" (rather than a single ring around a "bowl-shaped" depression). The ejecta blanket of Lyot crater contains both linear grooves and bulbous lobes, which implies emplacement through both ballistic and fluid processes. Fluid processes result from the excavation and entrainment of water as a liquid or vapor. The morphologic details of impact craters of various sizes on Mars provide important information not only about impact-related processes but also about the character and composition of the Martian surface and subsurface.

Coastal disaster zone

If the northern plains of Mars had been infilled by an ocean due to catastrophic floods more than 3 billion

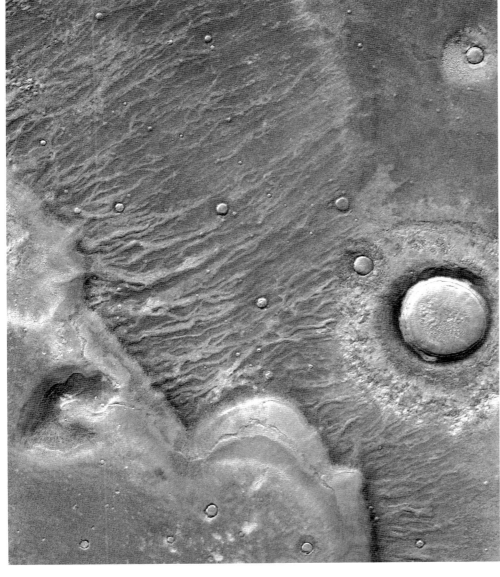

FIGURE 5.G View of an area of Deuteronilus Mensae in western Arabia Terra, showing dense parallel channel systems that are carved into darker material. These occur in various locations upstream of the older tsunami deposits (forming brighter surfaces) that are mapped out in Figure 5.F. (CTX image P16_007426_2266_XN_46N341W, 6 m/pixel, view 17 km by 20 km, north at top, centered at 46.4° N, 18.9° E, NASA/JPL-Caltech/MSSS).

years ago, as some investigators have proposed, the ocean could have been the target of huge impacts. Such impacts might generate enormous tsunami waves that could alter the coastal landscape. Along the southern margins of Chryse and Acidalia Planitiae, such waves may explain the lack of shoreline features observed along the margins of plains

deposits (Rodriguez et al., 2016; Costard et al., 2017). Instead, lobes of debris have been documented along two successive fronts (Figures 5.E, 5.F). Each of these is interpreted as being deposited by a tsunami that traveled up coastal plains, tens to hundreds of kilometers from possible paleo-shorelines, toward the southern highlands. In some

FIGURE 5.H Close-up view of bar in channeled surface shown in Figure 5.G, near northern terminus of channels. The surface material consists of extensive boulder deposits, including concentrations of rounded clasts; largest boulders are 10 m across (Rodriguez et al., 2016; from HiRISE ESP_028537_2270_RED, 25 cm/pixel, view about 350 m by 450 m, north at top, NASA/JPL-Caltech/University of Arizona).

areas, upslope of these lobes, parallel channel systems are observed (Figure 5.G) that might have resulted from tsunami backwash toward the ocean body. In addition, patches of boulders, which are as much as several meters across, occur along parts of the outer, older set of debris lobes (Figure 5.H). The boulders may represent the coarse debris, entrained and deposited by the lobate flows. Impacts that resulted in craters about 30–50 km in diameter could have generated tsunami waves, 50–75 m high, which would have had enormous energy, explaining the observed resurfacing features.

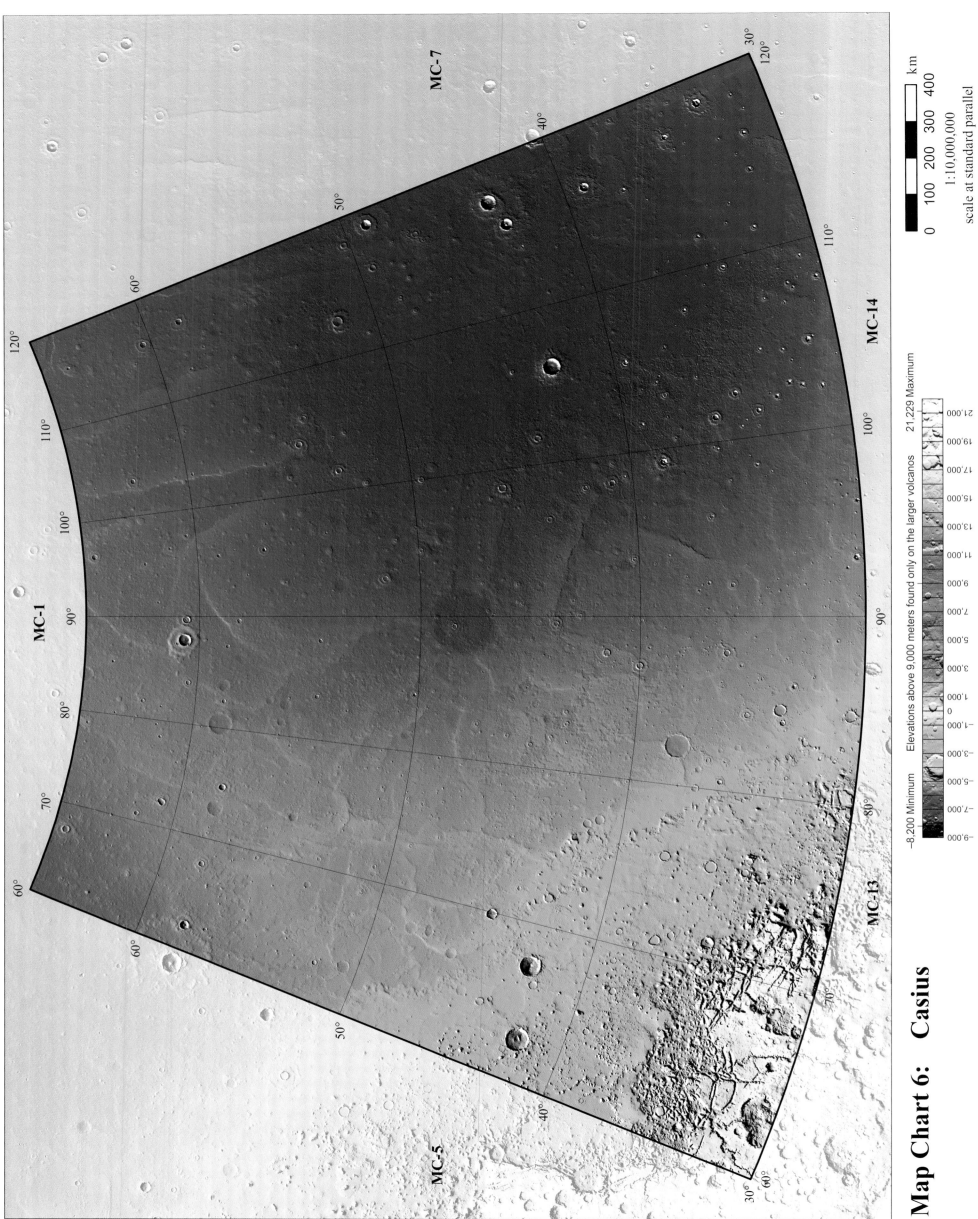

Map Chart 6: Casius

MC-7

MC-1

MC-14

MC-5

MC-13

1:10,000,000
scale at standard parallel

km

0   100   200   300   400

−8,200 Minimum   Elevations above 9,000 meters found only on the larger volcanos   21,229 Maximum

Elevation in meters

−9,000   −7,000   −5,000   −3,000   −1,000   0   1,000   3,000   5,000   7,000   9,000   11,000   13,000   15,000   17,000   19,000   21,000

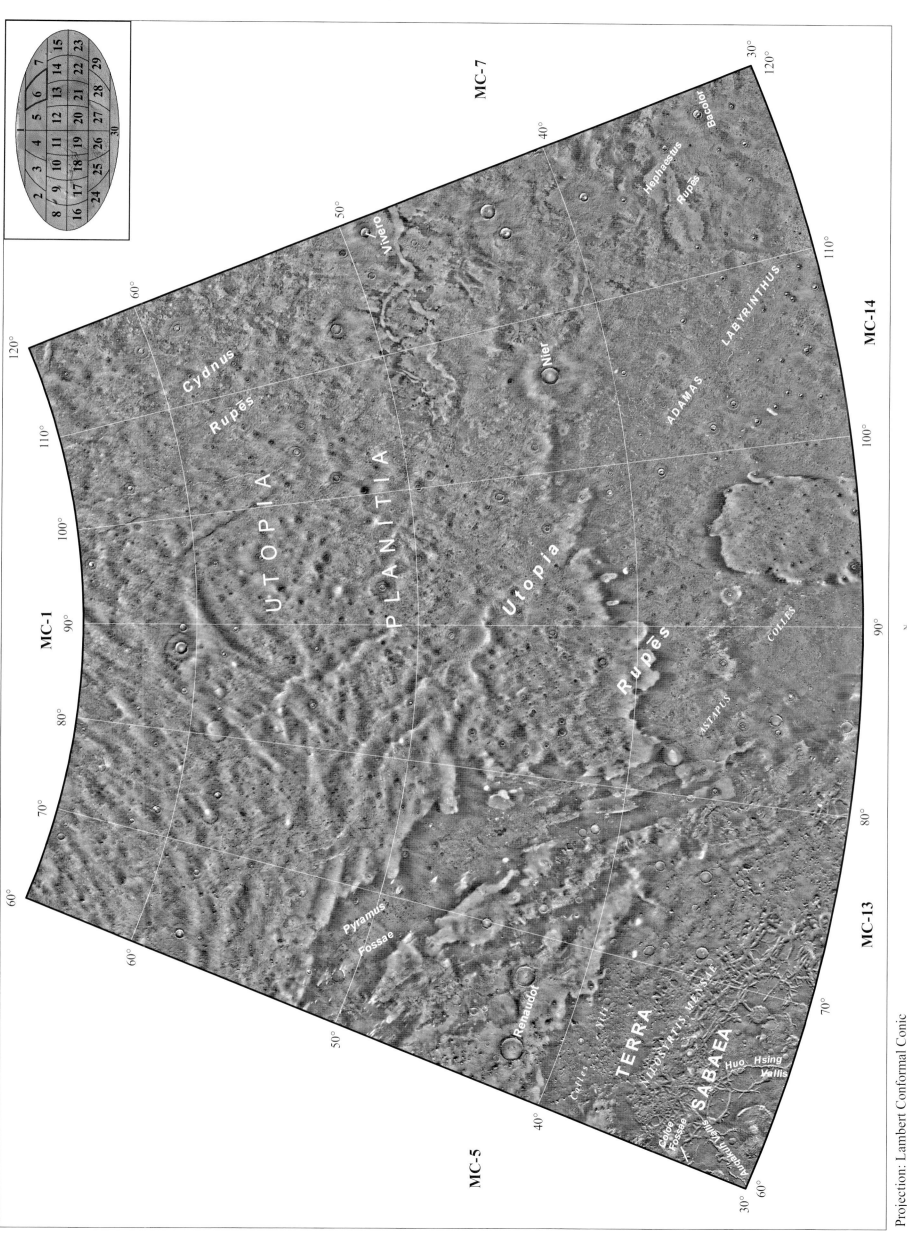

MC-1

MC-5

MC-7

MC-13

MC-14

120°   110°   100°   90°   80°   70°   60°

30°

Cydnus

Rupēs

U T O P I A

P L A N I T I A

Utopia

Rupēs

Vivero

Nier

ADAMAS

LABYRINTHUS

COLLES

ASTAPUS

Hephaestus

Rupēs

Bacolor

Pyramus

Fossae

Renaudot

Nili

Carius

T E R R A

NILOSYRTIS MENSAE

S A B A E A

Huo. Hsing

Vallis

Auqakuh Vallis

Oltis Valles

Colce

Fossae

Projection: Lambert Conformal Conic

Datum: Mars 2000 Sphere

Central Meridian: 90.0000

Standard Parallel 1: 36.1516

Standard Parallel 2: 59.4669

1:10,000,000

scale at standard parallel

km

0   100   200   300   400

N

S

E

W

Casius (MC-6)

Geography

The Casius quadrangle contains a bit of ancient cratered highlands of Terra Sabaea in the southwest corner but is mostly dominated by the pervasively flat, and more youthful, northern lowlands of Utopia Planitia. The east half of the quadrangle is below −4,000 m, while the highlands are close to the 0 m datum. The highland–lowland transition is represented in the Casius quadrangle by broken plateaus and knobs, including the Colles Nili and Nilosyrtis Mensae, which are surrounded by undulating terrain of intermediate roughness. The northern lowlands are extremely smooth at regional scales but are marked locally by various types of topographic scarps, ridges, troughs, mounds and depressions, including Utopia, Cydnus, and Hephaestus Rupēs, Adamas Labyrinthus, and Astapus Colles.

Geology

The terrains exposed in the Casius quadrangle are really, really flat. To illustrate the planar nature of the northern lowlands, the topographic differential from west to east across the center of the quadrangle is approximately 800 meters over 1,700 kilometers. That equates to 0.027°, or comparable to the flattest places on the surface of the Earth! How were these materials deposited? It is likely that the surface of the northern lowlands in Casius is composed of sheets of sediments that were originally deposited in oceans, lakes, or playas, creating a nearly horizontal surface that effectively approaches equilibrium across large areas – that is, containing no significant highs or lows. When this equilibrium surface is disrupted – for example, due to the formation of an impact crater – strong winds and perhaps settling of icy substrate eventually re-establish the horizontal equilibrium surface by reducing local highs and filling in local lows. This long-lived process is demonstrated in Casius by the existence of many shallow, circular depressions, and other subdued features, of various sizes. These are subtle remnants of impact craters that have been – or are in the process of being – scoured and filled, leading to a naturally horizontal surface (Figure 6.A).

Scarps, grooves, and depressions (Oh my!)

A formal process is used to identify spatially unique features on the surface of Mars so that scientists can have a common context for orienting and conducting research. Names that have been approved by the International Astronomical Union are intended to provide context and are not intended to provide specific indications about

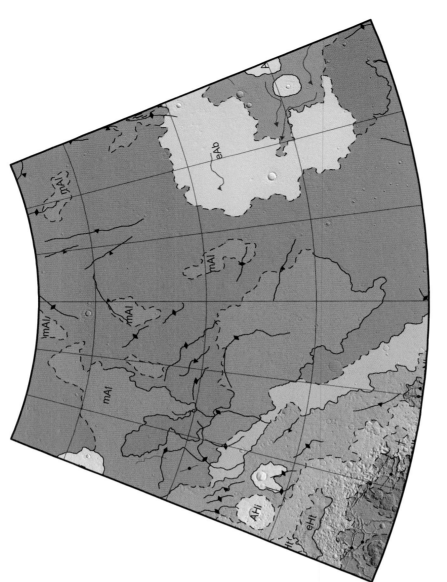

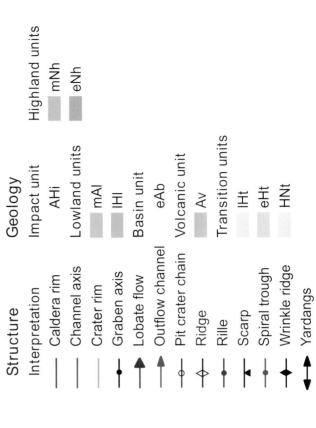

Structure

Interpretation
— Caldera rim
— Channel axis
— Crater rim
●— Graben axis
▲ Lobate flow
▲ Outflow channel
⊖ Pit crater chain
◇ Ridge
● Rille
◀ Scarp
● Spiral trough
◆ Wrinkle ridge
↕ Yardangs

Geology

Impact unit
AHi

Lowland units
mAl
lHl

Basin unit
eAb

Volcanic unit
Av

Transition units
lHt
eHt
HNt

Highland units
mNh
eNh

FIGURE 6.C Excerpt of THEMIS V27279025, showing the very fluid lavas (dark-colored), which extend into the Casius quadrangle and surround low mounds (mostly light-colored), and partly fill the troughs that make up Adamas Labyrinthus (18 m/pixel, view 15 km across, north at top, centered at 32.5° N, 110.2° E, NASA/JPL-Caltech/Arizona State University).

Smooth versus smooth

While the northern lowlands of Mars are very flat, it is important to note that this "flatness" is based on long distances (hundreds of kilometers). At very local scales (hundreds of meters), the northern plains can be either more or less "flat," depending both on the types of rocks and sediments and the geologic processes that occupy and occur within a particular region (Kreslavsky and Head, 2000). In the southeastern corner of the Casius quadrangle in the vicinity of Hephaestus Rupēs, a subtle change in topographic texture exists, which denotes a slight smoothing of the local surface. This smooth, flat surface formed not because of long-lived erosion but rather by the eruption of lava that flows very readily (Tanaka et al., 2005). This very fluid lava surrounded and filled even the subtlest local topographic highs and lows, including the labyrinth-like troughs of Adamas Labyrinthus (Figure 6.C). To underscore how fluid these lavas actually were, consider the fact that they were erupted from volcanic cracks, located over 1,200 km to the east.

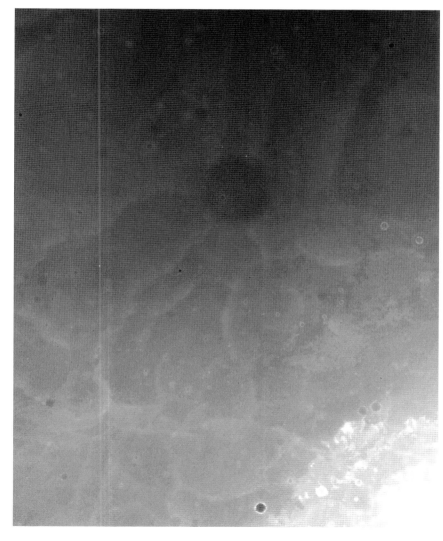

FIGURE 6.A (top) MOLA grayscale elevation image (view 1,000 km by 1,200 km, darker shades denote lower elevations) showing subdued, asymmetrical ridges of Utopia Rupēs, near a "ghost" or degraded crater.

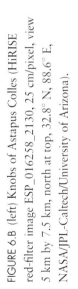

FIGURE 6.B (left) Knobs of Astapus Colles (HiRISE red-filter image ESP_016258_2130, 25 cm/pixel, view 5 km by 7.5 km, north at top, 32.8° N, 88.6° E, NASA/JPL-Caltech/University of Arizona).

the origin of a feature. As such, there are several different Latin "descriptor" terms that can be applied to similar landforms. In the Casius quadrangle, where subtle highs and lows are common, we can see this application of different terms. "Rupes" ("rupēs," plural form) is a scarp that separates a topographic high from an adjacent low. "Labyrinthus" is a set of intersecting valleys or ridges. "Collis" ("colles," plural) is a small hill or knob (Figure 6.B; Russell et al., 1992).

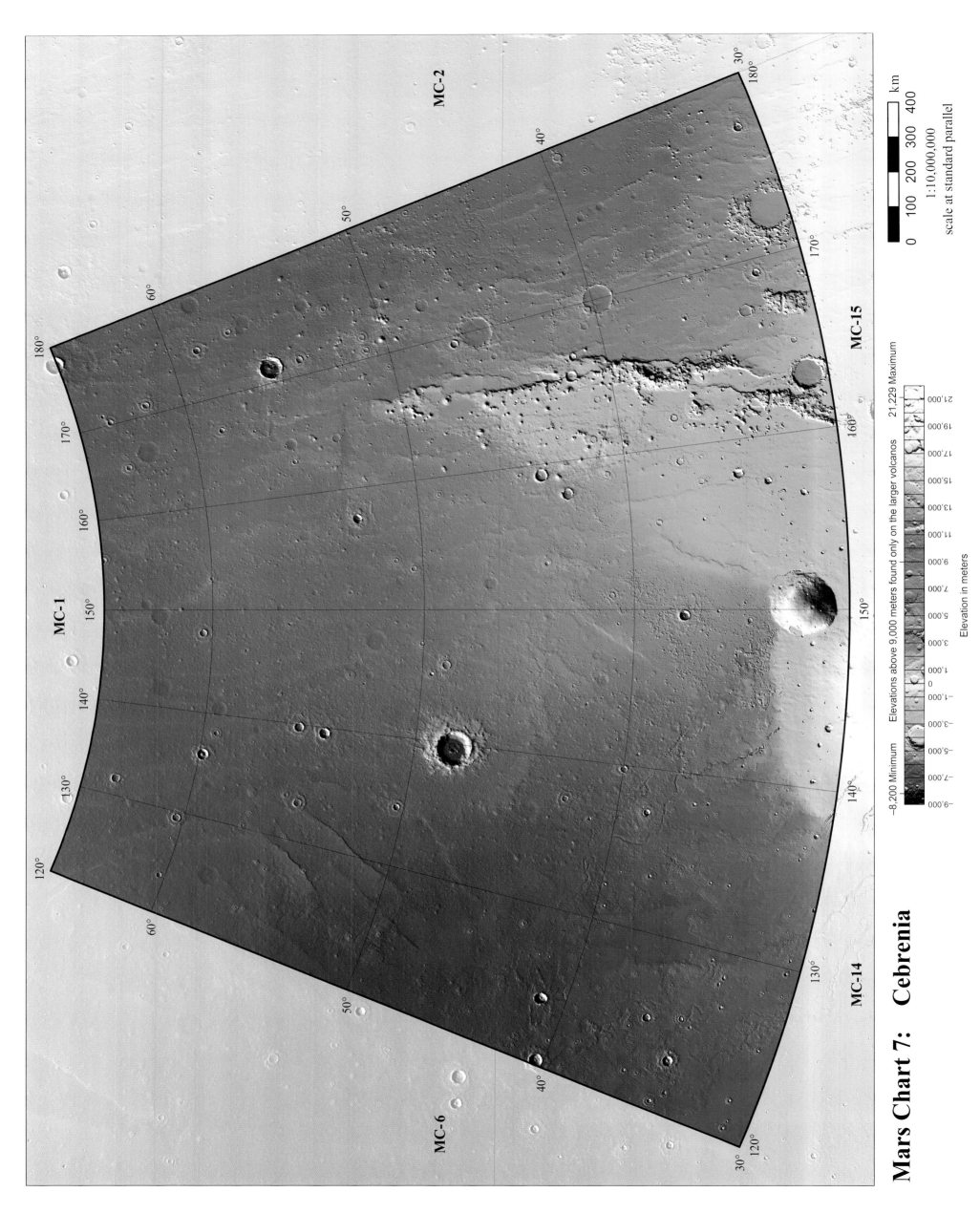

Mars Chart 7:   Cebrenia

1:10,000,000
scale at standard parallel

km
0   100   200   300   400

−8,200 Minimum   Elevations above 9,000 meters found only on the larger volcanos   21,229 Maximum

Elevation in meters

−9,000   −7,000   −5,000   −3,000   −1,000   0   1,000   3,000   5,000   7,000   9,000   11,000   13,000   15,000   17,000   19,000   21,000

MC-2
MC-1
MC-15
MC-14
MC-6

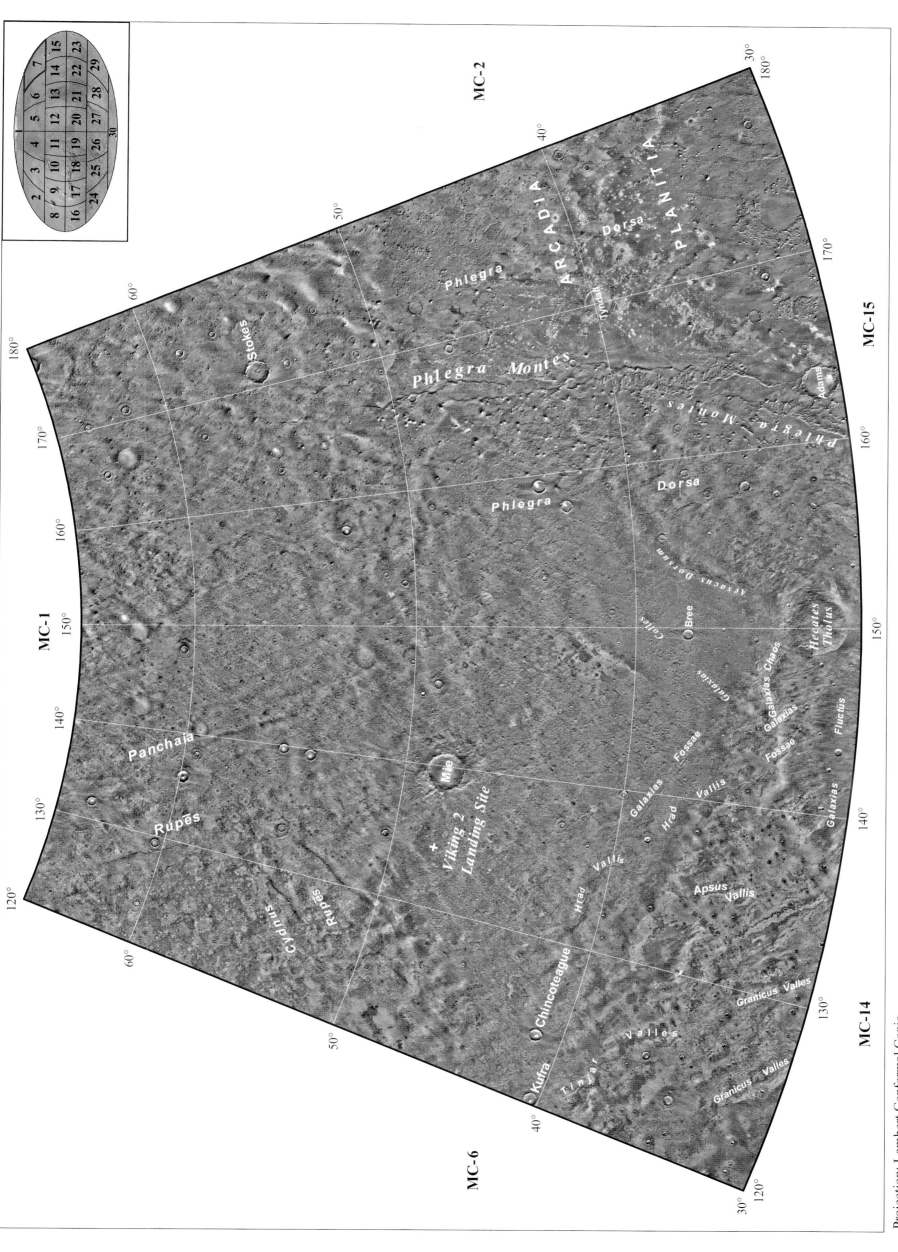

MC-1

MC-2

MC-15

MC-14

MC-6

180°   170°   160°   150°   140°   130°   120°

60°   50°   40°   50°   60°

30°   40°   160°   170°   180°   30°

Stokes

Phlegra

Phlegra   Montes

ARCADIA   Dorsa

PLANITIA

Tyndall

Phlegra   Dorsa

Phlegra   Montes

Adams

Aesacus Dorsum

Panchaia   Mie   Coloe   Bree   Hecates Tholus

Rupēs   Viking 2   Galaxias Chaos   Galaxias

Landing Site   Galaxias   Fossae   Fluctus

Cydnus   Galaxias   Fossae   Vallis

Rupēs   Hrad   Vallis   Galaxias

Chincoteague   Hrad   Vallis

Kufra   Apsus   Vallis

Tinjar   Valles   Granicus Valles

Granicus   Valles

N   E   W   S

Projection: Lambert Conformal Conic
Datum: Mars 2000 Sphere
Central Meridian: 150.0000
Standard Parallel 1: 36.1516
Standard Parallel 2: 59.4669

km

0   100   200   300   400

1:10,000,000
scale at standard parallel

Cebrenia (MC-7)

Geography

The Cebrenia quadrangle is mostly covered by the plains of eastern Utopia and western Arcadia Planitiae, which are split by the prominent, north-trending Phlegra Montes ridge belt. The south-central margin of the quadrangle includes the northern part of the Elysium rise, upon which Hecates Tholus forms a domical mountain that includes the highest point in the quadrangle, more than 8,000 m above the adjacent plains to the north. This edifice includes a series of nested summit calderas and extensive fluvial valleys (Figure 7.A). Lesser ridge and scarp systems in the plains include north-trending Phlegra Dorsa in Arcadia and northwest-trending Panchaia Rupēs and northeast-trending Cydnus Rupēs in Utopia. Several systems of sinuous channel systems, including Tinjar, Granicus, Apsus, and Hrad Valles, extend hundreds of kilometers north-westward from the Elysium rise into the deeper, central floor of Utopia basin, where the lowest

regional elevations (~5,000 m below the Martian datum) occur. The most prominent crater, 100-km-diameter Mie, occurs near the quadrangle's center. The Viking 2 landing site is more than 150 km west of Mie.

Geology

Vestiges of Noachian terrains form the Phlegra Montes and, nearby, elevated systems of knobs, some of which outline crater forms that are tens of kilometers across. The degradation forming the knobs and the crustal contraction that built Phlegra Montes both likely began in the Noachian Period and extended into the Hesperian Period (Tanaka *et al.*, 2005). Meanwhile, eruption of basaltic lava flows during the Hesperian and Amazonian Periods formed Elysium rise and spawned channelized flows, originating from the rise that covered the floor of Utopia basin during the Early Amazonian Epoch (Tanaka *et al.*, 2005). Volcanic interactions with groundwater may explain the flows, which may be volcanic debris flows, known as lahars (see below; Christiansen, 1989). Hecates

Tholus represents centralized volcanism, with its nested calderas formed by collapse, resulting from magma withdrawal within the volcano (Figure 7.B). The dense valley systems on Hecates Tholus testify to fluvial erosion, relating to geothermal heating and melting of summit snow-pack (Fassett and Head, 2006). Arcadia and northern Utopia are largely covered by Late Hesperian sediments, which are tectonically contracted into systems of sinuous wrinkle ridges

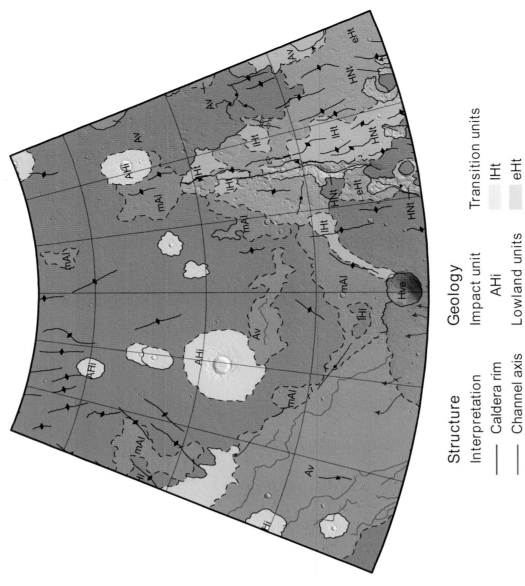

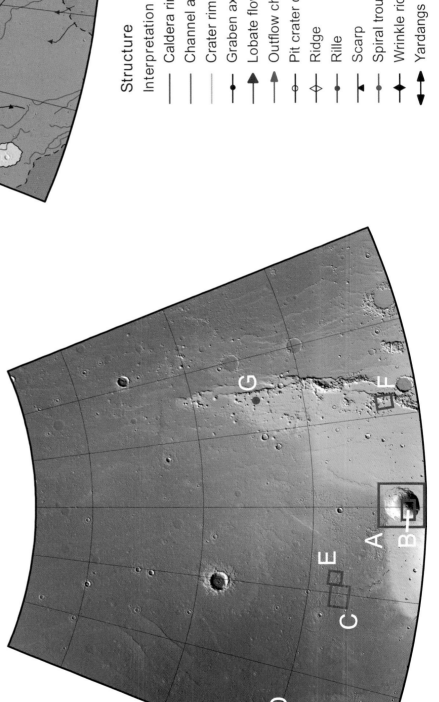

Structure

Interpretation

—	Caldera rim
—	Channel axis
—	Crater rim
—•—	Graben axis
▲	Lobate flow
▲	Outflow channel
—o—	Pit crater chain
—◇—	Ridge
—•—	Rille
—◀	Scarp
—•	Spiral trough
—◆—	Wrinkle ridge
↕	Yardangs

Geology

Impact unit
- AHi

Lowland units
- mAl
- lHl

Basin unit
- eAb

Volcanic units
- AHv
- lAv
- Hve
- Av

Transition units
- lHt
- eHt
- HNt

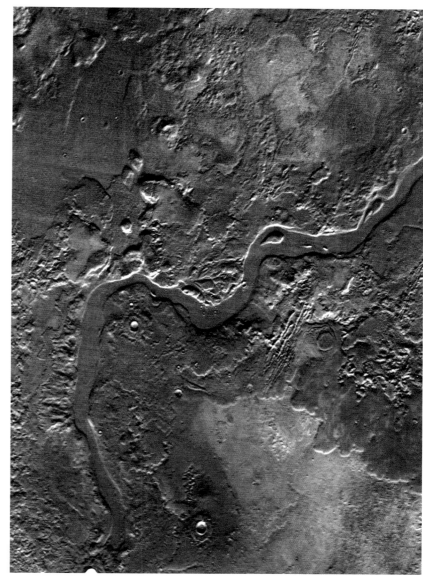

FIGURE 7.B This color image of Hecates Tholus shows the summit region at center of Figure 7.A. Features related to lava tube collapse and/or ice removal (pits) and fluvial action (narrow channels) radiate downslope from the summit caldera (HRSC orbit 32-color view, 25 m/pixel, 90 km by 60 km, centered at 32° N, 150° E, ESA/DLR/FU-Berlin).

103

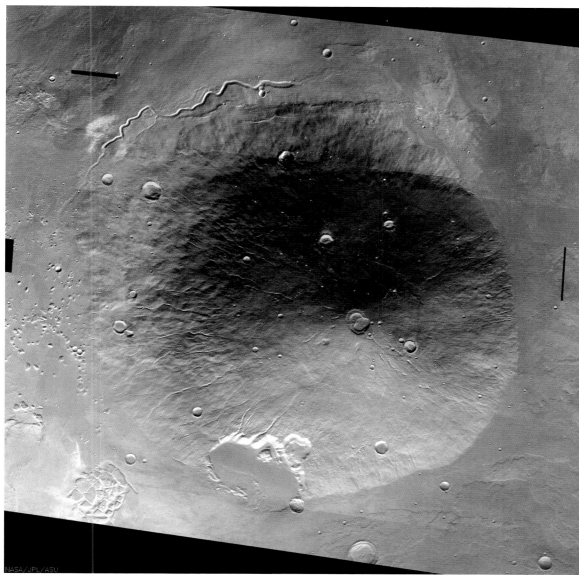

FIGURE 7.A The oval depression on the northwest flank of Hecates Tholus has features consistent with glacial activity (de Pablo and Centeno, 2012; de Pablo et al., 2013; THEMIS daytime infrared mosaic, 100 m/pixel, 250 km across, north at top, 30° N to 34° N, 148° E to 153° E, NASA/JPL-Caltech/Arizona State University).

and partly buried by remnants of a Middle Amazonian ice and dust mantle (Tanaka et al., 2005; Skinner et al., 2012).

Giant lahars on Mars?

The western margin of Elysium rise is marked by trough systems that are radial to the rise, the more prominent of which occur in the Elysium quadrangle (MC-15) to the south. From these features emanate channeled, lobate flows, tens of

kilometers across and hundreds of kilometers long, forming a total deposit about 900 km across, extending for 2,400 km in MC-6, MC-7, MC-14, and MC-15 (Christiansen, 1989). Individual flows are tens of meters to about a hundred meters thick.

FIGURE 7.C (right) Part of Hrad Vallis (THEMIS daytime infrared mosaic, 100 m/pixel, 98 km by 112 km, centered near 36.9° N, 139.8° E, NASA/JPL-Caltech/Arizona State University).

The Atlas of Mars

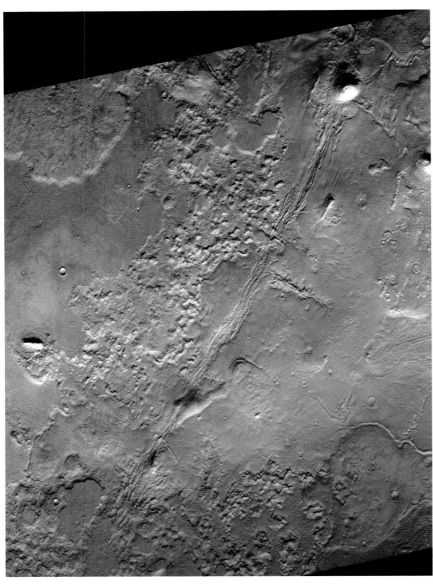

FIGURE 7.E Galaxias Fossae and associated deposits and mounds (CTX image P17_007738_2162_XN_36N218W, 6 m/pixel, about 30 km across, centered near 36.9° N, 141.2° E, NASA/JPL-Caltech/MSSS).

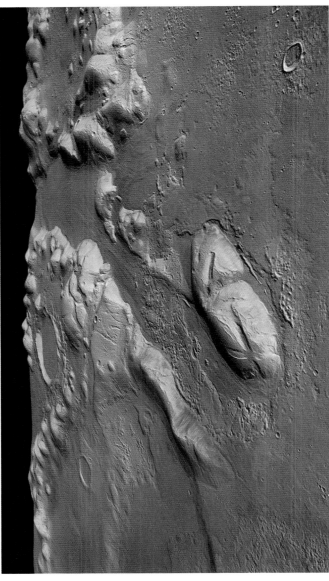

FIGURE 7.F View toward the east of southern Phlegra Montes centered at 33° N, 162° E (HRSC image, orbit 9465, 16 m/pixel, about 80 km across, ESA/DLR/FU-Berlin).

FIGURE 7.D Lahar deposits downslope of Granicus Valles and near Tinjar Valles. Note bright, rugged material with branching ridges, partly buried by mostly dark, flat-lying, complexly overlapping deposits. The deposits both infill, and in places dissect, sinuous channels (CTX image G19_025632_2213_XI_41N237W, 6 m/pixel, about 30 km wide, centered near 39.4° N, 123.2° E, NASA/JPL-Caltech/MSSS).

The lahars have complex surfaces, indicative of a blend of multiple flow surges, fluvial erosion and sedimentation, variable material competence, and a mixture with lava flows (Figures 7.C, 7.D). The channels within them include teardrop-shaped bars, and in places their floors are lower than adjacent plains, both indicative of erosive flow within the channels. If the flows are lahars, they are much larger than their counterparts on Earth. The lahars may include material that has eroded

from the trough systems as well as erupted pyroclastic material. Water for the lahars may have been impounded within the Elysium rise behind a permafrost barrier.

Iced or hot?

Near Galaxias Fossae, the terrain consists of a mixture of complex landforms that seem to indicate a complicated geologic history, involving

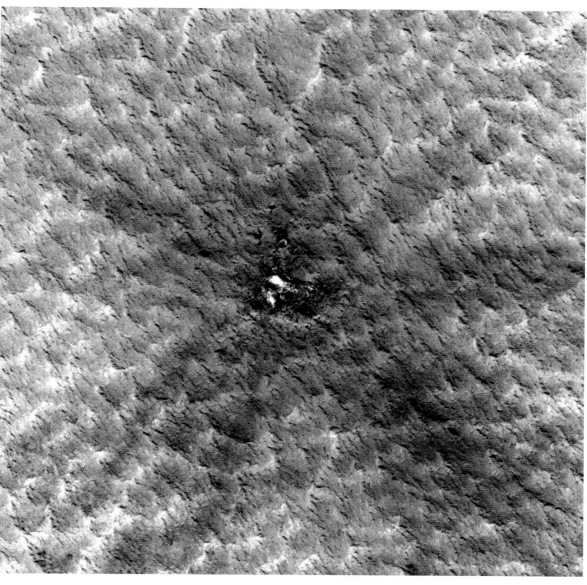

FIGURE 7.G A new crater, betrayed by the dark material excavated onto a surface, otherwise covered with light-colored dust (HiRISE image ESP_016954_2245, 30 cm/pixel, 3.5 km across, north at top, 44.2° N, 164.2° E, NASA/JPL-Caltech/University of Arizona).

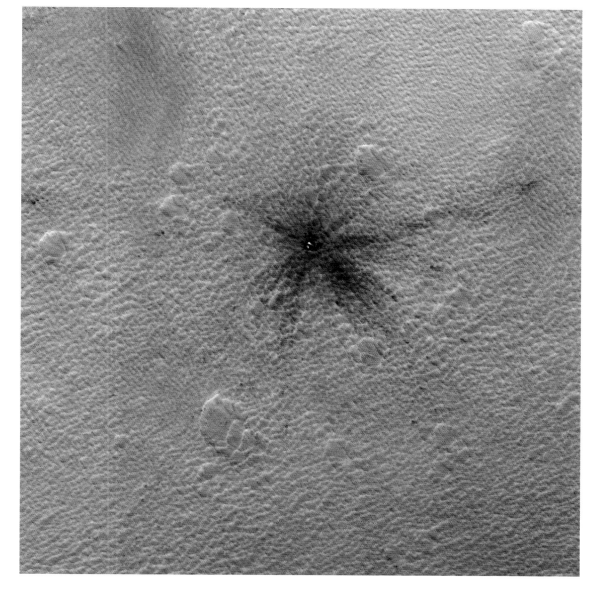

FIGURE 7.H The detail of Figure 7.G view shows the excavated ice, pure enough to be easily visible, in the 20-m crater (detail of HiRISE image ESP_016954_2245, NASA/JPL-Caltech/University of Arizona).

both hot volcanic eruptions and icy surface modi-fication. In Figure 7.E, the linear system of cracks follows the main trend of the Galaxias Fossae trough system. A pitted cone, aligned with the cracks, may have resulted from magma erupting along the crack system. Surrounding materials appear to be muted, eroded, and warped, possibly due to ice accumulation, removal, and deform-ation. Magmas associated with Elysium rise vol-canic activity could have traveled along buried fracture systems such as these, locally emerging and interacting with ground ice to create a host of features.

What goes up must come down

Phlegra Montes form the most prominent ridge system in the Martian lowlands, extending 1,400 km and rising as much as 4,000 m above adjacent Arcadia Planitia. The north–south trend parallels trends of adjacent sinuous ridges. These features are thought to result from crustal contraction, driven by a combination of planetary shrinkage that was caused by cooling of Mars as well as tectonic forces induced by the great mass of the enormous Tharsis volcanic rise to the east (e.g. Tanaka *et al.*, 1991). Phlegra Montes also are

concentric to Utopia basin and possibly align with a buried impact basin ring. The perspective image (Figure 7.F, produced from Mars Express High Resolution Stereo Camera [HRSC] data) shows a portion of the southern Phlegra Montes, made up of domical mounds that are dissected by troughs and valleys. The trough at center is partly infilled by lineated material, likely underlain by ice that has flowed west down the trough, carrying with it eroded debris from the mounds. Smaller but simi-lar debris aprons ring the rounded knobs and scarps.

A chilly surprise beneath the surface

A handful of freshly formed craters appear in images taken from Mars orbit. A spectacular example (Figures 7.G, 7.H) shows that the impact excavated down to water ice present in the shal-low subsurface. This is consistent with the findings of the Phoenix Lander at 68° N (see Chapters 4 and 5), where ice was found a few centimeters beneath the surface (Smith *et al.*, 2009).

105

The Atlas of Mars

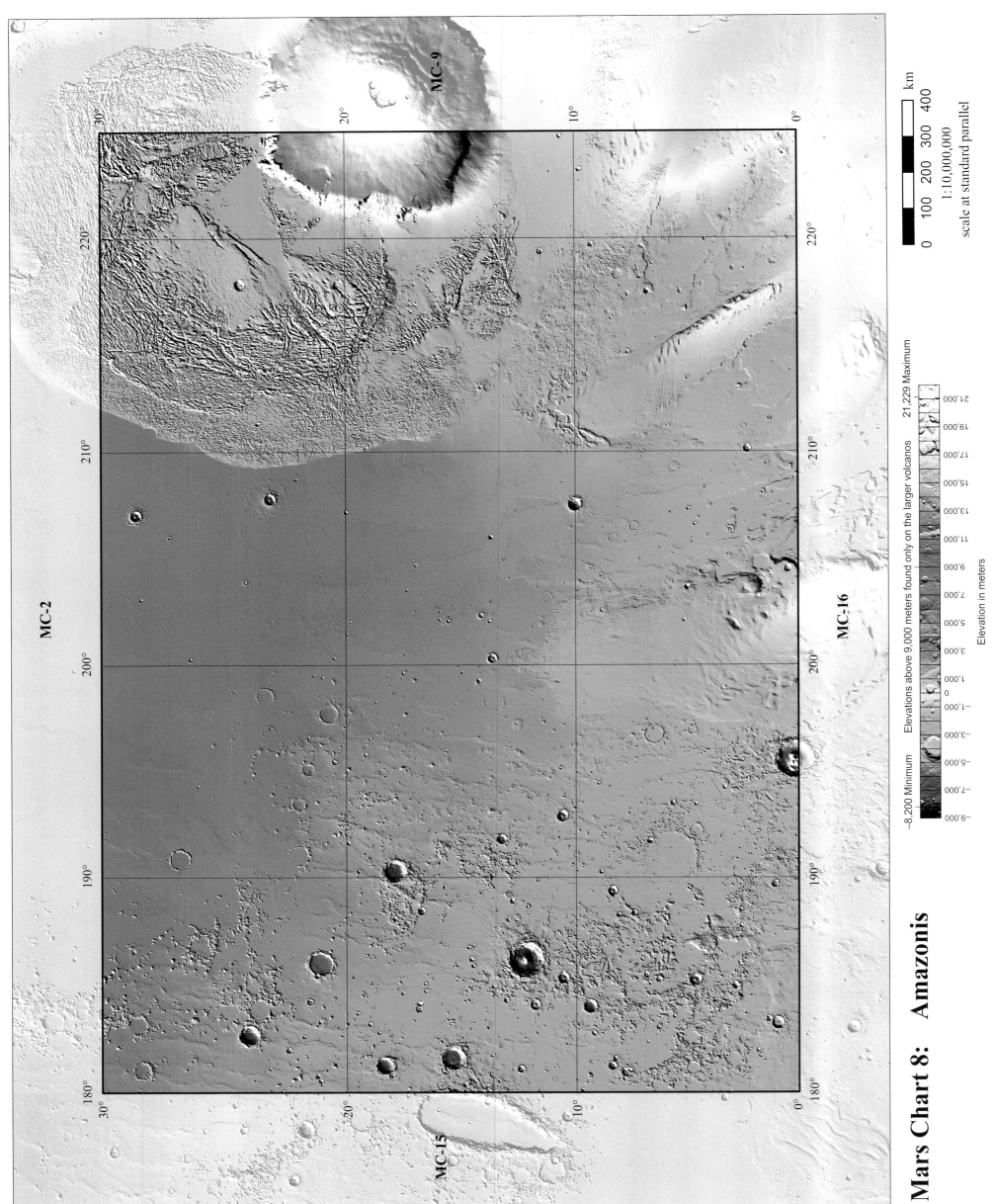

Mars Chart 8: Amazonis

MC-2

MC-9

MC-15

MC-16

-8,200 Minimum   Elevations above 9,000 meters found only on the larger volcanos   21,229 Maximum

| -9,000 | -7,000 | -5,000 | -3,000 | -1,000 | 0 | 1,000 | 3,000 | 5,000 | 7,000 | 9,000 | 11,000 | 13,000 | 15,000 | 17,000 | 19,000 | 21,000 |

Elevation in meters

0   100   200   300   400

km

1:10,000,000
scale at standard parallel

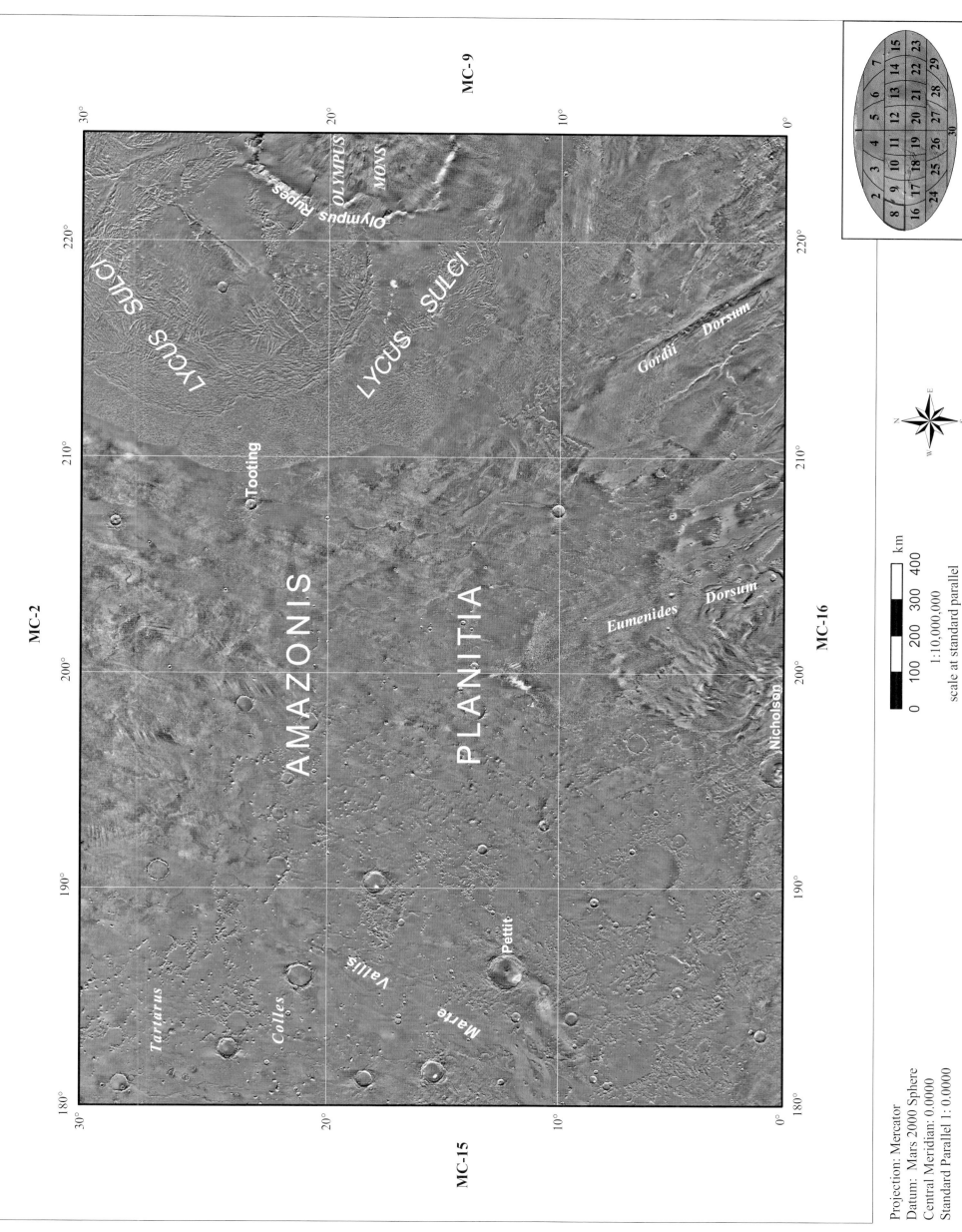

MC-2

MC-9

MC-15

MC-16

30°   20°   10°   0°

180°   190°   200°   210°   220°   30°

LYCUS SULCI

LYCUS SULCI

OLYMPUS MONS

Olympus Rupes

Gordii Dorsum

Tooting

AMAZONIS

PLANITIA

Eumenides Dorsum

Nicholson

Tartarus

Colles

Marte Vallis

Pettit

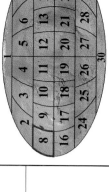

N
W   E
S

km

0   100   200   300   400

1:10,000,000
scale at standard parallel

Projection: Mercator
Datum: Mars 2000 Sphere
Central Meridian: 0.0000
Standard Parallel 1: 0.0000

Amazonis (MC-8)

Geography

Amazonis Planitia occupies the north-central part of the map region. The east-central and northeastern parts of the map are dominated by the western margin of Olympus Mons (MC-9) and its associated aureole deposits. This forms one of the most pronounced changes in relief in the solar system when considering a topographic transect, and involves more than 25,000 m in elevation change from the summit of the giant volcano to the regionally flat, lava-flow-covered plains of Amazonis Planitia. Olympus Mons is the only part of the quadrangle above datum. The aureole deposits, which extend up to 750 km west from the flank of the volcano and lie 1–3 km below datum, are marked by the Lycus Sulci ridge systems (see Figure 8.A), which make up broad lobes, hundreds of kilometers across. In the southern part of the map, near the highland–lowland boundary, pronounced mesas, Gordii and Eumenides Dorsa, mark the landscape. Just south of the mesas is the debouchment site of Mangala Valles

(MC-16). In the southwestern part, knobby-looking terrain marks a transitional boundary that separates the northern plains, including Amazonis Planitia at –3 km to –4 km, from the cratered highlands to the south. Included in this rugged terrain, located in the west-central part of the map, is Marte Vallis, a distinct but shallow valley system that connects Elysium (MC-15) and Amazonis Planitiae.

Geology

The Amazonis quadrangle records far-reaching geologic and hydrologic histories. Because Mars is half the radius of Earth, its ratio of surface area to volume is about twice as large. Given a similar internal composition, Mars should lose internal heat faster than Earth, just as a small baked potato cools off faster than a large one. Thus, one enduring paradigm about Mars is that, relative to Earth, it is internally cooled and inactive. Evidence does suggest an overall, long-term decline in resurfacing of Mars, although the record of older activity is likely underestimated. Given this context, Amazonis Planitia and its surroundings record

surprisingly extensive, geologically recent activity (Fuller and Head, 2002; Dohm *et al.*, 2008), including some of the youngest and flattest lava plains on Mars (Tanaka *et al.*, 2005; Dohm *et al.*, 2008).

In the south, mesas Gordii and Eumenides Dorsa and valleys indicate dynamic geologic and hydrologic activities. These include volcanism, ancient floods (Dohm *et al.*, 2001a), wind, and possibly ice-stream activity (Kite and Hindmarsh, 2007). The Mangala Valles system of channels

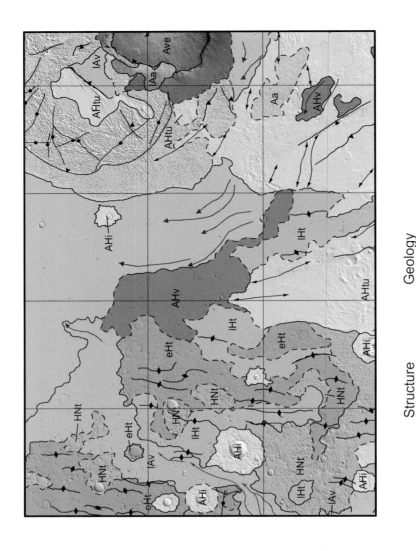

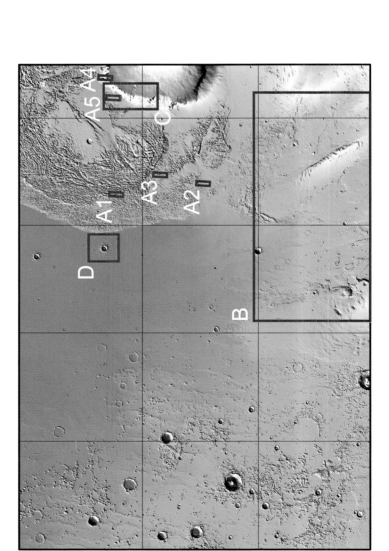

Structure

Interpretation
— Caldera rim
— Channel axis
— Crater rim
•— Graben axis
▲ Lobate flow
↑ Outflow channel
⊶ Pit crater chain
◇— Ridge
•— Rille
◂ Scarp
•— Spiral trough
◆— Wrinkle ridge
↕ Yardangs

Geology

Impact unit   Highland unit
AHi   INh

Volcanic units
Ave
AHv
IAv

Apron units
Aa
IAa

Transition units
IHt
eHt
AHtu
HNt
Htu

FIGURE 8.A THEMIS visible images (18 m pixel size) showing details in the Olympus Mons aureole (Lycus Sulci) and along the volcano's basal scarp. Images are 18 km wide and have north at the top (all images NASA/JPL-Caltech/Arizona State University). Figure continued on next page.

A1 Ridges and grooves of aureole showing dark streaks on slopes (V02589003, 22.1° N, 212.8° E).

A2 Wind-modified aureole, showing narrow grooves aligned with the dominant wind direction and ridges that are sculpted into yardangs (V02539007, 14.7° N, 213.7° E).

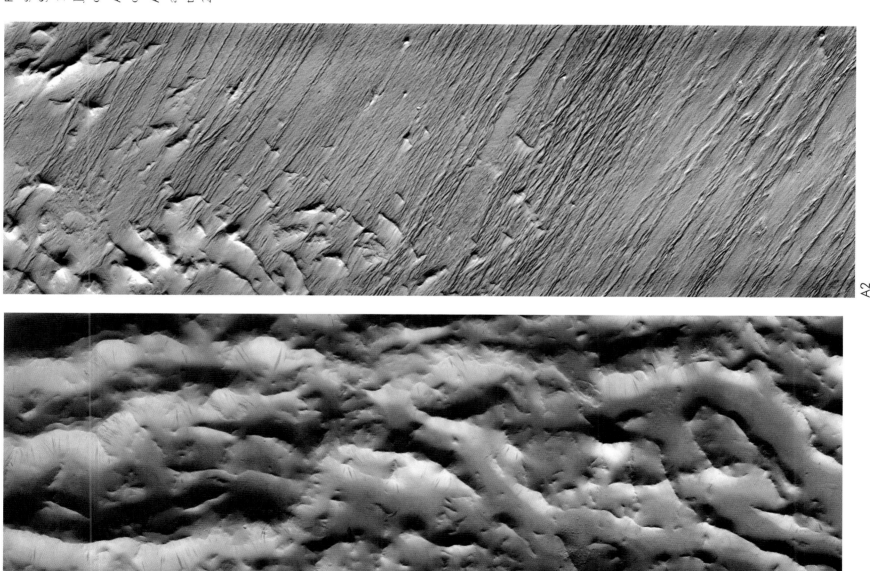

A2

A1

A3 Crescent-shaped cavities, possibly created by wind erosion (V08506017, 18.6° N, 214.6° E).

A4 Modification at base of scarp: lobate landslides (V1713005, 23.2° N, 136.2° E).

A5 Flow textures near base of scarp, suggestive of rock or ice glacier flow (V40102005, 22.5° N, 137.9° E).

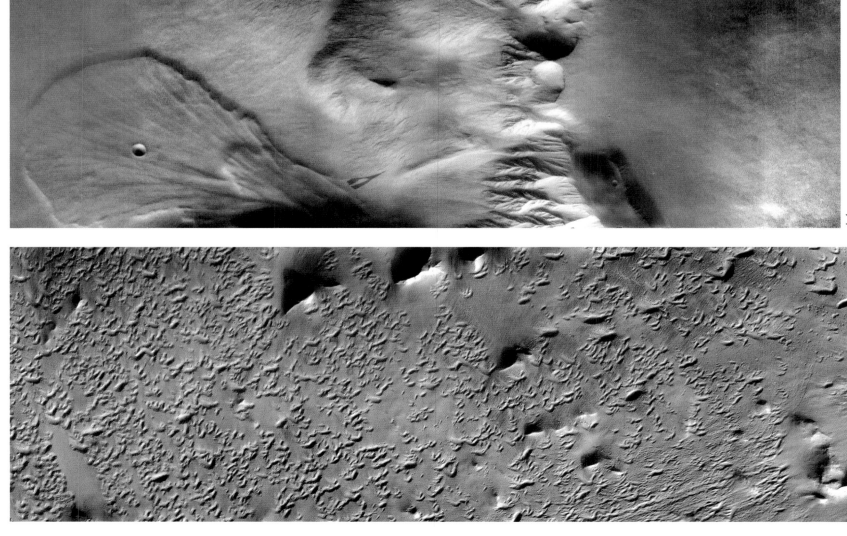

A5

A4

A3

Mysterious trimmings of Olympus Mons

The aureole deposits of Olympus Mons are marked by systems of ridge patterns, organized into broad lobes on all sides of the volcano that seem to record mass movements (e.g. Lucchitta, 1981; Scott and Tanaka, 1986). These lobes occur below the basal scarp, named Olympus Rupes, which also surrounds Olympus Mons. The scarp reaches a height of 8 km on the volcano's northwest flank (see Figures 8.C1, 8.C2). Many mechanisms have been proposed for the emplacement of the aureole deposits, including pyroclastic flows (Morris, 1982) and/or the degradation of a once-broader volcano as evidenced by its basal escarpment (Carr et al., 1977). The latter activity may have involved either catastrophic landslides (Lopes et al., 1982) or slow-moving, gravity-driven spreading along basal detachment zones, possibly lubricated by ground ice, groundwater, and/or clay-rich materials (Tanaka, 1985; McGovern et al., 2004; McGovern and Morgan, 2009). Well-formed lobes and deposits, marked by concentric ridge systems along the northwestern Olympus Rupes cliffs (Figures 8.A4, 8.A5, 8.C1), may result from glacier formation during the Late Amazonian Epoch (as proposed by Head et al., 2005).

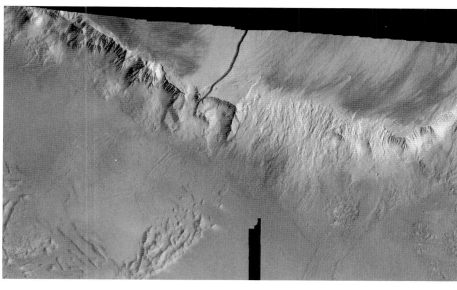

FIGURE 8.C1 HRSC nadir image shows west part of Olympus Mons, centered on the Olympus Rupes escarpment, which reaches a height approaching 8 km in this area (HRSC image H0143_0009_ND4, 13 m/pixel, view 150 km by 225 km, 21° N, 222° E, north at top, ESA/DLR/FU-Berlin).

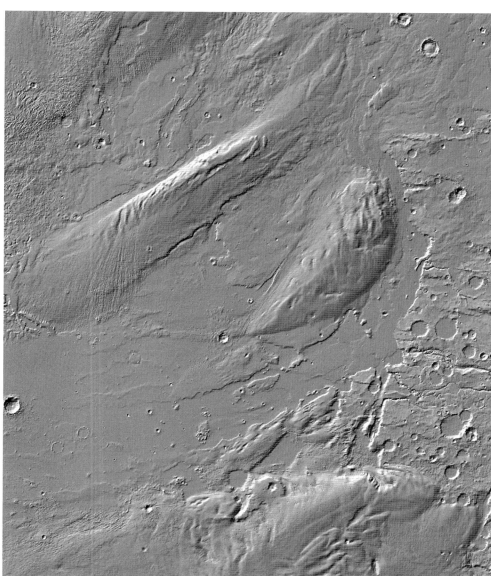

FIGURE 8.B Eumenides Dorsum (along left edge of image) and Gordii Dorsum (prominent in top center) are huge, streamlined plateau forms, made of relatively soft materials of uncertain origin (perhaps wind-deposited volcanic ash and/or other fine particles eroded from the Martian surface) that have undergone deep, long-lived wind erosion. The materials overlie the cratered highland boundary scarp, which is locally dissected by troughs that are aligned with the scarp as well as by small valleys perpendicular to the scarp. Volcanic flows in the lower right part of the image bury the cratered highlands and cover the lower parts of the large, streamlined plateaus (MOLA shaded relief view including part of MC-16, view about 1,000 km by 1,200 km, 7° S to 10° N, 202° E to 222° E).

records flood discharges from a graben of Sirenum Fossae, located within a north-trending basin (see MC-16; Chapman and Tanaka, 1993; Craddock and Greeley, 1994; Zimbelman et al., 1994; Anderson et al., 2012). Knobby-looking terrain in the southwest indicates erosion that occurred mainly since the end of the Noachian Period (Tanaka et al., 2005). Marte Vallis may have

carried discharges from the Cerberus Fossae fracture system to the west (MC-15). However, the vallis is filled with Late Amazonian flood lavas. Amazonis Planitia, for which the Amazonian Period is named, is also largely covered by Late Amazonian flood lavas emanating from buried sources in the southern and perhaps northeastern parts of the plain. Other Amazonian activity

includes long-lived and variable wind erosion of relatively soft, layered materials that make up Eumenides and Gordii Dorsa (Figure 8.B) and parts of Lycus Sulci to form streamlined ridge forms known as yardangs (Figures 8.A2, 8.A3).

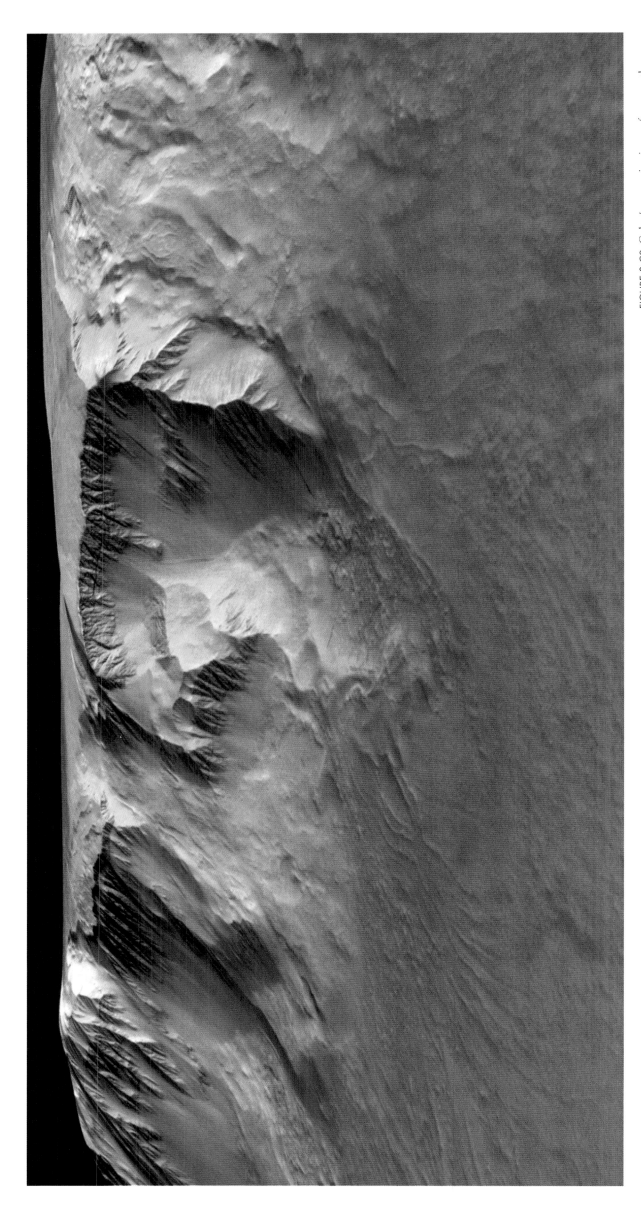

FIGURE 8.C2 Color/perspective image from northern half of area in Figure 8.C1 shows the typical form of the cliff: faceted upper slope, smooth debris on middle slope, and flow-like features at the base (HRSC image 037–160404–0143, press release 034, field of view about 75 km across, view to east, ESA/DLR/FU-Berlin).

Splat! at Tooting crater: Wet flow during impact

Some, but not all, impact craters on Mars show fluid flow features in their ejecta; these are not seen in lunar craters. The presence of water or ice near the surface is a probable factor. In the spectacular example in Figure 8.D, multiple overlapping lobes of ejecta display a variety of surfaces, including smooth, hummocky, or radially striated. Elevated ridges ("distal ramparts") are common along the terminal edges of the ejecta lobes (Mouginis-Mark, 2015). Tooting occurs on Late Amazonian lava flows and is one of the youngest craters on Mars for its size (27 km in diameter).

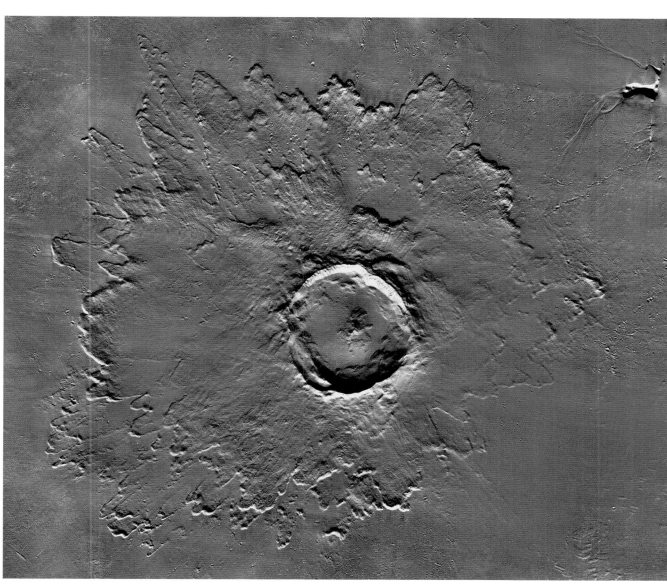

FIGURE 8.D Tooting crater, view 130 km by 148 km (18 m/pixel, north at top, 23.2° N, 207.8° E), shown in a mosaic of THEMIS visible images, with additions from infrared images. The multiple lobes of ejecta show flow ridges and terminal ramparts, consistent with flow by debris containing water (NASA/JPL-Caltech/Arizona State University).

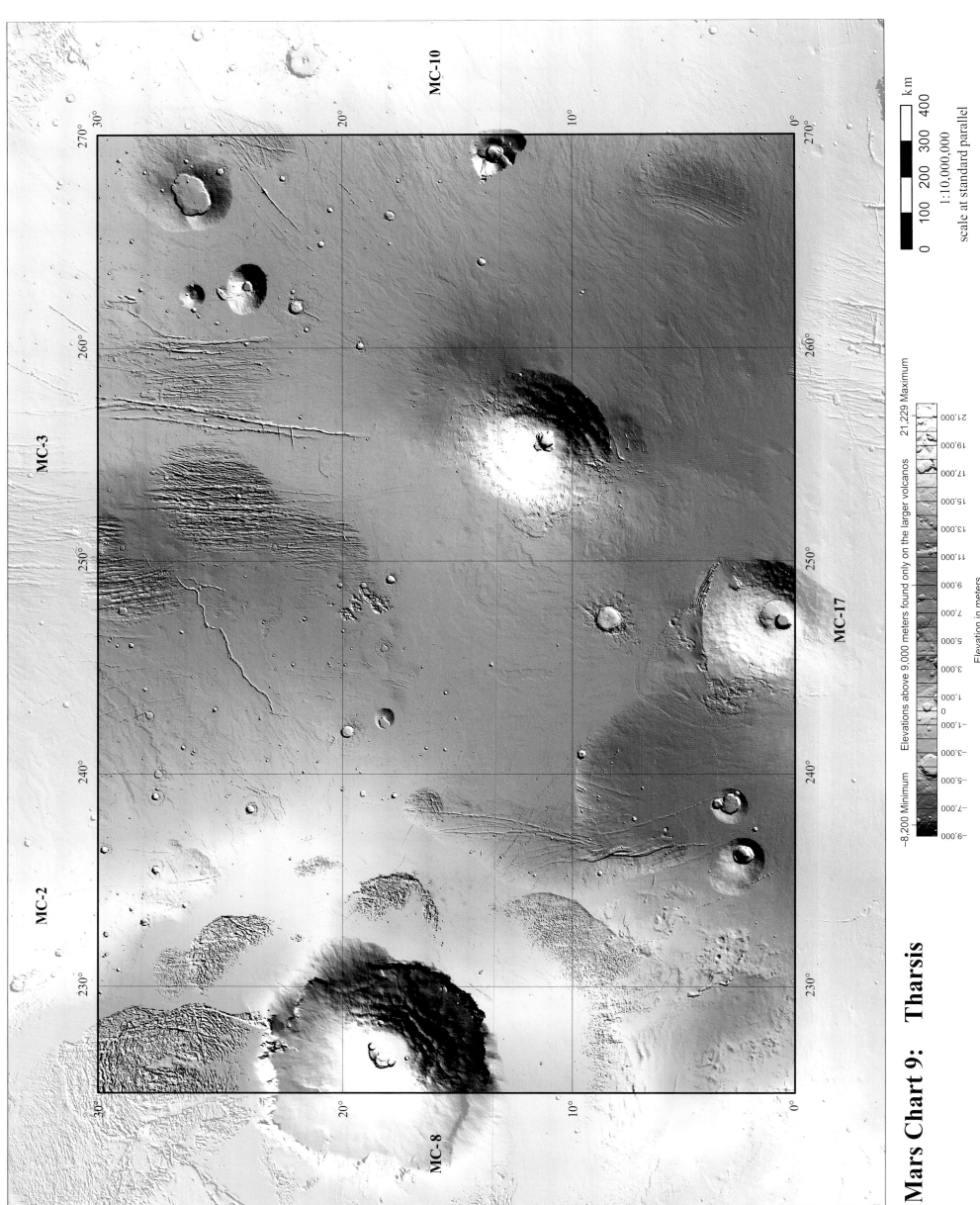

Mars Chart 9:   Tharsis

MC-10

MC-3

MC-2

MC-8

MC-17

270°

260°

250°

240°

230°

0°

30°

20°

10°

1:10,000,000
scale at standard parallel

km
0   100   200   300   400

−8,200 Minimum   Elevations above 9,000 meters found only on the larger volcanos   21,229 Maximum

−9,000 −7,000 −5,000 −3,000 −1,000 0 1,000 3,000 5,000 7,000 9,000 11,000 13,000 15,000 17,000 19,000 21,000

Elevation in meters

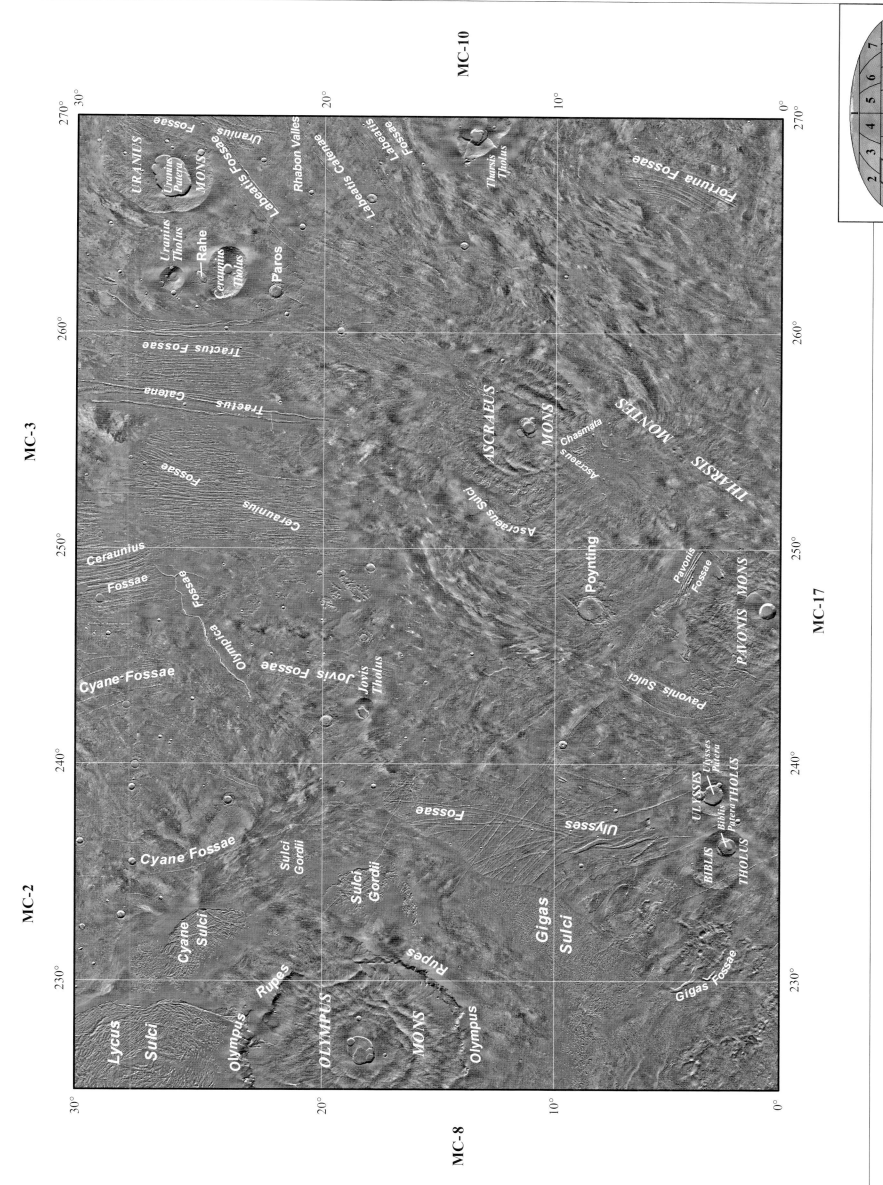

MC-3

MC-10

MC-2

MC-8

MC-17

URANIUS MONS
Uranius Patera
Uranius Tholus
Ceraunius Tholus
Rahe
Paros
Uranius Fossae
Labeatis Fossae
Labeatis Catenae
Rhabon Valles
Labeatis Fossae
Tharsis Tholus
Fortuna Fossae

Tractus Fossae
Tractus Catena
Tractus
Ceraunius Fossae
Ceraunius Fossae
ASCRAEUS MONS
Ascraeus Chasmata
Ascraeus Sulci
THARSIS MONTES

Cyane Fossae
Olympica Fossae
Jovis Fossae
Jovis Tholus
Poynting
Pavonis Fossae
PAVONIS MONS
Pavonis Sulci

Cyane Fossae
Sulci Gordii
Sulci Gordii
Fossae
Ulysses
ULYSSES
Ulysses Patera
BIBLIS THOLUS
Biblis Patera
Gigas Sulci
Gigas Fossae

Lycus Sulci
Olympus Rupes
OLYMPUS MONS
Olympus Rupes
Cyane Sulci

270° 30°
20°
MC-10
10°
0° 270°

30° MC-2
20° MC-8
10°
0°

230° 20° 10° 0°
230° 240° 250° 260° 270°

N
W E
S

0 100 200 300 400 km
1:10,000,000
scale at standard parallel

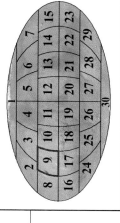

1	2	3	4	5	6	7
8	9	10	11	12	13	14	15
16	17	18	19	20	21	22	23
24	25	26	27	28	29
30

Projection: Mercator
Datum: Mars 2000 Sphere
Central Meridian: 0.0000
Standard Parallel 1: 0.0000

Tharsis (MC-9)

Geography

The Tharsis quadrangle spans the northern half of the highest region on the surface of Mars. High plains cover most of the map region, interrupted by immense shield volcanoes as well as grooved, ridged, and hummocky terrains. To the northwest, Olympus Mons constitutes the largest volcano in the solar system (Figure 9.A). Standing some 22 km above the immediately surrounding terrain, Olympus Mons towers two-and-a-half times as high as Mount Everest (8.8 km above sea level on Earth), while its volume is more than 50 times greater than Mauna Loa, the most massive volcano on Earth (Bargar and Jackson, 1974; Plescia, 2004; Carr, 2006; see Table 5.2). Furthermore, Pavonis Mons and Ascraeus Mons are two of the three Tharsis Montes – giant shield volcanoes that form a northeast-trending chain; also on this trend are Ceraunius Tholus, Uranius Tholus, and Uranius Mons (Scott and Tanaka, 1986; Plescia, 2004). Other substantial volcanoes in the quadrangle include Biblis, Ulysses, Tharsis, and Jovis Tholi. Large, lobate, ridged aprons surround parts of Olympus, Pavonis, and Ascraeus Montes. Local swells expose systems of linear troughs, cracks, and pit chains. Where undeformed, the quadrangle ranges from 6,000 m elevation near the volcanoes to 1,000 m above datum and is surfaced by lobate lava-flow forms, except for a patch of hummocky deposits in the southwestern corner of the quadrangle. A lower plain surrounds the base of Olympus Mons, as described below.

Geology

The present surfaces of this volcanic region represent the most youthful phases of an extremely long-lived (Noachian through Amazonian, or from before 3.5 billion years ago to within the past few million years), and perhaps episodically active, volcanic province of Mars (Dohm and Tanaka, 1999; Anderson *et al.*, 2001; Dohm *et al.*, 2001b; 2001c). The Tharsis quadrangle displays huge volcanoes with central calderas and some flank vents as well as local clusters of small shields and fissure vents that have issued forth vast lava-flow fields. The volcanic plains surrounding the volcanoes display a varied accumulation of lavas in time and space. The Hesperian plains tend to be relatively elevated and more highly dissected by graben systems (Figure 9.B), indicative of broad areas of crustal extension. Younger, Amazonian plains overlap the older plains, as well as the gigantic aureoles and basal scarp of Olympus Mons that likely resulted from collapse of outer parts of a broader, earlier Olympus shield edifice (see MC-8). Locally, some fissures in the quadrangle east of Olympus Mons and Ceraunius Fossae are sources of channel systems that may have resulted from groundwater discharges (Figure 9.C). Among the youngest geologic landforms in the quadrangle are huge lobes of thin, ribbed material

on the northwestern flanks of Pavonis and Ascraeus Montes that may be relicts of former Martian ice-age glaciers.

Fire and ice: Glaciation

The northwest flanks of the Tharsis volcanoes, Ascraeus, Pavonis, and (in MC-17) Arsia Montes display fan-shaped deposits of unusual form and

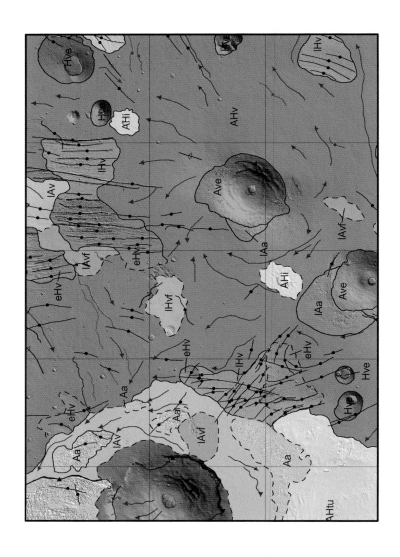

Structure

Interpretation
— Caldera rim
— Channel axis
— Crater rim
●— Graben axis
▲ Lobate flow
▲ Outflow channel
⊕— Pit crater chain
◇ Ridge
●— Rille
◀ Scarp
●— Spiral trough
◆ Wrinkle ridge
↕ Yardangs

Geology

Impact unit	Transition unit
AHi	AHtu
Volcanic units	Highland unit
IHvf	eHh
Ave	
eHv	
AHv	
IAvf	
IAv	
Hve	
IHv	
Apron units	
Aa	
IAa	

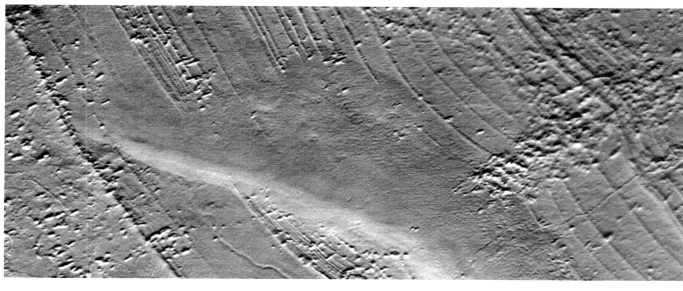

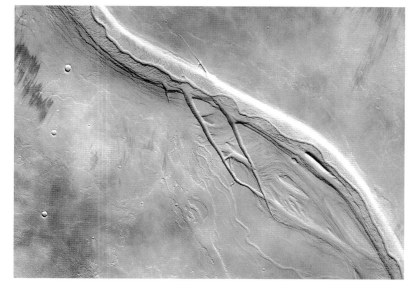

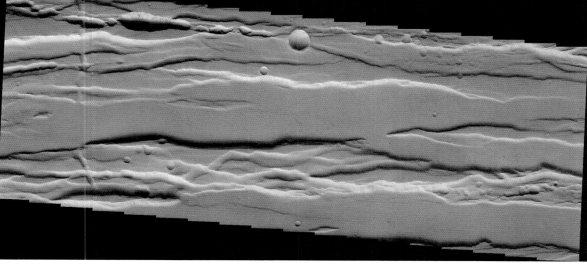

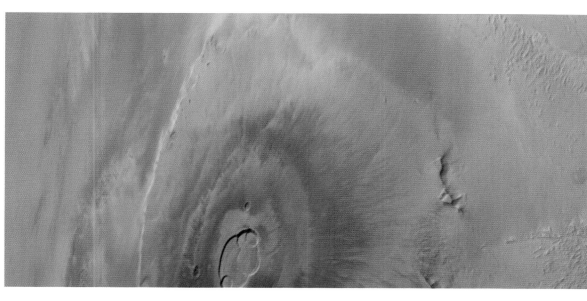

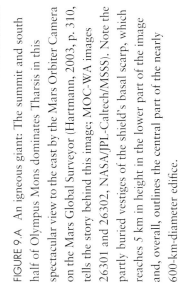

FIGURE 9.A An igneous giant: The summit and south half of Olympus Mons dominates Tharsis in this spectacular view to the east by the Mars Orbiter Camera on the Mars Global Surveyor (Hartmann, 2003, p. 310, tells the story behind this image; MOC-WA images 26301 and 26302, NASA/JPL-Caltech/MSSS). Note the partly buried vestiges of the shield's basal scarp, which reaches 5 km in height in the lower part of the image and, overall, outlines the central part of the nearly 600-km-diameter edifice.

origin (Ascraeus, Pavonis, and Arsia Sulci, respectively; Scott and Zimbelman, 1995; Scott et al., 1998; Shean et al., 2005; Kadish et al., 2008). While gravity-driven landslides, volcanic activity,

FIGURE 9.B Cracking up: Ceraunius Fossae are just some of the many fractures arrayed around the Tharsis region (seen also on maps MC-3, -10, -17; THEMIS visible image V44669007, 19 m/pixel, 20 km by 70 km, centered near 24.78° N, 251.2° E, NASA/JPL-Caltech/ Arizona State University).

or erosion have been proposed as explanations for the deposits, their features seem more consistent with a glacial origin (Shean et al., 2005; Kadish et al., 2008). Ridges parallel to the margins of the deposit are up to 100 m high and several kilometers wide and resemble the terminal deposits of cold-base glaciers on Earth. Where the ice sublimates, or changes from solid to vapor form, it

FIGURE 9.C The surface surrounding Olympica Fossae shows widespread channelized erosion, likely due to flow of water to the southwest prior to the opening of oblique fractures. Additional flooding then caused the main valley to cut down through these features into the lava plain. Later, lava flows created the texture on the floor of the valley and left the narrow channel (Plescia, 2013). This history is similar to that of young channels elsewhere on Mars (MC-15, MC-16) (CTX image G23_027342_2051_XN_25N116W, 6 m/pixel, view 24 km by 33 km, north at top, 23.4° N, 243.8° E, NASA/JPL-Caltech/MSSS).

leaves rock debris. This could also explain the knobby terrain as the remnants of the interior of a sublimated ice sheet. Smooth, elevated areas overlap the other deposits, as would be expected for a debris-covered remnant glacier (Figure 9.D). The fan-shaped deposits are Middle to Late Amazonian. Models of Mars climate suggest that times of greater tilt of its orbital axis (obliquity) can promote the accumulation of water ice at low latitudes and intermediate elevations (Head et al., 2003; Fastook et al., 2008). According to this hypothesis, the prevailing wind patterns led to ice accumulation northwest of the volcanoes (see also the west flank of Olympus Mons in MC-8).

FIGURE 9.D Ice amidst fire: The fan-shaped deposit northwest of Pavonis Mons shows three types of terrain. The parallel ridges mark locations where a glacial sheet terminated for a time, dropping debris as it sublimated, while the overlying knobby material marks the interior parts of also-vanished ice. The smooth hill at center is interpreted as thick debris covering existing ice (THEMIS daytime infrared image I01739006, 100 m/pixel, view 32 km by 77 km, north toward top, centered at 5.2° N, 243.9° E, NASA/JPL-Caltech/Arizona State University).

The Atlas of Mars

117

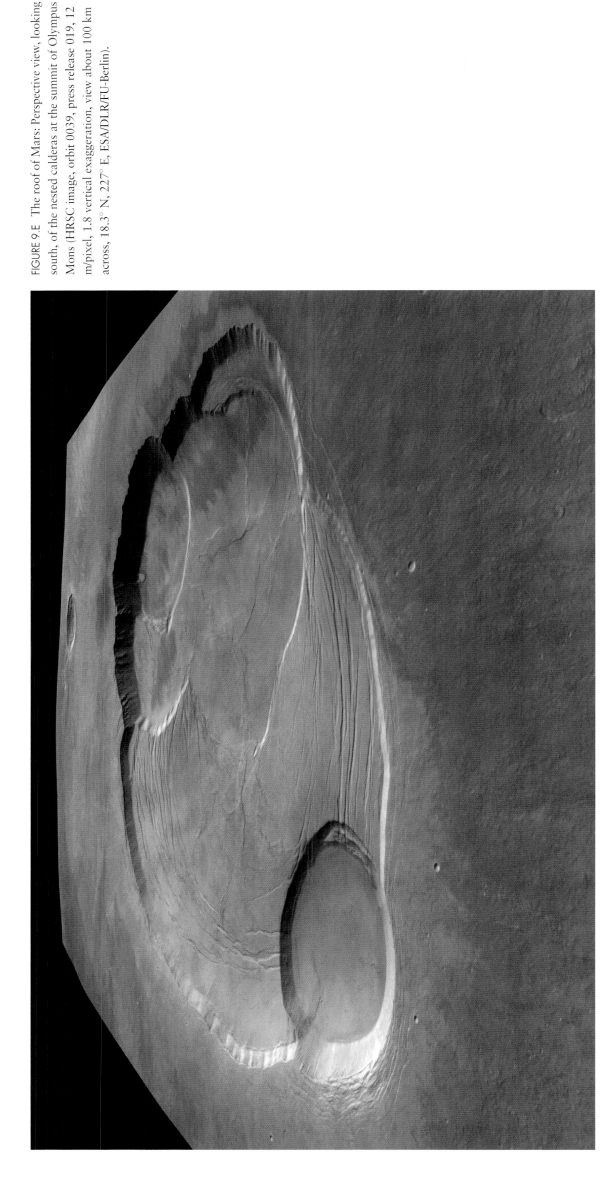

FIGURE 9.E The roof of Mars: Perspective view, looking south, of the nested calderas at the summit of Olympus Mons (HRSC image, orbit 0039, press release 019, 12 m/pixel, 1.8 vertical exaggeration, view about 100 km across, 18.3° N, 227° E, ESA/DLR/FU-Berlin).

Shaped on the top and sides

Olympus Mons has inspired fascination and numerous studies since its discovery. The summit of a shield volcano typically features a caldera, or circular collapse basin, formed when a magma chamber empties through vents or fissures (Mouginis-Mark et al., 2007). Olympus Mons features half a dozen superposed calderas (Figure 9.E; Mouginis-Mark et al., 1992; Crumpler et al., 1996; Neukum et al., 2004). The older of these show wrinkle ridges and concentric fractures. The density of impact craters on the caldera floors suggests that they were active on the order of hundreds of millions of years ago (Neukum et al., 2004; Robbins et al., 2010).

The flanks of Olympus Mons are not uniformly sloping but display concentric, less-sloping terraces, separated by steeper slopes (Figure 9.F). The terraces may represent thrust faulting, related to settling of the load of the volcano on the Martian lithosphere (Carr, 2006; Robbins et al., 2010). A more striking sign of the huge load of the volcano is the level moat, evident for several hundred kilometers beyond the flanks, especially on the south and east. Just as the mass of a diver at the end of a diving board bends it downward, the volcano has flexed the lithosphere downward (Watts, 2001; Chadwick and McGovern, 2011; Musiol and Neukum, 2012; Isherwood et al., 2013). The moat subsequently filled with younger lava flows, some of which flowed down the volcano's flanks (Figure 9.G).

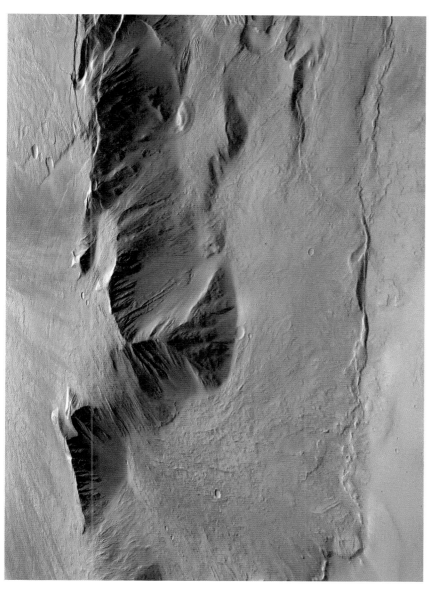

FIGURE 9.G Basal scarp or cliff (Olympus Rupes) of the eastern flank of Olympus Mons, looking west. Lava flows that postdate the cliff have flowed over it in several places (HRSC perspective from orbit 1089, 11 m/pixel, view about 50 km across, 17.5° N, 230.5° E, ESA/DLR/FU-Berlin).

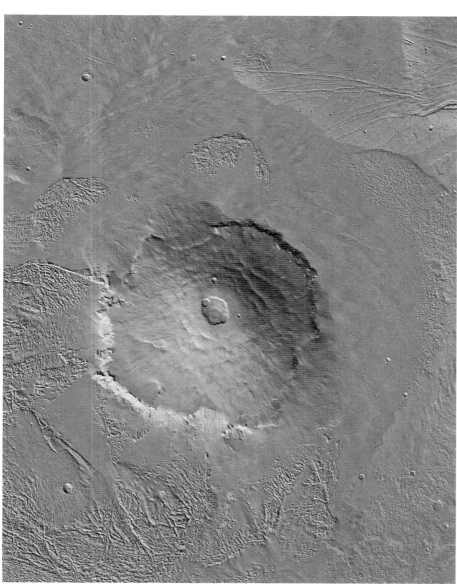

FIGURE 9.F Mega-washboard!: The impressive Olympus Mons edifice, nearly 600 kilometers in diameter at its imposing basal scarp, rises more than 21 km above the mean elevation of the planet. The summit caldera complex is surrounded by a series of undulating terraces that indicate gravitationally induced deformation of the edifice (MOLA digital elevation model, 128 pixels/degree and THEMIS daytime infrared mosaic, 100 m/pixel; view centered near 17.8° N, 225.8° E).

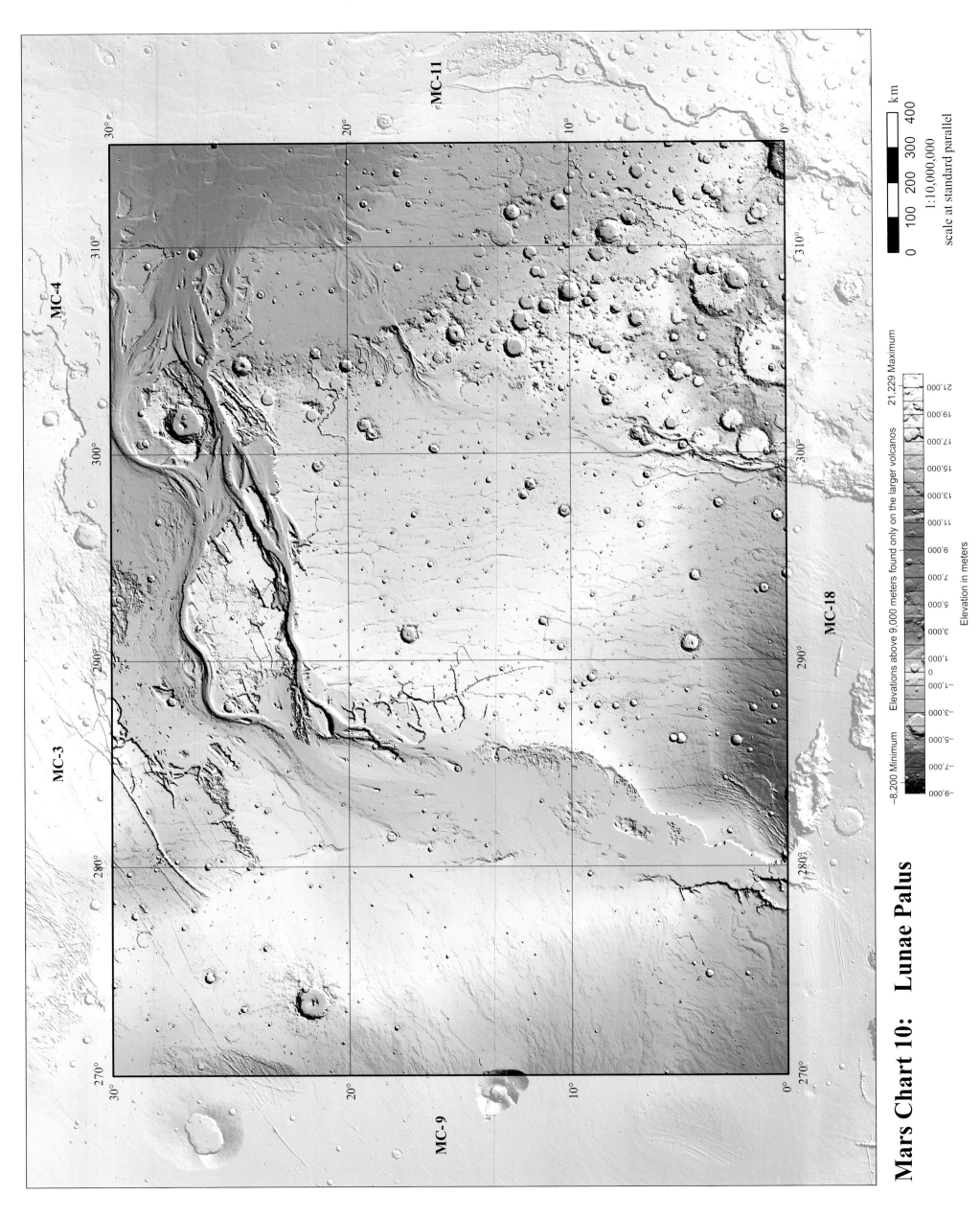

Mars Chart 10: Lunae Palus

MC-3

MC-4

MC-11

MC-18

MC-9

30°
20°
10°
0°

270°
280°
290°
300°
310°

km

1:10,000,000
scale at standard parallel

0   100   200   300   400

−8,200 Minimum   Elevations above 9,000 meters found only on the larger volcanos   21,229 Maximum

Elevation in meters

−9,000 −7,000 −5,000 −3,000 −1,000 0 1,000 3,000 5,000 7,000 9,000 11,000 13,000 15,000 17,000 19,000 21,000

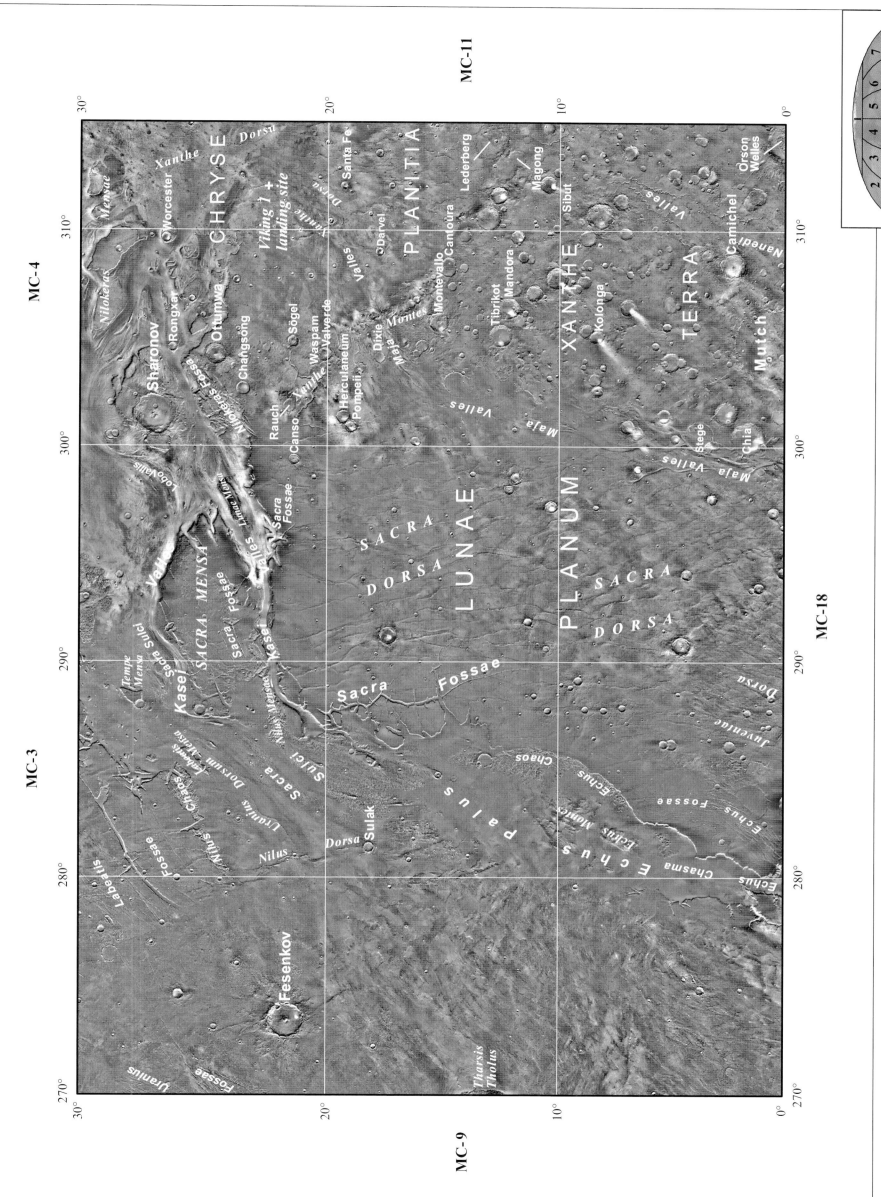

MC-4   MC-11

30°   310°   300°   290°   280°   270°

CHRYSE PLANITIA

XANTHE TERRA

LUNAE PLANUM

SACRA DORSA

SACRA MENSA

Kasei Valles

Xanthe Dorsa

Maja Valles

Echus Chasma

Echus Montes

Viking 1 + landing site

Sharonov
Worcester
Rongxar
Ofumwa
Changsöng
Sögel
Rauch
Canso
Waspam
Valverde
Herculaneum
Pompeii
Dixie
Maja Montes
Santa Fe
Darvel
Montevallo
Cantoura
Mandora
Tibrikot
Lederberg
Magong
Sibut
Kolonga
Camichel
Orson Welles
Nanedi Valles
Mutch
Stege
Chia
Sulak
Nilus Dorsa
Fesenkov
Tharsis Tholus

Xanthe Mensae
Nilokeras Mensae
Nilokeras Fossae
Lobo Vallis
Kasei Vallis
Lunae Mensa
Sacra Fossae
Tempe Mensa
Sacra Sulci
Sacra Fossae
Nilus Mensa
Nilokeras Chaos
Ledovnik Dorsum Mensa
Nilus Chaos
Labeatis Fossae
Uranius Fossae
Sacra Fossae
Sacra Sulci
Echus Palus
Juventae Dorsa
Echus Chaos
Echus Fossae
Juventae Dorsa

MC-3   MC-9

MC-18

km

0 100 200 300 400

1:10,000,000

scale at standard parallel

N
W E
S

Projection: Mercator
Datum: Mars 2000 Sphere
Central Meridian: 0.0000
Standard Parallel 1: 0.0000

Lunae Palus (MC-10)

Geography

This map spans diverse highland to lowland terrains from west to east. The western smooth region is the margin of the Tharsis rise that is centered to the west of the map, descending from 2 km to near datum. The elevated Lunae Planum and southern part of Tempe Terra occupy the central third of the map area. Lunae Planum rises to 3,500 m elevation along the southern border of the map as it approaches the Valles Marineris canyon system (south of the map area, MC-18). East of the planum lies the heavily cratered Xanthe Terra and Xanthe Montes. Cutting into these highlands are the large outflow channels Maja Valles and Kasei Valles. The latter originates from the irregular Echus Chasma depression that lies near datum. In the northeastern corner is Chryse Planitia, part of the vast northern lowlands, more than 3000 m below datum, and the destination for channel outflow as well as the Viking 1 lander.

Geology

A contrast in age is evident between the heavily cratered Xanthe Terra and Xanthe Montes (Noachian, eastern part of map), the flood basalts of the moderately cratered Lunae Planum (whose surface is of Hesperian age, center of map), and the relatively smooth, sparsely cratered Tharsis volcanic flows in the west (Amazonian).

Two sets of tectonic features in the Hesperian-age units are related to the Tharsis province, west of the map area. Lunae Planum displays north-trending wrinkle ridges, here indicative of shortening in an east–west direction. Narrow, faulted troughs trend northeast in Hesperian units in the western part of the map. These are radial to the center of the huge Tharsis rise, in which the massive load of igneous rocks has resulted in immense weighting of the lithosphere and thus regional deformation.

The channel systems flowing north and east toward Chryse Planitia developed from multiple discharge events. Older flow events were eroded or overprinted by subsequent activity. The contrast in character between the streamlined forms, dominating the lower part of the Kasei Valles in the north, versus the irregular, less distinct upper valley in the south (Echus Chasma) indicates distinctive origins and histories, as highlighted below.

A dry cataract in Kasei Valles

The northern part of the map displays Kasei Valles, a series of dry channels showing varied features, likely made by water and perhaps ice flowing to the north and then east. The northwestern two-thirds of the area in Figure 10.A was shaped by early, massive flood events (Baker and Milton, 1974; Tanaka and Chapman, 1992).

During a late flow event, two large waterfalls migrated upstream by erosion to the positions shown by the arrows. Each has a vertical drop of about 500 m and extends horizontally for 100 km. On Earth, some modern waterfalls have a greater vertical drop, but all are dwarfed by the width and implied flow rate of the Kasei falls. Several fresh-looking impact craters along Kasei indicate that significant time has elapsed since the flow events.

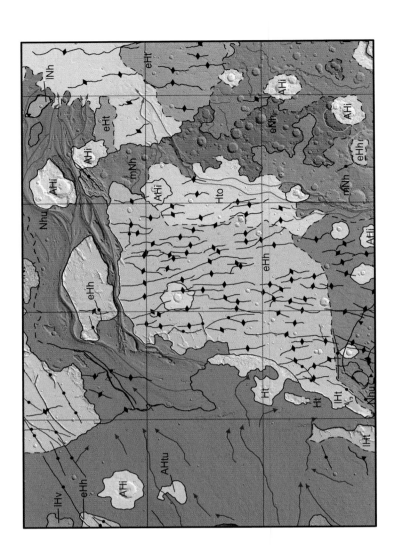

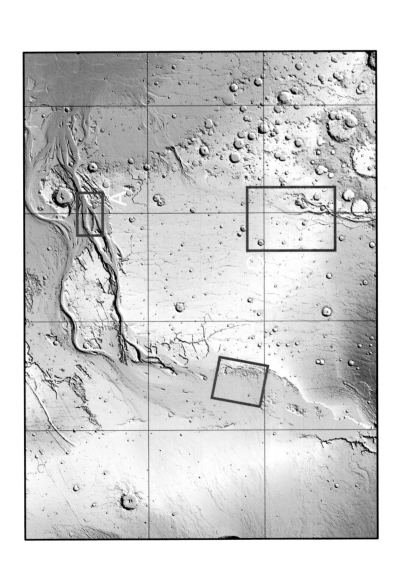

Structure

Interpretation

——	Caldera rim
——	Channel axis
——	Crater rim
•—•—	Graben axis
▲	Lobate flow
➤	Outflow channel
⊖—⊖	Pit crater chain
◇—◇	Ridge
•—•—	Rille
◄—◄	Scarp
•—•—	Spiral trough
◆—◆	Wrinkle ridge
↕	Yardangs

Geology

Impact unit
AHi

Volcanic units
eHv
AHv
Hve
lHv

Transition units
lHt
Hto
Ht
eHt
AHtu
HNt

Highland units
mNh
lNh
eHh
eNh
Nhu

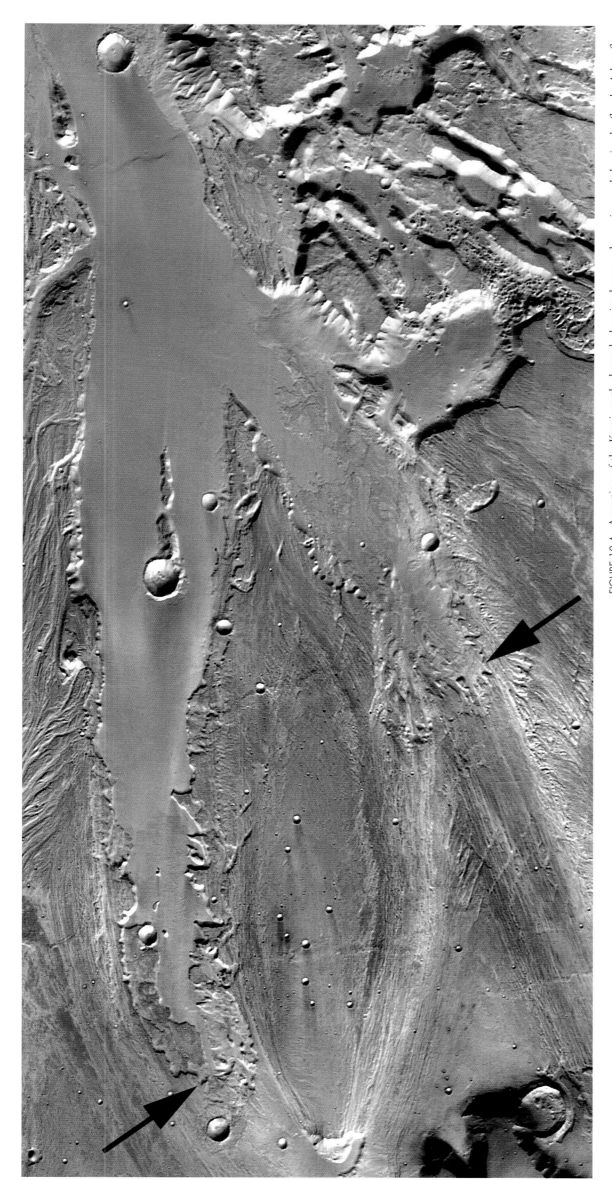

FIGURE 10.A A portion of the Kasei south channel, showing large-scale erosion and shaping by floods. A late flow event has carved two huge waterfalls, marked by the arrows. View is 95 km by 200 km, at 24° N, 300° E. (THEMIS daytime infrared mosaic, 100 m/pixel, north at top, NASA/JPL–Caltech/Arizona State University.)

that formed the channel and waterfalls. The channel east of the falls has an extremely smooth floor that shows few traces of the torrent it once carried. The smooth floor may be a younger lava flow that followed the deepest part of the channel and partially buried the craters (Dundas and Keszthelyi, 2014), though mudflow has also been suggested (Williams and Malin, 2004).

Echus Chaos: source of a flood

Following the Kasei Valles system upstream leads to this region, where channels and streamlined islands no longer appear. Instead, a large depression may be the source of the last of several giant floods to sweep down Kasei Valles. This depression has since partially filled with younger lavas (the smooth region in Figure 10.B), except for the rough-looking chaotic terrain near the eastern edge of the depression (Chapman *et al.*, 2010a; 2010b).

FIGURE 10.B The Echus Chaos in the eastern half of this image evidently was part of the adjacent Lunae Planum to the east before it collapsed and released flood water or ice. View is 200 km wide, centered at 12° N, 284° E. (Viking orbiter 1 image 858A33, 200 m/pixel, north at top, NASA/JPL-Caltech.)

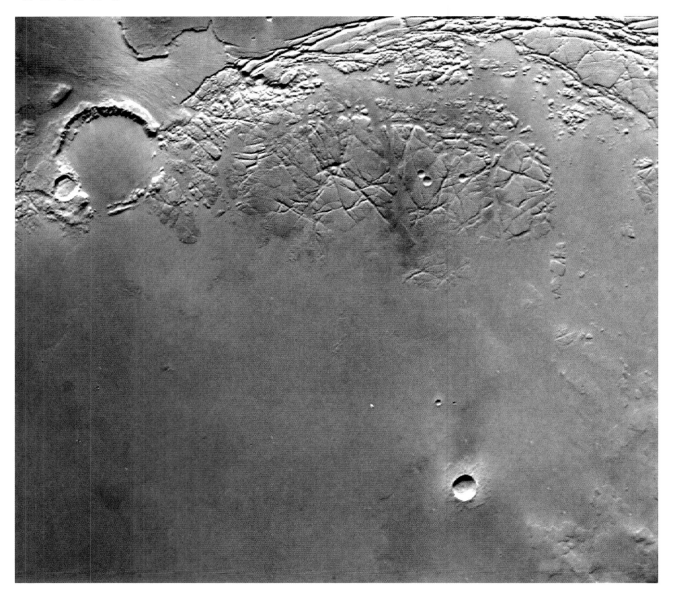

Aqueous nocturne

In the nighttime, thermal infrared image of part of the Maja Valles outflow channel system shown in Figure 10.C, bright areas are relatively warm, indicative of high thermal inertia materials, including bedrock, coarse sand, and afternoon-sunfacing rock faces (e.g. some crater walls and floor deposits). Dark areas represent finer-grained materials, including crater ejecta (see dark rays around some craters). Note the sinuous, north-trending, branching Maja Valles system of channels, which includes streamlined bars and longitudinal ridges and troughs within more pronounced channel reaches. For the most part, channel floors have an intermediate overall brightness and are relatively bright versus adjacent surfaces. Some of the bright patches in the left half of the image have lobate margins. Likely, the brighter, channel-related patches mark the extents of thin, fine-grained fluvial deposits, including lobate sheet flows that are produced by sediment-rich floods (Voelker *et al.*, 2012; 2013; T. Platz and M. Voelker, personal communication, 2015). In addition, evaporation of the water could have led to mineral precipitation and cementation, which may have also increased the thermal inertia of the deposits.

FIGURE 10.C THEMIS nighttime infrared view of Maja Valles. The areas of medium brightness include channels and lobate fluvial deposits (100 m/pixel mosaic, 333 km by 464 km, centered at 7.8° N, 299.5° E, NASA/JPL-Caltech/Arizona State University).

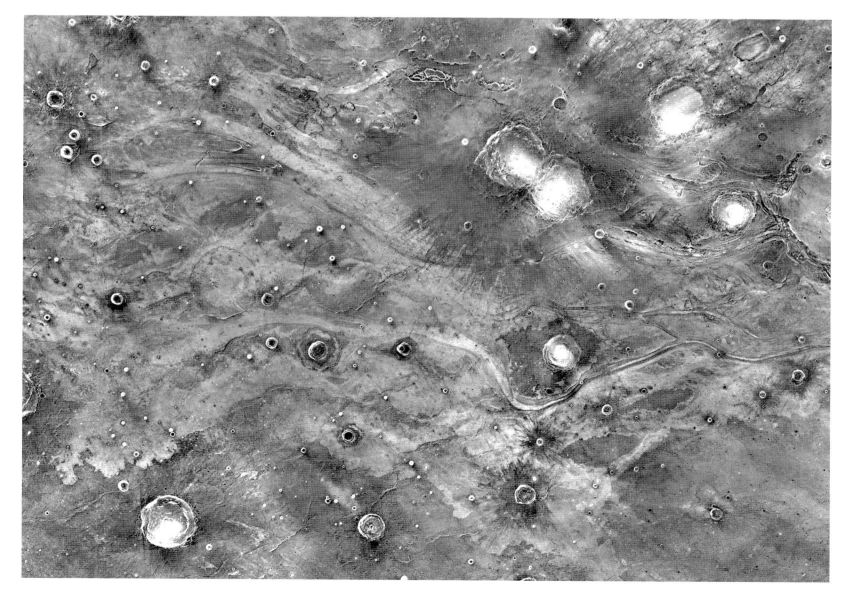

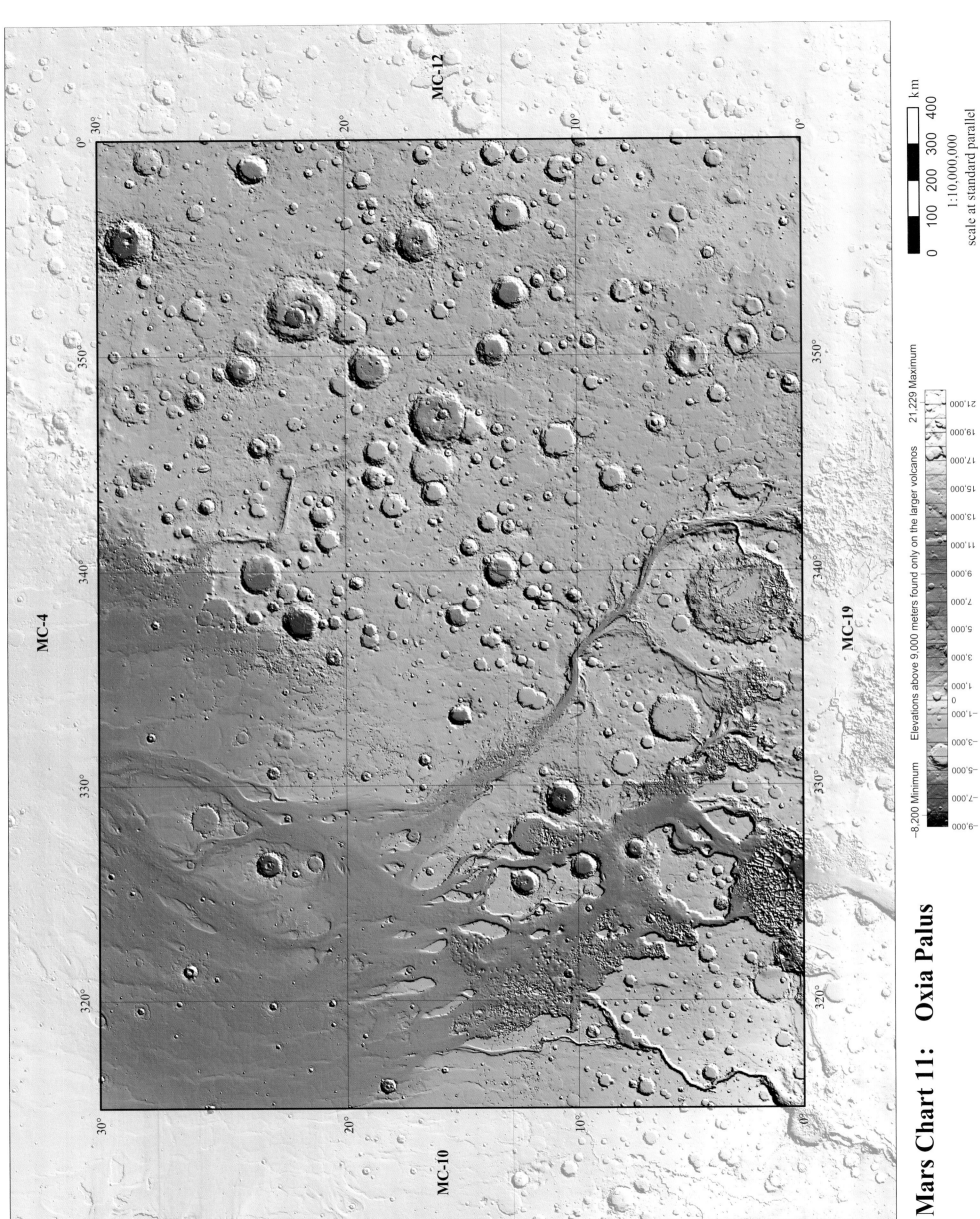

Mars Chart 11:   Oxia Palus

MC-4

MC-10

MC-12

MC-19

km

1:10,000,000

scale at standard parallel

0   100   200   300   400

−8,200 Minimum   Elevations above 9,000 meters found only on the larger volcanos   21,229 Maximum

Elevation in meters

−9,000   −7,000   −5,000   −3,000   −1,000   0   1,000   3,000   5,000   7,000   9,000   11,000   13,000   15,000   17,000   19,000   21,000

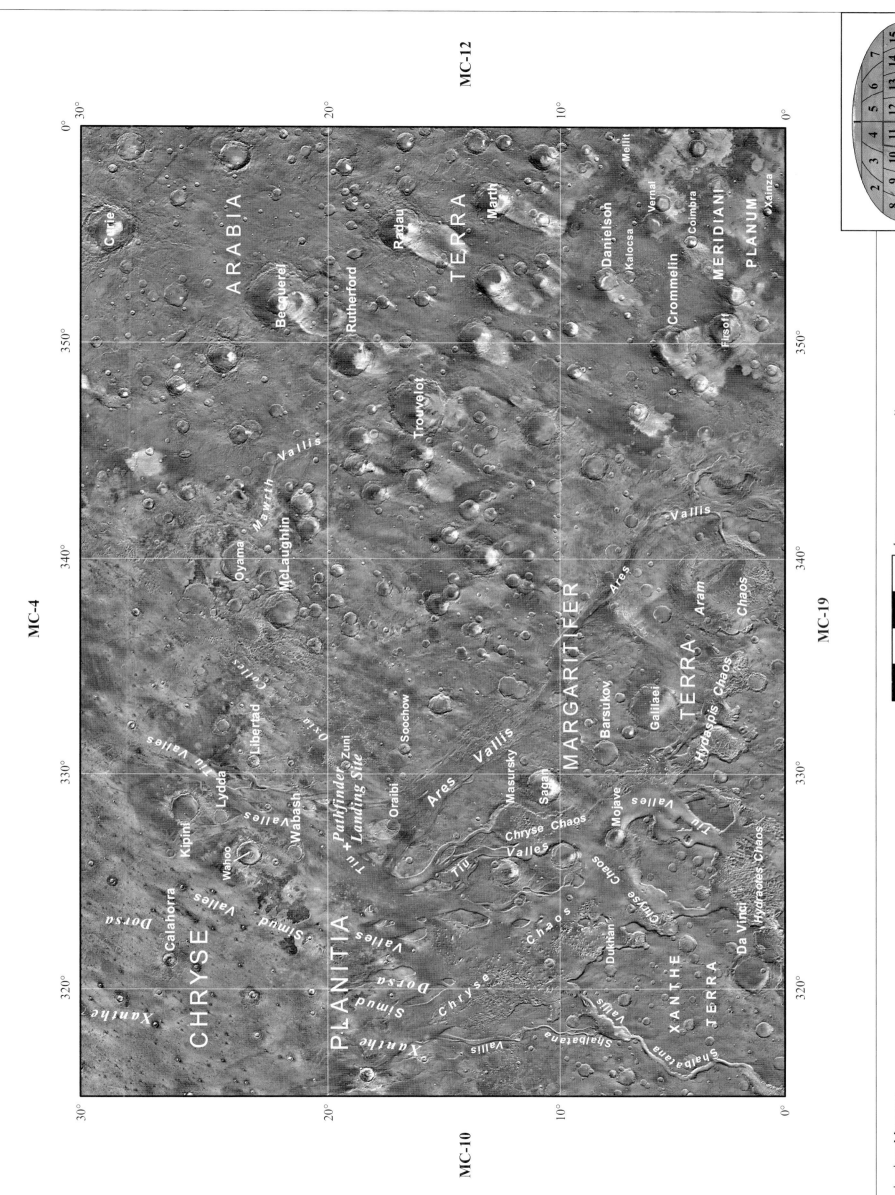

MC-4

MC-12

MC-10

MC-19

0° 30°

350°

340°

330°

320°

30° 20° 10° 0°

ARABIA

Curie

Becquerel

Rutherford

Radau

TERRA

Marth

Trouvelot

Danjelsoh

Kalocsa

Vernal

Mellit

Coimbra

Crommelin

MERIDIANI

Firsoff

PLANUM

Xainza

Oyama

McLaughlin

Mawrth Vallis

Coltes

Liberad

Oxia

Lydda

Kipini

Wahoo

Wabash Valles

Simud

Tiu Valles

Calahorra

Dorsa

Xanthe

CHRYSE

PLANITIA

Simud Dorsa

Xanthe Dorsa

Chryse

Soochow

Oraibi

Pathfinder
Landing Site

Zuni

Ares Vallis

Masursky

Sagan

Tiu

Chryse Chaos

Valles

Chryse Chaos

Chaos

Mojave

Tiu Valles

Dukhan

Shalbatana Vallis

Vallis

MARGARITIFER

Ares

Vallis

Barsukov

Galilaei

TERRA

Aram

Chaos

Hydaspis Chaos

Da Vinci

Hydraotes Chaos

XANTHE TERRA

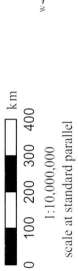

				7			
		5	6	13	14	15	
	3	4	12	20	21	22	23
	2	10	11	18	19	27	28	29
8	9	16	17	25	26	30

N
W E
S

km

0 100 200 300 400

1:10,000,000
scale at standard parallel

Projection: Mercator
Datum: Mars 2000 Sphere
Central Meridian: 0.0000
Standard Parallel 1: 0.0000

Oxia Palus (MC-11)

Geography

The Oxia Palus quadrangle is dominated in the east by the mid-elevation Arabia Terra (1,500–3,500 m below datum), with ancient cratered terrain transitioning from the southern highlands down to the northern plains. Arabia Terra includes a portion of Meridiani Planum, north of the Mars Exploration Rover-B (Opportunity) landing site (MC-19). Mawrth Vallis is the only large valley coming from this portion of Arabia Terra that has a mouth opening directly into the northern plains across the dichotomy boundary. The western half of the quadrangle includes portions of Margaritifer Terra and Xanthe Terra highland terrains, which are dissected by enormous outflow channels. These formed by catastrophic floods that left behind grooved channels, streamlined islands, chaotic terrain, braided-channel patterns, and the southern margin of the smooth northern plains deposits. From east to west, Ares, Tiu, Simud, and Shalbatana Valles carve curvilinear paths through the highland terrain until merging at the southern margins of Chryse Planitia. Chryse, at 3,000–4,000 m below datum, dominates the northwest portion of the quadrangle and lies within an ~1,600-km-diameter, ancient, largely degraded, and buried impact basin. The Mars Pathfinder landing site, which included the first rover (named Sojourner) to navigate on Mars, is located in Tiu Valles near the edge of the Ares Vallis deposits.

Geology

Arabia Terra is made up of Noachian highland materials, including a mélange of basement rocks, volcanic materials, ejecta materials, and wind- and water-transported deposits. Arabia Terra has many small depressions and impact basins, which likely were sites where water ponded, given the high incidence of layering inside these basins. An additional line of evidence for the prevalence of water is that orbiting imagers have detected minerals in the surface deposits that typically form in the presence of water in terrestrial environments. The remnants of Margaritifer and Xanthe Terrae appear more like the typical highland materials to the south, though they are more fragmented by extensive flood channels. Chaotic materials, hypothesized to be the result of large evacuations

Martian archipelago

In the western portion of the Oxia Palus quadrangle, there are isolated plateaus of cratered highland materials and smoothed, streamlined, teardrop-shaped mounds forming an archipelago in the middle of the outflow channels (Figure 11.A). These features are interpreted to be the result of subsurface water and sediment, which destabilized the overlying rock and led to collapse, are present at the head regions of, and interspersed within, the large outflow channels (Rotto and Tanaka, 1995). Sediments transported via the outflow channels are one of the main sources of the northern lowland materials, along with volcanic, impact, and wind-borne materials.

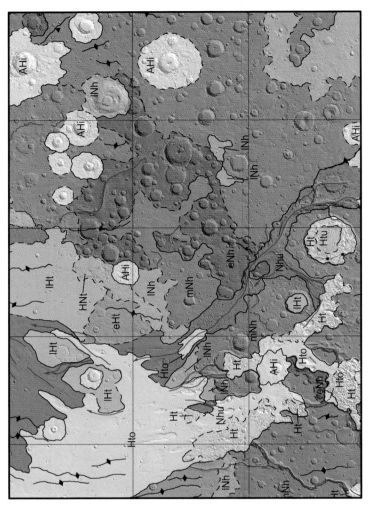

Structure

Interpretation

Caldera rim	
Channel axis	
Crater rim	
Graben axis	
Lobate flow	
Outflow channel	
Pit crater chain	
Ridge	
Rille	
Scarp	
Spiral trough	
Wrinkle ridge	
Yardangs	

Geology

Impact unit
- AHi

Lowland unit
- lHl

Transition units
- lHt
- Hto
- Ht
- eHt
- HNt
- Htu

Highland units
- mNh
- lNh
- eNh
- Nhu
- HNhu

FIGURE 11.A Streamlined, teardrop-shaped islands near the mouth of Ares Vallis. The shape of erosional remnants and features on the channel floor indicate flow to the northwest (upper left). Two features that postdate flooding and erosion are a wrinkle ridge, which trends north–south to the left of center, and the crater Lins, at lower right (which is 6 km across). (THEMIS visible image V01786010, 18 m/pixel, view 48 km by 51 km, north at top, 15.9° N, 330° E, NASA/JPL–Caltech/ Arizona State University.)

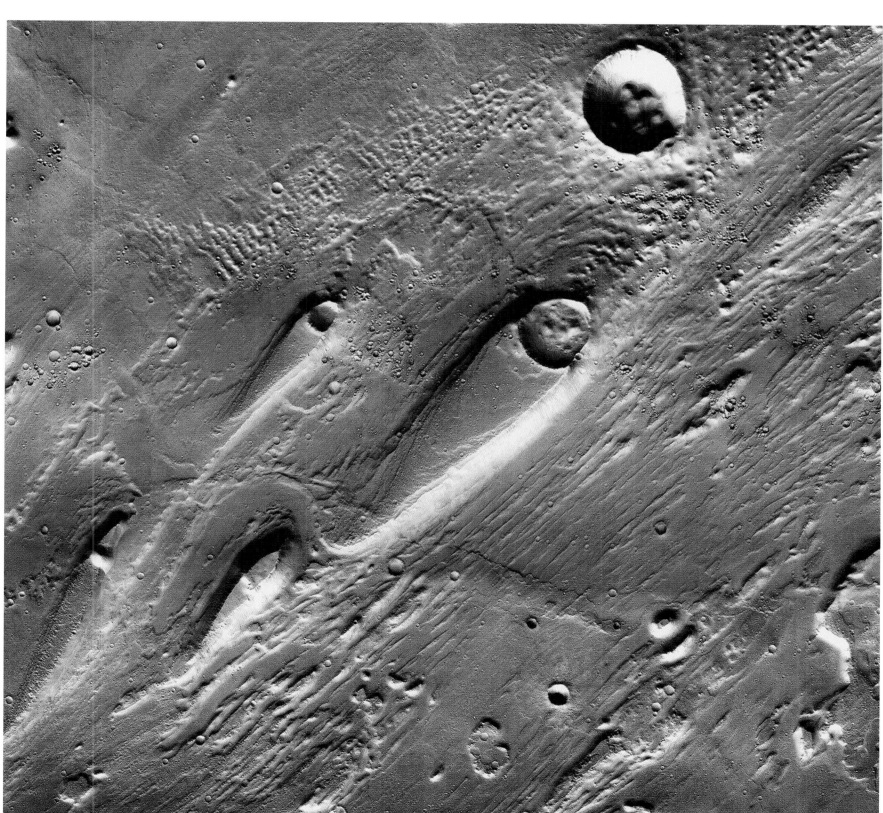

outflow channel, Ares Vallis. Spectral measurements indicate a large deposit of hematite, a form of iron oxide that likely formed in the presence of water, in parts of the basin (see also MC-19).

FIGURE 11.B Layering within the rim of an impact crater, 4 km in diameter, about 30 km west of Mawrth Vallis. The blue and white layers near the top are aluminum-bearing clays, while most of the image shows underlying iron- and magnesium-bearing clays. (Wray and Squyres, 2010; JPL, 2011; CRISM data, color exaggerated, overlain on HiRISE image PSP_004052_2045, 29 cm/pixel, view 1 km across, north at top, 24.3° N, 340.7° E, NASA/JPL-Caltech/University of Arizona/JHUAPL)

erosion. Above the ancient, cratered basement rocks, and exposed in the Mawrth Vallis walls, are layered materials with various compositions, detected using TES, OMEGA, and CRISM orbital spectrometers, but all indicating the role of water in the development of the region. Before Mawrth Vallis formed, the exposed bedrock was buried in sequences of sediment, likely from the materials that were being shed from the southern highlands, or alternatively in volcanic ash deposits, forming stacks of layered material (Wray et al., 2008; Loizeau et al., 2010). The layered compositions are composed of clay minerals that change throughout the canyon walls (Figure 11.B). Just above the bedrock, iron- and magnesium-rich clays form a >50-m-high wall of layered materials. On top of those clays are younger, aluminum-rich clay layers that are covered by a regional capping rock. These clay types also appear interlayered in nearby areas (Wray et al., 2008; Loizeau et al., 2010). The layered materials were deposited in the Middle to Late Noachian Epochs, and because Mawrth Vallis exposes these layers, it was likely active in the Late Noachian to Early Hesperian Epochs.

Aram Chaos: Fill, drain, fill, drain . . .

Aram Chaos is a crescent-shaped feature on the floor of an ancient, ~280-km-diameter impact basin. After the impact, the basin filled with material, and the pore space between clasts was occupied by volatiles, i.e. water and/or ice. It is hypothesized that evacuation of the subsurface volatiles caused instability and ultimately collapse of the ground surface (Glotch and Christensen, 2005). The resulting fractured and blocky morphology is called chaotic terrain (Figure 11.C). After the development of this chaotic terrain, the basin underwent multiple episodes of catchment and sediment deposition. The lake has a single outlet on its eastern margin, providing evidence that it drained into a north–south-trending stretch of the

megafloods that originated in the chaotic terrain to the south (see MC-19), much like features in the Channeled Scablands of the northwestern United States (see Chapter 5). These islands grade from larger, higher-standing plateaus in the south into the lower-relief, rounded, and smoothed mounds to the north where the floods spread across the topography. A great example of these streamlined mounds is located near the mouth of Ares Vallis

(Figure 11.A). In some cases, flood waters flowed around craters on their upstream sides, creating a teardrop shape (Moscardelli and Wood, 2011).

Layered clay cake

Mawrth Vallis is a 636-km-long valley cutting through surficial materials and exposing bedrock that reveals a history of repeated deposition and

FIGURE 11.C Parts of Aram Chaos: Denser, rockier materials, which occur at the surface, are in warm colors (brown, yellow) in this image, while dust-covered areas are in cool colors (purple, blue-gray). The outlet that partially drained Aram is at the eastern (right) edge of the image; dust is absent at the outlet. (THEMIS mosaic of visible and infrared bands, 100 m/pixel, view 290 km across, north at top, 3° N, 339° E, image by THEMIS team, NASA/JPL-Caltech/Arizona State University.)

131

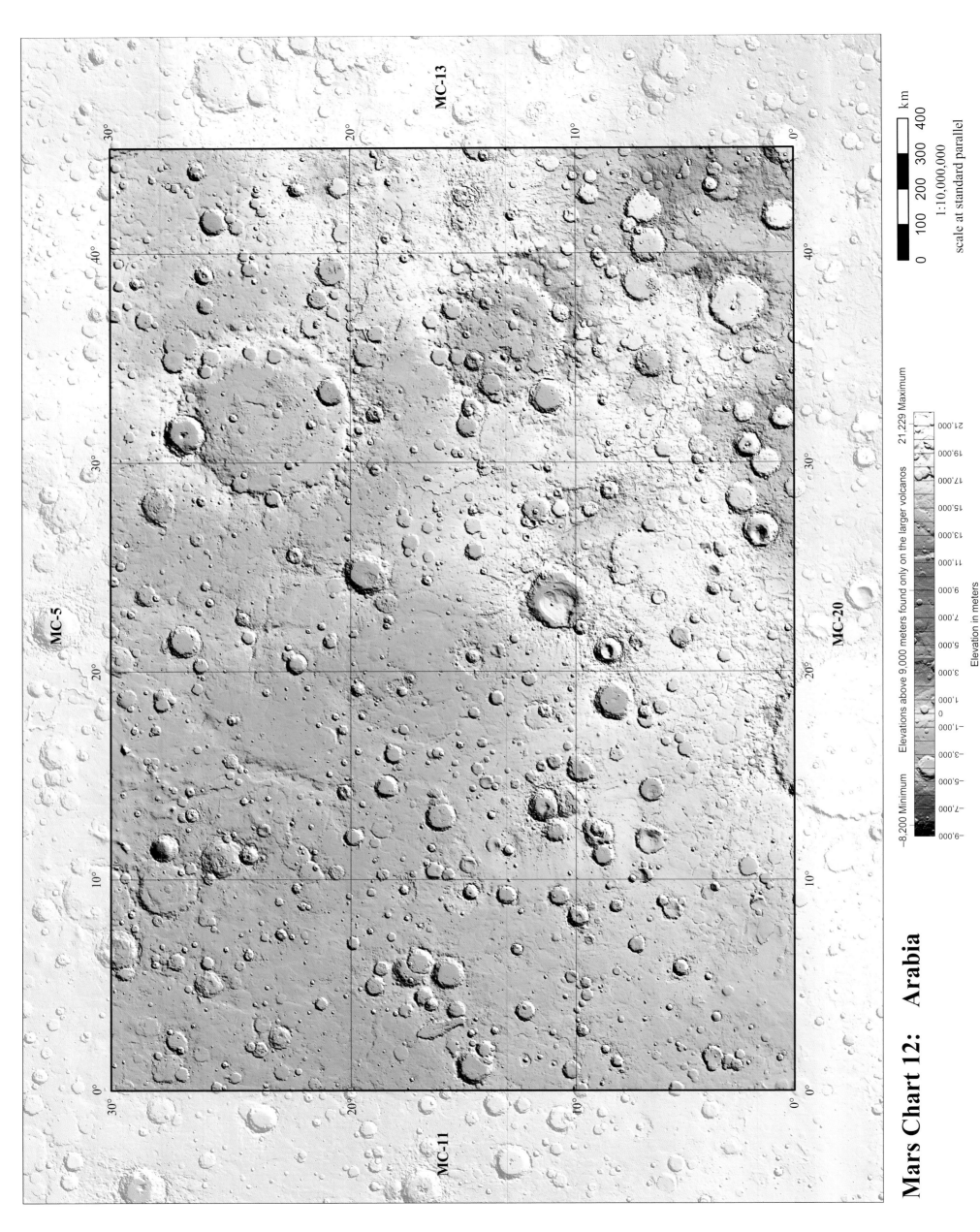

Mars Chart 12:   Arabia

MC-13

MC-5

MC-20

MC-11

1:10,000,000
scale at standard parallel

km
0   100   200   300   400

−8,200 Minimum   Elevations above 9,000 meters found only on the larger volcanos   21,229 Maximum

Elevation in meters

−9,000 −7,000 −5,000 −3,000 −1,000 0 1,000 3,000 5,000 7,000 9,000 11,000 13,000 15,000 17,000 19,000 21,000

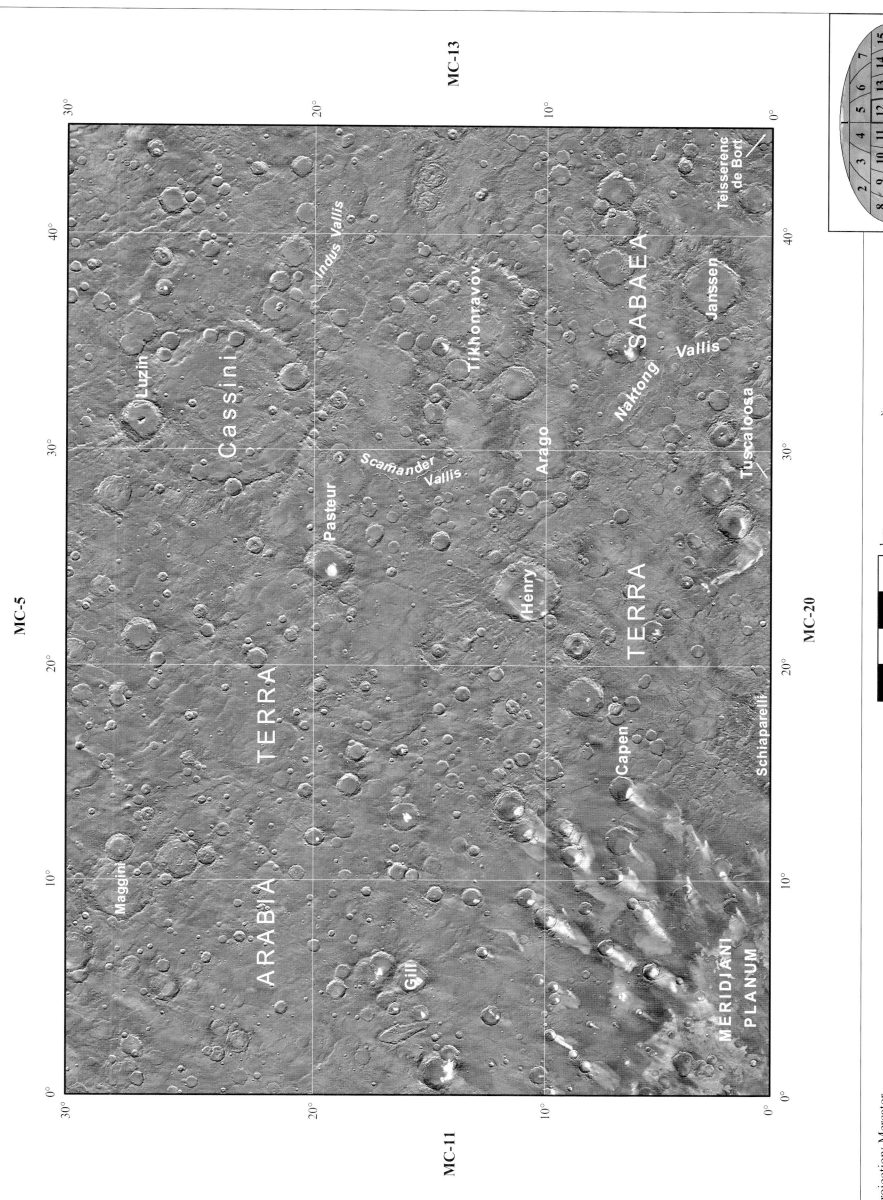

MC-5

MC-13

MC-11

MC-20

30° 40° 30° 20° 10° 0°

Luzin

Cassini

Indus Vallis

Tikhonravov

SABAEA

Janssen

Naktong Vallis

Tuscaloosa

Scamander Vallis

Arago

Pasteur

TERRA

Henry

TERRA

Maggini

ARABIA

Capen

Gill

Schiaparelli

MERIDIANI PLANUM

Teisserenc de Bort

N
W E
S

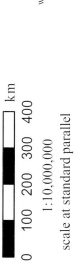

km

0 100 200 300 400

1:10,000,000

scale at standard parallel

Projection: Mercator
Datum: Mars 2000 Sphere
Central Meridian: 0.0000
Standard Parallel 1: 0.0000

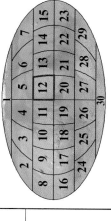

Arabia (MC-12)

Geography

The Arabia quadrangle has both highland materials of Terra Sabaea and transition materials of Arabia Terra, including a portion of Meridiani Planum. Craters dominate the landscape, with Arago, Capen, Cassini, Gill, Henry, Janssen, Pasteur, Teisserenc de Bort, Tikhonravov, and Tuscaloosa. The longest valleys are Naktong, Scamander, and Indus Valles. The topography slopes from 1 km above datum in the southeast down to 2 km below datum in the northwest. Across this transition, the southeastern, topographically irregular, densely cratered Terra Sabaea gives way to the moderately impacted, but smoother, intercrater plains of Arabia Terra. Degraded domical mounds are present in several craters, most notably Henry, but are not present in all the craters along the boundary between the higher-standing Terra Sabaea and Arabia Terra.

Geology

Terra Sabaea is a highland terrain with tectonic ridges, impact-related high-standing massifs, and valleys incised through the surface. The materials making up the area are a mixture of rocks, formed by volcanic, sedimentary, impact, and eolian processes. Large impacts as well as clusters of smaller impacts are present on the surface, and range from moderately to heavily degraded. Infilling of low-lying intercrater areas and the interiors of some craters occurred by a combination of fluvial transport, where valleys are present, and dust accumulation by airfall, where no valleys are present. Hypotheses on the depositional environment that formed layered strata in domical, intracrater mounds include groundwater, niveal (snow), glacial (ice), surface runoff processes, and volcanic ashfall. It is possible that craters with no mounds but with horizontal floors, indicative of infilling, may have gone through similar processes as those with mounds, but to a lesser degree. Arabia Terra also contains several large craters (>100 km in diameter); however, these are almost entirely degraded by erosional processes or buried by the influx of materials that were transported from the adjacent highland region. Craters with significant relief are rare in Arabia, and almost all of the craters have horizontal to sub-horizontal floors. A high density of partially buried or exhumed craters, as well as ghost craters (those with relief that is subdued to the point of complete burial) provide evidence that lower-lying surfaces were exposed and eroded for a long period before burial by an influx of materials.

A valley for the ages

Naktong Vallis is a typical dendritic, tree-like, valley system, similar to those seen on Earth. These systems are called dendritic because they have a trunk channel that is fed by branching channels, called tributaries, and, if sketched, they resemble the structure of a tree. Naktong Vallis cuts through Terra Sabaea, breaching and debouching into the southern margin of Arago crater (Irwin *et al.*, 2005). Crater densities indicate that this valley is Late Noachian to Early Hesperian in age. The age provides important chronological boundaries for a time when Mars had a thriving hydrologic cycle in which precipitation, overland flow, and groundwater were reshaping the surface (Fassett and Head, 2008).

In the lower reaches of the channel, landforms resemble those in terrestrial river valleys. A terrace (Figure 12.A) is an abandoned river bank of a channel, which remains as the flow reacts to changes in the water level at the mouth of the stream. As the water level of a lake drops, the river will cut down

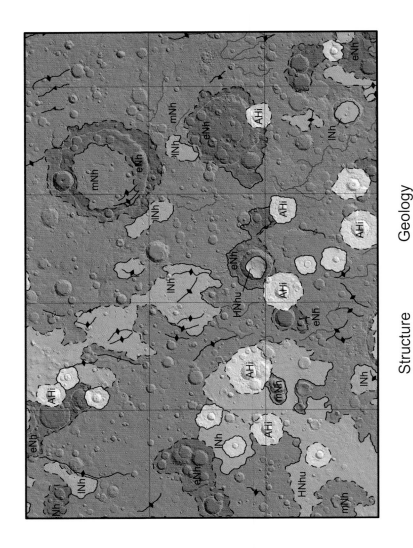

Structure

Interpretation

Caldera rim	
Channel axis	
Crater rim	
Graben axis	
Lobate flow	
Outflow channel	
Pit crater chain	
Ridge	
Rille	
Scarp	
Spiral trough	
Wrinkle ridge	
Yardangs	

Geology

Impact unit
AHi

Highland units
mNh
lNh
eHh
eNh
HNhu

through the surface to reach lake level. This process abandons the old bank, which becomes a terrace. Inverted channels (Figure 12.B) are remnants of the path taken by the water. Preservation of these surfaces can be achieved through several processes: (1) percolation of mineral-rich water into the subsurface, leaving behind the minerals as cementing agents as the water evaporates, (2) armoring of the channel by deposition of larger grains, or (3) filling in the channel by more resistant materials, i.e. lavas. Each process results in the channel fill ending up more resistant to erosion, relative to surrounding materials. As erosion removes the weaker material that surrounds the stronger material, a sinuous ridge, or inverted channel, is exposed.

Layered mounds in craters

The ~1.5–2.0-km-thick mound of materials in Henry crater, located near the Arabia Terra and Terra Sabaea boundary (Figure 12.C), consists of layered strata that are hypothesized to have been materials deposited in lakes, by glaciers and/or through eolian processes. In Henry crater, near the change from highland to transition terrain, these strata are exposed and provide a glimpse into the Martian past. The mound in Henry is not isolated; many craters in Arabia Terra have similar mounds and layered strata (see also Gale crater in MC-23). This observation suggests that this type of deposition was areally extensive and that large volumes of materials were transported to this region.

Windy residue

Light and dark streaks extend from some impact craters in the Arabia Terra portion of the quadrangle (Figure 12.D). Comparison with terrestrial analogs suggests that sand-sized material on the floors of craters is moved by prevailing winds to create the streaks, which have been observed to change over two decades of imaging. Bedforms on the streaks are consistent with wind transport. Light-colored material in the streaks may originate from lacustrine deposits in the craters upwind of the streaks (Rodriguez et al., 2010). Some craters in this region produce streaks and others do not; possible reasons for this variation include differences in crater depth, cementation of crater floor materials, ratio of sand to smaller grains, and age and exposure of the source deposits (see also MC-22).

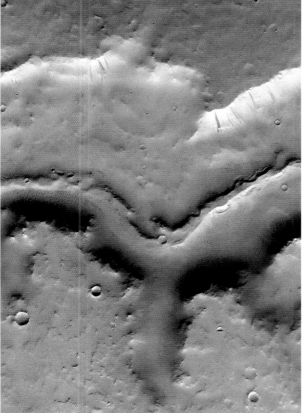

FIGURE 12.A A portion of a CTX image showing terraces (T1 and T2) along Naktong Vallis. The original surface (OS) represents the surface before the channel formed. The initial formation of the channel is preserved next to the OS and is the preserved level of the oldest terrace (T1). The river then downcut to a new level at an elevation equal to the youngest preserved terrace (T2). Another incision cut into the T2 surface and formed the channel preserved in the bottom of the valley (blue arrow). These incisions could be due to changes in rock strength, stream power, and/or base level, or could be due to responses to changes in slope due to tectonic processes (image P19_008375_1874_XN_07N329W, 6 m/pixel, view about 7 km across, north at top, centered at 7.3° N, 30.7° E, NASA/JPL-Caltech/MSSS).

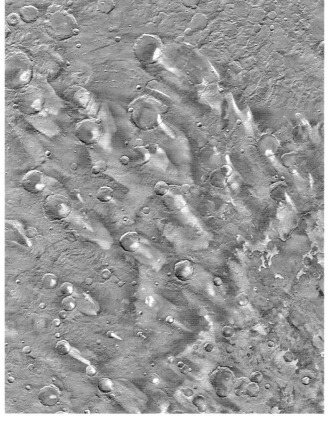

FIGURE 12.B CTX image showing an inverted channel in Naktong Vallis. The sinuosity (curvature) matches that of the walls of the valley. Image is about 13 km across (image P04_002758_1880_XI_08N329, 6 m/pixel, north at top, 7.8° N, 30.5° E, NASA/JPL-Caltech/MSSS).

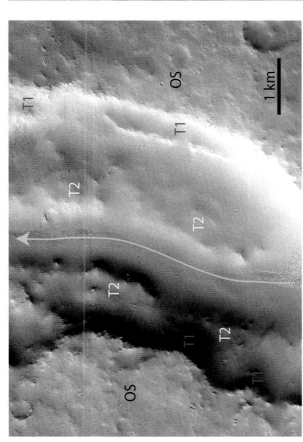

FIGURE 12.C HiRISE view (~1.5 km across) showing layering on the northeast side of the mound in Henry crater (image PSP_009008_1915, 25 cm/pixel, north at top, centered at 11.3° N, 23.7° E, NASA/JPL-Caltech/University of Arizona).

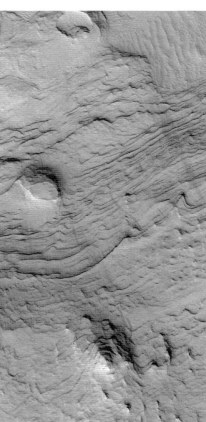

FIGURE 12.D THEMIS daytime infrared mosaic, view 700 km by 950 km (north toward top, 0° N to 12° N, 0° E to 14° E), showing streaks near craters. Note that some of the light steaks extend to the next crater in the downwind direction. The streaks likely formed by winds, moving from northeast to southwest (NASA/JPL-Caltech/Arizona State University).

The Atlas of Mars

Mars Chart 13: Syrtis Major

1:10,000,000
scale at standard parallel

km
0   100   200   300   400

Elevations above 9,000 meters found only on the larger volcanos

−8,200 Minimum   21,229 Maximum

−9,000 −7,000 −5,000 −3,000 −1,000 1,000 3,000 5,000 7,000 9,000 11,000 13,000 15,000 17,000 19,000 21,000
                              0
Elevation in meters

MC-14

MC-6

MC-5

MC-21

MC-12

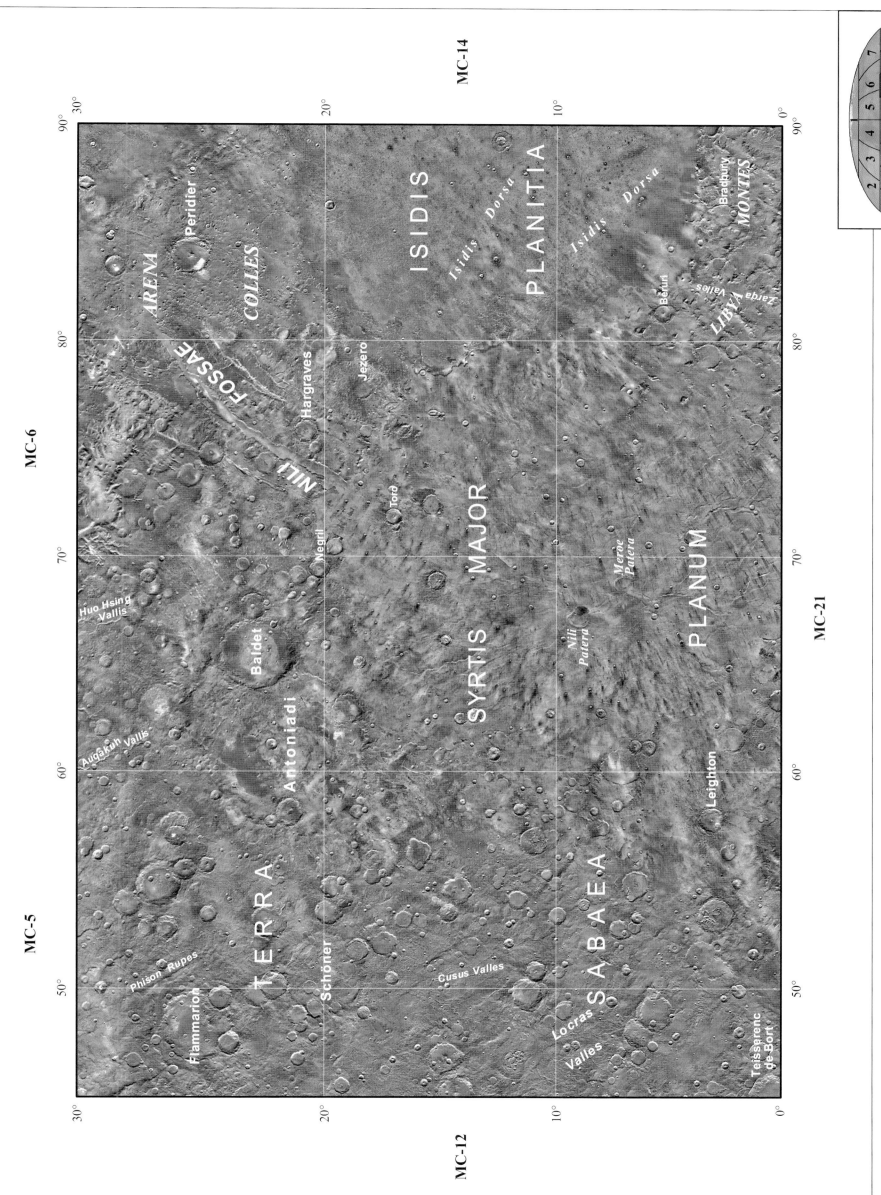

MC-6

MC-5

MC-14

MC-12

MC-21

30°

20°

10°

0°

90°   80°   70°   60°   50°   30°

ARENA

Peridier

COLLES

ISIDIS

Isidis Dorsa

NILI

FOSSAE

Hargraves

Jezero

PLANITIA

Isidis Dorsa

Bëuri

LIBYA

Zarda Valles

Bradbury

MONTES

Negril

Tord

SYRTIS   MAJOR

Meroe
Patera

Huo Hsing
Vallis

Baldet

Nili
Patera

PLANUM

Augakuh Vallis

Antoniadi

TERRA

Schöner

Cusus Valles

SYRTIS

SABAEA

Leighton

Phison Rupes

Flammarion

Locras
Valles

Teisserenc
de Bort

Projection: Mercator
Datum: Mars 2000 Sphere
Central Meridian: 0.0000
Standard Parallel 1: 0.0000

1:10,000,000
scale at standard parallel

km
0   100   200   300   400

N
W   E
S

Syrtis Major (MC-13)

Geography

The Syrtis Major quadrangle is named after a prominent dark feature, discovered by Christiaan Huygens, in 1659, as the first surface feature recognized on another planet (Figure 13.A). The name "Syrtis Major," however, originated more than 200 years later, with the astronomer Giovanni Schiaparelli. The quadrangle contains the volcanic province, defined by Syrtis Major Planum, a large but gently sloping dome that rises to almost 2,100 m in elevation. Centered on Syrtis Major Planum are two rimless depressions named Nili and Meroe Paterae. West of Syrtis Major Planum lie the ancient cratered highlands of Mars, named Terra Sabaea. This ancient terrain descends to near datum and is pockmarked with multiple ancient impact craters and basins, including Antoniadi (400 km in diameter), Baldet (181 km in diameter), and Flammarion (174 km in diameter). Isidis Planitia resides to the east of Syrtis Major Planum, and represents the smooth, infilled surface of a very ancient multi-ring impact basin that was at least 1,500 km in original diameter. The floor of Isidis Planitia is close to 4 km below

datum. Large-scale landforms, characteristic of giant impact basins, are circumferential to Isidis Planitia. To the south, Libya Montes forms an arcuate ring of large, angular mountains, some of which rise >2 km above the surrounding plains. Northwestern Isidis Planitia is bounded by the elongate and arcuate troughs of Nili Fossae, which are up to 40 km wide and 1,300 m deep. Arena Colles forms the best-preserved highland–lowland transition zone within the Syrtis Major quadrangle, and it is composed of a field of rugged knobs as well as their intervening plains.

Geology

The geologic history of the terrain within the Syrtis Major quadrangle is diverse, complex, and long-lived. The Isidis-forming impact likely occurred ~4 billion years ago and has since been a topographic reservoir for lavas that poured in from Syrtis Major Planum, for sediments that were eroded from the adjacent, higher-standing terrain, as well as for sediment emplacement, perhaps by oceans, lakes, and playas from the northeast and by possible glaciation (Ivanov et al., 2012). Syrtis Major Planum formed through the eruption of very fluid lavas, both in the vicinity of Nili and

Meroe Paterae as well as from large cracks that radiate away from the shield-like, though very low relief, volcanic edifice (Hiesinger and Head, 2004). Through time, the volcano grew in height and breadth as the lavas extended across adjacent terrains. Nili Patera is a particularly complex volcanic area, having features indicative of both effusive and explosive eruptions and basaltic to silica-rich volcanic rock compositions (Fawdon et al., 2015).

In addition, the margins of a late-stage cone in the patera display hydrated silica mineralogic signatures that may indicate deposits produced by volcanically driven hydrothermal activity (Skok et al., 2010). Jezero crater, on the western edge of Isidis Planitia, is the planned landing site for the Mars 2020 rover mission. Jezero contains remnants of a river delta and displays minerals likely to have been altered by water.

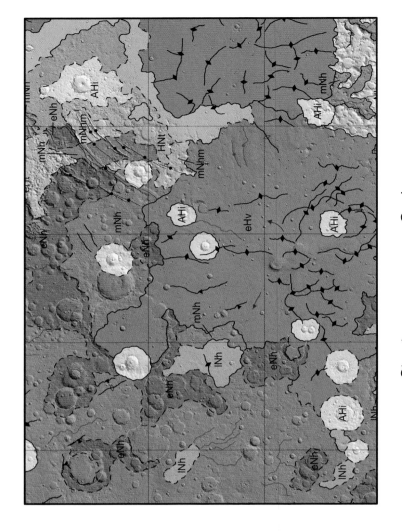

Structure

Interpretation

——	Caldera rim
——	Channel axis
——	Crater rim
•——	Graben axis
▲	Lobate flow
▲	Outflow channel
⊖——	Pit crater chain
◇——	Ridge
•——	Rille
◀——	Scarp
•——	Spiral trough
◆——	Wrinkle ridge
↕	Yardangs

Geology

Impact unit
AHi

Lowland unit
lHl

Volcanic unit
eHv

Transition units
lHt
eHt
HNt

Highland units
mNh
lNh
eNh
mNhm

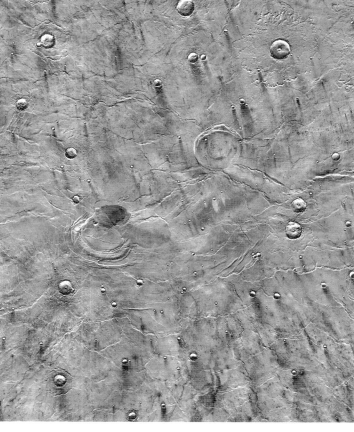

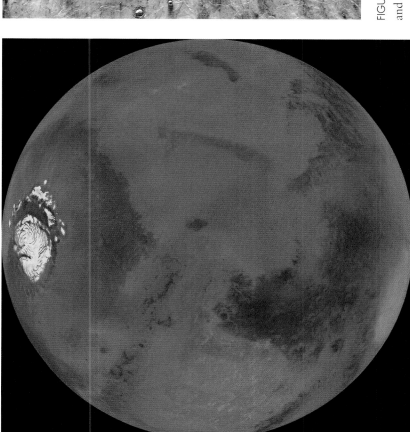

FIGURE 13.A Syrtis Major was originally the name of the prominent dark feature just left of center in this view of Mars. Isidis basin is the lighter, smooth area near the center of the image, while Hellas basin is the lighter area near the bottom left, and the north polar cap is at the top. Because of the contrast with surrounding terrain, Syrtis Major helped early observers measure the rate of rotation of Mars (Mars Digital Image Map, from Viking orbiter imagery, image credit NASA/JPL-Caltech/USGS).

FIGURE 13.B THEMIS daytime infrared mosaic showing the summit of Syrtis Major. Nili and Meroe Paterae (at upper left center and lower right center, respectively) are calderas whose locations align with other features, circumferential to the Isidis basin. Light and dark streaks at small craters show prevailing wind direction (100 m/pixel, view 350 km by 475 km, north at top, 5° N to 11° N, 64° E to 72° E, NASA/JPL-Caltech/Arizona State University).

Volcanoes love (giant impact) basins

The locations of Syrtis Major Planum and – perhaps more specifically – Nili and Meroe Paterae (Figure 13.B) relative to Isidis Planitia are conspicuous. Notice that Nili Fossae and Libya Montes are circumferential to Isidis Planitia. Extend those arcuate features and it becomes clear that they cut through Syrtis Major Planum. This is not a coincidence. Large-scale impacts like the one that formed the Isidis basin were so large and powerful that they pierced deep into the ancient crust, which responded by forming systems of large faults and fractures, concentric to the excavated, inner basin (Hiesinger and Head, 2004). Such pronounced concentric systems of structures identify basins like Isidis on Mars and other

planetary bodies as "multi-ring impact basins." Magma that is generated deep within the interior of the planet can exploit these crustal fractures and ascend to the surface, eventually to form large volcanic centers, after prolonged activity. Compare the location of other large volcanic centers on Mars such as Elysium Mons (MC-14, MC-15) and Malea Planum (MC-28) and notice that they too form at the margins of large multi-ring impact basins (Wichman and Schultz, 1989; Figure 5.11 in Chapter 5 is a global map of impact basins). Thus, at least some of the largest volcanic provinces on Mars have a special geologic affinity to the ring systems of the larger impact basins.

Pool and riffle (and repeat)

Many locations within the Martian highlands are scoured by dense networks of fluvial channels, and

the terrain in Syrtis Major quadrangle is no different (Figure 13.C). With close examination, there is a pattern in the occurrence of these channel networks. Often, they scour highland surfaces and then merge into an irregularly shaped basin that has a comparatively smooth surface. Move downslope and you may see another network of channels, which individually converge into a smooth basin. This shape and association can be thought of as "pool and riffle" morphology. The "pool" refers to the basin into which the channels carry their sediment. When the basin filled up, it was overtopped and the water then migrated to the next depression. This overtop and scour is the "riffle." We see this pattern repeated in multiple locations on the surface of Mars, which is not unlike the shape of terrestrial channels as they course from high-standing mountains to low-standing coastal regions, though on a much smaller scale.

FIGURE 13.C CTX image of Locras Valles. Channels form pools as they run through basins, such as the old crater in the bottom half of the image, before continuing downslope as a "riffle" (image G04_019688_1883, 6 m/pixel, view 28 km by 140 km, north at top, 8.3° N, 47.8° E, NASA/JPL-Caltech/MSSS).

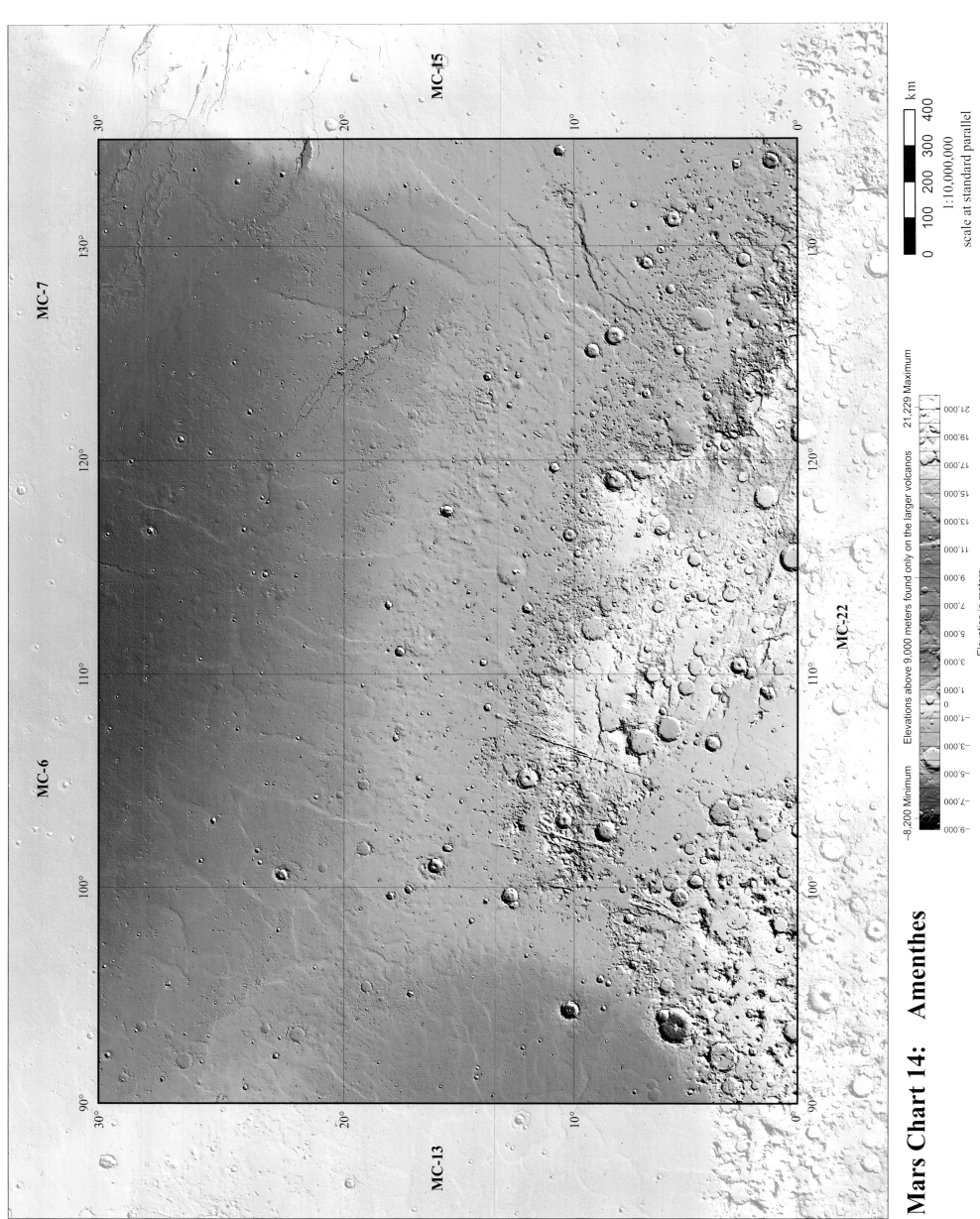

Mars Chart 14:   Amenthes

MC-7

MC-6

MC-15

MC-13

MC-22

1:10,000,000
scale at standard parallel

0   100   200   300   400
km

−8,200 Minimum   Elevations above 9,000 meters found only on the larger volcanos   21,229 Maximum

−9,000 −7,000 −5,000 −3,000 −1,000 0 1,000 3,000 5,000 7,000 9,000 11,000 13,000 15,000 17,000 19,000 21,000

Elevation in meters

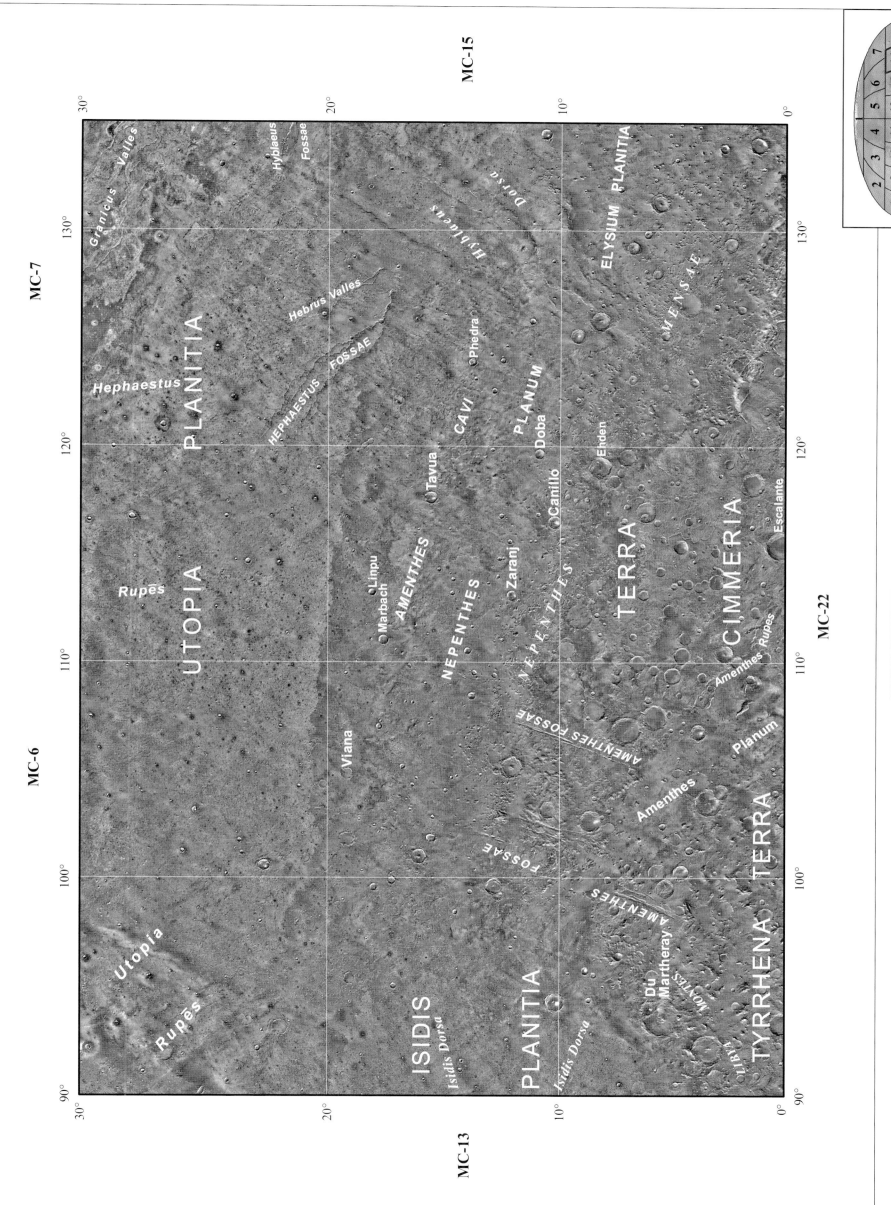

MC-7

MC-15

MC-6

MC-22

MC-13

30°

20°

10°

0°

30°

20°

10°

0°

130°

120°

110°

100°

90°

130°

120°

110°

100°

90°

Granicus Valles

Hyblaeus Fossae

Hebrus Valles

HEPHAESTUS FOSSAE

Hyblaeus Dorsa

Hephaestus

PLANITIA

UTOPIA

Rupes

Utopia

Rupēs

ISIDIS

PLANITIA

Isidis Dorsa

Isidis Dorsa

Viana

Marbach

Linpu

AMENTHES

NEPENTHES

NEPENTHES

NEPENTHES FOSSAE

FOSSAE

AMENTHES

Du
Martheray

LIBYA MONTES

TYRRHENA TERRA

Tavua

CAVI

PLANUM

Phedra

Doba

Zaranj

Canillo

Ehden

TERRA

CIMMERIA

Amenthes Rupes

Planum

Amenthes

ELYSIUM PLANITIA

MENSAE

Escalante

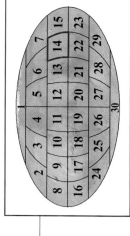

km

0   100 200 300 400

1:10,000,000

scale at standard parallel

Projection: Mercator
Datum: Mars 2000 Sphere
Central Meridian: 0.0000
Standard Parallel 1: 0.0000

Amenthes (MC-14)

Geography

The Amenthes quadrangle contains parts of the Martian southern highlands and northern lowlands, as well as the transition between the two. In the southern part of the quadrangle, Cimmeria and Tyrrhena Terrae form rugged, cratered plateaus as high as 1 km above datum, which are gouged by the long, linear depressions of Amenthes Fossae. Highland terrae in this quad are separated by Amenthes Planum, an elongate, topographic basin that is located as much as 1 km below the highlands. The northern part of the Amenthes quadrangle consists of southern Utopia Planitia, over 4 km below datum. The western part of the quadrangle is made up of eastern Isidis Planitia, which is nearly as low in elevation. Both Utopia and Isidis Planitiae – which are centered outside of the Amenthes quadrangle – are plains of sediments that fill very ancient impact basins. Various scarps and depressions mark the surface of these lowland planitiae. From south to north, the highland–lowland boundary is defined by distributed fields of knobs and intervening plains of Nepenthes Mensae, rolling plains of Nepenthes Planum, and isolated and coalesced depressions of Amenthes Cavi.

Geology

The terrains exposed in the Amenthes quadrangle exemplify the morphologic differences between young and old surfaces. To the south, surfaces are higher-standing, rugged, and contain an abundance of impact craters, pointing toward a long history of material erosion. In contrast, to the north, surfaces are lower-standing, smooth, and contain a relative paucity of impact craters, pointing toward a history of material deposition. The quadrangle also contains the margins of two giant basins that were formed by impacts so large that the crust of the planet was shredded by faults that circle such basins. Though the impacts were ancient, affecting Mars when it was a planetary infant, the scars are still visible as elevated mountains (Libya Montes) and circumferential troughs (Amenthes Fossae), each of which form outer rings of the gigantic Isidis and Utopia impact basins. The highland–lowland boundary in this quadrangle may be related to the annular troughs of these ancient basins, wherein sediment and, likely, groundwater accumulated to form isolated aquifers. Some curious surface features in these basin zones may have resulted from deformation and expulsion of ice- and water-rich sediment.

Eruptions of curious origin

The plain in the Amenthes quadrangle is quite different from most other highland–lowland boundary plains of Mars. Its low latitude suggests the region was amenable to the past presence of liquid water at or near the planetary surface. The Amenthes quadrangle is also home to two populations of cone-shaped features that formed from the eruption of subsurface material. The first population occurs in a roughly east–west band in the center of Nepenthes Planum. Therein, cones several kilometers in diameter, with large central depressions, emanate lobate flows with rugged surfaces. The cones extend across and partly form the highland–lowland boundary plain (Figure 14.A). The features look similar to terrestrial volcanoes (Broz and Hauber, 2013), though the extended distance from volcanic sources and their occurrence above sedimentary sequences open the possibility

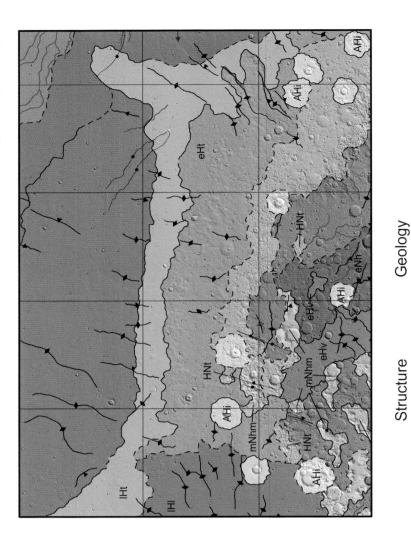

Structure

Interpretation
— Caldera rim
— Channel axis
— Crater rim
●— Graben axis
▲ Lobate flow
▲ Outflow channel
◇— Pit crater chain
●— Ridge
|— Rille
◀— Scarp
●— Spiral trough
◆— Wrinkle ridge
↕ Yardangs

Geology

Impact unit
AHi

Lowland unit
lHl

Volcanic units
eHv
AHv
Av

Transition units
lHt
eHt
HNt

Highland units
mNh
eNh
mNhm

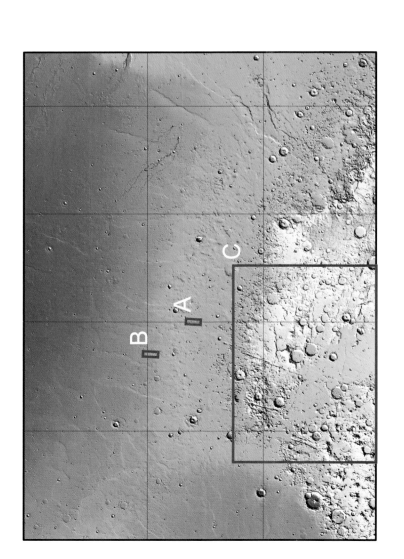

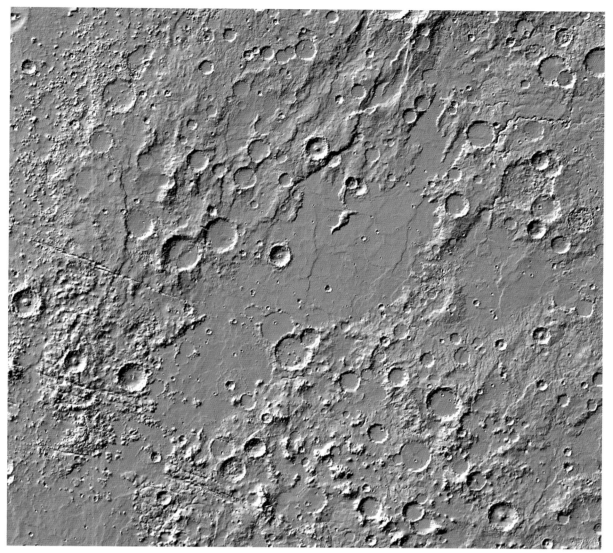

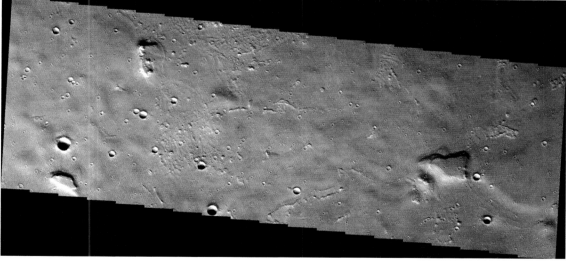

FIGURE 14.B Small-diameter cone-shaped features (smallest features visible) occur in clusters and linear alignments at the southern margin of Utopia Planitia (THEMIS visible image V27678030, 18 m/pixel, view 18 km by 50 km, north at top, centered at 19.7° N, 117° E, NASA/JPL-Caltech/Arizona State University).

also discussion in MC-4 and additional examples in Figure 4.D). Though characteristics indicate that the eruptive source is close to the surface, perhaps formed through the de-gassing of near-surface sediments, their origin remains uncertain.

Down in the valley

Amenthes Planum extends over 1,000 km, from Hesperia Planum south of the map area to Isidis Planitia, and has served as a conduit for fluvial

FIGURE 14.A Examples of large, cone-shaped landforms that emanate rugged, lobate flows. Such emanations make up at least part of the highland–lowland boundary plain (THEMIS visible image V14113011, 18 m/pixel, view 18 km by 55 km, north at top, centered at 16.3° N, 110° E, NASA/JPL-Caltech/Arizona State University).

that the features resulted from the eruption of sub-surface mud and gas (Skinner and Tanaka, 2007).

The second population of cone-shaped landforms occurs in the lowland plains at the southern margin of Utopia Planitia. These features have much smaller diameters (a few hundred meters) than the first population, have small central depressions, and occur in lineated or arcuate groups (Figure 14.B; see

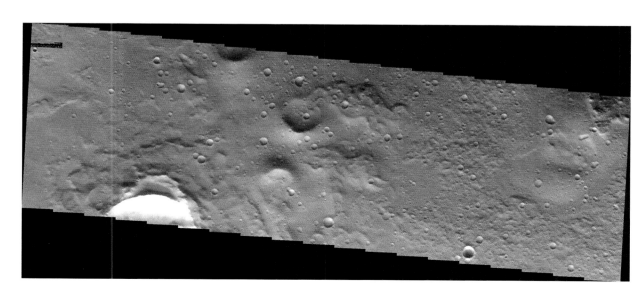

FIGURE 14.C Amenthes Planum crosses the center of this view, which extends south into MC-22. At the lower right, a sinuous channel enters Palos crater from the south (see Figure 22.C; MOLA hillshade image, 1,100 km by 1,200 km, north at top, 7° S to 13° N, 97° E to 115° E).

Planum, and then to Isidis Planitia. Overlapping the fluvial activity in time were Hesperian-age volcanic flows, perhaps originating in Hesperia Planum (MC-22), which formed ridged plains in Amenthes Planum. The faults of Amenthes Fossae were reactivated as extensional valleys (grabens), no later than the Early Hesperian Epoch (Erkeling et al., 2011). The steep slopes of Amenthes Rupes have been interpreted as either extensional or contractional faults (Schultz, 2003; Watters, 2003; Capprarelli et al., 2007). Small valleys cut into the Amenthes ridged plains during the Amazonian Period.

and volcanic material moving northwest (down-slope) since the Noachian Period (Erkeling et al., 2011). Amenthes Planum occupies a trough that formed across older highland terrain, at or soon after the impact that formed the Isidis basin to the west (Figure 14.C). The faults responsible for Amenthes Fossae are probably also related to the impact formation of one or both of the Isidis and Utopia basins. To the south of the map area (see MC-22, Tyrrhenum) dendritic valley networks and distinctly younger sinuous channels directed fluvial material through Palos crater, into Amenthes

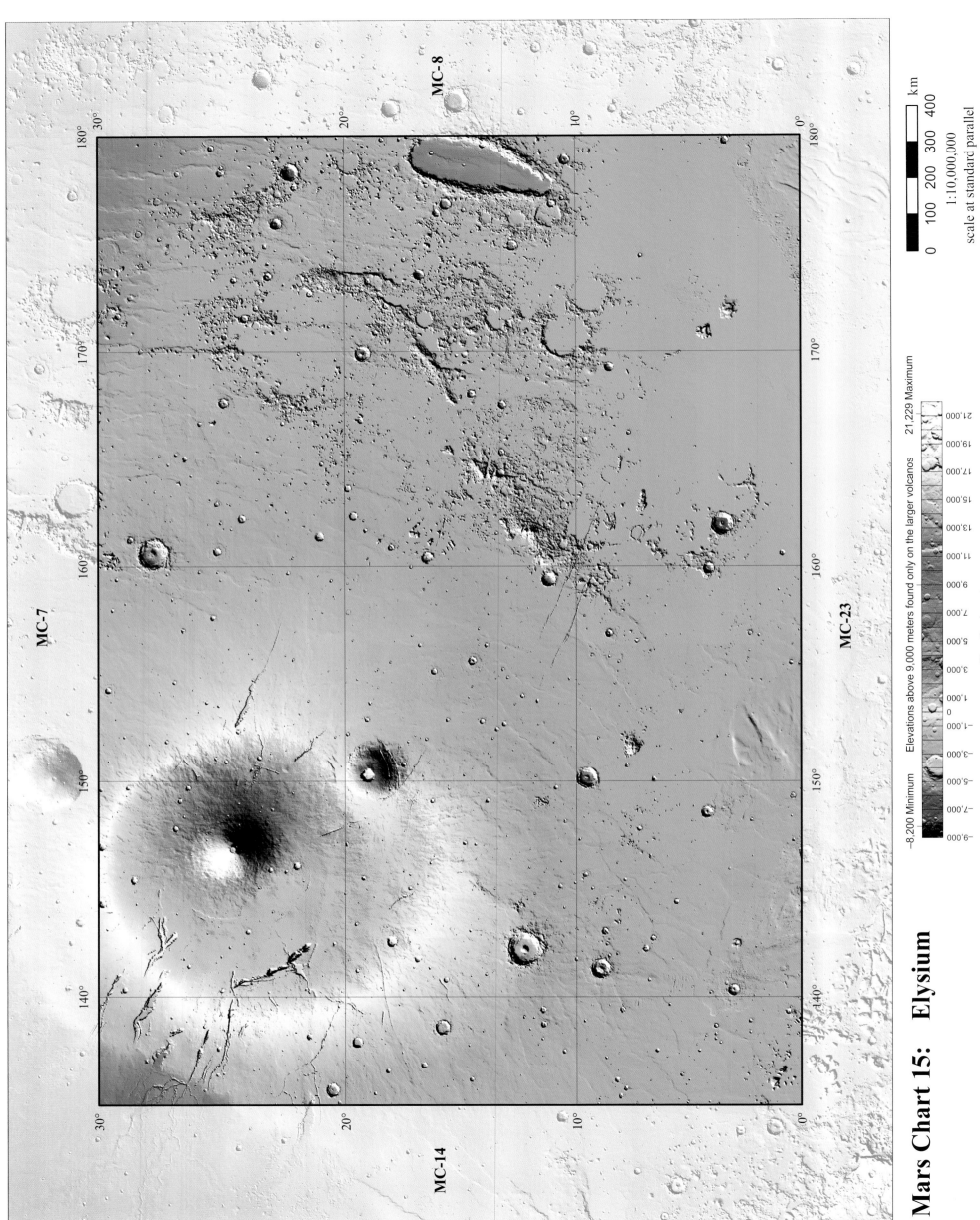

Mars Chart 15: Elysium

MC-7

MC-8

MC-14

MC-23

180° 170° 160° 150° 140°

30°

20°

10°

0°

180°

170°

160°

150°

140°

0°

10°

20°

30°

1:10,000,000
scale at standard parallel

km

0   100   200   300   400

−8,200 Minimum   Elevations above 9,000 meters found only on the larger volcanos   21,229 Maximum

−9,000 −7,000 −5,000 −3,000 −1,000 0 1,000 3,000 5,000 7,000 9,000 11,000 13,000 15,000 17,000 19,000 21,000

Elevation in meters

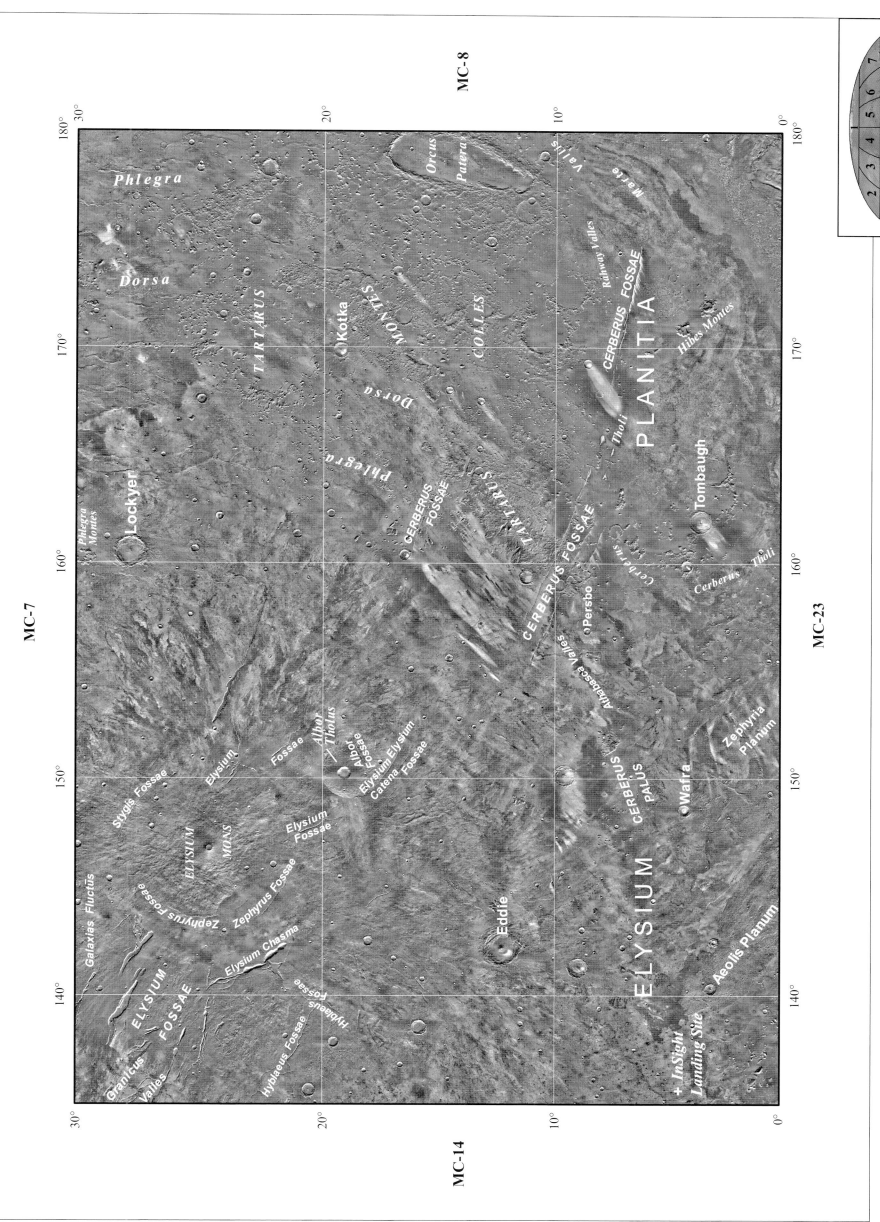

MC-7

MC-8

Phlegra

Dorsa

30°

180°

Orcus
Patera

MC-8

Marte
Vallis

20°

Rahway Valles

CERBERUS FOSSAE

170°

T A R T A R U S

Kotka

M O N T E S

PLANITIA

Hibes Montes

Dorsa

Lockyer

Phlegra

COLLES

Tholi

Phlegra Montes

160°

CERBERUS
FOSSAE

T A R T A R U S

Tombaugh

Cerberus Tholi

Stygis Fossae

150°

CERBERUS FOSSAE

Persbo

MC-7

MC-23

Elysium

Fossae

Albor
Tholus

Athabasca Valles

1:10,000,000
scale at standard parallel

Galaxias Fluctūs

ELYSIUM

MONS

Zephyrus Fossae

Albor
Fossae

Elysium
Catena

Elysium
Fossae

CERBERUS
PALUS

Wafra

150°

Zephyria
Planum

Elysium Chasma

Eddie

km

0   100 200 300 400

ELYSIUM

140°

ELYSIUM

Hyblaeus Fossae

ELYSIUM
FOSSAE

Hyblaeus Fossae

+ InSight
Landing Site

Aeolis Planum

140°

30°

20°

MC-14

10°

0°

MC-23

1				
2	5	6	7	
3	4	11	12 13 14 15
	10	18	19 20 21 22 23
8 9	17		26 27 28 29
16		24 25	30	

N

W ✦ E

S

Projection: Mercator
Datum: Mars 2000 Sphere
Central Meridian: 0.0000
Standard Parallel 1: 0.0000

Elysium (MC-15)

Geography

Elysium Mons rises 14 km above the surrounding plains, while nearby Albor Tholus is 4 km high. Much of the quadrangle consists of plains near datum to −3,000 m, but in the east, Tartarus Montes, Tartarus Colles, and the rimmed depression, Orcus Patera, constitute a more rugged region, largely made up of knobs and low plateaus and ridges that separate Elysium Planitia from Amazonis Planitia to the east (MC-8). To the south of Orcus, Marte Vallis extends from Elysium Planitia into the Amazonis basin. Elysium Planitia includes the landing site of the InSight mission, which is exploring the interior of Mars using geophysical measurements.

Geology

The Elysium rise has contributed significantly to the regional geologic record in the eastern hemisphere of Mars. Magmatic activity associated with the growth of the Elysium rise dates back at least to the Hesperian Period, according to crater statistics. This activity led to the construction of the prominent Elysium Mons and Arbor Tholus shield volcanoes, and vast aprons of sheet lavas surrounding them to form a broad rise. The rise shows subsequent sculpturing by tectonism and erosion. Processes included the formation of both radial and concentric fault and ridge systems as well as pit crater chains. The development of structurally controlled canyons was especially prevalent along the northwestern flank. Lahar-like flow materials partly drape the northwestern flank and extend

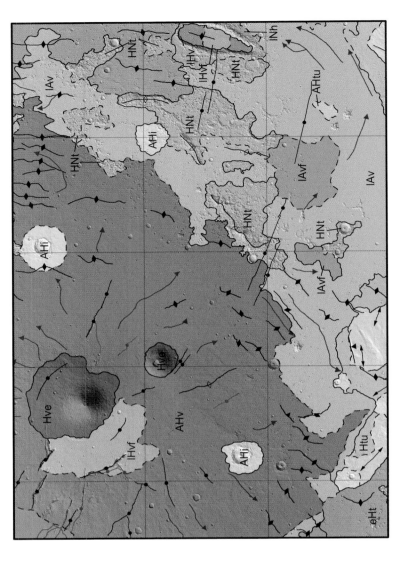

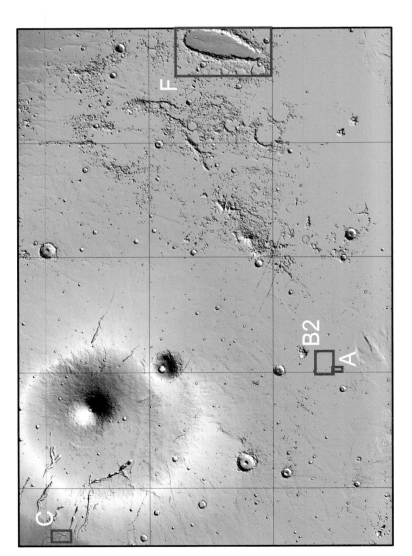

Structure

Interpretation

——	Caldera rim
—	Channel axis
	Crater rim
●—	Graben axis
▲	Lobate flow
↑	Outflow channel
○—	Pit crater chain
◇—	Ridge
●—	Rille
◀—	Scarp
●—	Spiral trough
◆—	Wrinkle ridge
↕ ↑	Yardangs

Geology

Impact unit

- AHi

Volcanic units

- IHvf
- AHv
- IAvf
- IAv
- Hve
- Av
- IHv

Transition units

- IHt
- eHt
- AHtu
- HNt
- Htu

Highland unit

- INh

Highland unit

- INh

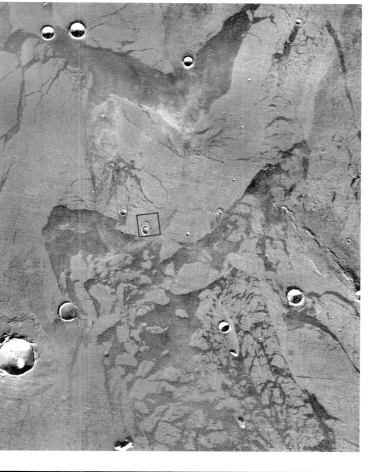

FIGURE 15.B2 A THEMIS daytime infrared mosaic of the region of Figure 15.B1 shows the importance of resolution in interpreting geology. The crater (2 km in diameter) is just visible, but only the HiRISE image suggests that the material next to the crater is not impact ejecta. Box shows location of Figure 15.B1 (100 m/pixel, view 100 km by 125 km, north at top, 5.0° N to 6.7° N, 149.7° E to 151.8° E, NASA/JPL-Caltech/Arizona State University).

MRO/HiRISE

ESP_018537_1860_RED

NASA/JPL/University of Arizona

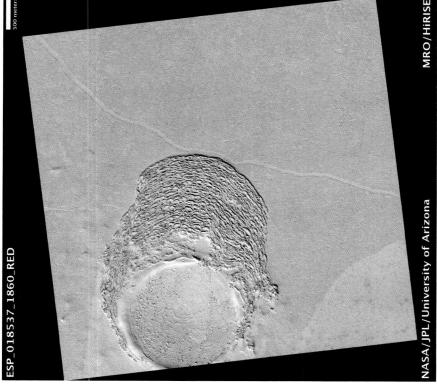

FIGURE 15.B1 In Cerberus Palus, west of Athabasca Valles, material flowing east to west, and interpreted as lava, has piled up against a 2-km-diameter crater (HiRISE image ESP_018537_1860, 28 cm/pixel, 6 km by 6 km, north at top, 5.98° N, 150.74° E, NASA/JPL-Caltech/University of Arizona).

Formation materials (see MC-23). Magmatic, tectonic, and hydrologic activity continued into the most recent Late Amazonian Epoch. One example of such youthful activity is the Athabasca Valles system of channels and their infilling lava flows, which source from Cerberus Fossae. Also young are scattered low, shield-like constructs, typically tens of kilometers across, in the central and eastern parts of Elysium Planitia. Crater statistics date much of these features to within the last few hundred million years (Vaucher et al., 2009).

Rotations and ridges near the equator

Athabasca Valles originated from the faults and fractures of Cerberus Fossae, near the southern margin of the Elysium rise, possibly within the last few tens of millions of years (Plescia, 1990; Murray et al., 2005). Athabasca Valles exhibits features indicating erosion by moving water, with subsequent partial infill through the emplacement of lava flows, similar to several other outflow channels on Mars. The scarcity of impact craters on the lava flows indicates that they are quite recent in Mars history. Spiral patterns, or coils with widths ranging from 5 to 30 m, occur among polygonal-patterned ground in the plains of Cerberus Palus (Figure 15.A). Similar features form on Earth when lava flows cool and form solid surfaces above flowing lava, creating plates that break apart as the flow moves. The troughs between the plates develop polygonal texture as they cool and gas escapes through cracks at the edges of the polygons.

As the plates move past one another, the intervening troughs shear and the polygons rotate, forming the coils (Ryan and Christensen, 2012).

Also in this region is an old impact crater, where lava moving from right to left (east to west) piled up against the crater rim (Figures 15.B1 and 15.B2). The lava spilled over the eastern rim and partly filled the crater.

Troughs, valleys, and mud on Elysium Mons

The Elysium Fossae are troughs or rilles on the northwestern and southeastern slopes of Elysium Mons, interpreted to be grabens or fault-

FIGURE 15.A Part of HiRISE image of Athabasca Valles (ESP_028084_1845_RED, 50 cm/pixel, 170 m by 390 m, north at top, 4.53° N, 150.25° E, NASA/JPL-Caltech/University of Arizona). Note coil forms on surface, suggesting rotation (about a vertical axis) of material between surface blocks.

some 2,000 km into the Utopia basin (MC-7) to the northwest of the quadrangle. Elongated promontories, visible along the southwestern margin of the quadrangle, named Aeolis Planum and Zephyria Planum, are partly composed of Medusae Fossae

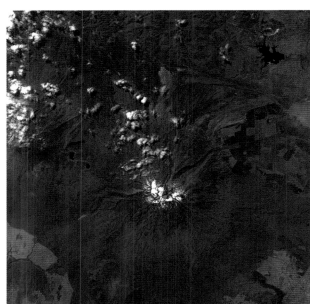

FIGURE 15.D Lahar flow on Earth. Before (left, February 2002) and after (right, March 2007) images of a Mount Ruapehu, New Zealand, showing a new lahar (channel fill in gray, initially extending eastward from summit, formed 7 days before the right image was taken) on the flank (view about 35 km across, north at top, Terra-ASTER images, NASA).

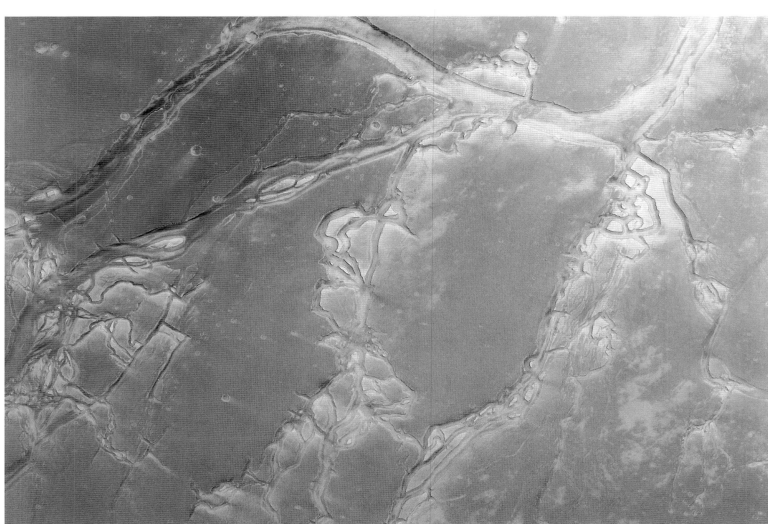

FIGURE 15.C The head of the Granicus Valles system of channels is in the southeast corner of this HRSC image, at the mouth of a tectonic trough. Different channels likely were active at different times, and some formed and became active as volcanic flows caused local melting. The largest channels in the image are several kilometers wide and several hundred meters deep (HRSC press release 272, orbit 1383, 24 m/pixel, view 55 km by 80 km, rotated so north is at top, 27° N, 136° E, ESA/DLR/FU-Berlin).

controlled erosional valleys. At the downslope end of one trough are the intersecting valleys of Granicus Valles (Figure 15.C). Activity at the nearby volcano included the interaction of injected magma, water (as ice and possibly groundwater), and host-rock materials. When these water-charged lavas and rock materials reached the end of the grabens, they deposited broad fans of sediment dissected by distributary channels. While the number of flow events and relative importance of water and ice is unclear (Nussbaumer, 2008), the resulting landforms are similar to those from volcanic mudflows, or lahars, on Earth (Figure 15.D). Maps of elemental abundances, such as for potassium (Boynton *et al.*, 2007, paragraph 56; Figure 15.E), suggest that the resting place for the flows is in the Utopia basin (Christiansen, 1989; see MC-7).

Orcus Patera: mysterious cavity

Orcus Patera, at the eastern edge of the quadrangle, is an elongate, Late Noachian depression with a relatively flat floor, flooded by younger, Late Hesperian lavas, and a raised rim, consistent with an impact origin (Figure 15.F; van der Kolk *et al.*, 2001). Its elongate outline (more than three times as long as it is wide), however, likely requires a very shallow angle of impact. Explaining Orcus Patera as a volcanic feature has also been proposed, but this is problematic when matching its morphology to known examples. Contraction of the crust between the Elysium and Amazonis Planitiae has also been suggested to account for the shape. East–west-trending grabens, cutting the patera floor, apparently were activated by and were conduits for the later volcanic activity.

Potassium

Low K   High K

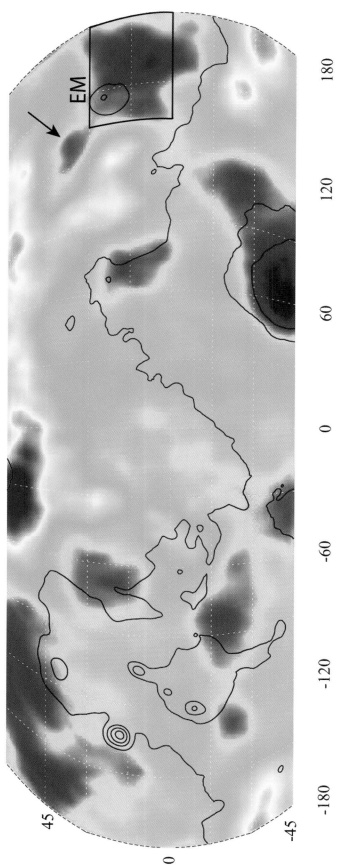

FIGURE 15.E (top) Global map of potassium (K) with contours of constant surface elevation. Location of Elysium Mons (EM) within MC-15 is shown. Note that the low-potassium region, northwest of Elysium Mons (at arrow), is consistent with the lahar deposits having an origin at the volcano, which also shows low potassium (NASA/JPL-Caltech/University of Arizona).

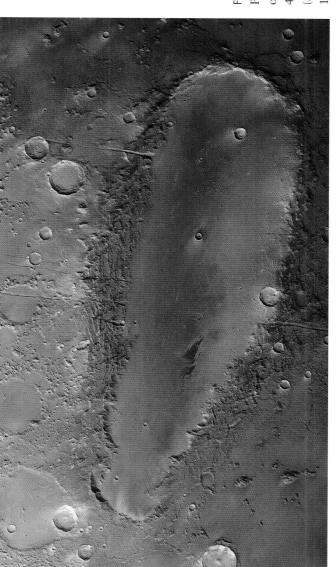

FIGURE 15.F (left) HRSC mosaic of part of Orcus Patera, a 380 km by 140 km depression of uncertain origin. Relative to the surrounding plain, the floor is 400–600 m lower while the rim rises as much as 1800 m (30 m/pixel, view 260 km by 450 km, north to right, 14° N, 178° E, ESA/DLR/FU-Berlin).

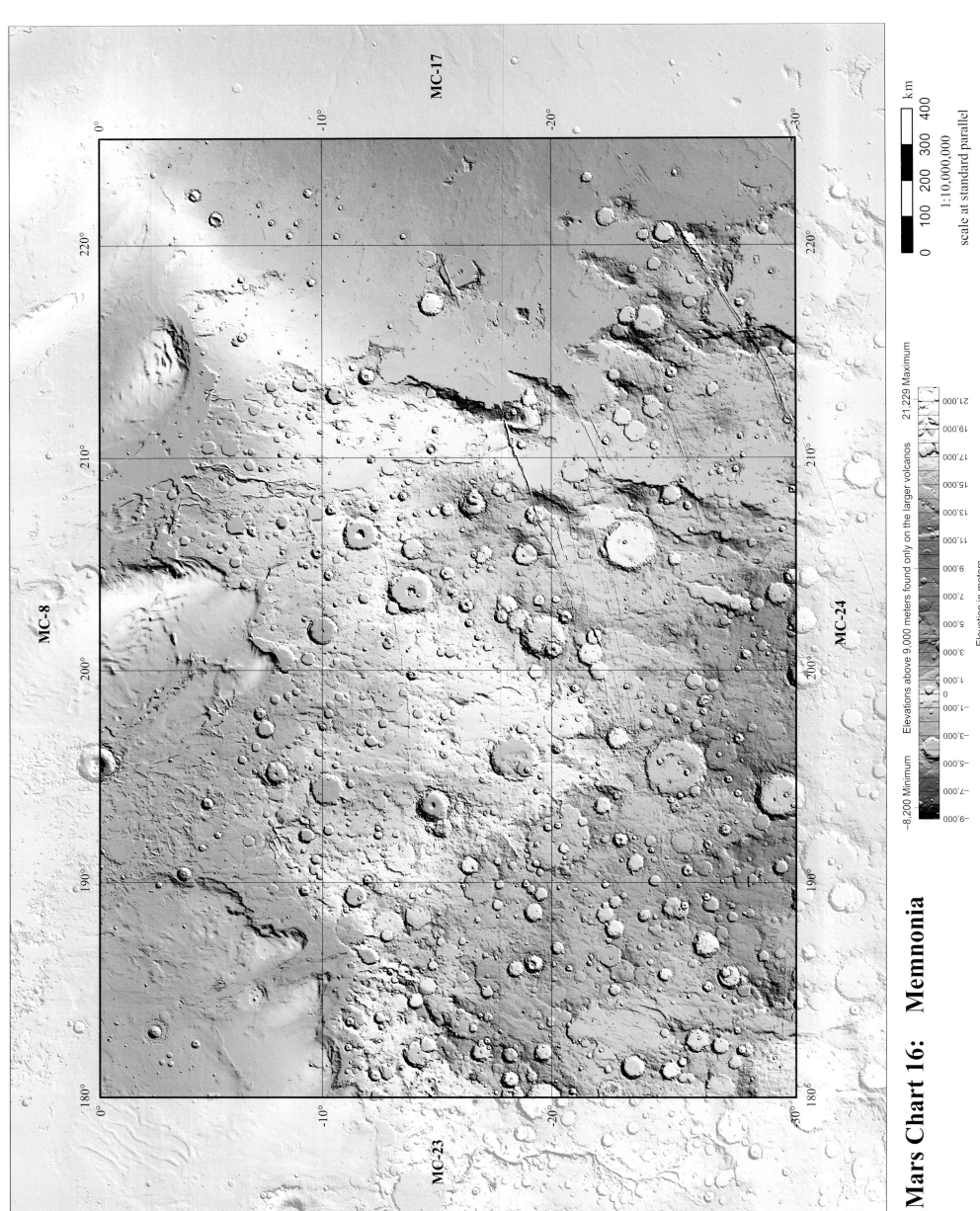

Mars Chart 16: Memnonia

−8,200 Minimum Elevations above 9,000 meters found only on the larger volcanos 21,229 Maximum

1:10,000,000
scale at standard parallel

Elevation in meters

km
0 100 200 300 400

MC-17

MC-8

MC-24

MC-23

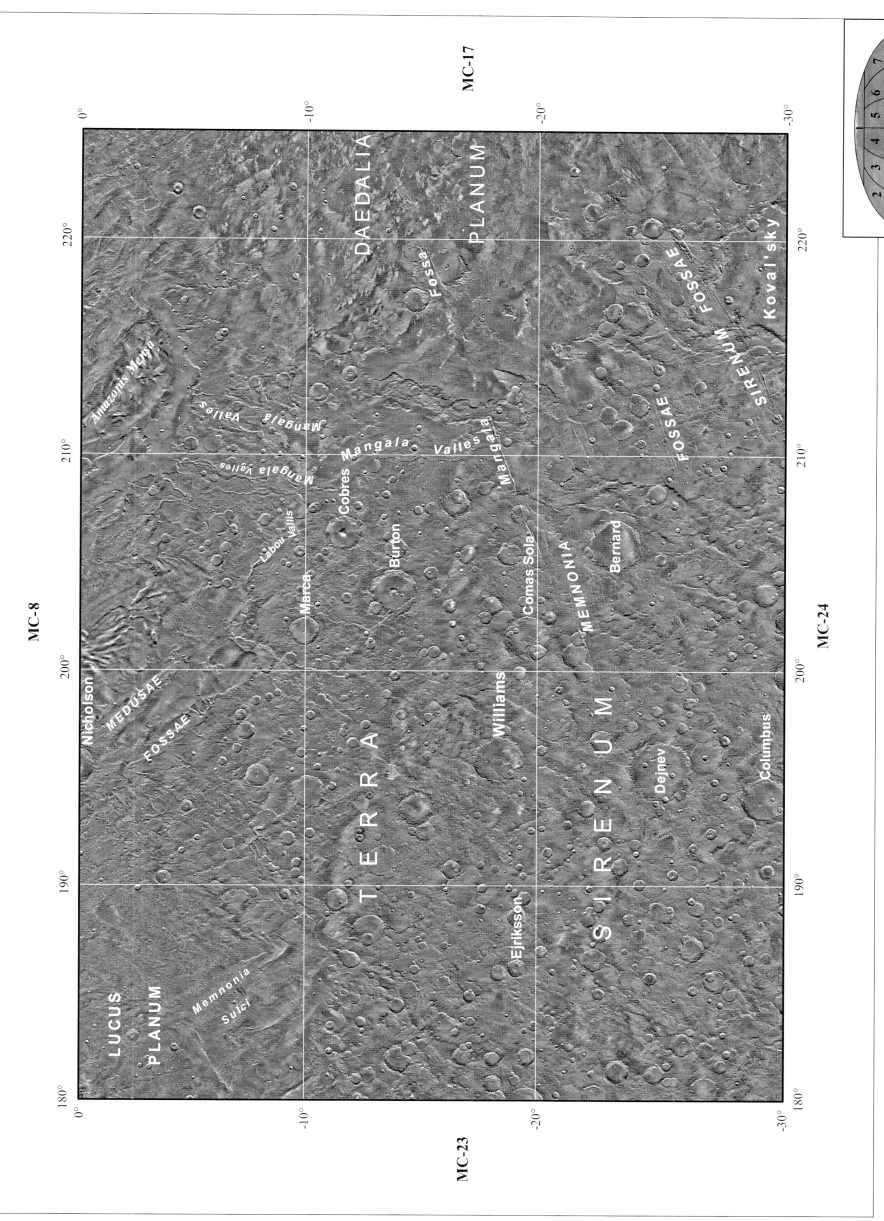

MC-17

MC-8

MC-23

MC-24

0°
-10°
-20°
-30°

180°
190°
200°
210°
220°
0°

LUCUS
PLANUM

Memnonia
Sulci

Nicholson

MEDUSAE

FOSSAE

Amazonis Mensa

Labou Vallis

Mangala Valles

Mangala Valles

Cobres

Marca

Burton

Mangala Valles

Mangala

Fossa

DAEDALIA

PLANUM

T E R R A

Ejriksson

Williams

Comas Sola

MEMNONIA

Bernard

FOSSAE

SIRENUM FOSSAE

Koval·sky

S I R E N U M

Dejnev

Columbus

Projection: Mercator
Datum: Mars 2000 Sphere
Central Meridian: 0.0000
Standard Parallel 1: 0.0000

km
0   100 200 300 400
1:10,000,000
scale at standard parallel

Memnonia (MC-16)

Geography

Memnonia occurs on the western margin of the Tharsis volcanic province and on the southern margin of Amazonis Planitia. Its elevation ranges from a few kilometers above datum in Terra Sirenum to a few kilometers below datum along the northern edge of the map sheet. The cratered and faulted terrain of Terra Sirenum gives way to Tharsis lava flows to the east within Daedalia Planum. Valleys and channels mark the northern margin of the highlands, where they grade into low regions interrupted by elongate plateaus like Amazonis Mensa and the broader Lucus Planum.

Geology

Faults of the Memnonia Fossae and Sirenum Fossae trend east-northeast, radial to the Tharsis volcanic province. They cut diverse rock materials of varying ages, ranging from the Noachian to the Amazonian (Anderson and Dohm, 2011;

Anderson *et al.*, 2012). The eastern margin of Terra Sirenum includes very large and ancient tectonic structures, several thousands of kilometers in length, including north-trending, structurally controlled basins. These features are among those cut by the younger fossae. The ancient, highly cratered and deformed terrain of Terra Sirenum records distinct magnetic signatures (Anderson and Dohm, 2011; Anderson *et al.*, 2012; Karasozen *et al.*, 2012; Figure 3.5 in Chapter 3). This ancient terrain contrasts with the younger, onlapping Tharsis lava flows of Daedalia Planum. The wind-etched materials of the Medusae Fossae Formation (Figure 16.A; see MC-23) form a broad but discontinuous east-west band along the highland–lowland boundary (e.g. Scott and Tanaka, 1986; Greeley and Guest, 1987; Scott and Chapman, 1991).

Tharsis ridge belt and structurally controlled basins

Some of the large, north-trending structures in eastern Terra Sirenum could be markers of the

these structures reflect regional extension across the narrow mountain ranges. Farther east in the Tharsis ridge belt, however, structures suggest the opposite kind of distortion: shortening across the belt (MC-18, MC-25). If different parts of the Tharsis ridge belt share a common history, this history is likely to be complex and/or to have multiple phases.

early development of Tharsis, also identified as the Tharsis ridge belt (see Figure 4.2; Schultz and Tanaka, 1994; Karasozen *et al.*, 2012), and/or they may represent pre-Tharsis activity (Anderson *et al.*, 2012). The systems of basins are reminiscent of basin and range topography of the southwest United States (Figures 16.B1 and 16.B2; Anderson *et al.*, 2012; Karasozen *et al.*, 2012). On Earth,

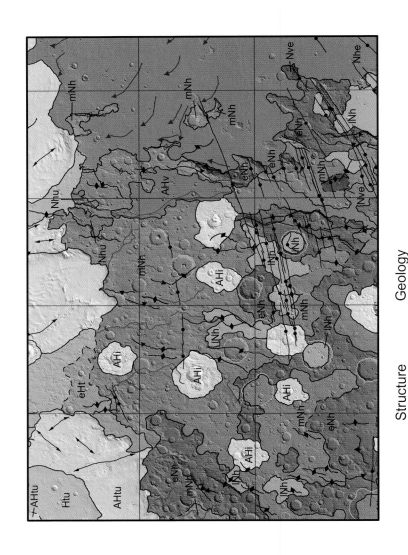

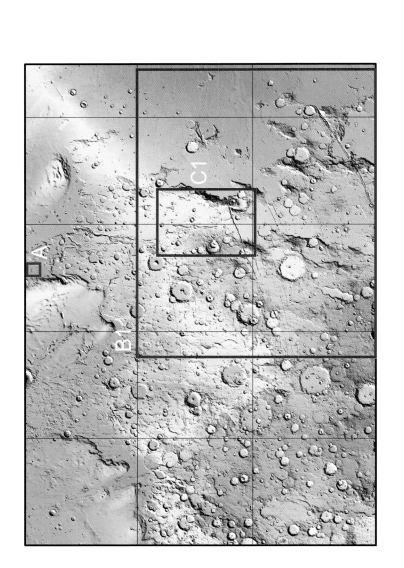

Structure

Interpretation

	Caldera rim
	Channel axis
	Crater rim
	Graben axis
	Lobate flow
	Outflow channel
	Pit crater chain
	Ridge
	Rille
	Scarp
	Spiral trough
	Wrinkle ridge
	Yardangs

Geology

Impact unit
- AHi

Volcanic units
- eHv
- Nve
- AHv

Transition units
- lHt
- eHt
- AHtu
- HNt
- Htu

Highland units
- mNh
- lNh
- eNh
- Nhe
- Nhu

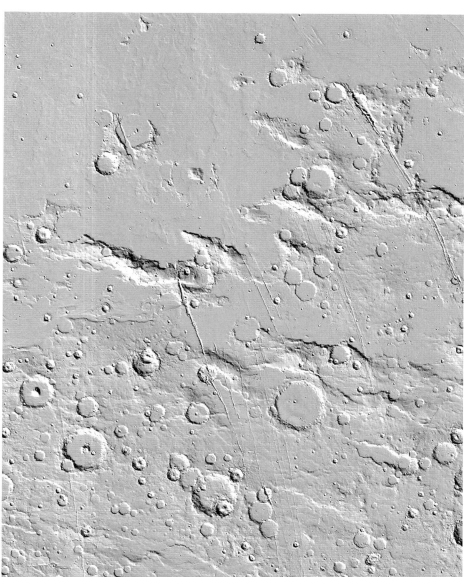

FIGURE 16.A Perspective view to the south–southeast across the sculpted deposits of the Medusae Fossae Formation (MC-23). Wind and/or water have modified friable materials that may be mostly volcanic ash deposits (HRSC press release 423, orbit 5114, 13 m/pixel, view spans about 50 km, centered at 206° E, 1° S, ESA/DLR/FU-Berlin).

Mangala Valles

An interplay among Tharsis igneous activity, large faults, and water-enriched, structurally controlled basins is particularly evident through Mangala Valles (Anderson et al., 2012). An outflow channel system, Mangala Valles originates within a distinct, north-trending, structurally controlled basin (Anderson et al., 2012) at the intersection of the

basin with one of the northeast-trending faults of Medusae Fossae, named Mangala Fossa (Figures 16.C1, 16.C2). The outflow channel system extends north to Amazonis Planitia and the northern plains, nearly 900 km from the headwaters at Mangala Fossa. Magma was injected along a fault in Mangala Fossa. The heating released floodwaters from the subsurface, which flowed through

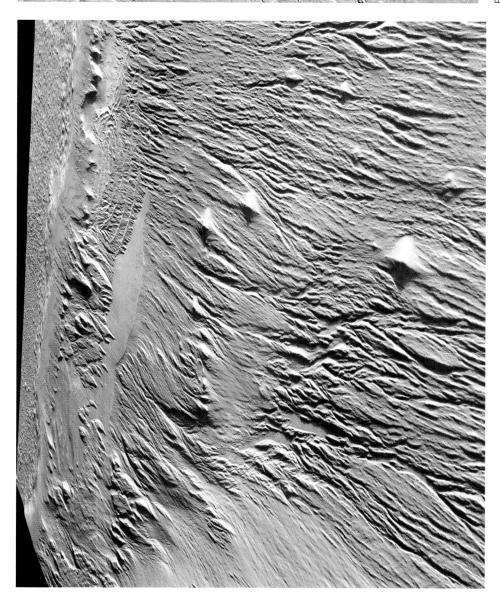

FIGURE 16.B1 MOLA gray hillshade image of Terra Sirenum (view about 1200 km by 1500 km; north at top).

the north-trending, structurally controlled basin and created Mangala Valles. The distinct outflow system had more than one flood episode, ranging possibly from the Early Hesperian Epoch to the Amazonian Period, resulting in valleys, streamlined islands, and terraces (Figures 16.C2–16.C4) (Chapman and Tanaka, 1993; Craddock and Greeley, 1994; Zimbelman et al., 1994). The later Amazonian floods have been interpreted as lava flows rather than water floods (Leverington, 2007; Basilevsky et al., 2009; McEwen et al., 2012).

Alternatively, pits that resemble thermokarst, in smooth deposits in the channel of upper Mangala Valles, suggest a sediment-rich ice cap that formed as floodwaters receded (Figure 16.C5; Levy and Head, 2005). The interplay among tectonic structures such as faults (as a conduit for magma and water and other volatiles), heat (from the magma), and water (groundwater and possibly ice) is of interest to both geologists and biologists, as it often results in interesting landforms, minerals, and prime habitats for life here on Earth.

153

The Atlas of Mars

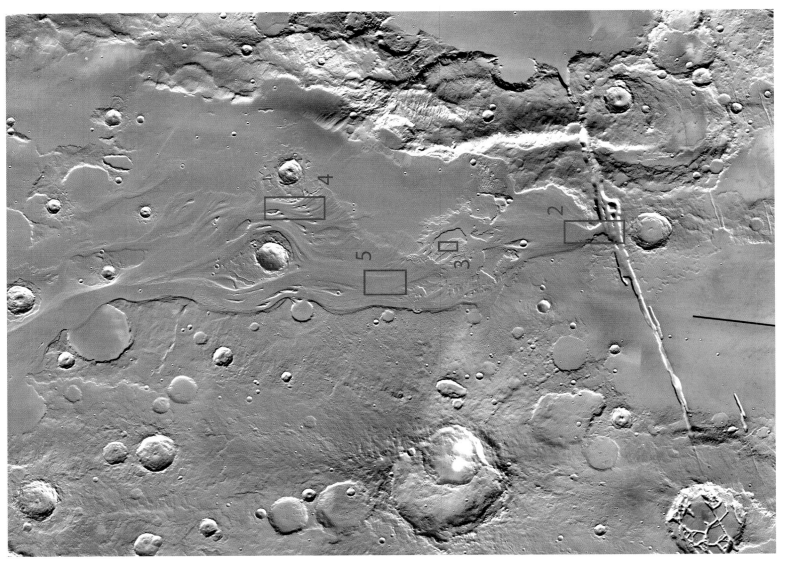

FIGURE 16.C1 This THEMIS daytime infrared mosaic captures the narrow, linear fault valley, Mangala Fossa, as it slashes through a north-trending, structurally controlled valley. The valley displays smooth plains materials along its floor and tectonically derived ranges along its margins. Mangala Valles originate at a breach in Mangala Fossa and extend to the north. Boxes show locations of Figures 16.C2–16.C5 (100 m/pixel, view 348 km by 474 km, north at top, 12° S to 20° S, 207° E to 213° E, images and captions NASA/JPL-Caltech/Arizona State University).

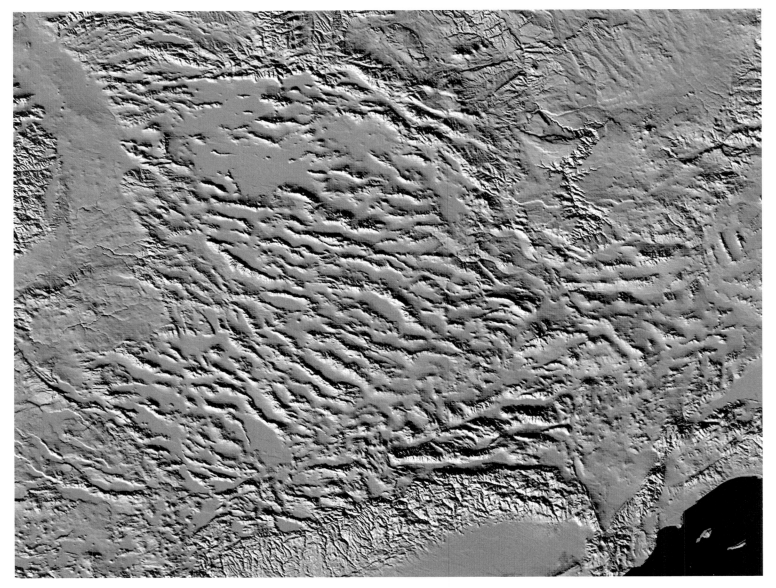

FIGURE 16.B2 Shaded relief image of the basin and range terrain of the western USA, for comparison with Terra Sirenum. Both images extend about 1,200 km north to south. The form, if not the exact scale, of the ranges and basins in the center of the image is reminiscent of those in Terra Sirenum. (Shaded relief data for USA from Thelin and Pike, 1991; north at top, courtesy US Geological Survey.)

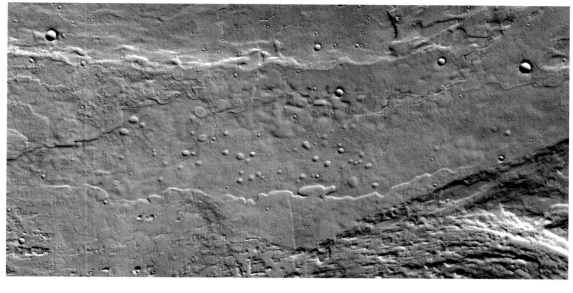

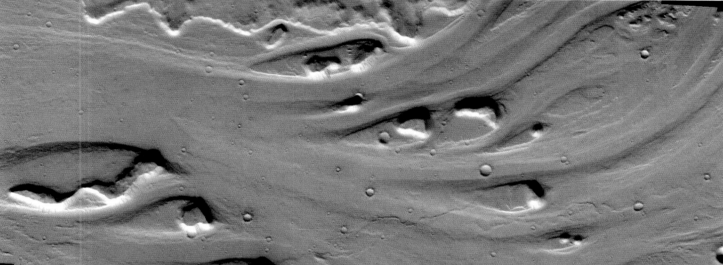

FIGURE 16.C5 Overlying the scoured channel floor is a smooth deposit (center of image) with many circular pits. The deposit is similar in age to the other channel features (Late Hesperian or Early Amazonian; Levy and Head, 2005) and may represent the remains of a sediment-rich ice cap on the flowing water. After the flow ended, ice in the deposit may have sublimated, leaving the pits and arcuate margins. This scenario is consistent with conditions (cold and dry) currently found on Mars. View 13 km by 25 km, from THEMIS visible image V04400003 (17 m/pixel, NASA/JPL-Caltech/Arizona State University).

FIGURE 16.C4 Constricted as it flowed between two impact craters (outside of the scene), the Mangala flood scoured the channel to a greater depth and carved hills and knobs into graceful streamlined shapes. This scene, 14 km wide, is part of THEMIS visible image V18628003 (17 m/pixel, NASA/JPL-Caltech/Arizona State University).

155

FIGURE 16.C3 Downstream, along the valley floor, the floods could have thinned the material covering an aquifer, releasing additional subsurface waters that joined the main flood. Left behind is a tangle of mesas and intervening troughs, called a "chaos." The scene, about 4.5 km wide, is part of image ESP_018469_1630, taken by the HiRISE camera on the Mars Reconnaissance Orbiter (26 cm/pixel, NASA/JPL-Caltech/University of Arizona).

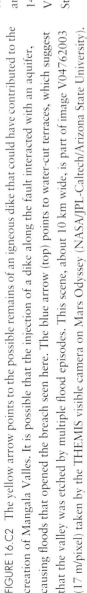

FIGURE 16.C2 The yellow arrow points to the possible remains of an igneous dike that could have contributed to the creation of Mangala Valles. It is possible that the injection of a dike along the fault interacted with an aquifer, causing floods that opened the breach seen here. The blue arrow (top) points to water-cut terraces, which suggest that the valley was etched by multiple flood episodes. This scene, about 10 km wide, is part of image V04762003 (17 m/pixel) taken by the THEMIS visible camera on Mars Odyssey (NASA/JPL-Caltech/Arizona State University).

The Atlas of Mars

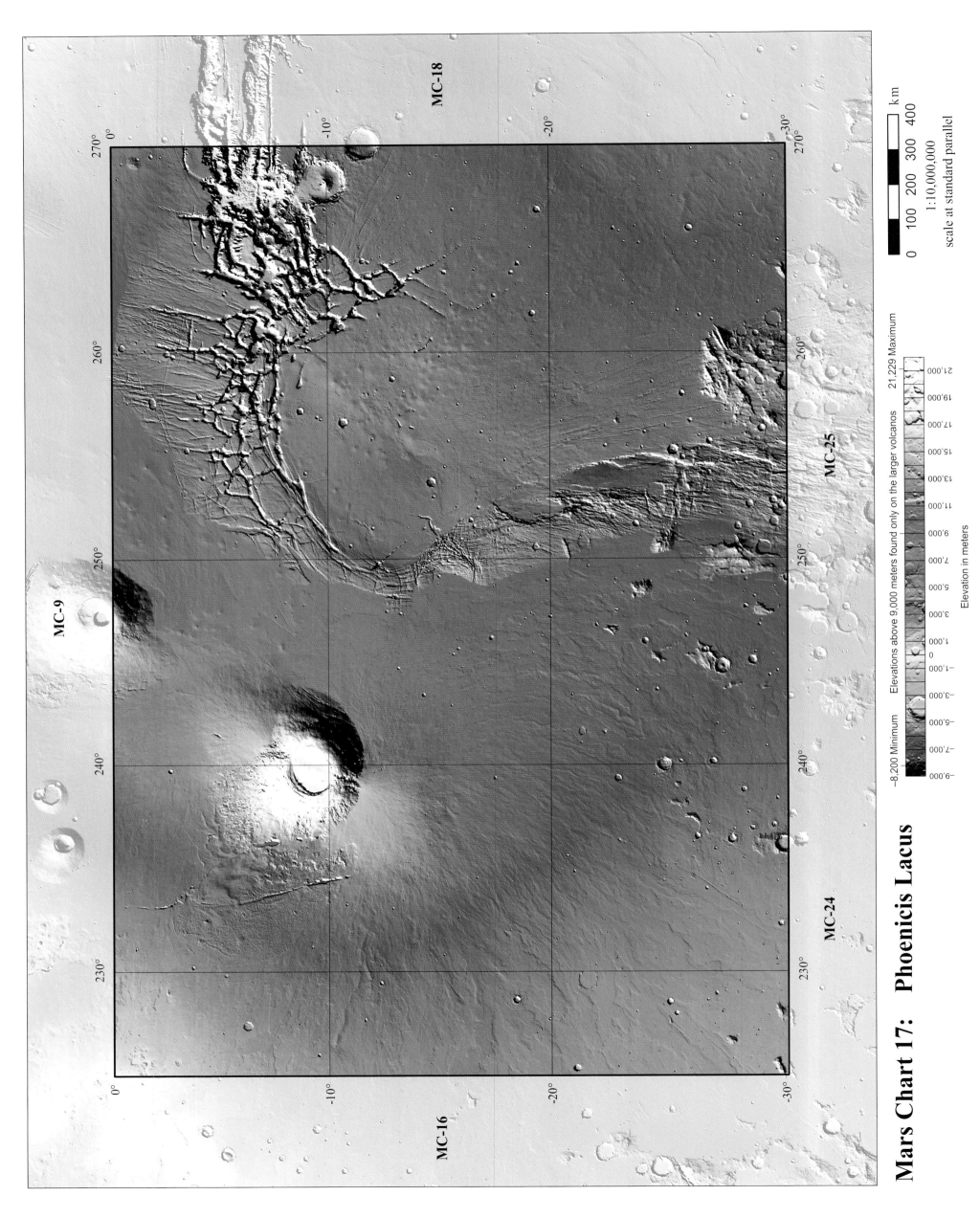

Mars Chart 17:   Phoenicis Lacus

1:10,000,000
scale at standard parallel

km

Elevations above 9,000 meters found only on the larger volcanos

-8,200 Minimum

21,229 Maximum

Elevation in meters

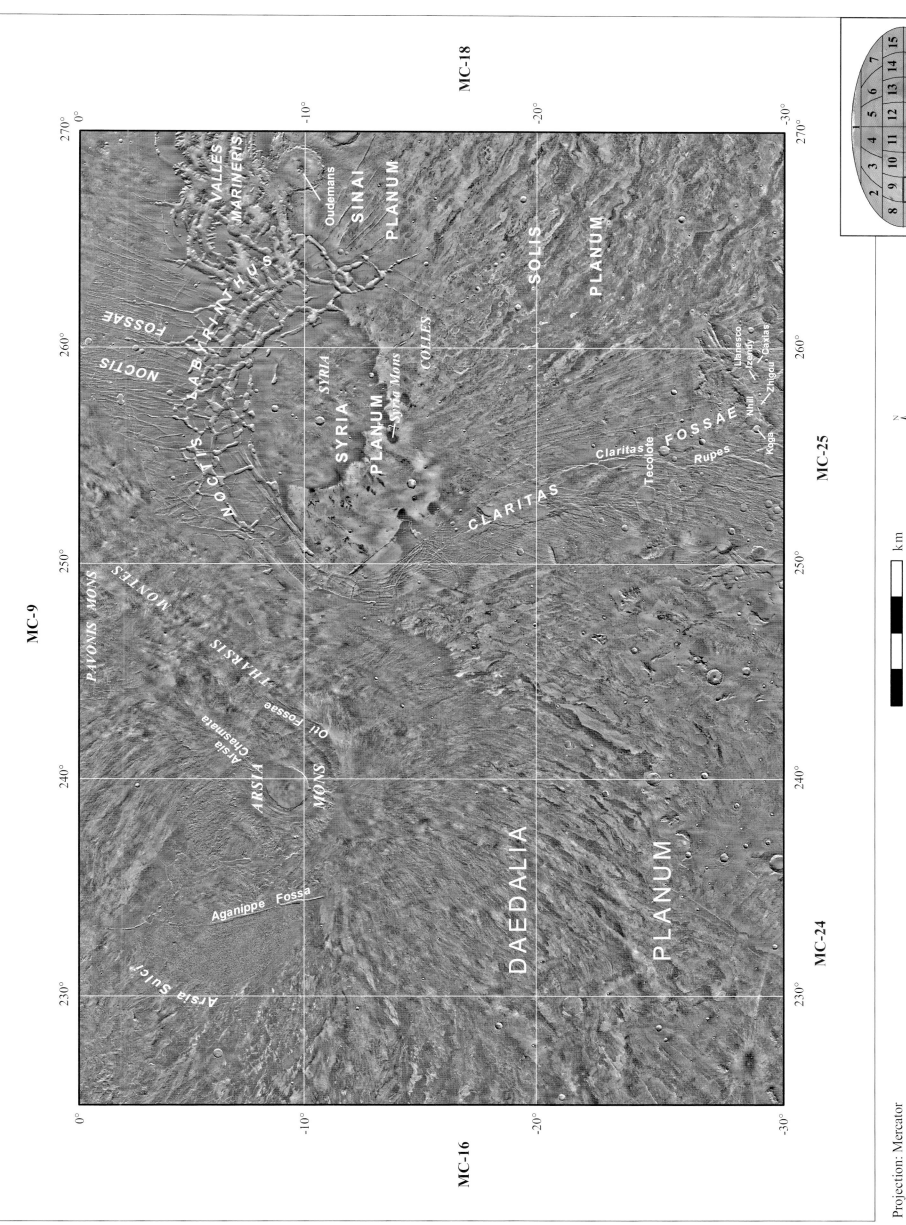

MC-9

MC-18

MC-16

MC-25

MC-24

270°

260°

250°

240°

230°

VALLES

MARINERIS

Oudemans

SINAI

PLANUM

NOCTIS LABYRINTHUS

FOSSAE

NOCTIS

SYRIA

SYRIA
PLANUM

Syria Mons

COLLES

SOLIS

PLANUM

CLARITAS

Claritas

FOSSAE

Claritas
Tecolote

Nhill

Rupes

Koga

Zhigou

Llanesco

Izendy

Claxias

PAVONIS MONS

MONTES

THARSIS

Oti Fossae

ARSIA

Arsia
Chasmata

MONS

Aganippe Fossa

Arsia Sulci

DAEDALIA

PLANUM

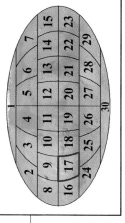

1		
5	6	7
4	13	14	15
3	12	21	22	23
2	11	20	29
9	10	19	28
8	17	18	27
16	25	26
24		30

N
W · E
S

0   100   200   300   400   km

1:10,000,000

scale at standard parallel

Projection: Mercator
Datum: Mars 2000 Sphere
Central Meridian: 0.0000
Standard Parallel 1: 0.0000

0°

-10°

-20°

-30°

Phoenicis Lacus (MC-17)

Geography

The Phoenicis Lacus quadrangle shows the heart of the Tharsis region that dominates the western hemisphere of Mars. Elevations are high: except for the floor of Valles Marineris, essentially the entire quadrangle lies above datum. Arsia Mons rises over 11 km from the surrounding plain and is more than 400 km across. It marks the southwest end of the northeast-trending Tharsis Montes, which also include Pavonis Mons and Ascraeus Mons (MC-9). The eastern half of the map is dominated by Syria, Sinai, and Solis Plana, high plateaus that are capped by a broad field of dozens of smaller volcanic shields. From Arsia Mons across Syria Planum the elevation is 6,000 m or more, descending to 2,000 m at the southwest corner of the map. Wrapping for 1,000 km around the north and west margin of Syria Planum is the Noctis Labyrinthus, where numerous canyons and depressions intersect in a maze-like pattern (see Figure 2.3 in Chapter 2). A large, rugged promontory, informally referred to as Claritas rise, lies along the southern margin of the quadrangle and east of Claritas Fossae (Dohm et al., 2009a). Valles Marineris extends east of Noctis Labyrinthus for several thousand kilometers (across MC-18).

Geology

The quadrangle, located in the heart of the Tharsis rise, includes volcanic and tectonic structures of wide-ranging sizes and geometric patterns. The development of such features demonstrates that the geology of this part of Mars has been dominated by long-term magmatic activity. The resulting landscapes in the quadrangle range from giant shield volcanoes (e.g. Arsia Mons) to shield fields (on Syria Planum and south of Pavonis Mons), and the igneous and tectonic plateaus contain lava-flow fields (the planum areas). In turn, these broader features are marked by various local fracture, fault, rift, pit-crater chain, channel, and canyon systems (forming fossae, catenae, valles, chasmata as well as Noctis Labyrinthus). Syria Planum is a major center of Tharsis tectonic activity, possibly due to a dynamic mantle plume. The planum is mostly of Noachian and Hesperian age and is surrounded by dense systems of concentric and radial narrow graben systems, including Noctis Labyrinthus (Sakimoto et al., 2003; Baptista et al., 2007; Baptista and Craddock, 2010). Overlapping lava-flow field sequences intermingle with sets of narrow grabens that make up Claritas Fossae, thereby establishing several successive stages of extensional deformation, related to the growth of Tharsis (Tanaka and Davis, 1988). Northeast of Syria Planum, the Noctis Labyrinthus canyon maze appears to result from the extension of Valles Marineris canyon development into a pre-existing, cross-hatched, Syria-centered tectonic fabric.

Ancient promontory: product of early plate tectonics?

The Claritas rise at the southern edge of the quadrangle developed during the Early Noachian Epoch prior to and/or as part of the incipient development of Tharsis. The promontory is associated with the development of the large mountain range that is distinct in the Thaumasia quadrangle

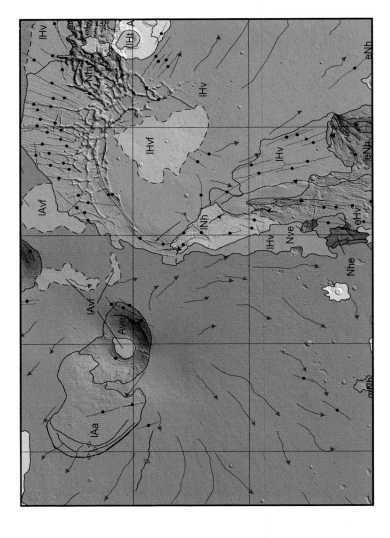

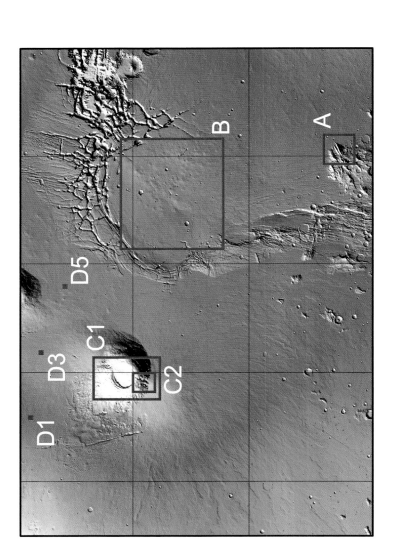

Structure

Interpretation
— Caldera rim
— Channel axis
— Crater rim
● Graben axis
▲ Lobate flow
▲ Outflow channel
⊖ Pit crater chain
⬦ Ridge
● Rille
◀ Scarp
● Spiral trough
◆ Wrinkle ridge
↕ Yardangs

Geology

Impact unit
AHi

Volcanic units
IHvf
Ave
eHv
Nve
AHv
IAvf
IHv

Apron unit
IAa

Transition units
IHt
AHtu
Htu

Highland units
mNh
INh
eNh
Nhe
Nhu

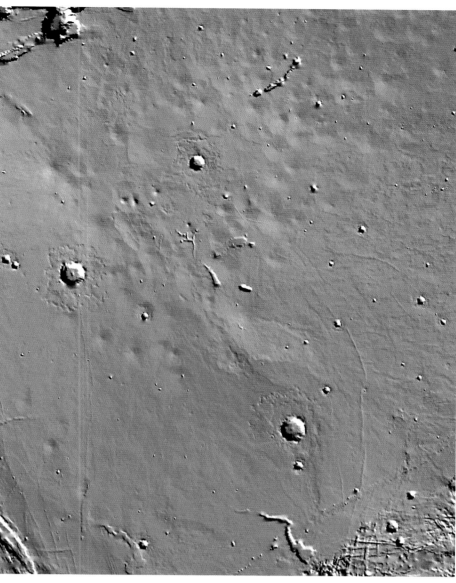

FIGURE 17.A   The eastern part of the Claritas rise is shown in this HRSC view looking toward the west. The rugged hills in the western (background) half of the image are onlapped by lavas to the east (foreground), while the entire region is cut by south–southeast-trending grabens, representative of Tharsis-related extension. Thin clouds are present in the northern part of the image (HRSC press release 095, 40 m/pixel, foreground about 120 km wide, center near 28° S, 260° E, ESA/DLR/FU-Berlin).

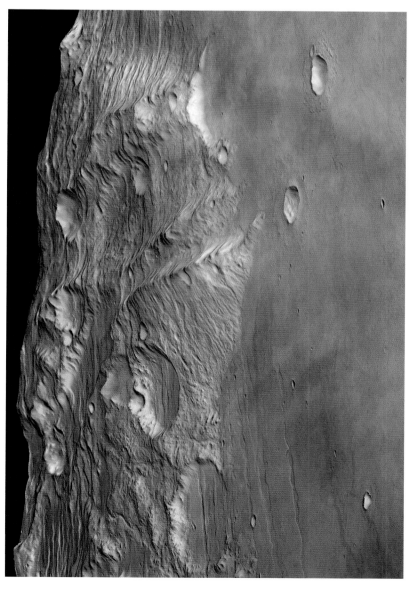

FIGURE 17.B   Volcanic centers in Syria Planum, forming low shields and elongate depressions (MOLA color hillshade image, 600 km by 480 km, north at top, view 9° S to 17° S, 251° E to 261° E).

(MC-25), interpreted as having been produced by major crustal shortening (Dohm and Tanaka, 1999; Dohm et al., 2001c). Claritas rise is one of only two locales on Mars where the hydrated mineral serpentine has been identified, the other being Nilosyrtis Mensae in MC-6 (Ehlmann et al., 2010; Figure 17.A). These geologic associations among ancient Claritas and Thaumasia features have been proposed as evidence for an episode of plate-tectonic activity on Mars more than 4 billion years ago (Dohm et al., 2015).

An elevated plain with a fractured margin

Syria Planum is an oval-shaped region, speckled by dozens of low volcanic shields and fissure vents, and it dominates the eastern part of the Phoenicis Lacus quadrangle (Dohm et al., 2001b; Figure 17.B). Related to its episodic stages of volcanic

and tectonic activity, dating back at least to the Late Noachian Epoch, thousands of faults and wrinkle ridges occur radial to and concentric about this high plain (Anderson et al., 2001). The lavas of Syria Planum extend hundreds of kilometers to the south and southeast from the broad shield complex. Syria lavas flowed onto the Noachian rock materials and tectonic structures, which form rugged mountain ranges of the Thaumasia quadrangle (MC-25) to the south and southeast (Dohm et al., 2001c). Syria Planum and the partly encircling Noctis Labyrinthus, along with Alba Mons and possibly a dozen or so other sites in the Tharsis region, are ovoid centers of volcanism and tectonism, hundreds of kilometers across, which may result from underlying mantle plumes; such structures are more extensively developed on Venus, where they are called "coronae" (Watters and Janes, 1995; Tanaka et al., 1996).

Volcanoes of wide influence

Also visible in the western parts of the quadrangle are Arsia Mons (Figures 17.C1, 17.C2) and associated concentrically ridged apron deposits along its northwest flank, the southern part of Pavonis Mons, and sheet lavas of a relatively young stage of Tharsis development (see details in similar features in MC-8 and MC-9). The apron deposits may be related to major climate swings (and/or volcanic activity), which promoted growth and retreat of glaciers on the volcanoes' northwestern flanks. Computerized models of the Martian climate predict transfer of ice from the polar regions to these immense mountains during high obliquity (that is, a high degree of tilt of the planet's axis of rotation), in spite of their lower-latitude locations (Scott et al., 1998; Forget et al., 2006).

A variety of cavernous pits

Examples of three types of cave openings occur north of Arsia Mons (Cushing, 2011). Lava tubes, which occur on Earth and in volcanic regions on other planets, commonly empty out when the lava flow ceases. The tube can collapse completely to form a rille (sinuous valley), or the roof can collapse locally to make a pit or depression. Figures 17.D1 and 17.D2 show one such 60-m-long skylight into a lava tube where wind-blown dust has partially filled the opening. Another type of cave opening is associated with long, narrow fault valleys, or grabens, along which associated fissures may have once filled with magma and then drained, causing collapse and pit formation. These caves may be deeper than lava tubes. Figures 17.D3 and 17.D4 show an example, 60 m long and perhaps 50 m deep, partially filled with material

The Atlas of Mars

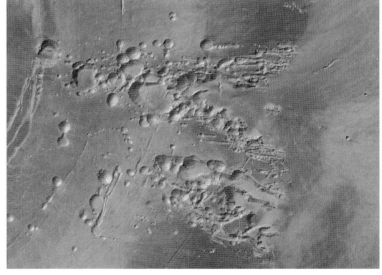

FIGURE 17.C2 HRSC nadir image (HRSC press release 046, orbit 0263, 20 m/pixel, north at top) of the southern flank of Arsia Mons, showing the numerous collapse pits on the slope of the volcano. The pits are in turn partly filled or covered by lavas on the basal plain and in the summit caldera (ESA/DLR/FU-Berlin).

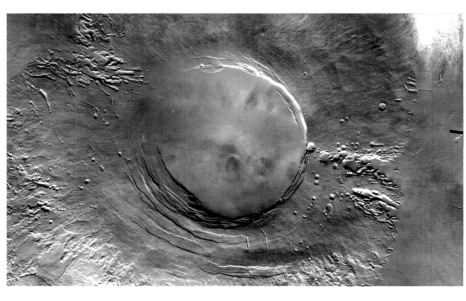

FIGURE 17.C1 THEMIS mosaic of Arsia Mons summit caldera and flanks (daytime infrared, 100 m/pixel, view 207 km by 335 km, north at top, centered on 9° S, 239° E). The pits and collapse features on the north-northeastern and southern slopes align with the trend of Pavonis and Ascraeus Mons, suggesting major deep-seated basement structures that controlled the location of all three volcanoes (NASA/JPL-Caltech/Arizona State University).

FIGURE 17.D1 (top) HiRISE image of likely lava tube with several openings winding across image from lower right to upper left. Figure 17.D2 shows opening at lower right (image ESP_016767_1785_RED, 3 km by 4 km, all five images are 25 cm/pixel with north at top).

FIGURE 17.D2 Opening into lava tube (HiRISE ESP_016767_1785, near 1° S, 236° E).

by wind or gravity. A few pit openings (Figure 17.D5, opening about 50 m across) are not clearly associated with lava tubes or faults (see Cushing, 2012).

Caves and pits have been suggested as targets in the search for microbial life on Mars (European Space Agency, 2012). Caves on Mars could be associated with subsurface water or ice. Caves would also be more hospitable for human visitors than the Martian surface, because the overlying rock would protect them from exposure to radiation, temperature extremes, and wind-blown dust (Cushing, 2012).

FIGURE 17.D3 (left) Overview HiRISE image showing graben, revealed by surface morphology, with several cave openings and pits. Detailed image (Figure 17.D4) shows opening at left center (image ESP_014380_1775_RED, 3 km by 5 km).

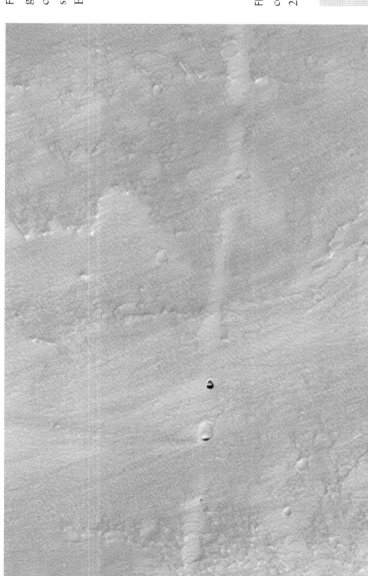

FIGURE 17.D4 (bottom left) Cave opening formed by collapse along graben (HiRISE ESP_014380_1775, near 2° S, 242° E).

FIGURE 17.D5 (below) Cave opening of uncertain origin (HiRISE ESP_023531_1840, near 4° S, 248° E, all images NASA/JPL-Caltech/University of Arizona).

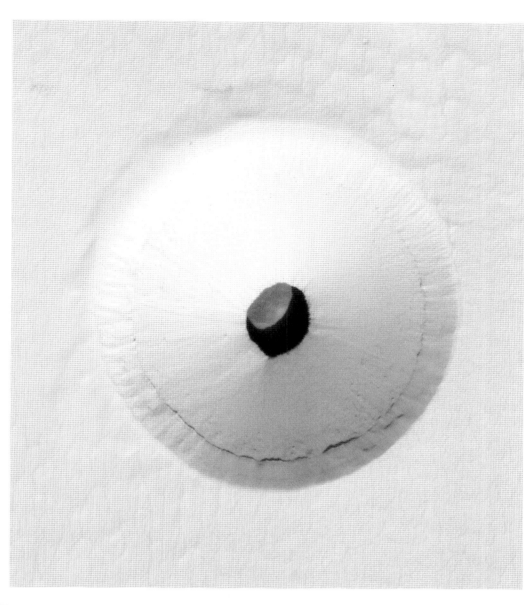

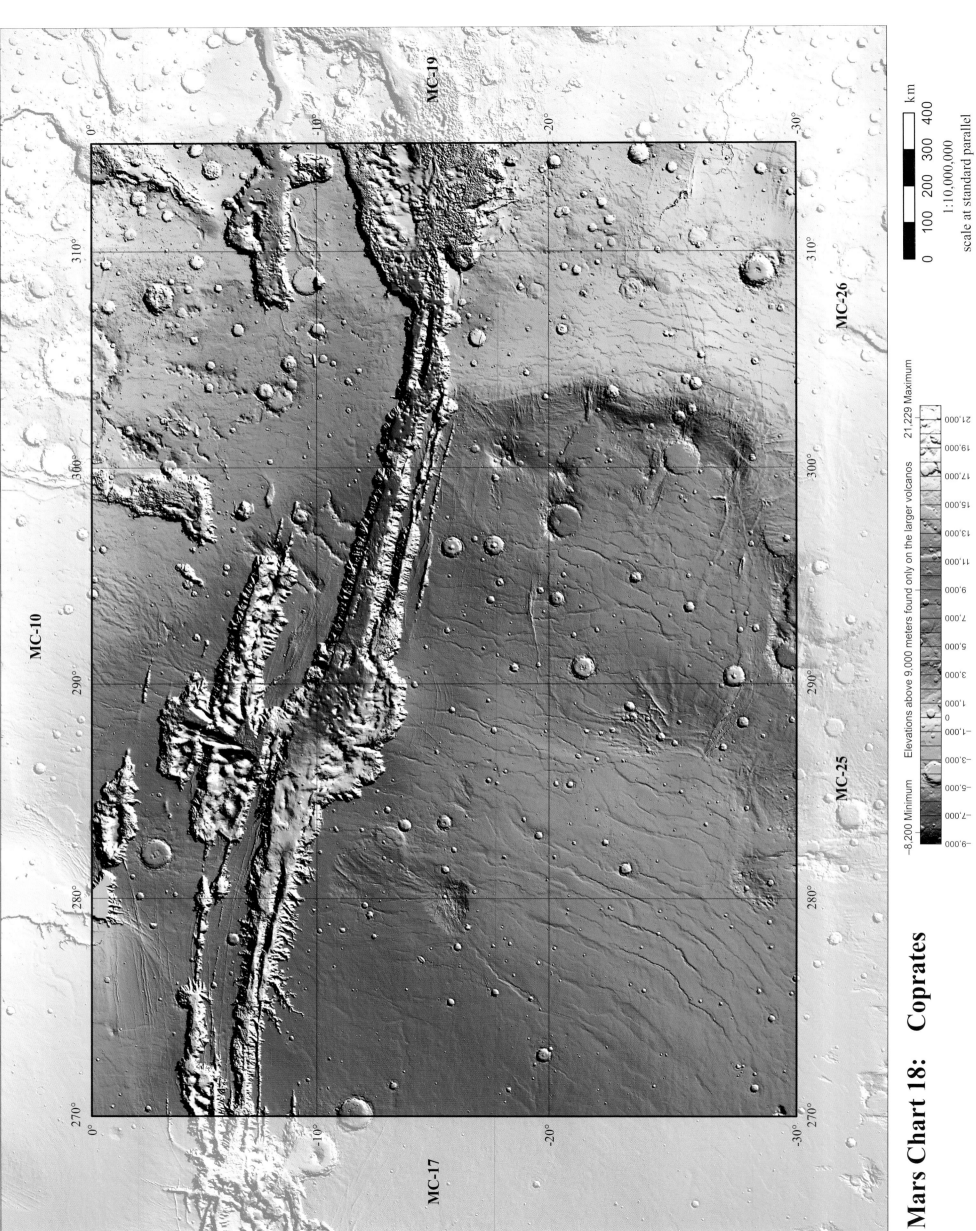

Mars Chart 18:   Coprates

Elevation in meters

–8,200 Minimum   Elevations above 9,000 meters found only on the larger volcanos   21,229 Maximum

–9,000 –7,000 –5,000 –3,000 –1,000 0 1,000 3,000 5,000 7,000 9,000 11,000 13,000 15,000 17,000 19,000 21,000

1:10,000,000
scale at standard parallel

km
0   100   200   300   400

MC-10   MC-19   MC-17

MC-26   MC-25

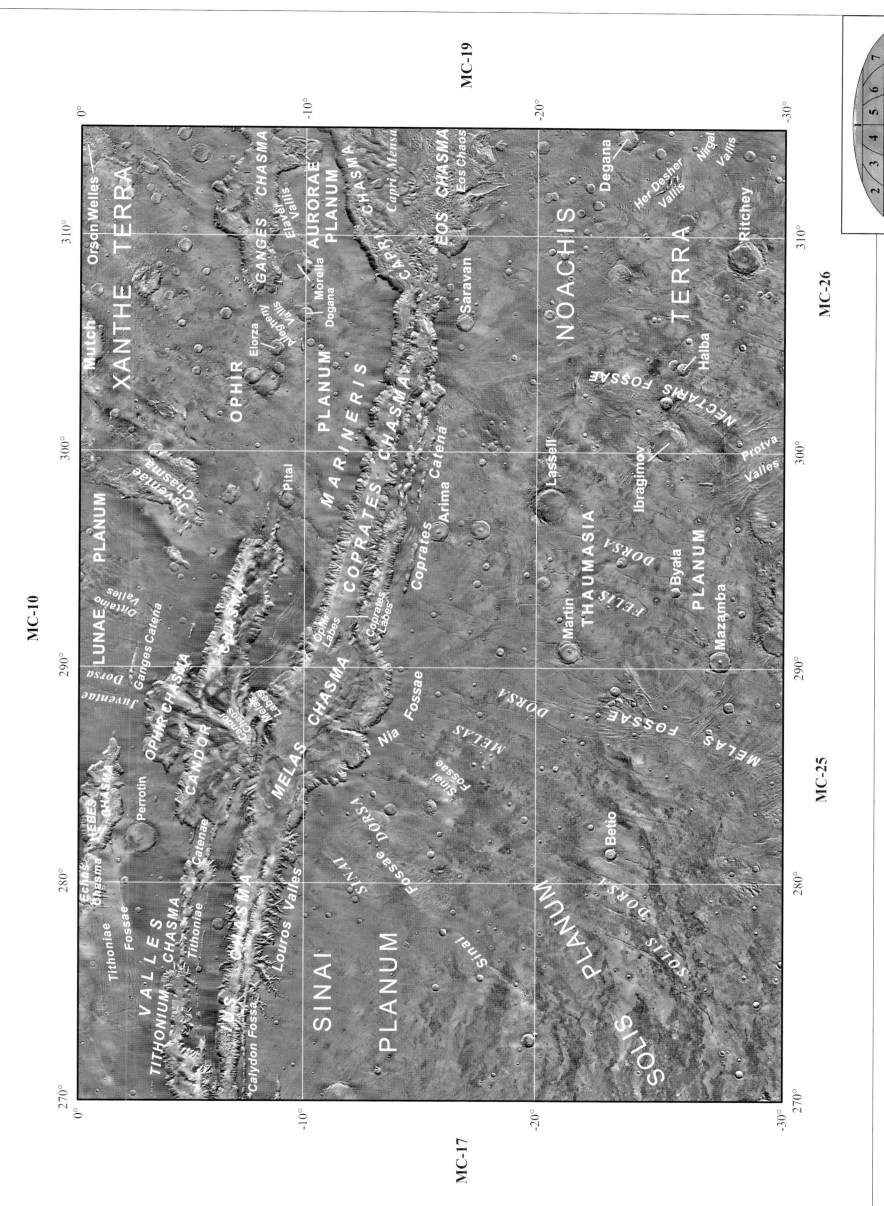

MC-10

MC-19

MC-26

MC-25

MC-17

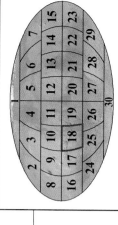

1			
2	3	4	5	6	7
8	9	10	11	12	13	14	15
16	17	18	19	20	21	22	23
24	25	26	27	28	29
		30		

270°   280°   290°   300°   310°   0°

0°

−10°

−20°

−30°

XANTHE TERRA
Mutch
Orson Welles
Ophir
Elorza
Eloza
OPHIR PLANUM
GANGES CHASMA
Allegheny Vallis
Elaver Vallis
AURORAE PLANUM
Morella
Dogana
CAPRI CHASMA
Capri Mensa
EOS CHASMA
Eos Chaos
Saravan
Degana
Her Desher Vallis
Nirgal Vallis
NOACHIS TERRA
Ritchey
Halba
NECTARIS FOSSAE
Protva Valles
Ibragimov
Lassell
THAUMASIA
Martin
FELIS DORSA
Byala
THAUMASIA PLANUM
Mazamba
MELAS FOSSAE
Betio
SOLIS PLANUM
Solis Dorsa

LUNAE PLANUM
Juventae Chasma
Dittaino Valles
Ganges Catena
Juventae Dorsa
OPHIR CHASMA
Pital
MARINERIS PLANUM
COPRATES CHASMA
Ophir Labes
Coprates Labes
Coprates Catena
Arima
Coprates
Perrotin
CANDOR CHASMA
CANDOR CHASMA
Melas Labes
Candor Chaos
ECHUS CHASMA
HEBES CHASMA
Tithoniae Fossae
TITHONIUM CHASMA
Tithoniae Catenae
Calydon Fossa
VALLES MARINERIS
IUS CHASMA
Louros Valles
MELAS CHASMA
Nia Fossae
SINAI DORSA
SINAI PLANUM
Sinai
MELAS DORSA

1:10,000,000
scale at standard parallel

0   100 200 300 400
km

N
W   E
S

Projection: Mercator
Datum: Mars 2000 Sphere
Central Meridian: 0.0000
Standard Parallel 1: 0.0000

Coprates (MC-18)

Geography

In this quadrangle, the deepest canyon system on Mars and one of the most spectacular in the solar system, Valles Marineris, extends more than 2,500 km across the map, cutting the northern part of the Thaumasia plateau (informal name). Parts of the floor are more than 5,000 m below datum and as much as 10,000 m below the plateau rim. The system of canyons includes Melas, Candor, and Ophir Chasmata, prominent at its central part, Coprates along its eastern part, and Ius and Tithonium Chasmata, forming its western arm. Echus, Hebes, Juventae, and Ganges Chasmata form additional, separate canyons, north of the main canyon system. The canyons of Noctis Labyrinthus, west of the map (MC-17), join Valles Marineris, which connects eastward with the chaos-filled canyons of Capri and Eos Chasmata. Some of the other chasmata are linked to outflow channels north (MC-10) and east of the map (MC-19). Sinai, Solis, and Thaumasia Plana make up high plains that span the western to south-central parts of the Coprates quadrangle. Together they consti-tute much of the Thaumasia plateau, a dish-

shaped, elevated region at 3,000–5,000 m. At their eastern edges, the Thaumasia plateau and Ophir Planum drop as much as 3,000 m down to variably cratered plains of Noachis and Xanthe Terrae, respectively. The informally named Coprates rise is a prominent north-trending mountain range, forming the eastern margin of the Thaumasia plateau, south of Coprates Chasma. Crossing this entire region are north- to northeast-trending wrinkle ridges as well as local sets of narrow grabens and pit chains.

Geology

The Thaumasia plateau is apparently made up largely of Noachian lava flows, at least several thousand meters thick (McEwen *et al.*, 1999). It formed during the Middle to Late Noachian Epochs as one of the earlier stages of Tharsis volcanic and tectonic development, perhaps the result of mantle plume activity (Frey, 1979). The vast canyon system of Valles Marineris began to form within the northern part of the Thaumasia plateau at least as far back as the Late Noachian Epoch (Dohm *et al.*, 2001c). The origin of Valles Marineris is uncertain; some have suggested that it may be the result of local plume-driven

magmatism, but others consider that strike–slip deformation was a primary instigator (Andrews-Hanna, 2011; Yin, 2012; see the Tectonics section in Chapter 5). In addition to the canyons, exten-sional tectonism, perhaps driven by rifting and uplift along Valles Marineris as well as by regional Tharsis volcanic loading upon the lithosphere, resulted in systems of narrow grabens, including

Sinai, Nia, Melas, and Nectaris Fossae, and others (Dohm *et al.*, 2001c). Contractional deformation, due both to regional Tharsis loading and to global contraction as the result of planetary cooling, may account for the wrinkle ridge systems.

The Thaumasia plateau and central Tharsis region may have contained a vast aquifer system, the groundwater of which, perhaps trapped

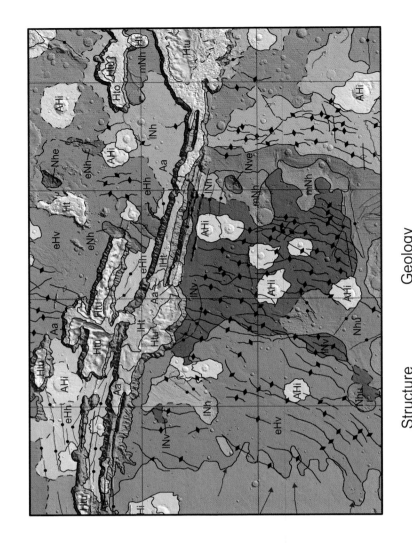

Structure

Interpretation

———	Caldera rim
———	Channel axis
———	Crater rim
●—	Graben axis
▲	Lobate flow
▲	Outflow channel
⊖—	Pit crater chain
◇—	Ridge
●—	Rille
◄—	Scarp
●—	Spiral trough
◆—	Wrinkle ridge
↕	Yardangs

Geology

Impact unit
AHi

Volcanic units
lNv   eHv
eHv   Nve
AHv   lHv

Apron unit
Aa

Transition units
lHt   Hto
Ht   Htu

Highland units
mNh   lNh
eHh   eNh
Nhe   Nhu

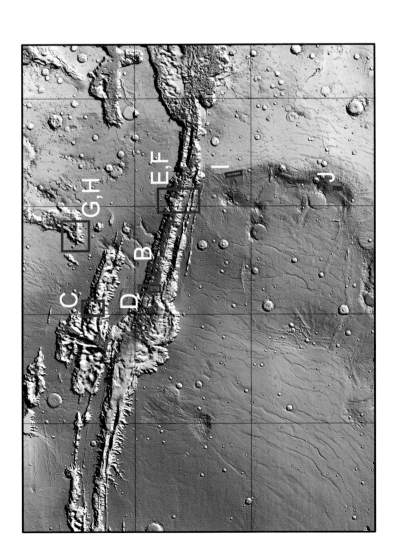

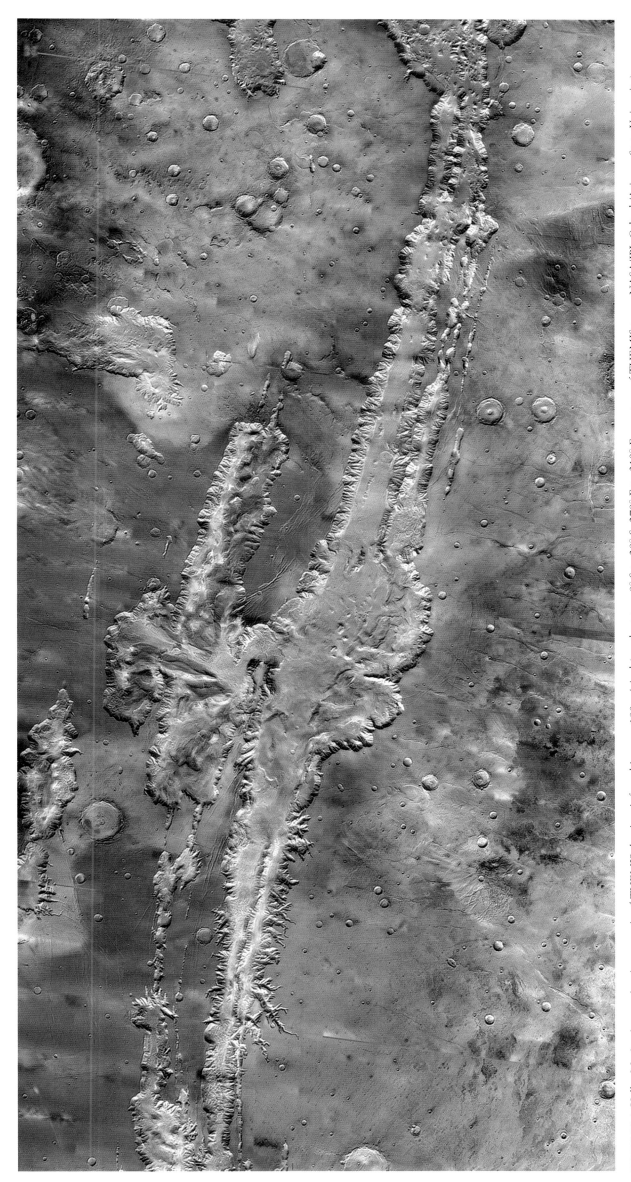

FIGURE 18.A Valles Marineris overview in a mosaic of THEMIS daytime infrared images, 100 m/pixel (north at top, 0° S to 20° S, 270° E to 310° E, courtesy of THEMIS team, NASA/JPL-Caltech/Arizona State University).

beneath a permafrost cap, was later tapped when the development of Valles Marineris canyons and fault systems provided conduits for its release during the Late Hesperian Epoch (Carr, 1979; Dohm *et al.*, 2001b). Gargantuan groundwater discharges may have further enlarged the Valles Marineris canyons. These could also have produced the associated chaotic terrains and outflow channels that extend north and east from Valles Marineris. The pit chains that parallel Coprates Chasma may be indicative of collapse due to magma withdrawal or even subterranean erosion. Also during the Late Hesperian Epoch, the canyons

became partly infilled by stratified, moderately bright, sulfate-rich rocks, commonly referred to as "interior layered deposits" (e.g. Witbeck *et al.*, 1991; Murchie *et al.*, 2009). Their origin seems to require a water-rich environment, perhaps within lakes, and volcanic and sedimentary explanations for their accumulation have been proposed. The deposits are susceptible to wind erosion, as evidenced by streamlined ridges on their steeper slopes, where wind velocities tend to be increased.

Little evidence for Amazonian volcanism or tectonism in the Coprates quadrangle has been documented. However, enormous hollows scar many

sections of Valles Marineris canyon walls, from which extend dozens of lobate and grooved landslide deposits, upwards of tens of kilometers in length. Crater-density analysis indicates that they formed throughout the Amazonian Period (Quantin *et al.*, 2007). Wind erosion and formation and movements of local dune and dust patches have continued to alter the landscape over time.

A view beneath the surface

The Valles Marineris system is not only spectacular in scale (Figure 18.A), it also shows superb

examples of major faults (Figure 18.B), exposures of layered rocks that partly compose Thaumasia (Figure 18.C), and gravity-collapse processes (Figure 18.D). The underground-collapse features that probably led to the formation of the chasm are found nearby (Figure 18.E), while younger interior layered deposits, with water-bearing minerals, occupy the floor of the valley (Figures 18.F–18.H).

An ancient plain and its margins

Thaumasia Planum is a tectonic province along the eastern part of the shield volcano complex of

165

The Atlas of Mars

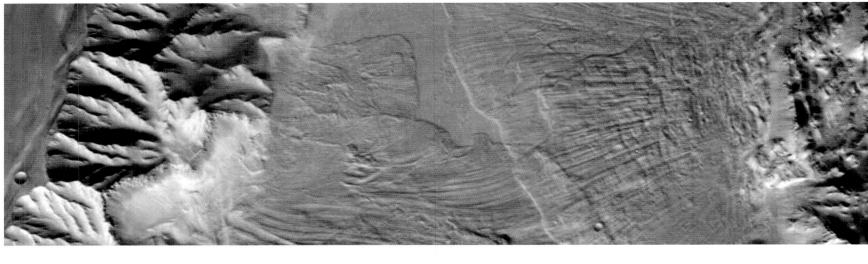

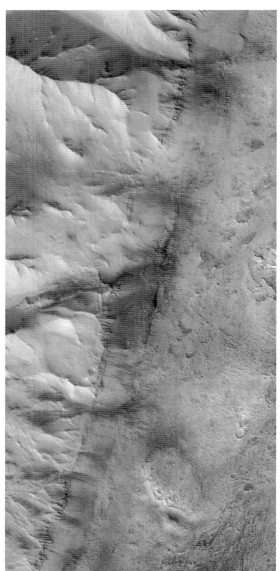

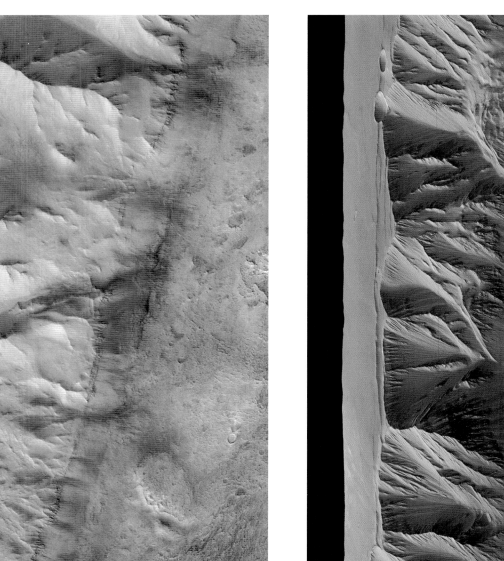

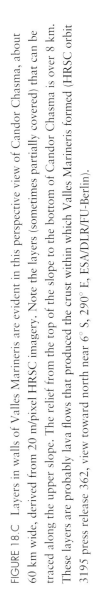

FIGURE 18.B Coprates wall fault: This view shows a remarkably linear portion of the base of the north wall of Coprates Chasma. The walls of Valles Marineris elsewhere are modified by landslides and have an irregular outline, making it unclear whether all the walls originally had such alignment (CTX image P16_007258_1686_XN_11S066W, 5.2 m/pixel, 25 km by 10 km, north at top, 11.5° S, 293.9° E, NASA/JPL-Caltech/MSSS).

FIGURE 18.D (right) Landslides are common features in Valles Marineris, given the very high valley walls. Coprates Labes, shown in this THEMIS daytime infrared image, records several overlapping deposits. The long slide from the upper, or north, wall of the canyon was overlapped later by the ends of two landslides from the south wall. The grooves or lineations on the landslides also occur in examples on Earth and indicate direction of motion (THEMIS image I51935002, 100 m/pixel, view 30 km by 110 km, north at top, center at 11.5° S, 292.25° E, NASA/JPL-Caltech/Arizona State University).

FIGURE 18.C Layers in walls of Valles Marineris are evident in this perspective view of Candor Chasma, about 60 km wide, derived from 20 m/pixel HRSC imagery. Note the layers (sometimes partially covered) that can be traced along the upper slope. The relief from the top of the slope to the bottom of Candor Chasma is over 8 km. These layers are probably lava flows that produced the crust within which Valles Marineris formed (HRSC orbit 3195 press release 362, view toward north near 6° S, 290° E, ESA/DLR/FU-Berlin).

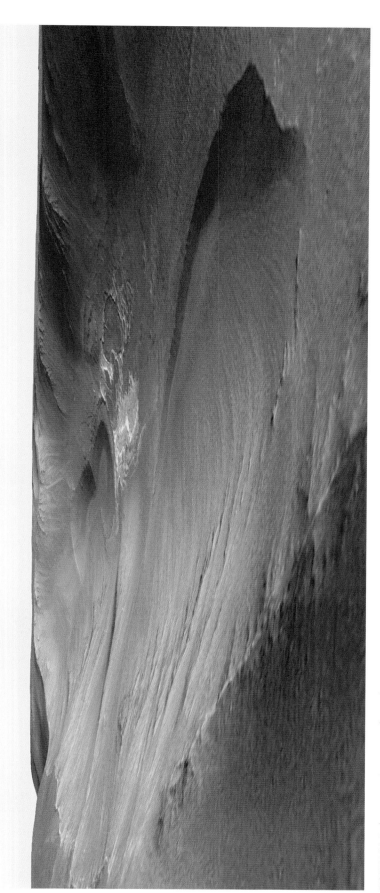

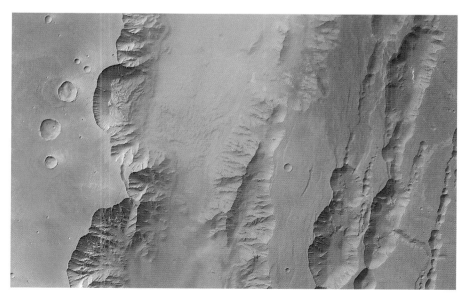

FIGURE 18.E Catenae: In this section of Coprates Chasma, several parallel lines of closed depressions ("catenae"), are seen, south of and parallel to the main canyon. These are only a couple of kilometers deep, much less than the main canyon. The catenae do not connect to outflow channels or low regions by surface drainage, so their collapse represents removal of subsurface material, probably magma, rock debris, water, and/or ice. The trend suggests that this process was promoted by fractures parallel to the Valles Marineris system (HRSC press release 191/orbit 449, 48 m/pixel, view about 120 km by 200 km, north at the top, 13.5° S, 300° E, ESA/DLR/FU-Berlin).

FIGURE 18.F A perspective view of one depression at the lower left of Figure 18.E shows a feature resembling a delta, formed in a former lake, in the foreground, and a white patch of sulfate minerals that indicate evaporation of surface water (Panorama about 10 km across from THEMIS visible image at 18 m/pixel, view toward the southeast, 15° S, 300° E, courtesy THEMIS team, NASA/JPL-Caltech/Arizona State University).

Syria Planum (MC-17). The western margin of Thaumasia Planum, including faults related to the development of Melas Chasma, is overlapped by younger lava-flow materials of Syria, Sinai, and Solis Plana, forming a distinct geologic contact that extends south from Melas Chasma to the mountain range of the northern part of the Thaumasia quadrangle (MC-25), informally named the Thaumasia highlands. Melas Fossae, which occur along the western margin of Thaumasia Planum and corresponding geologic contacts, also experienced onlap of lava flows from the west (Dohm et al., 2001c).

The eastern margin of Thaumasia Planum abuts a north-trending mountain range, informally named the Coprates rise. The mountain range extends from near the Coprates Chasma southward, forming the southeast margin of Thaumasia plateau. Cuestas, plateaus made up of tilted, terraced materials, are distinct along its eastern flank, as well as a system of faults with trends generally paralleling the range, named Nectaris Fossae.

Nearly perpendicular to the trend of the mountain range and Nectaris Fossae, rift systems mark the activity of Tharsis, including uplift of the Thaumasia plateau (Figures 18.I, 18.J). The mountain range has been variously interpreted to result from reverse faulting resulting from the development of the Thaumasia plateau (Schultz and Tanaka, 1994), rotation of the Thaumasia crustal block, and/or a megaslide originating from what is now Syria Planum (Montgomery et al., 2009).

167

The Atlas of Mars

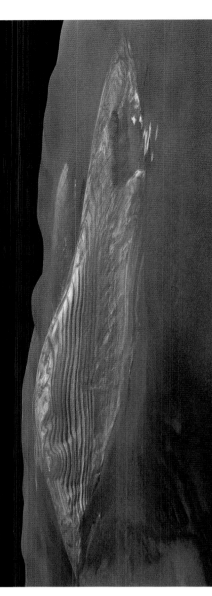

FIGURE 18.H A perspective view, looking east, of the elongated ridge on the valley floor at "d" in Figure 18.G, from HRSC imagery. The ridge extends 30 km north–south and is 2,500 m in height (press release 053, orbit 0243; 23 m/pixel, ESA/DLR/FU-Berlin). Deposits on the floor of Valles Marineris have layering that is distinct from the much older rocks in the valley walls and formed after the valley system itself.

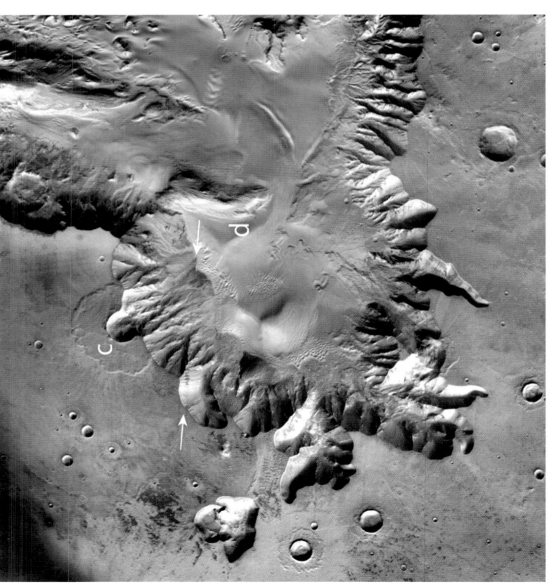

FIGURE 18.G Juventae Chasma is 5 km deep and shows a number of interesting features in this thermal infrared image. The contrasting colors show variations in thermal inertia. Materials that cool rapidly at night, such as sand and dust, are green and blue, while the bedrock, which retains heat for a longer time, is red (THEMIS infrared, 100 m/pixel, 150 km × 140 km, north at top, 5° S, 297° E, NASA/JPL–Caltech/Arizona State University). Three features of interest are associated with the canyon slope at the upper center of the image. A fault, shown by the arrows, runs nearly east–west across the view. Contrasting colors along the fault suggest that it controls where the bedrock lies at the surface. An impact crater at the top of the same slope (at letter "c") has been partially destroyed by slope erosion. This crater has an ejecta pattern that indicates that subsurface water or ice was present when it formed. An elongated ridge on the valley floor at the east end of the same slope (at letter "d") consists of interior layered deposits of calcium and magnesium sulfate minerals, which indicate an origin in water. These are also shown in Figure 18.H. Also, on the western margins of the canyon floor, patches of rippled ridges are made up of wind-blown sand dunes.

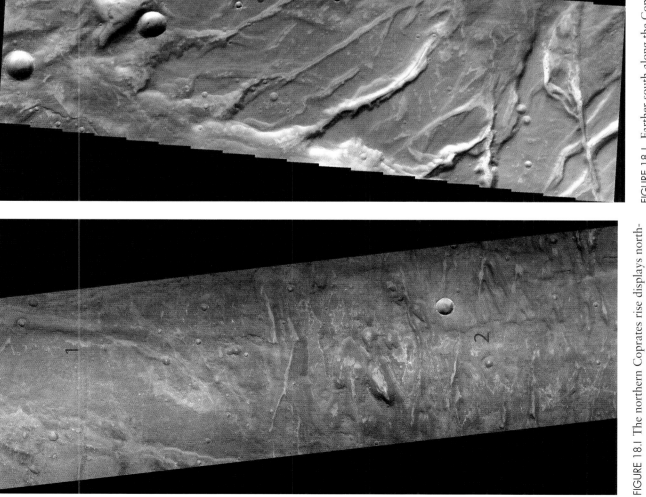

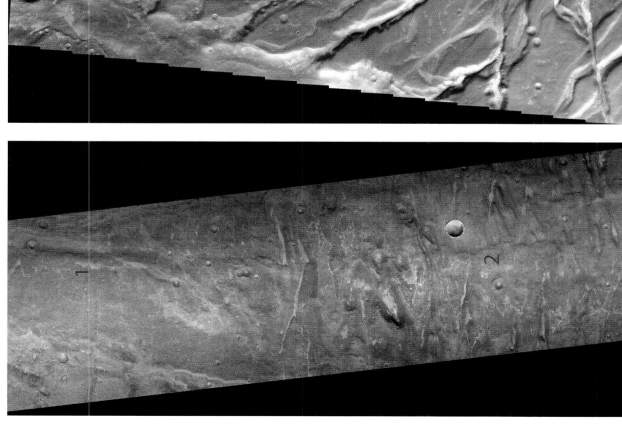

FIGURE 18.I The northern Coprates rise displays north-trending fractures (Nectaris Fossae, at 1), possible contractional (fold-and-thrust) wrinkle ridges that also trend northerly, and east–west fractures that are related to the formation of the Tharsis rise (both at 2; CTX image G09_021894_1608, 6 m/pixel, 20 km by 70 km, north at top, centered at 18.5° S, 303° E, NASA/JPL-Caltech/MSSS).

FIGURE 18.J Farther south along the Coprates rise, an area sloping down to the southeast (lower right) displays several sets of cross-cutting fractures and tilted plateaus with uphill-facing erosional cliffs, known as "cuestas" (THEMIS visible image V24016003, 18 m/pixel, 18 km by 63 km, north at top, centered at 27.1° S, 301.6° E, NASA/JPL-Caltech/Arizona State University).

The Atlas of Mars

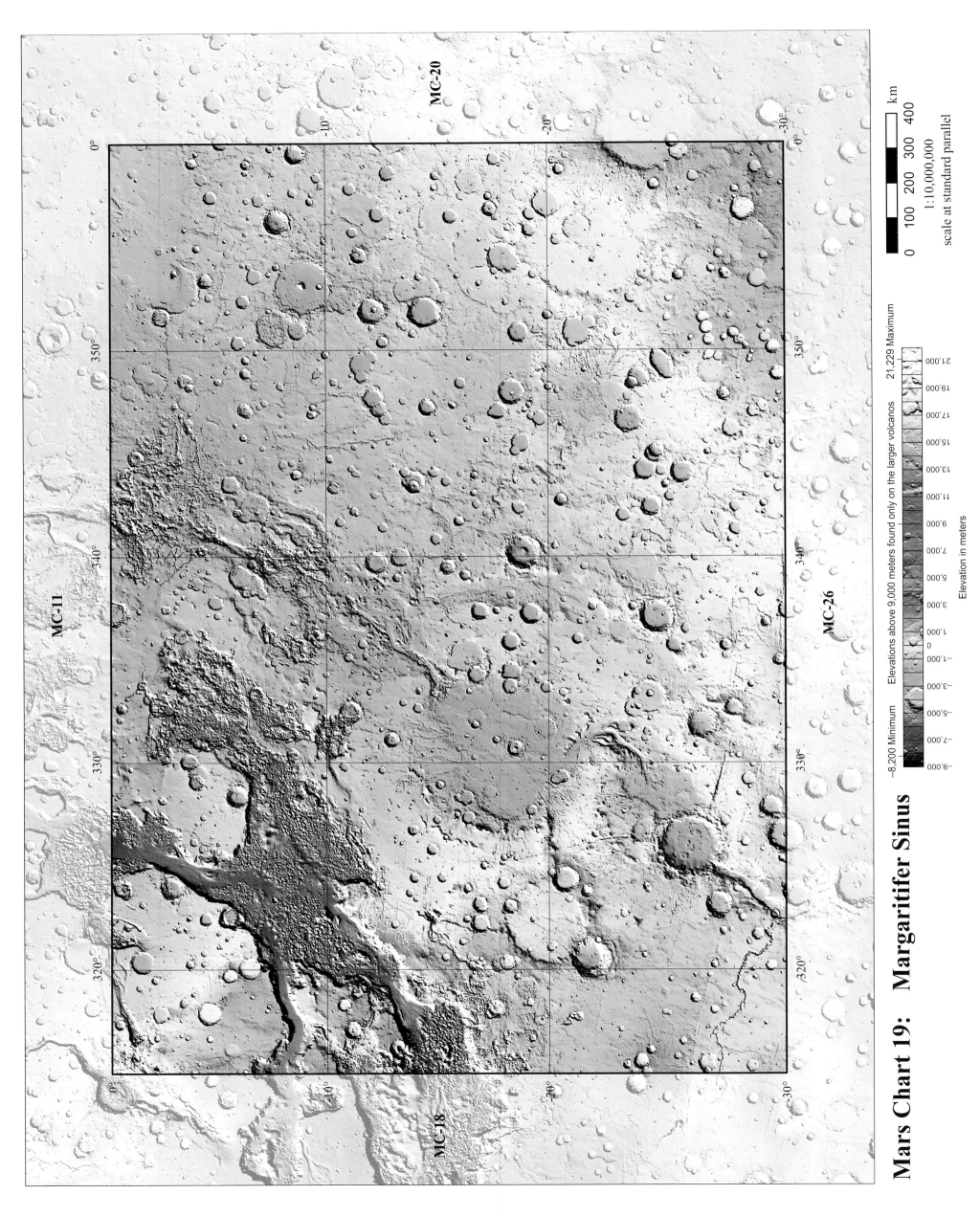

Mars Chart 19: Margaritifer Sinus

MC-20

MC-11

MC-26

MC-18

km

1:10,000,000
scale at standard parallel

0   100   200   300   400

−8,200 Minimum   Elevations above 9,000 meters found only on the larger volcanos   21,229 Maximum

Elevation in meters

−9,000 −7,000 −5,000 −3,000 −1,000 0 1,000 3,000 5,000 7,000 9,000 11,000 13,000 15,000 17,000 19,000 21,000

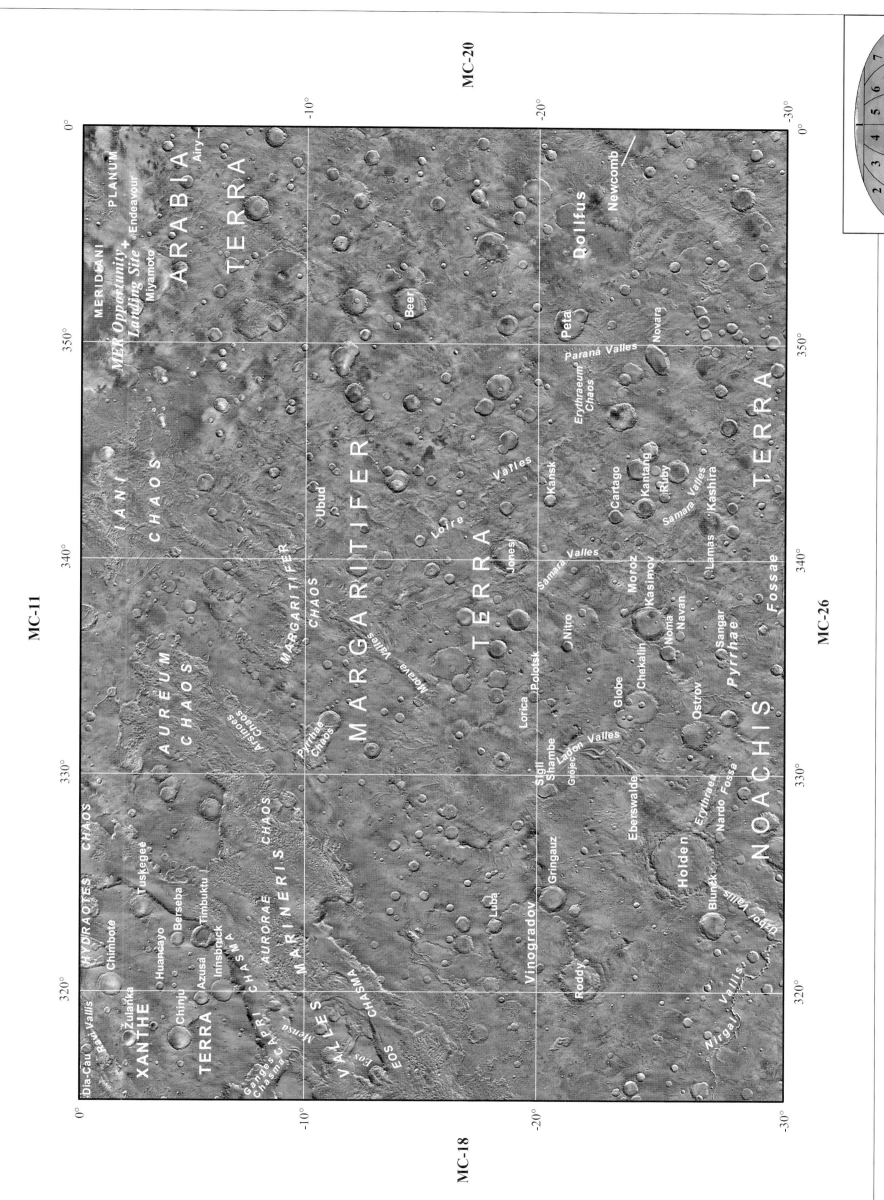

MC-11

MC-20

MC-18

MC-26

0°   320°   330°   340°   350°   0°

XANTHE TERRA

HYDRAOTES CHAOS

Dia-Cau
Raul Vallis
Žulanka
Chimbote
Tuskegee
Chinju
Huancayo
Berseba
Azusa
Timbuktu
Innsbruck

CAPRI CHASMA
AURORAE CHAOS
MARINERIS
Mensa

EOS CHASMA
EOS CHASMA
VALLES

ARSINOES CHAOS

AUREUM CHAOS

IANI CHAOS

MARGARITIFER CHAOS

MARGARITIFER

Pyrrhae Chaos

Morava Valles

Pyrrhae Valles

ARABIA

MERIDIANI PLANUM

MER Opportunity
Landing Site
Miyamoto
Endeavour
Airy

TERRA

Ubud

Beer

Loire

Jones

Valles

MARGARITIFER

TERRA

Samara Valles

Kansk

Cartago
Kantang
Ruby
Samara Valles
Kashira

Erythraeum Chaos

Paraná Valles

Peta

Novara

Dollfus

Newcomb

Vinogradov

Roddy

Gringauz

Luba

Lorica

Polotsk

Nitro

Sigli
Shambe
Ladon Valles
Grjotecon

Ebetswalde

Holden

Erythraea
Nardo Fossa

Blunck
Nirgal Vallis

Uzboi Vallis

Globe

Chekalin

Kasimov

Nona
Navan

Ostrov

Sangar

Lamas

Moroz

Pyrrhae
Fossae

NOACHIS

TERRA

N

W   E

S

km

0   100   200   300   400

1:10,000,000

scale at standard parallel

Projection: Mercator
Datum: Mars 2000 Sphere
Central Meridian: 0.0000
Standard Parallel 1: 0.0000

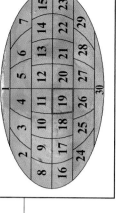

		5	6	7	
	4	12	13	14	15
3	10	11	20	21	22	23
2	9	17	18	19	27	28	29
8	16	24	25	26	30	

Margaritifer Sinus (MC-19)

Geography

The Margaritifer Sinus quadrangle comprises the distal eastern portions of the Valles Marineris canyon floor, including Eos, Ganges, and Capri Chasmata. Those features transition into the Iani, Margaritifer, Aurorae, and Aureum Chaoses that dominate much of the remaining northwestern and north–central region of the quadrangle. The heavily cratered, high-standing plains of Noachis Terra, at 1 km above datum, in the southern portion of the map, transition into the moderately cratered, middle-relief plains of Arabia Terra to the northeast. The elevation descends to 3,000 m below datum in parts of Xanthe and Margaritifer Terrae and drops another 1,000 m or more into eastern Valles Marineris. Located in a portion of Arabia Terra, Meridiani Planum was the hematite-rich landing site of the Mars Exploration Rover-B, Opportunity (Figure 19.A; Christensen and Ruff, 2004; Christensen *et al.*, 2005). A portion of the

Xanthe Terra highlands in the northwest shows how the formation of Valles Marineris isolated some of the highland terrains. The large channel and lake system of the Uzboi–Ladon–Morava (ULM) Valles runs diagonally from the south–southwest to the north–northeast, through several large ancient crater basins, thought to have held lakes, including Holden crater.

Geology

The ancient highland crust dominates the landscape, with tectonic and erosional features leaving their scars across the area. The large ULM channel and lake system carried large volumes of water, draining as much as ~9 percent of the planet (Grant and Parker, 2002). Some of the water from the ULM and many smaller channel systems, along with precipitation and subsequent runoff water, seeped back into the ground, where it collected. This water was held until it violently discharged due to one or more of the following:

the opening of Valles Marineris, Tharsis tectonic activity, or impact events. Whichever event(s) is responsible, a catastrophic discharge of the underground reservoirs ensued, leaving void space in the subsurface. Overlying materials collapsed into the voids, resulting in the large blocky fields that remain. Water discharged from the chaotic terrain is the likely source of the floods that carved large outflow channels, north of the chaotic regions (see MC-11).

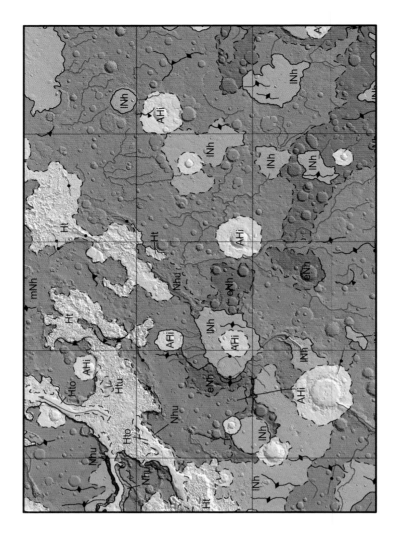

Structure

Interpretation
— Caldera rim
| Channel axis
|| Crater rim
•— Graben axis
▲ Lobate flow
▲ Outflow channel
⬦ Pit crater chain
◇ Ridge
•— Rille
◀ Scarp
•— Spiral trough
◆ Wrinkle ridge
↕ Yardangs

Geology

Impact unit
AHi

Transition units
Hto
Ht
Htu

Highland units
mNh
lNh
eNh
Nhu
HNhu

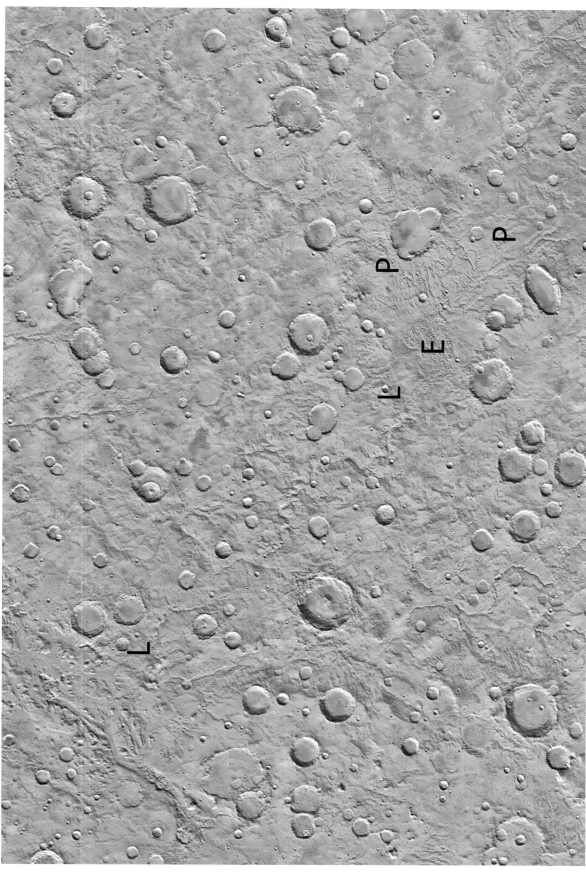

FIGURE 19.B THEMIS daytime infrared and MOLA color elevation mosaic of the Paraná–Loire Valles drainage system in eastern Margaritifer Terra. The dendritic Paraná Valles (P) flow north and west to Erythraeum Chaos (E) and then into the Loire Valles system (L; view spans about 1,400 by 1,000 km; north at top, 10° S to 26° S, 332.5° E to 357.5° E, NASA/JPL-Caltech/Arizona State University).

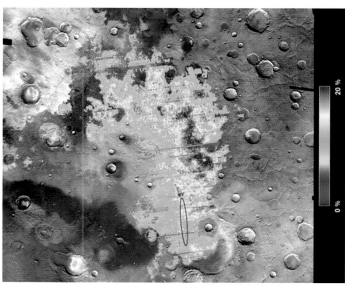

FIGURE 19.A Distribution of the mineral hematite as mapped by the TES instrument on Mars Global Surveyor, in Meridiani Planum in the northeast corner of the quadrangle (percentage derived from analysis of spectra; Christensen and Ruff, 2004). Hematite, an oxide of iron, commonly forms in the presence of water. The discovery of this hematite-rich area motivated its choice as the Opportunity landing site (in the black oval, see Figure 4.14 of the Opportunity traverse in Chapter 4; the base is the THEMIS daytime infrared mosaic, 640 km by 700 km, north at top, 7° S to 5° N, 350° E to 3° E, NASA/JPL-Caltech/Arizona State University).

The Atlas of Mars

FIGURE 19.C Drainage pattern in upper part (bottom, or south part of image) to middle part of Paraná Valles (center of image); the eastern part of Erythraeum Chaos is at the top. Near the center of the image a fault (arrow) appears to offset the valleys and plateaus. A meandering channel (at M) is pronounced on the floor of a valley, just north of the fault but not south of it, suggesting development after the tectonic activity (CTX image G02_018820_1574, 6 m/pixel, 30 km by 110 km, north at top, centered at 22.7° S, 348.9° E, NASA/JPL-Caltech/MSSS).

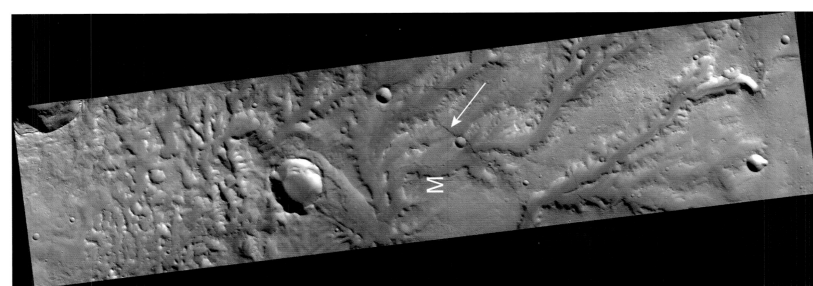

FIGURE 19.D Delta in Eberswalde crater with meandering distributary channels, comparable to those formed on Earth where rivers meet lakes or oceans. An incised channel that probably contributed sediment to the delta is visible at the lower left (MOC-NA high-resolution image mosaic MOC2–543a with color added from THEMIS data; 3 m/pixel, view about 15 km across, north at top, 326.4° E, 23.8° S; credit NASA/JPL-Caltech/MSSS).

Valleys, big and small

Some of the most interesting regional-scale geomorphologic features in this area are the dense, preserved channel forms, both large and small. While the central portion of the quadrangle is dominated by the large ULM outflow (Irwin and Grant, 2013), Noachis and Arabia Terrae are grooved by long, largely individual narrow channels as well as by densely networked tributary channel systems. Many of the channel networks are similar to those found on the surface of the Earth, associated with precipitation-fed water runoff, which typically form dendritic, or branching, tree-like patterns. An example of a dendritic pattern is Paraná Valles (Figure 19.B). The tributaries, or "branches," collect water and sediment from a large area, and transport the effluent into the main channel, or "trunk" (Figure 19.C). The fluid transported by the Paraná Valles branches first ponded in the Erythraeum Chaos, before continuing westward into the Loire Valles trunk. The Loire Valles system has few small tributaries along its ~1,000 km path through the highlands, before it intersects Morava Valles in the ULM system (Grant and Parker, 2002; Grant et al., 2009).

On the delta

The Margaritifer Sinus quadrangle contains multiple examples of deltas, which mark the intersection of running water with standing water. On Earth, this happens where rivers run into lakes or into the oceans, e.g. the Mississippi river delta. These features have a fan shape, indicative of the spreading out of water and the material carried by the water as the river escapes the narrow confines of its valley. A wonderful example of preserved delta morphologies is located in Eberswalde crater, a small basin located northnortheast of Holden crater (Figure 19.D). This delta has inverted linear features, interpreted to be the paths of the distributary channel network, surrounded by lobes of sediment, deposited as the running water slows and mixes with the standing water (Malin and Edgett, 2003; Wood, 2006; Malin et al., 2010). The antitheses of the tributaries in the upper portions of river systems, the distributaries, dispense the water and material suspended in the water, spreading it over the basin floor. As the moving water slows, the materials settle out of the water, and build up a lobate deposit. Eventually, the height of an active lobe becomes too tall, and the water veers off to an easier path to make deposits. As this process repeats, multiple overlapping lobes of materials create a fan shape. The multiple lobes and layering are indications that water endured and was cyclic in the Martian past – in this case about 3.5 billion years ago. Erosion at the distal margins of the fan exposes layering that may include clues to climatic conditions, source materials, and even possible past life on Mars.

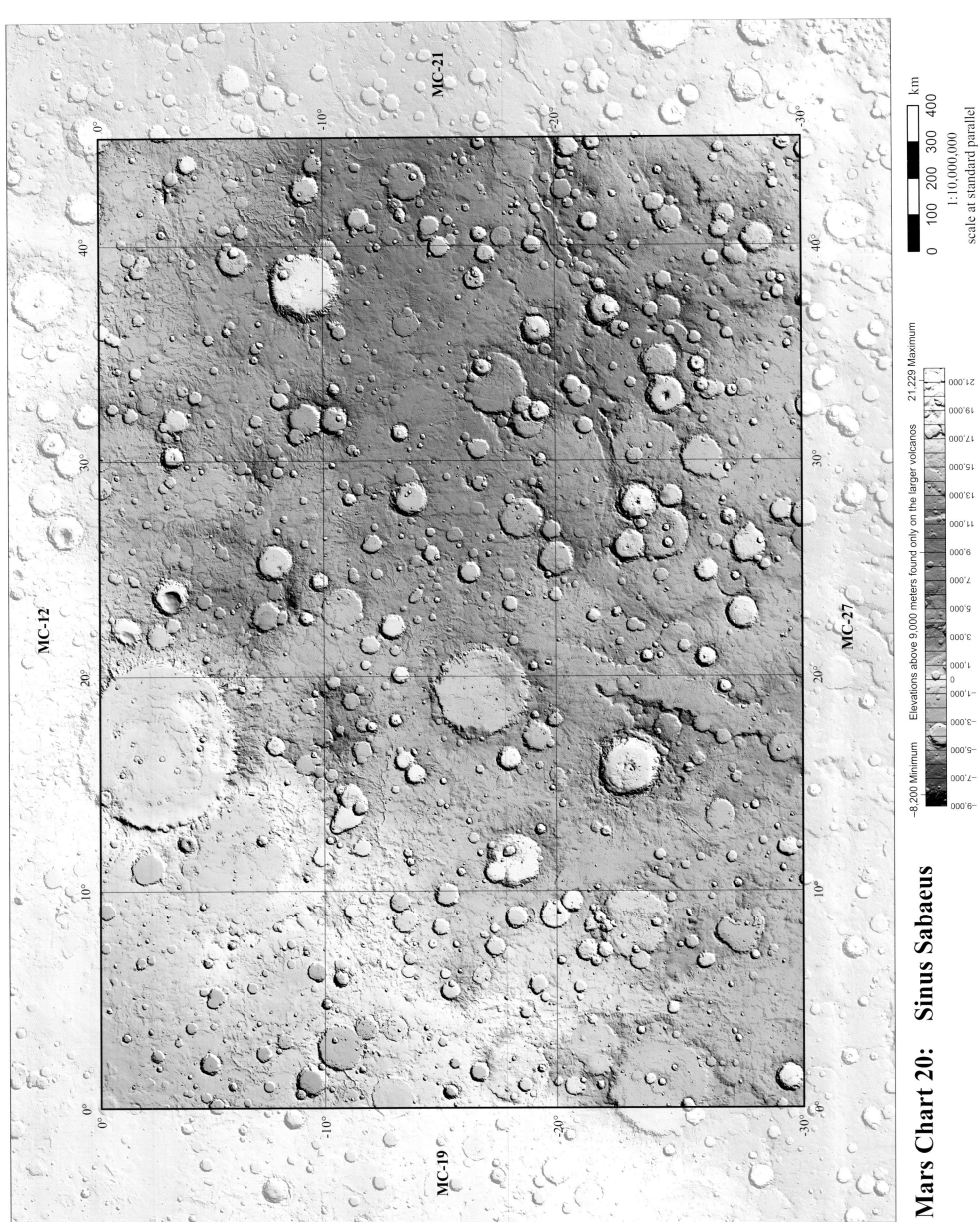

Mars Chart 20: Sinus Sabaeus

1:10,000,000
scale at standard parallel

0   100   200   300   400
km

−8,200 Minimum   Elevations above 9,000 meters found only on the larger volcanos   21,229 Maximum

Elevation in meters

-9,000 -7,000 -5,000 -3,000 -1,000 0 1,000 3,000 5,000 7,000 9,000 11,000 13,000 15,000 17,000 19,000 21,000

MC-21

MC-12

MC-19

MC-27

0°   -10°   -20°   -30°

0°   10°   20°   30°   40°

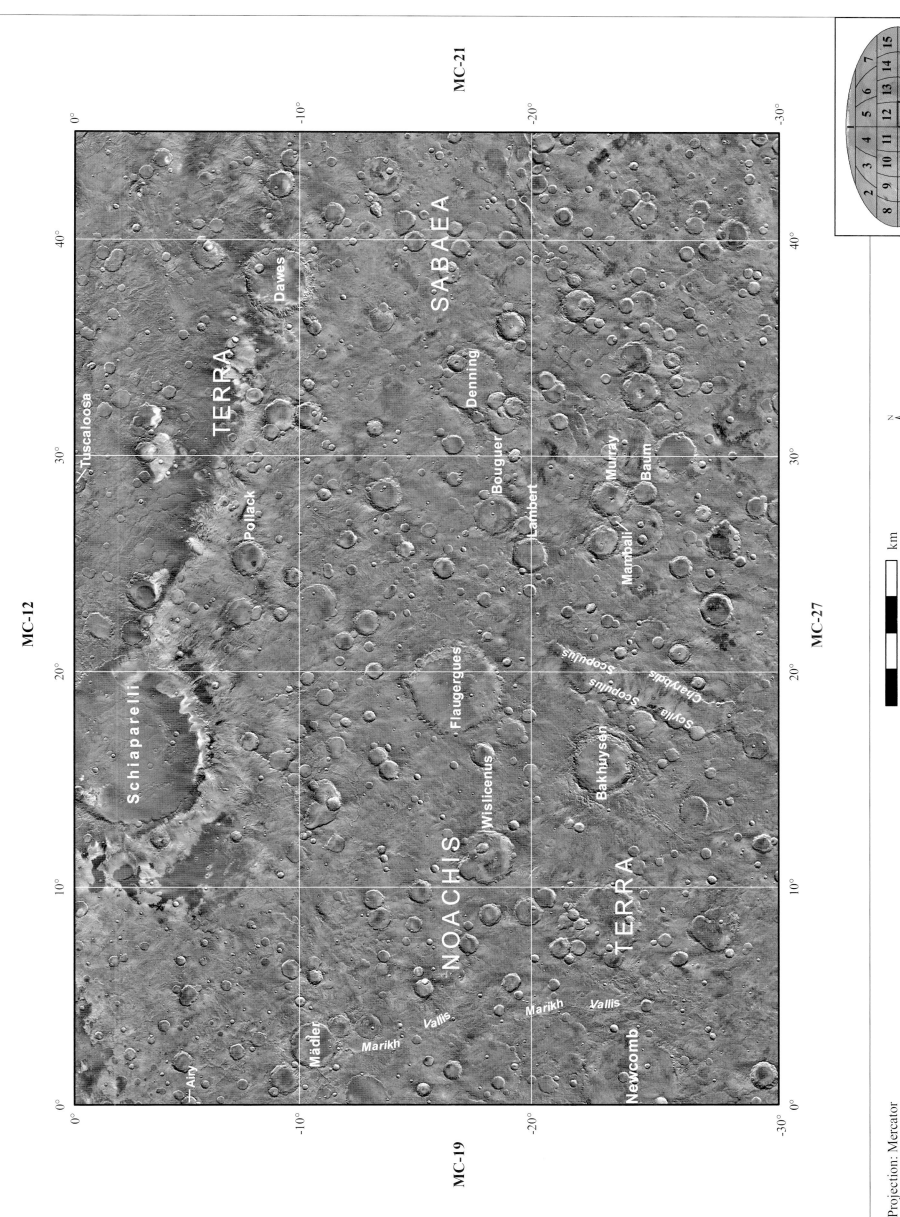

MC-21

MC-12

MC-19

MC-27

Tuscaloosa

Dawes

TERRA

SABAEA

Denning

Pollack

Bouguer

Murray

Lambert

Baum

Mambali

Schiaparelli

Flaugergues

Scopulus

Scopulus

Charybdis

Scylla

Bakhuysen

Wislicenus

NOACHIS

TERRA

Mädler

Marikh

Vallis

Marikh

Vallis

Newcomb

Ainy

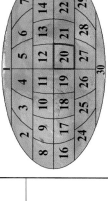

0   100   200   300   400
                                    km

1:10,000,000

scale at standard parallel

Projection: Mercator
Datum: Mars 2000 Sphere
Central Meridian: 0.0000
Standard Parallel 1: 0.0000

Sinus Sabaeus (MC-20)

Geography

The rugged highland terrains of Noachis Terra and Terra Sabaea dominate the region. The higher-standing, tectonically deformed, and densely cratered Terra Sabaea contains Scylla and Charybdis Scopuli. Also present are Denning, Bouguer, Lambert, Dawes, Pollack, Schiaparelli, Tuscaloosa, and Bakhuysen craters. To the west, the relatively subdued, but still rugged highland region of Noachis Terra has Newcomb, Wislicenus, and Mädler craters; and Marikh and a portion of Evros Valles. Numerous other moderately to highly degraded craters are scattered throughout the area. Valley networks ranging from tens of kilometers to thousands of kilometers in length dissect much of the topography. Wide grabens scar parts of Terra Sabaea. The region slopes from close to 3,000 m above datum in Terra Sabaea to as low at −1,500 m in the northwest. The northeast region is a portion of the zone that occurs between highland terrains to the south and transition terrain of Arabia Terra to the north (MC-12).

Geology

Similarly to other highland regions, the deposits in the area likely include volcanic, sedimentary, impact, and eolian-derived materials. The highland units span the Noachian Period and are interspersed with Amazonian/Hesperian crater materials. Large grabens are present in the area and are thought to be due to stresses created by impacts that formed Hellas and Isidis Planitiae, which lie to the southeast (MC-27, MC-28) and east-northeast (MC-13, MC-14), respectively. Schiaparelli crater is a multi-ring impact, overlapping the northern border of the quadrangle. The floor of this crater contains smaller impacts with layered deposits.

A crater within a crater within a crater

Schiaparelli crater is an interesting impact with not one, but at least two, and possibly three visible rims. Craters formed by one impact event but with multiple, concentric rims are common on Mercury, the Moon, and Mars and are called multi-ring impact basins (Spudis, 1993). Sometimes the impactor, or meteorite, is so large and moving so fast that when it hits, it forms a different kind of

crater from the normal bowl shape. Schiaparelli has two readily visible rings; the inner ring has a diameter of ~200 km, while the outer one is ~415 km across. This overlaps what appears to be an older, highly degraded, multi-ring impact basin to its west-southwest (Figure 20.A).

On the floor of Schiaparelli there are several craters with layered deposits, but one of these stands out above the rest. This unnamed, 2.3-km-wide crater has multiple, nearly circular layers, exposed in an outstanding example of "layer-cake" stratigraphy (Malin Space Science Systems, 2003; Figure 20.B). The rhythmic layers indicate a repetitive process, which may represent lake, volcanic ash, and/or eolian deposits. The layers likely covered the bottom of the crater and have the current form thanks to eolian erosion, indicated by the yardang development on the surface.

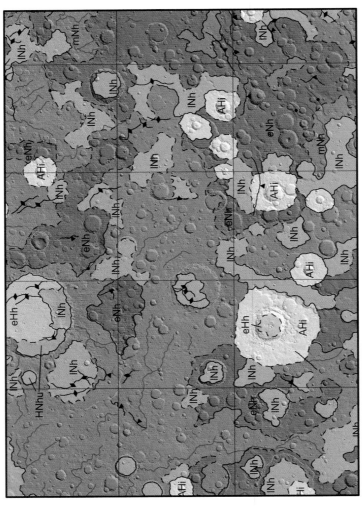

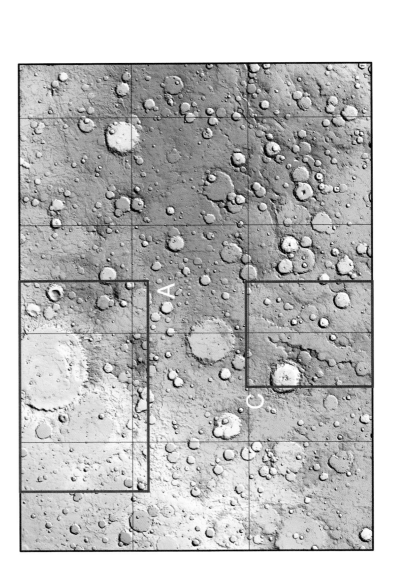

Structure	Geology
Interpretation	**Impact unit**
—— Caldera rim	AHi
—— Channel axis	**Highland units**
—— Crater rim	mNh
•—— Graben axis	lNh
▲ Lobate flow	eHh
▲ Outflow channel	eNh
◦—— Pit crater chain	HNhu
◇—— Ridge	
●—— Rille	
◄—— Scarp	
●—— Spiral trough	
◆—— Wrinkle ridge	
↕ Yardangs	

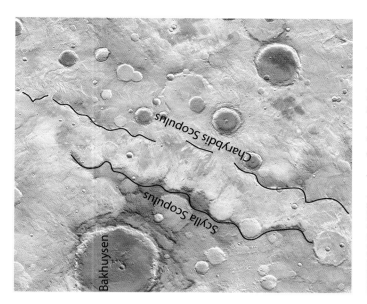

FIGURE 20.C MOLA color shaded relief (local rather than global color scheme) over THEMIS daytime infrared, showing Scylla and Charybdis Scopuli. A broad, apparently fault-bounded graben, outlined by the scopuli, is ~75 km wide (north at top, view spans 20° S to 30° S, 15° E to 25° E, NASA/JPL-Caltech/ Arizona State University).

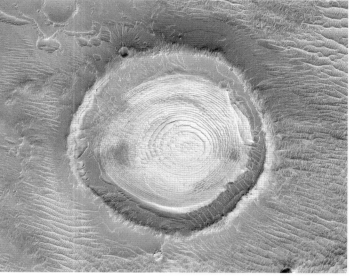

FIGURE 20.B Stack of layered rocks in an unnamed crater within Schiaparelli crater (see location in Figure 20.A). The illumination is from the left (west), and it appears that the center is the highest-standing region (MOC-NA image R0600195, 2 m/pixel, crater is 2.3 km wide, north at top, 0.9° S, 13.85° E, NASA/JPL-Caltech/MSSS).

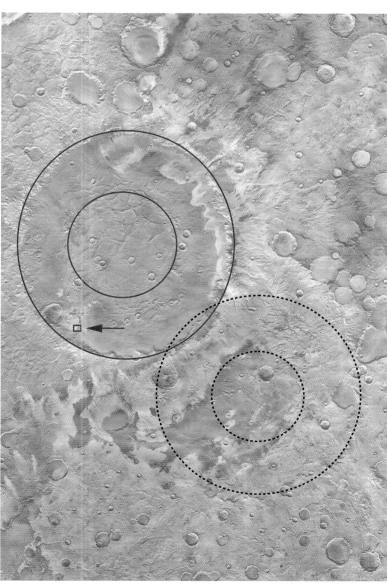

FIGURE 20.A MOLA color shaded relief (local rather than global color scheme) over THEMIS daytime infrared, showing the multi-ring impact Schiaparelli (solid red lines) and an unnamed multi-ring basin to its southwest (broken black lines). The inner ring of Schiaparelli is ~200 km in diameter and the second ring is ~415 km in diameter. Note the location of Figure 20.B (small box shown by arrow; north at top, view spans 11.5° S to 1.5° N; 5° E to 25° E, NASA/JPL-Caltech/Arizona State University).

The scope of the Scopuli

Scylla and Charybdis Scopuli are scarps on either side of a depression that averages ~75 km wide along its ~500-km-long scar, to the southeast of Bakhuysen crater (Figure 20.C). This type of feature is called a "graben." It is the surficial expression of extensional (or pulling apart) strain, and its enormous size indicates a lot of stress relief. The graben's orientation, circumferential to Hellas Planitia (southeast of the quadrangle), indicates that it likely formed as a response to the

excavation of the largest well-preserved impact structure on Mars – Hellas basin. The basin is believed to be ~4 billion years old, and therefore, this graben is the same age. Scylla Scopulus, the westernmost scarp, has a relief of ~1 km, and Charybdis Scopulus has a relief of ~800 m. For the past 4 billion years, this graben has undergone erosion and has been filled with deposits, indicated by both the degraded craters and the circular alcoves along the scarps. Currently, the relief is only ~1 km but it very likely exhibited much higher relief when it was a younger feature.

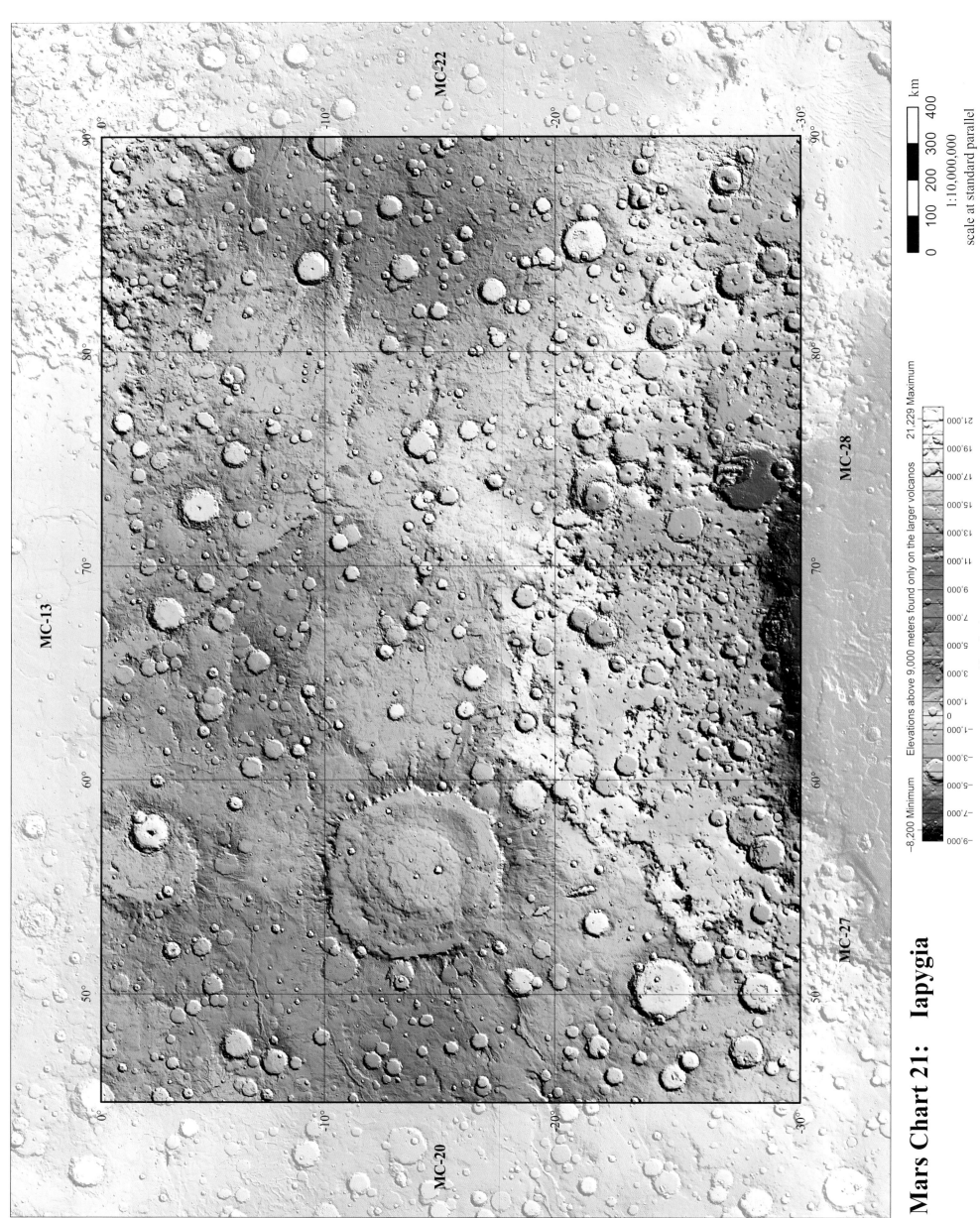

Mars Chart 21: Iapygia

1:10,000,000
scale at standard parallel

km
0 100 200 300 400

−8,200 Minimum   Elevations above 9,000 meters found only on the larger volcanos   21,229 Maximum

−9,000 −7,000 −5,000 −3,000 −1,000 0 1,000 3,000 5,000 7,000 9,000 11,000 13,000 15,000 17,000 19,000 21,000

Elevation in meters

MC-13

MC-22

MC-20

MC-27

MC-28

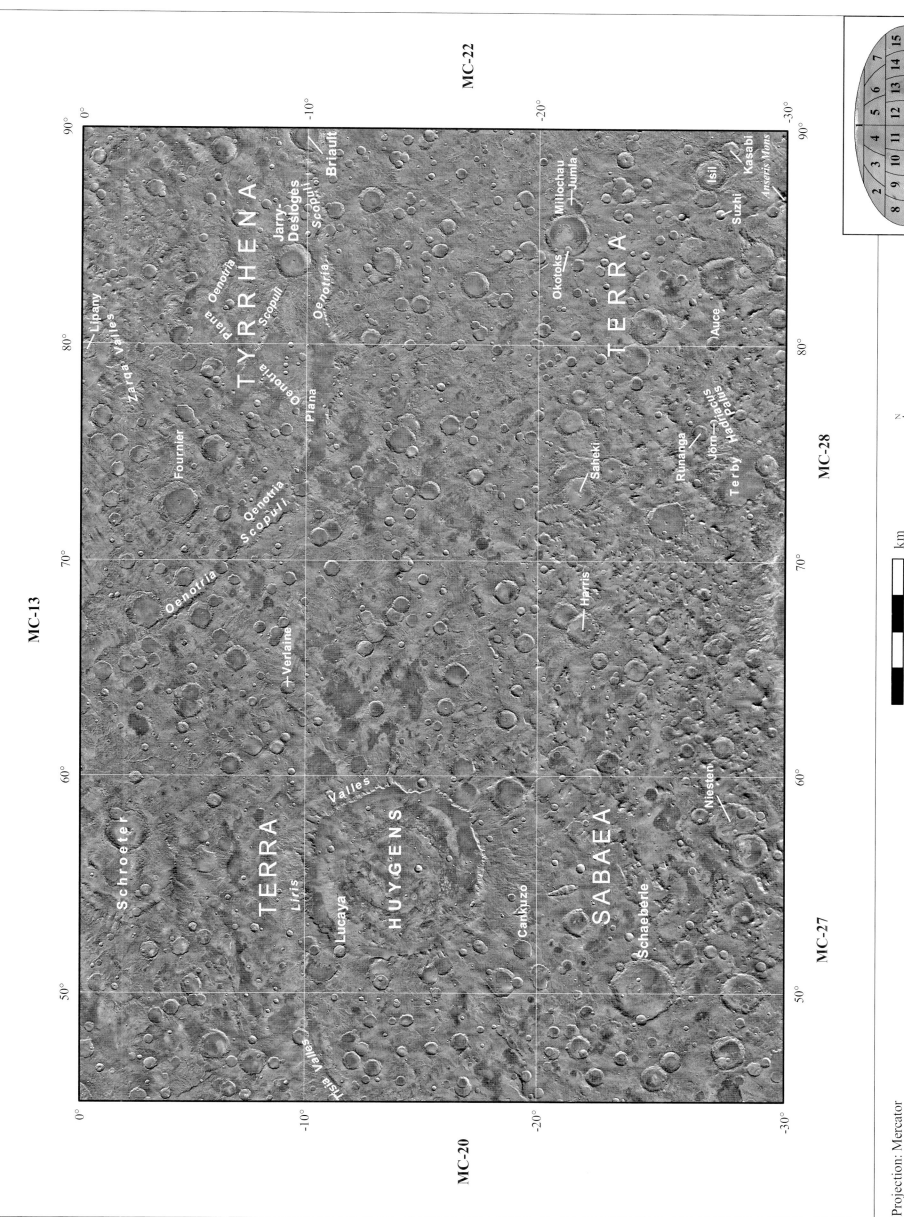

MC-13

MC-22

MC-20

MC-28

MC-27

0°

90°   80°   70°   60°   50°   0°

-10°

-20°

-30°

TYRRHENA

Lipany
Zarqa Valles
Fournier
Oenotria Plana
Oenotria Scopuli
Oenotria
Oenotria Scopuli
Verlaine

Jarry-
Desloges
Scopuli
Plana
Briault

TERRA
Schroeter
Liris
Lucaya
Tisia Valles

HUYGENS
Valles
Cankuzo

SABAEA
Schaeberle
Niesten

Harris

Saheki

Okotoks
Millochau
Jumla

TERRA

Runanga
Jörn—Dacus
Terby Hadriacus
Terby

Auce

Isil
Suzhi
Kasabi
Anseris Mons

N
W   E
S

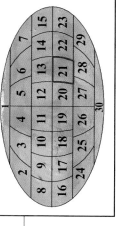

km
0   100 200 300 400
1:10,000,000
scale at standard parallel

Projection: Mercator
Datum: Mars 2000 Sphere
Central Meridian: 0.0000
Standard Parallel 1: 0.0000

Iapygia (MC-21)

Geography

The Iapygia quadrangle consists almost entirely of heavily cratered highlands, as high as 3 km above datum, descending to the northern basin rim (0–3 km below datum) and floor (3 to over 5 km below datum) of Hellas, and a piece of the southwestern rim of Isidis basin. Terra Sabaea makes up the western two-thirds of the quadrangle, whereas Tyrrhena Terra makes up the third that is east of a topographic divide, at ~75° E. An arcuate, north-facing series of scarps, Oenotria Scopuli, crosses this divide and appears to be concentric to Isidis basin. Huygens forms a prominent impact basin, with an outer rim of ~470 km in diameter, and it has an inner (250-km-diameter) and partial intermediate (350-km) ring.

Geology

The Hellas basin impact is the dominant, Early Noachian geologic event that shaped this region (Tanaka and Leonard, 1995; Leonard and Tanaka, 2001). Along with the basin cavity, the impact appears to have deposited a thick,

extensive ejecta blanket that accounts for the higher plateau areas in the terrae. The inner Hellas rim consists of scattered massifs and intervening plains, attesting to extensive mass wasting of this part of the rim. The main scarp of Oenotria Scopuli appears to be an outer ring structure of Isidis basin, which cuts what may be dominantly Hellas ejecta, thereby demonstrating that the Isidis impact postdated the Hellas impact (Tanaka et al., 2014). Ridges and scarps in the western half of the quadrangle mostly trend northeast to east, roughly concentric to Hellas and locally radial to Isidis and Huygens; these features are likely fault systems, resulting from the impacts, and rejuvenated later by other tectonic activity, including planetary contraction due to cooling (Watters, 1993). During the Middle Noachian to Late Noachian Epochs, precipitation led to dissection of higher terrain, depositing sediments in lower areas. Along the northern margin of Hellas, eolian accumulations filled in some local catchment areas.

Over the top!

Terby crater, 170 km across, includes a rather unusual sequence of layered deposits (Wilson

et al., 2007; Ansan et al., 2011). Most of Terby is filled to a level that rises to the lowest, southern point of the crater's rim, about 4,500 m below datum. However, the northeastern sector is complex (Figures 21.A, 21.B) and includes rounded depressions, as much as 500 m deep, and a couple of irregular plateaus that rise as high as 1,700 m below datum – above most of the crater rim!

Likely, the eroded deposit is heavily dust-laden and poorly cemented. Some or most of the lower part of the deposit may have formed by runoff – a valley system crosses the crater rim on its east side. The upper section seems to require air-fall deposition, perhaps related to volcanism and/or a climate with a more active, possibly denser, atmosphere.

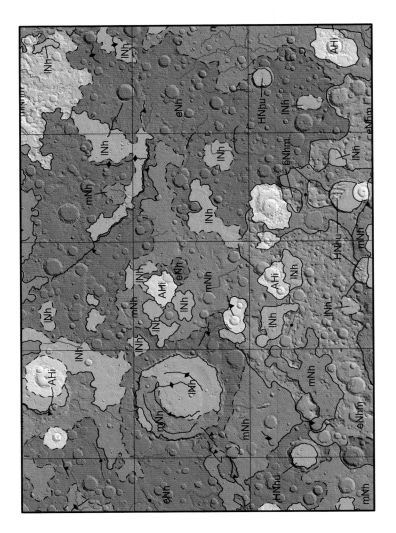

FIGURE 21.A View to the north–northwest of layered deposits within Terby crater (HRSC perspective of image from orbit 4199, 13 m/pixel, view about 50 km across, centered at 27.5° S, 74.5° E, ESA/DLR/FU-Berlin). Figure 21.B shows the small ridge at the center of the left edge of this image.

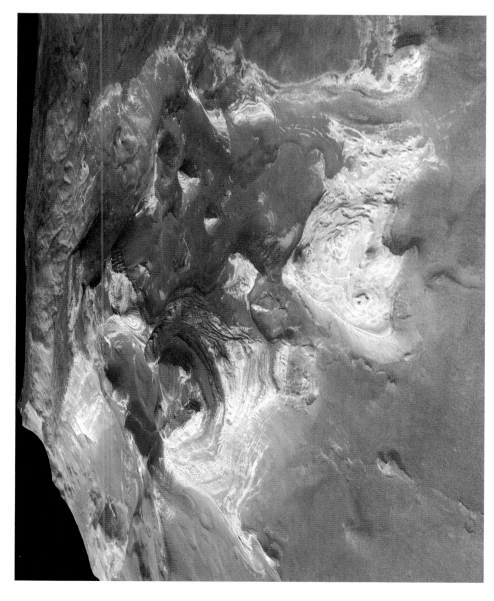

Hadriacus Palus

Just to the east of Terby, in Hadriacus Palus, are clues to the origin of layered deposits that fill parts of both Terby and surrounding areas. Hadriacus Cavi are irregular depressions, with walls formed by steep cliffs that show a cross-section of the rock layers (Figure 21.C). Figure 21.D shows a cliff of columnar-jointed rock that may be lava or an ignimbrite/ash deposit (Fortezzo and Skinner, 2012). On Earth, columnar joints form when a cooling lava or ash flow contracts, creating regular, usually vertical, fractures. The light-gray layer above the columnar-jointed rock shows channel-shaped forms, suggesting fluvial transport and

reworking during the filling of the region (Figures 21.E, 21.F). Above this, on the cliff in Figure 21.F, can be seen outcrops of darker layers that slope gently to the north in the cliff face (Skinner and Fortezzo, 2012).

Double "whammy" digs up the past

The oldest rocks on Mars tend to be deeply buried over time, as younger rocks cover them up. Occasionally, however, a large impact may excavate deep into the crust and reveal otherwise hidden materials. Such is the case for a spot that was uplifted by Huygens crater, then further exposed by the 35-km-diameter Lucaya crater on Huygens'

FIGURE 21.B Detail of layered deposits in Terby crater (HiRISE PSP_001662_1520, red filter, 25 cm/pixel, 5 km by 7.5 km, north at top, NASA/JPL-Caltech/University of Arizona).

FIGURE 21.C Hadriacus Cavi show dunes on the floors while surrounding cliffs display details of the rocks that once filled the region (part of HiRISE image PSP_006198_1525_RED, 25 cm/pixel, view 1.25 km across, north at top, centered at 27.31° S, 78.07° E, NASA/JPL-Caltech/University of Arizona). The boxes indicate detailed views in Figures 21.D (north box), 21.E (southwest box), and 21.F (east box).

rim. On the floor of Lucaya are light-toned rocks, rich in carbonate (Figure 21.G), which may have previously been buried to a depth of 5 km. Scientists think that the carbon-dioxide-rich atmosphere was perhaps a hundred times thicker on early Mars than at present, when precipitation of water from the atmosphere was once possible. However, at the end of the Noachian Period, precipitation became much less common, probably due to thinning of the atmosphere. If carbonate deposits like these are widespread within the crust of Mars, this would explain where that early carbon dioxide atmosphere went (Wray et al., 2011).

FIGURE 21.D Detail showing columnar jointed lavas (along the cliff in the northwest part of image), the tops of which are also evident on the slope farther to the southeast (view 270 m by 325 m, north at top).

FIGURE 21.E (left) Detail of layered rocks, showing cross-section of asymmetrical fluvial channel (light, arcuate shape at center of image; view 270 m by 230 m, north at top).

FIGURE 21.F (above) Detail with dunes on valley floor, columnar lava, light layer with channel forms, and dark layered section in which some light layers are contiguous with benches on the top of the mesa (view 270 m by 325 m, north at top).

FIGURE 21.G (left) Detail of a color HiRISE image (ESP_012897_1685, IRB version, 50 cm/pixel, view 1 km across, north at top, 11.6° S, 51.9° E, NASA/JPL-Caltech/University of Arizona), showing light-colored layers at right center, having iron and/or calcium-rich carbonate minerals.

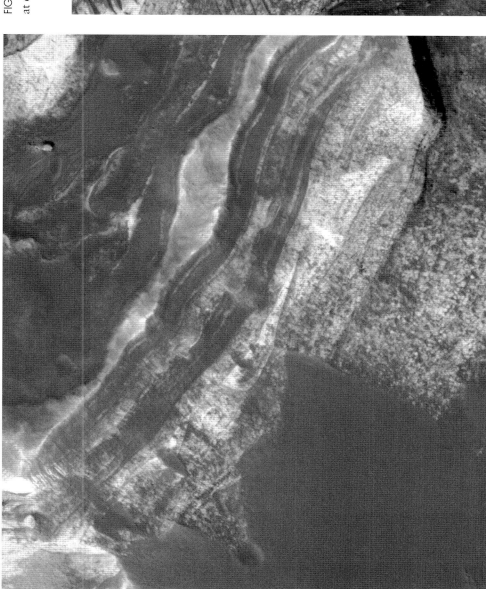

185

The Atlas of Mars

Mars Chart 22:   Mare Tyrrhenum

MC-23

MC-14

MC-21

MC-29

MC-28

km

1:10,000,000

scale at standard parallel

−8,200 Minimum   Elevations above 9,000 meters found only on the larger volcanos   21,229 Maximum

Elevation in meters

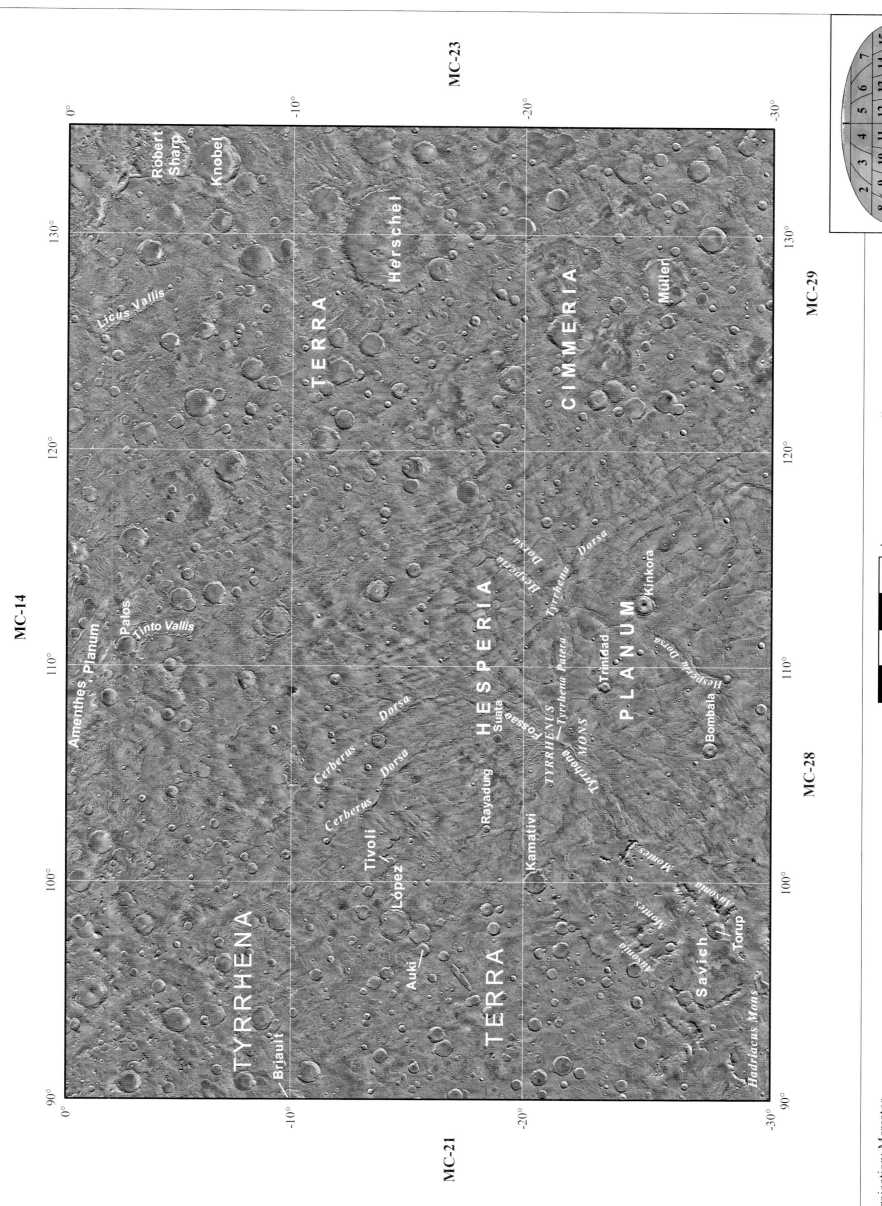

MC-23

MC-29

MC-14

MC-28

MC-21

0° -10° -20° -30°

TYRRHENA

Briault

Robert Sharp

Knobel

Licus Vallis

Amenthes Planum

Palos

Tinto Vallis

TERRA

Herschel

TERRA

Auki

López

Tivoli

Cerberus

Cerberus Dorsa

Dorsa

HESPERIA

Rayadurg

Suata

Fossae

Kamativi

Kinkora

CIMMERIA

Müller

Hesperia Dorsa

Tyrrhena Dorsa

Tyrrhena Patera

TYRRHENUS MONS

Tyrrhena

Trinidad

Bombala

PLANUM

Hesperia Dorsa

Savich

Torup

Ausonia Montes

Ausonia Montes

Hadriacus Mons

0° 130° 120° 110° 100° 90°

N
W E
S

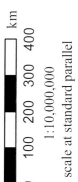

km

0   100 200 300 400

1:10,000,000

scale at standard parallel

Projection: Mercator
Datum: Mars 2000 Sphere
Central Meridian: 0.0000
Standard Parallel 1: 0.0000

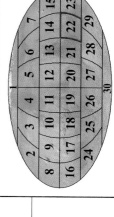

Mare Tyrrhenum (MC-22)

Geography

This quadrangle, most of which lies 1–3 km above datum, consists of northern and central Hesperia Planum, a wrinkle-ridged volcanic plain, bordered by the cratered highlands of Tyrrhena Terra to the west and Terra Cimmeria to the east. In the center of Hesperia Planum lies the broad Tyrrhenus Mons shield. The northern part of Hadriacus Mons and a few outer massifs of Hellas basin, Ausonia Montes, crop out in the southwest corner of the quadrangle. Herschel crater forms a 275-km-diameter double-ring impact basin along the eastern margin of the quadrangle.

Geology

The building of Tyrrhenus Mons occurred during the Noachian Period along with impact bombardment of the surrounding highland surface. Subsequent erosion of the volcanic materials led to the furrows on the volcano and indicate that it is made up of ash produced by explosive eruptions, as a result of water within or interacting with the rising magma (Greeley and Crown, 1990; Gregg *et al.*, 1998). Eruptions from Tyrrhenus and

perhaps surrounding buried fissures became dominated by lavas, by the Hesperian Period, likely accounting for the planar infilling of Hesperia Planum. These materials were chosen as a reference for the timing of the beginning of the Hesperian Period on Mars (Scott and Carr, 1978). The Hesperian plains are moderately covered by generally well-preserved impact craters, indicating that they formed at an early time, after which the overall erosion of the Martian surface was greatly reduced. In contrast, older Noachian-age impact craters and other landforms display signs of more substantial degradation and even obliteration (e.g. Craddock and Howard, 2002; Irwin *et al.*, 2013). For example, the rim of Herschel crater has been deeply incised by water runoff valleys and bludgeoned by subsequent impact craters (Figure 22.A). The activity of Tyrrhenus Mons, though, was not over, and from it a distinctive field of tongue-shaped lavas flowed southwest in the direction of Hellas basin, toward the end of the Hesperian Period (Greeley and Crown, 1990).

Old and wrinkled

Alas, Hesperia Planum (Figure 22.B) is showing its age of more than 3.5 billion years. It is wrinkled

by eons of planetary contraction since lavas and ash originally had produced a once relatively smooth, flat surface, by burying nearly all of the underlying Noachian rolling and cratered landscape. Alternatively, Noachian volcanic activity, interacting with groundwater, together with

enhanced water runoff, contributed to planation of the Hesperia region prior to volcanic infilling. In any case, the pile of lavas amounts to a brittle skin, overlying the crushed rock that makes up the buried cratered terrain. Contraction of the lavas due to cooling of the planet led to faulting and

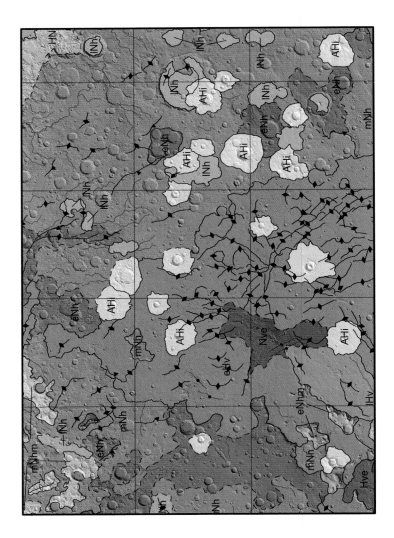

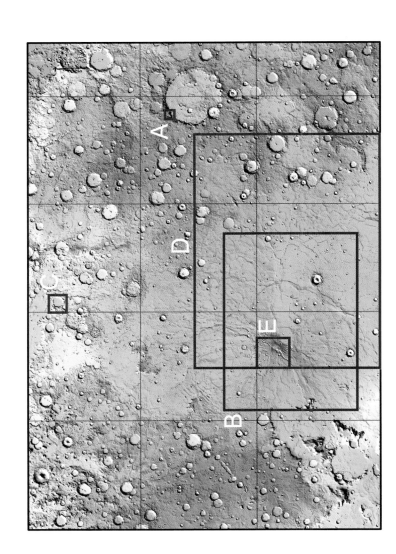

Structure	Geology
Interpretation	Impact unit
Caldera rim	AHi
Channel axis	Volcanic Units
Crater rim	eHv
Graben axis	Nve
Lobate flow	Hve
Outflow channel	lHv
Pit crater chain	Transition Units
Ridge	eHt
Rille	HNt
Scarp	Highland Units
Spiral trough	mNh
Wrinkle ridge	lNh
Yardangs	eNh
	eNhm
	mNhm

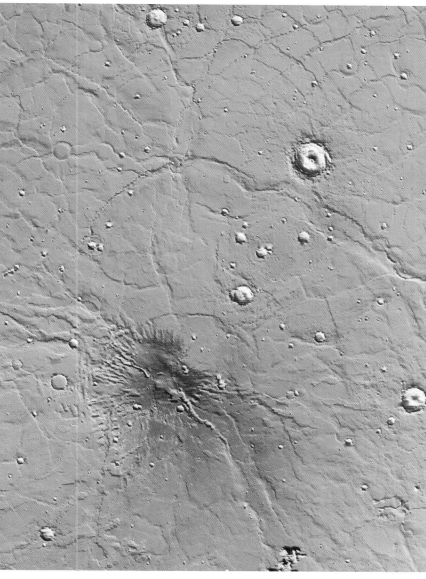

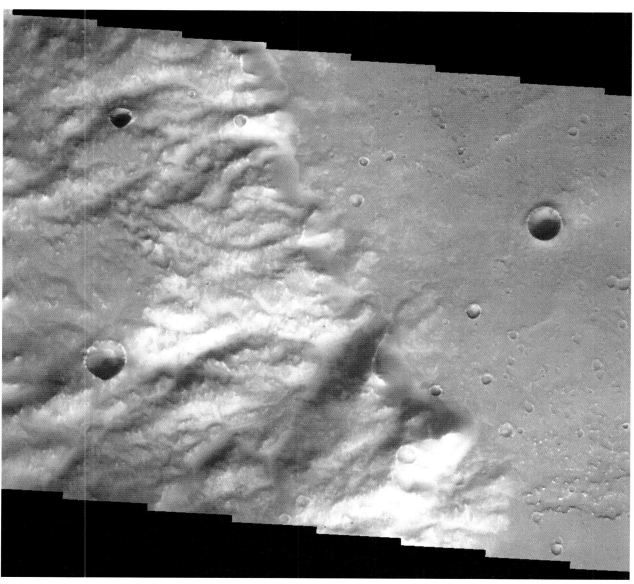

FIGURE 22.B MOLA Colored shaded relief of Tyrrhenus Mons and wrinkle ridges of Hesperia Planum; view spans about 600 km by 800 km (north at top, 18° S to 28° S, 102° E to 117° E).

sediments (Carr, 2006), as is the case at the Spirit landing site at Gusev (see MC-23).

A windy place

Eastern Hesperia Planum shows both light and dark streaks, associated with craters and other features. These show a pattern of prevailing winds (Figure 22.D), consistent with computer models of motions in the atmosphere of Mars. The bright streaks appear to be accumulations of very fine dust, while dark streaks are made of coarser sand and gravel, perhaps exposed by removal of fine dust (Hartmann, 2003; Carr, 2006).

FIGURE 22.A Rim of Herschel crater, pummeled by impact craters and carved by runoff channels (THEMIS visible image V07249002, 18 m/pixel, 18 km by 21 km, north at top, 12.5° S, 128.7° E, NASA/JPL-Caltech/Arizona State University).

folding, which produced sinuous, crossing, broad ridges, marked by narrower, sharply crested crenulations known as "wrinkle ridges" – like the wrinkles on the skin of a shriveled apple (Watters, 1993). The largest wrinkle ridges in Hesperia Planum extend for more than 500 km and exceed 30 km in width and 300 m in height.

Palos crater: what kind of flooring?

Palos shows channels entering and leaving through the rim (Figure 22.C). While the crater must have contained water when the channels were forming, the layered deposits on the floor could be primarily lava flows rather than

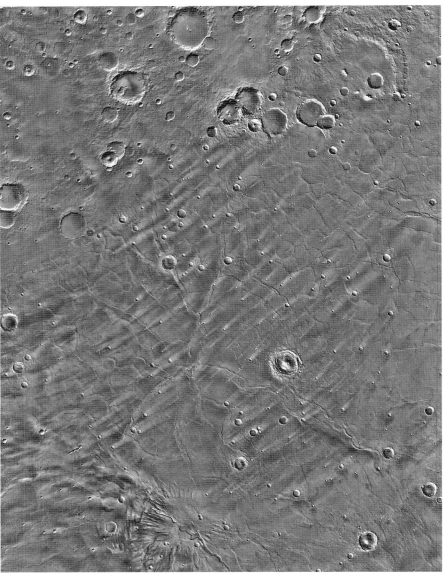

FIGURE 22.D   Eastern Hesperia Planum, showing streaks created by the prevailing wind (MDIM, 900 km by 1,200 km, north at top, 15° S to 30° S, 105° E to 127° E, USGS).

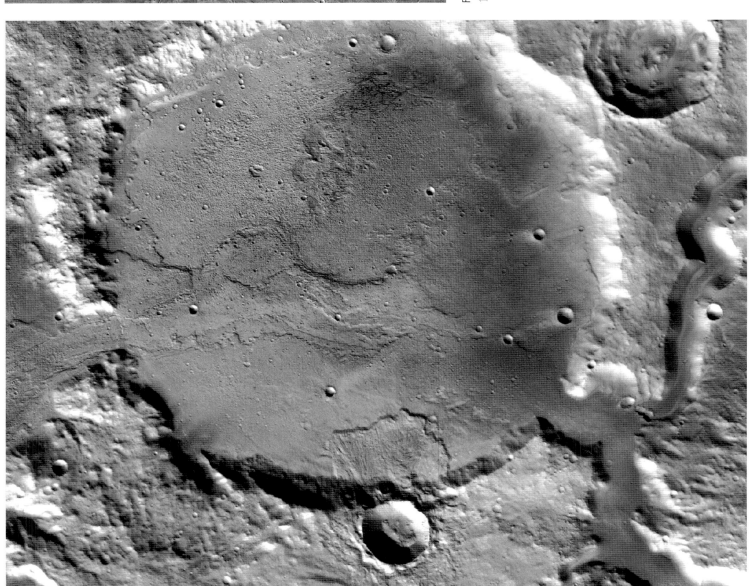

FIGURE 22.C   HRSC image of Palos crater, 55 km in diameter. Tinto Vallis enters from the south, while the breach in the northern rim leads to Amenthes Planum (H0962_0000_ND4, 12.5 m/pixel, north at top, 110.8° E, 2.7° S, ESA/DLR/FU-Berlin).

Two are better than one

Tyrrhena Patera lies at the summit of Tyrrhenus Mons. While other volcanoes on Mars show multiple calderas (circular regions of collapse due to withdrawal of magma), Tyrrhenus Mons displays a pair of possible calderas. These lie within a broad oval of fractures and pit chains that are connected by a broad trough (this composite depression makes up the patera). The floors of these features are covered by a continuous deposit of relatively young, presumably volcanic material. Channels with flat floors also run downslope from the caldera, dissecting layered flank materials, presumably made up largely of relatively easily eroded volcanic ash deposits (Figure 22.E).

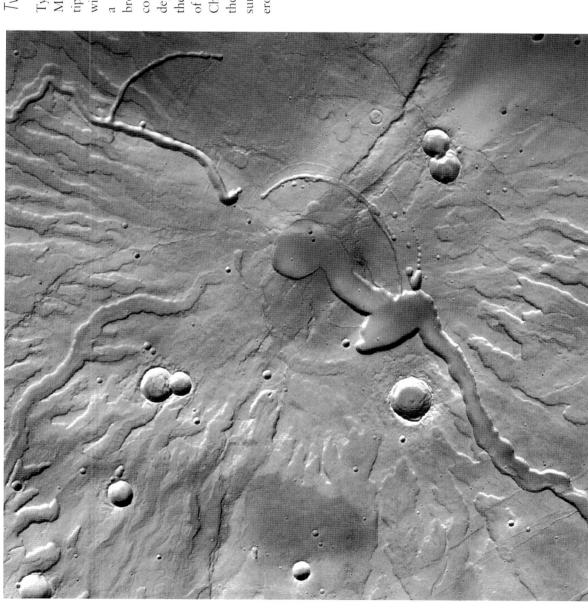

FIGURE 22.E Tyrrhenus Mons summit region, showing dual calderas, oval system of fractures and pit chains, and radiating channels cutting into eroded, layered flank materials (HRSC H1920_0000_ND4, 22 m/pixel, view 150 km across, north at top, centered at 22° S, 106° E, ESA/DLR/FU-Berlin).

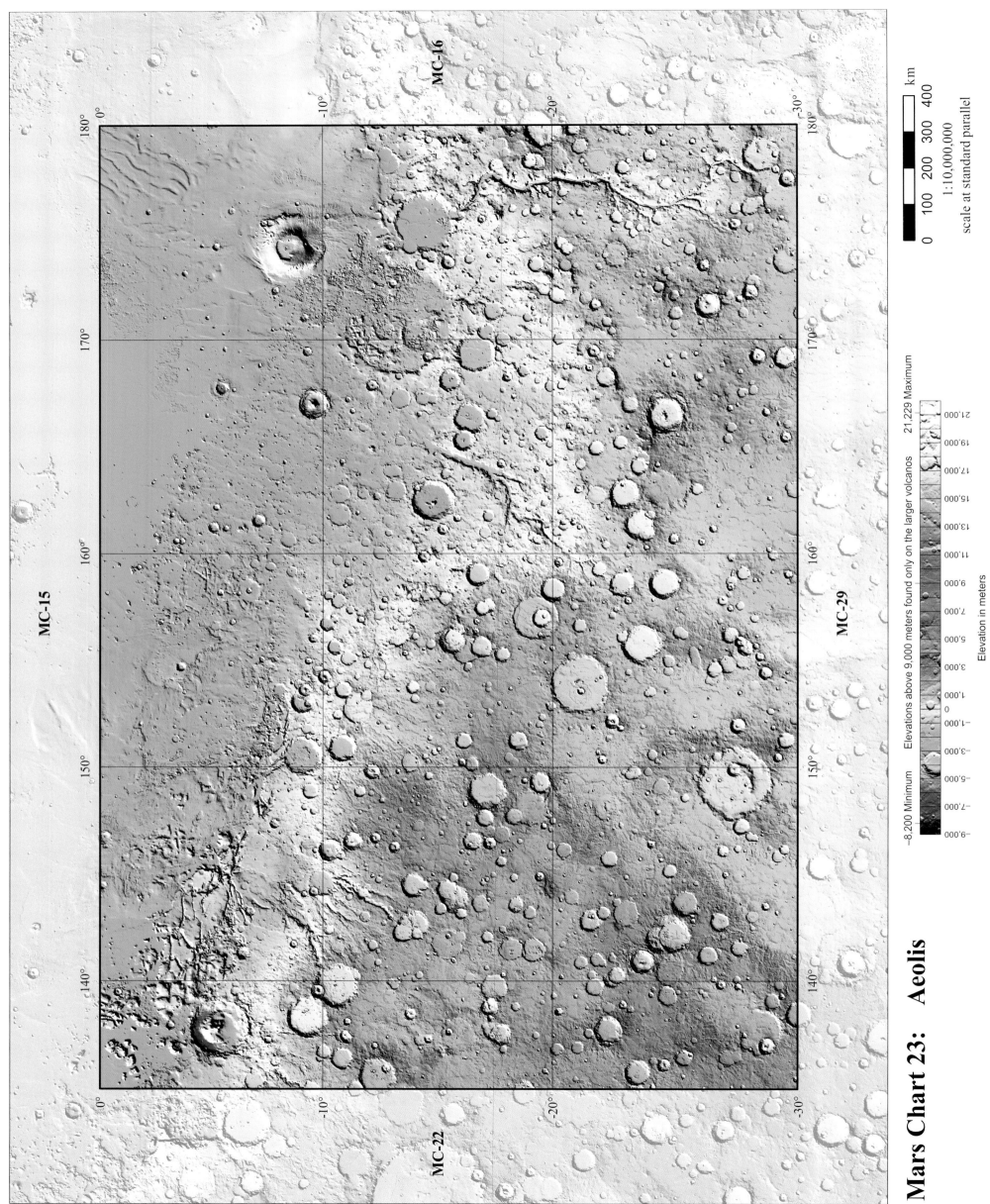

Mars Chart 23: Aeolis

MC-15
MC-16
MC-22
MC-29

km

1:10,000,000
scale at standard parallel

−8,200 Minimum   Elevations above 9,000 meters found only on the larger volcanos   21,229 Maximum

Elevation in meters

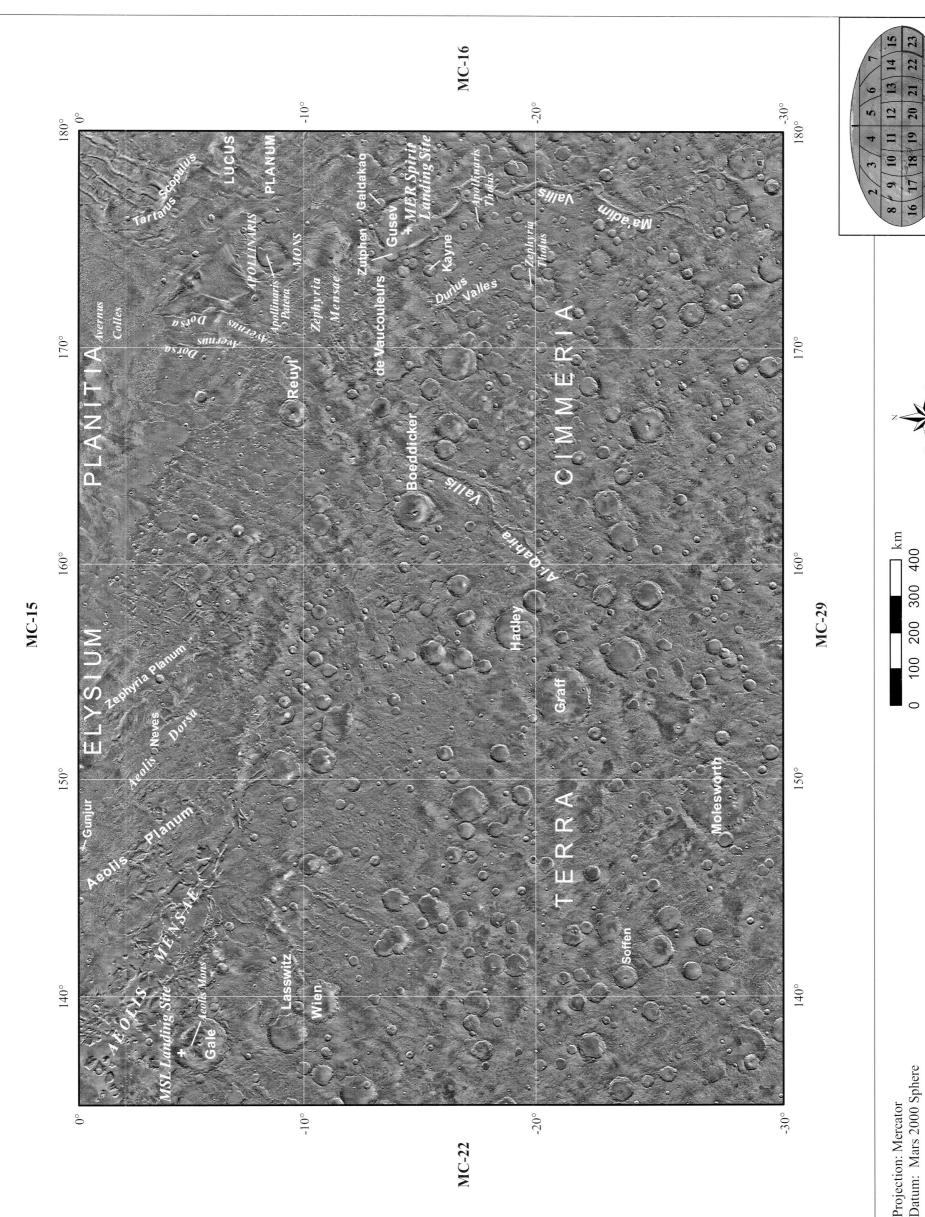

MC-15

MC-16

MC-22

MC-29

0° 180°

−10° 170°

−20° 170°

−30° 180°

ELYSIUM PLANITIA

LUCUS PLANUM

AEOLIS MENSAE

TERRA CIMMERIA

MSL Landing Site
Gale
Aeolis Mons
Lasswitz
Wien
Gunjur
Aeolis Planum
Zephyria Planum
Aeolis Neves Dorsa
Reuyl
Avernus Colles
Tartarus Scopulus
Avernus Dorsa
Avernus Dorsa
Apollinaris Patera
APOLLINARIS MONS
Zephyria Mensae
Zutphen Galdakao
Gusev
MER Spirit Landing Site
Kayne
Durius Valles
de Vaucouleurs
Boeddicker
Al-Qahira Vallis
Hadley
Graff
Soffen
Molesworth
Apollinaris Tholus
Zephyria Tholus
Ma'adim Vallis

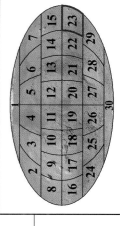

2 3 4 5 6 7
8 9 10 11 12 13 14 15
16 17 18 19 20 21 22 23
24 25 26 27 28 29
30

N
S
E
W

km
0 100 200 300 400

1:10,000,000

scale at standard parallel

Projection: Mercator
Datum: Mars 2000 Sphere
Central Meridian: 0.0000
Standard Parallel 1: 0.0000

Aeolis (MC-23)

Geography

The highland–lowland boundary transects the central part of the Aeolis quadrangle. Elevations range from 3 km above datum in the highlands to 3 km below datum in the lowlands at the north edge of the quadrangle. Aeolis Mensae and de Vaucouleurs display the rugged, gradational nature of the boundary in this region. Along the boundary is Gusev crater, the landing site of the Mars Exploration Rover (MER-A), Spirit (Squyres et al., 2004; Arvidson et al., 2006), while near the west edge of the map is Gale crater, location of the Mars Science Laboratory (MSL) rover, Curiosity (Golombek et al., 2012; Grotzinger et al., 2012). Ma'adim Vallis originates from a complex of proposed paleolake basins and enters Gusev crater from the south (MC-29; Irwin et al., 2002). About 200 km directly north of Gusev is a prominent, 200-km-wide and 5-km-high shield volcano, Apollinaris Mons.

Geology

The highland–lowland dichotomy in this quadrangle has degraded southward over time, through varying processes. These include fluvial modification such as spring-fed activity, glacial activity, solifluction of clay-enriched materials, and slumping (e.g. Squyres, 1989; Kargel et al., 1995; Tanaka et al., 2005; Fairén et al., 2011; Davila et al., 2013). Thus, the mélange of materials that partly infilled the northern plains includes sedimentary deposits from erosion of cratered highland materials. In addition, the far-reaching and diverse geologic and hydrologic activities recorded in the northern plains include (e.g., see both Tanaka et al., 2005 and Dohm et al., 2009b and references therein): impact cratering; magmatic- and impact-driven tectonism and related seismicity; periglacial and glacial modification; marine and paleolake activity; eolian, fluvial, debris-flow, and volcanic resurfacing; subterranean gas release such as sedimentary (mud) volcanism; possible burial of extensive ice; and possible impact-generated hydrothermal lakes.

Apollinaris Mons

This volcano is believed to have been constructed by both pyroclastic and effusive eruptions over time from the Late Noachian to the Late Hesperian Epochs, with several stages of development, including the formation of a multi-stage caldera, approximately 80 km in diameter, at its summit and late-stage emplacement of a distinct deposit on its southern flank (Figure 23.A; Robinson et al., 1993; Scott et al., 1993; Greeley et al., 2005). The volcano has been identified as a potential source

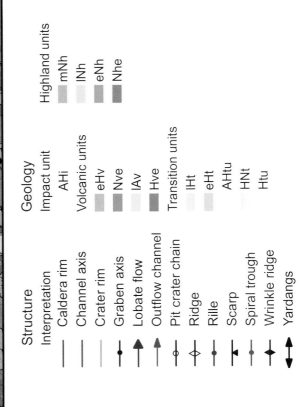

Structure

Interpretation

— Caldera rim
— Channel axis
— Crater rim
◆ Graben axis
▲ Lobate flow
→ Outflow channel
⊶ Pit crater chain
◇ Ridge
● Rille
◀ Scarp
● Spiral trough
◆ Wrinkle ridge
↕ Yardangs

Geology

Impact unit
AHi

Volcanic units
eHv
Nve
lAv
Hve

Transition units
lHt
eHt
AHtu
HNt
Htu

Highland units
mNh
lNh
eNh
Nhe

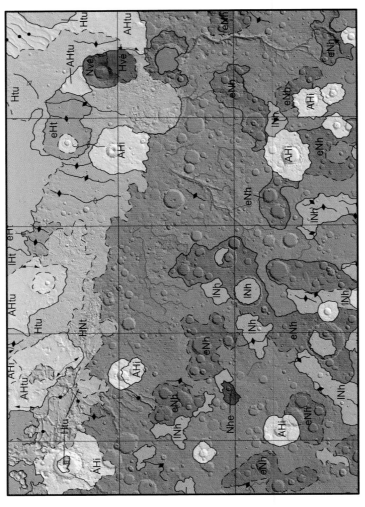

FIGURE 23.A Apollinaris Mons, showing clouds over the 80-km-wide summit caldera. The eruptive fan on the southern flank covers and postdates both the older flanks and the basal scarp of the volcano (MOC image PIA02006; composite of red and blue images, 500 m/pixel, view 200 km by 320 km, north at top, 6.5° S to 12° S, 172.3° E to 176° E, NASA/JPL-Caltech/MSSS).

for Medusae Fossae Formation materials[1] (Kerber and Head, 2010; El Maarry et al., 2012), as well as a contributor of materials and geochemical signatures that were encountered by Spirit within Gusev crater.

Chemical traces of water

Apollinaris Mons (Figure 23.A) developed in a water-enriched environment along the highland–lowland boundary (Scott et al., 1993). The volcano and its surroundings record extensive volcanism, movements of the ground, and movement of water, processes that likely interacted to create hydrothermal activity (Schulze-Makuch et al., 2007; El Maarry et al., 2012). Features that mark such activity include lava-flow materials, lahar-like flows, valleys, faults and fractures, mounds that possibly mark venting, and impact craters that are indicative of water-rich target materials, among others (El Maarry et al., 2012). Possible long-term magma–water interactions are consistent with GRS-based elemental information, which indicates elevated hydrogen and chlorine when compared to the rest of the equatorial region of Mars (Figure 23.B; Boynton et al., 2004; Dohm et al., 2008). In situ evidence of hydrothermal activity has also been identified by the Spirit rover at the Columbia Hills, within Gusev crater (Squyres et al., 2004). Such a history of water and energy interactions make Apollinaris Mons and its surroundings within the Aeolis quadrangle a prime target for future astrobiologic investigation (El Maarry et al., 2012; Farmer, 2000).

Extraterrestrial layer cake

Features imaged in Gale crater more than a decade before the arrival of the Curiosity rover show a key reason for choosing this landing site. A considerable thickness of horizontally stratified deposits forms a mountain, Aeolis Mons, within the crater (Figure 23.C). The height of Aeolis Mons (known informally as Mount Sharp) has led to suggestions that the layered rocks once completely filled Gale crater but that later much of the crater was exhumed by erosion (Malin and Edgett, 2000b). Alternatively, deposition in a mostly dry environment, dominated by downslope wind, could build the mound in the center of the crater without filling the surrounding moat with sediment (Kite et al., 2013). This thick section, comparable in scale to the Grand Canyon on Earth, led scientists to target it with the Curiosity rover in hopes of reading the geologic story that it records (see Figure 4.16 in Chapter 4).

Deposited far and wide?

Layered, wind-carved deposits have long been recognized along this part of the dichotomy boundary (Scott and Tanaka, 1982; 1986; Hynek et al., 2003; Mandt et al., 2008). A name commonly applied to these deposits is from one of the locations where they were first recognized: Medusae Fossae (MC-16). On the basis of its morphologic characteristics (for example, Figure 16.C in MC-16), the Medusae Fossae Formation has been interpreted to consist of ash-flow tuffs (Malin, 1979; Scott and Tanaka, 1982; 1986), ancient polar deposits (Schultz and Lutz, 1988), or pyroclastic and eolian materials (Greeley and Guest, 1987). The materials may have a source in the Tharsis volcanoes (Hynek et al., 2003) or Apollinaris Mons (Kerber et al., 2011). Inverted channel systems record an episode of fluvial activity during accumulation of the Medusae Fossae materials (Burr et al., 2009a). Delta-shaped landforms have been interpreted as river deposition by distributary channels into a water body such as an ocean in the northern plains (eastward flow in Figure 23.D; DiBiase et al., 2013), while others argue for a fluvial channel/alluvial fan setting (flow to west in Figure 23.D; Jacobsen and Burr, 2017; Di Pietro et al., 2018).

---

[1] The Medusae Fossae Formation is included in the "Amazonian and Hesperian transition undivided unit" of the global geologic map in Chapter 5.

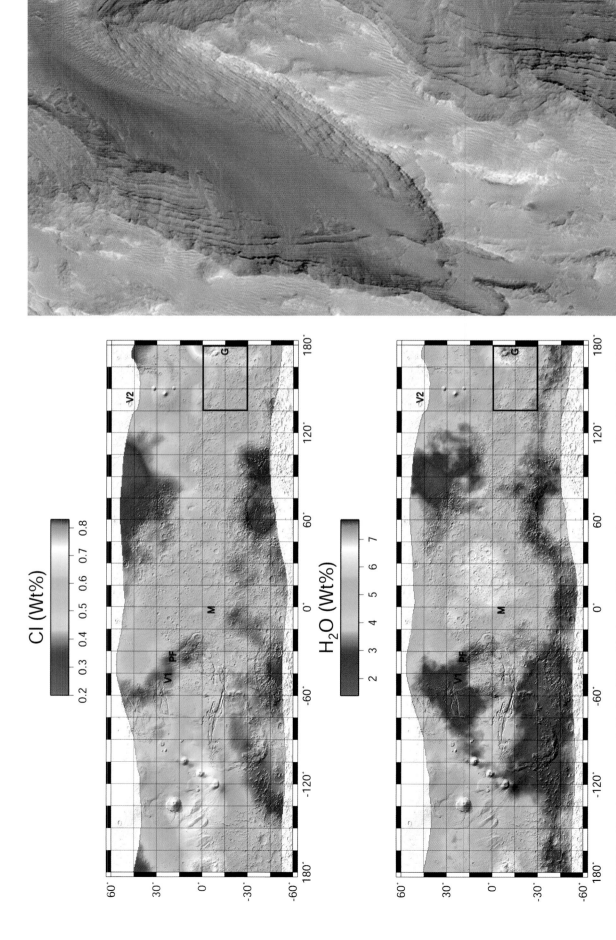

Cl (Wt%)

0.2 0.3 0.4 0.5 0.6 0.7 0.8

H₂O (Wt%)

2 3 4 5 6 7

FIGURE 23.B Abundance maps of chlorine (Cl) and hydrogen (H; expressed as H₂O), from Mars Odyssey GRS measurements of gamma rays emitted from the surface as a result of cosmic-ray bombardment. Landing sites: V1 and V2, Viking; PF, Pathfinder; M, Meridiani; G, Gusev. Note that the area near Gusev is enriched in Cl and H. Box outlines MC-23 Aeolis map area (NASA/JPL-Caltech/University of Arizona).

FIGURE 23.C Layers in Aeolis Mons, Gale crater, imaged by the MOC-NA (MOC release MOC2–480, 2003, 3 m/pixel, view 3 km wide, north at top, near 4.9° S, 138.3° E, NASA/JPL-Caltech/MSSS).

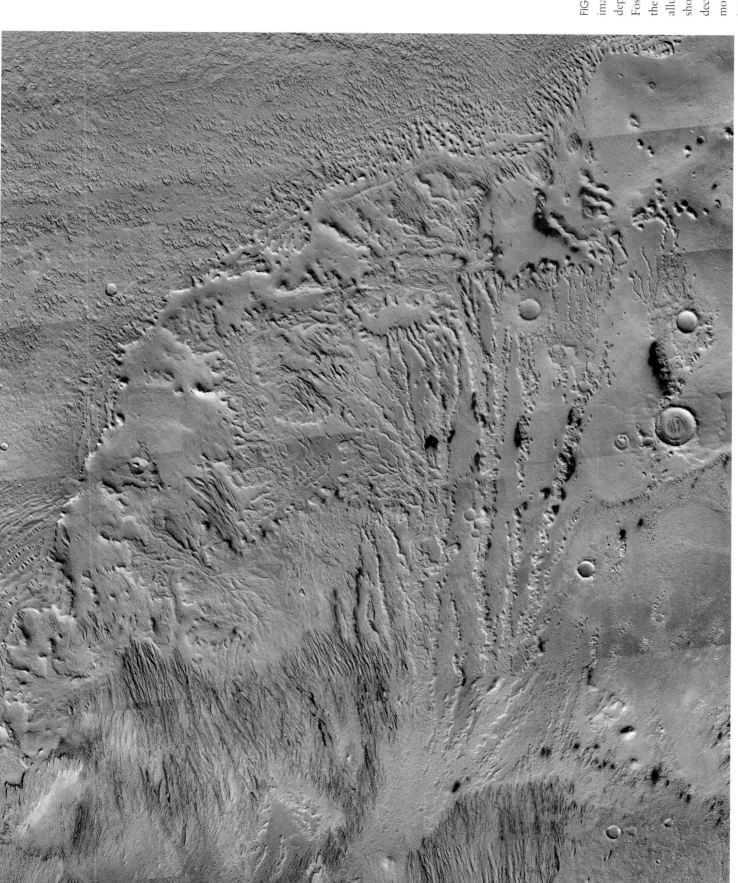

FIGURE 23.D Channels in a fan pattern dominate this image. They may indicate a possible east-directed deposition of a delta in Aeolis Dorsa in the Medusae Fossae Formation (DiBiase *et al.*, 2013). Alternatively, the channels could have formed by westward flow in an alluvial setting (Jacobsen and Burr, 2017). The region shown presently slopes down to the west (left), decreasing in elevation by several hundred meters (CTX mosaic, 6 m/pixel, view about 90 km by 75 km, north at top, 5° S, 155° E, NASA/JPL-Caltech/MSSS).

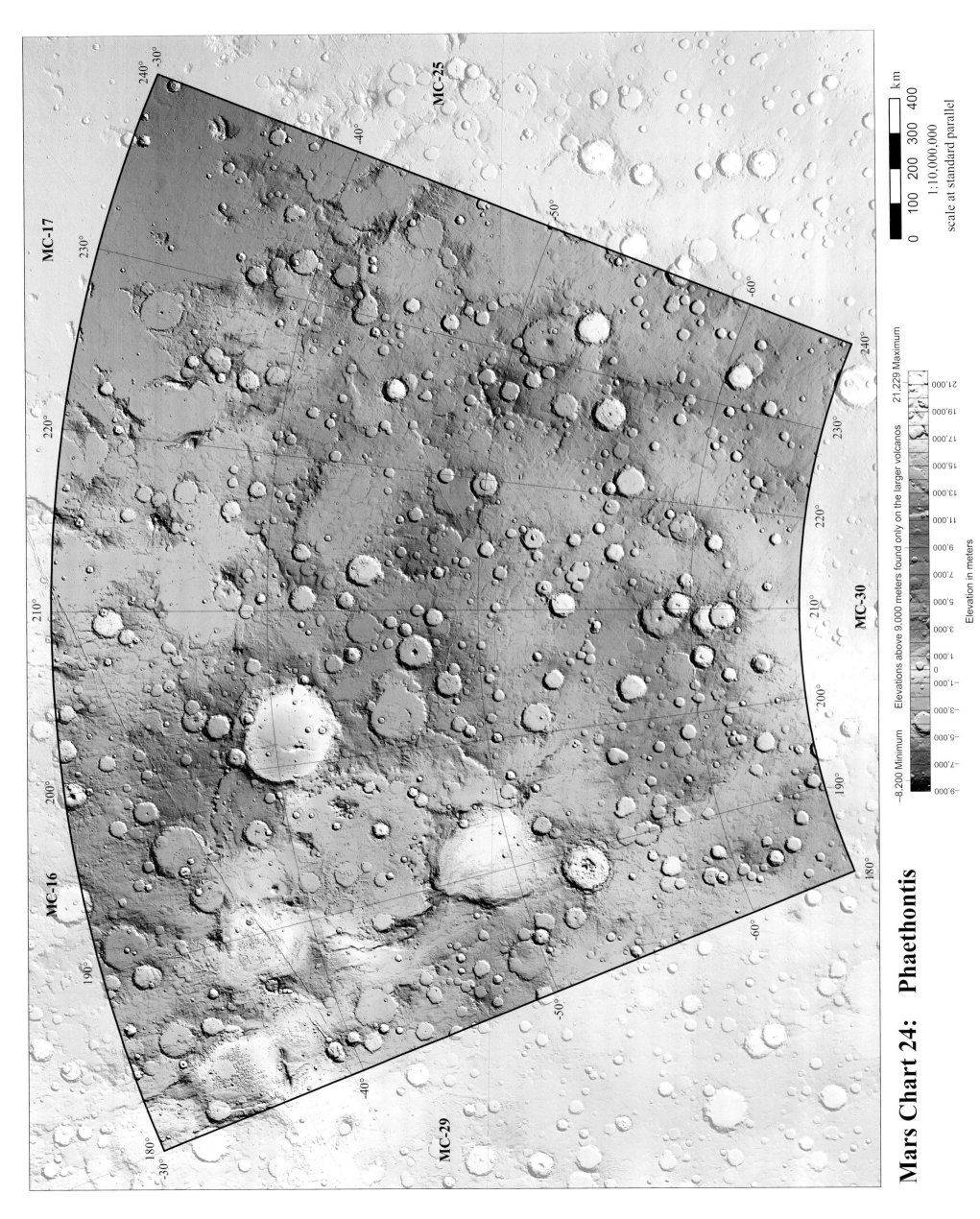

Mars Chart 24:   Phaethontis

MC-17

MC-25

MC-16

MC-30

MC-29

1:10,000,000
scale at standard parallel

km

0   100   200   300   400

−8,200 Minimum   Elevations above 9,000 meters found only on the larger volcanos   21,229 Maximum

Elevation in meters

−9,000 −7,000 −5,000 −3,000 −1,000 0 1,000 3,000 5,000 7,000 9,000 11,000 13,000 15,000 17,000 19,000 21,000

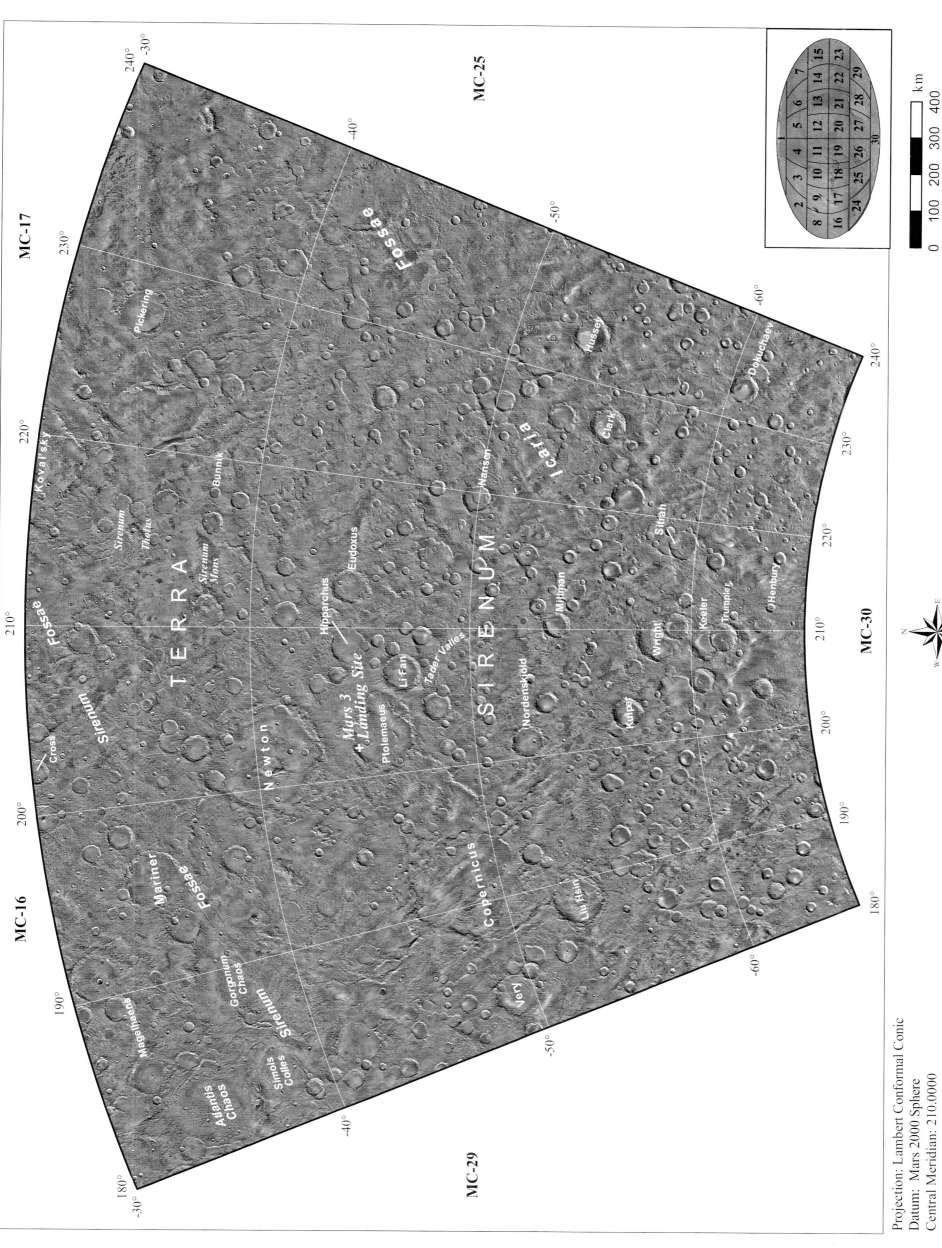

MC-17

MC-16

MC-25

MC-29

MC-30

240°

230°

220°

210°

200°

190°

180°

-30°

-40°

-50°

-60°

Fossae

Pickering

Kovalsky

Sirenum

Thollus

Sirenum Mons

Bunnik

Eudoxus

Hipparchus

Li Fan

Tader Valles

Ptolemaeus

Nansen

Icaria

Hussey

Clark

Sitah

Millman

Nordenskiöld

Wright

Keeler

Trumpler

Henbury

Dokuchaev

TERRA

SIRENUM

Fossae

Cross

Sirenum

Mariner

Fossae

Gorgonum Chaos

Magelhaens

Atlantis Chaos

Simois Colles

Sirenum

Newton

*Mars 3
+ Landing Site*

Copernicus

Very

Liu Hsin

Kuiper

Projection: Lambert Conformal Conic
Datum: Mars 2000 Sphere
Central Meridian: 210.0000
Standard Parallel 1: -36.1516
Standard Parallel 2: -59.4669

1:10,000,000

scale at standard parallel

km

0   100 200 300 400

Phaethontis (MC-24)

Geography

The Phaethontis quadrangle is dominated by the cratered highlands of Terra Sirenum, which display prominent, marginal basins and tectonic structures, reaching thousands of kilometers in length. Except for the interiors of larger craters, elevations are generally 1–3 km above datum. Tharsis lava-flow materials inundate and partially cover the rugged, ancient terrain in the northeast corner of the quadrangle. Some of the structural basins in the northwestern part of the quadrangle display disrupted floors, referred to as chaotic terrain, most notably Atlantis Chaos and Gorgonum Chaos. The segmented, narrow graben systems of Sirenum Fossae and Icaria Fossae extend southwestward from the Tharsis rise, northeast of the quadrangle, cutting both ancient cratered highland materials and some of the older Tharsis lava flows.

Geology

Phaethontis records both pre- (or proto-)Tharsis activity and the subsequent, volcanic and tectonic development of the southwestern margin of the Tharsis rise, at least since the Middle Noachian Epoch up to the present day. North-trending, fault-bounded basins in the northeast part of the quadrangle connect to similar ancient basins to the north (see MC-16). North-trending wrinkle ridges are prominent in the western part of the

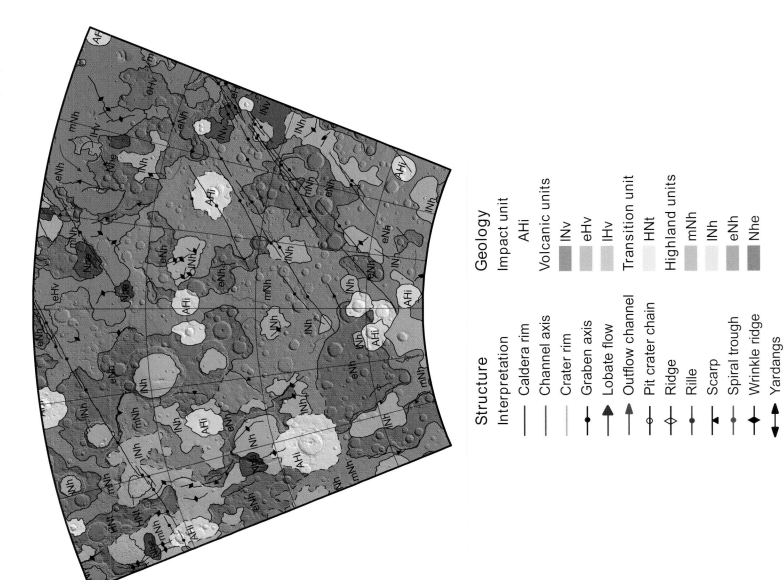

Structure

Interpretation

— Caldera rim
| Channel axis
| Crater rim
●| Graben axis
▲ Lobate flow
▲ Outflow channel
⊕ Pit crater chain
◆ Ridge
● Rille
◀ Scarp
● Spiral trough
◆ Wrinkle ridge
↕ Yardangs

Geology

Impact unit
AHi

Volcanic units
lNv
eHv
lHv

Transition unit
HNt

Highland units
mNh
lNh
eNh
Nhe

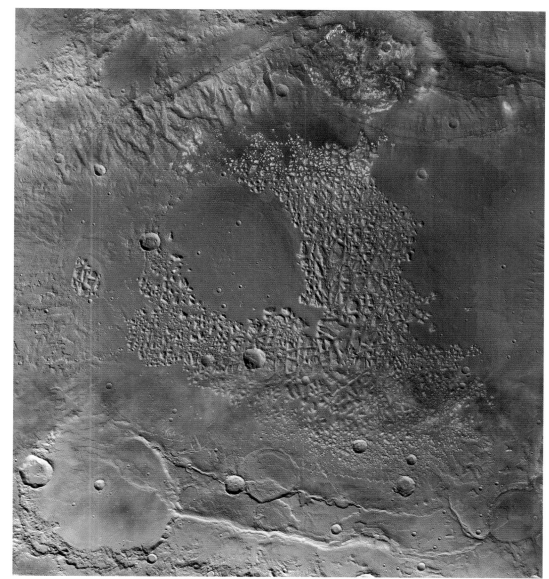

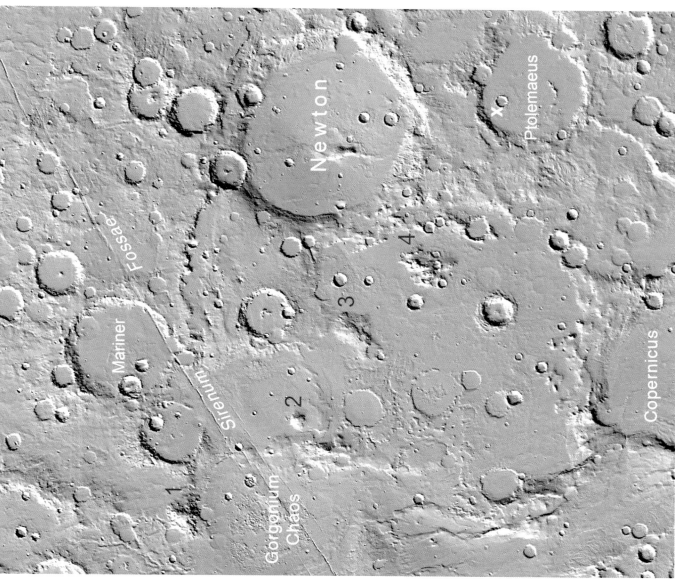

FIGURE 24.A Atlantis Chaos is the terrain at the center of this image, made up of numerous mounds and hills. These may be remnants of a once continuous layer that collapsed when it was partially eroded or removed. The cause may be similar to that of chaos regions in the northern hemisphere of Mars, which involve possible melting and/or sublimation of subsurface ground ice. Later emplacement of planar material partially buried the chaotic terrain (Adeli *et al.*, 2012; Golder and Gilmore, 2013). An alternate interpretation is that the mounds and hills are erosional remnants that were later partially covered (Wendt *et al.*, 2013; mosaic of four HRSC images, 14 m/pixel, view about 250 km across, north at top, 32.5° S to 36° S, 180° E to 185° E, ESA/DLR/FU Berlin).

FIGURE 24.B Mariner crater, the first crater imaged on Mars by a spacecraft, lies near the top of this MOLA hillshade mosaic, while Newton and Ptolemaeus are the two largest craters in the right half of the image. The "X" in Ptolemaeus marks the suspected 1971 landing site of the Mars 3 spacecraft. Features numbered 1 to 4 may be degraded volcanoes which lie along an alignment that results from a giant, south polar impact. North is slightly to the left of top (view 850 km by 1,000 km, centered at 40° S, 196° E).

quadrangle. Many of the high-standing rock outcrops are dissected by ancient valley networks. Basins in the northwestern part of the quadrangle may have had long-term water and/or ice enrichment, as evidenced by development of chaotic terrains (Figure 24.A). Tharsis rise-related deformation in the quadrangle includes a fold belt, concentric to the rise, as well as the radial narrow graben systems. The graben systems may be underlain by magmatic dikes, which could have been sources for the lavas in the

northeastern part of the quadrangle (see also MC-16). Some of these flows are relatively pristine, digitate, and undeformed when compared to both cratered terrains and older lava flows, indicating that tectonism and resurfacing declined over time.

Historic views

Mariner crater was the first large crater imaged on Mars (Figure 24.B; Figure 2.2 in Chapter 2). It

appears in the most-reproduced image from the Mariner 4 mission of 1965 (frame 11; Leighton *et al.*, 1965) and contributed to the first impression that Mars was more like the ancient, cratered

Moon than the imagined, more Earth-like world of pre-Space Age speculation. The likely landing site of the Soviet Mars 3 spacecraft (Hartmann, 2003, Ch. 16; Strooke, 2012) is nearby, in the crater Ptolemaeus. While it landed and sent signals back from the surface – the first robotic spacecraft to do so on Mars – signals were lost after a few seconds and never recovered.

Ancient eruption path?

Most recognized southern-hemisphere volcanoes on Mars occur along alignments that diverge from the south pole, perhaps resulting from mantle plumes that were caused by an ancient, hypothetical south polar giant impact (Leone, 2016). Dozens of other isolated mountains (montes) and domical features (tholii), tens of kilometers across and larger, also follow such alignments. Although these other features generally lack diagnostic volcanic landforms such as summit calderas and lava flows, it may be that they are so old that cratering and erosion have largely obliterated such details. In western Phaethontis, a few examples occur along a northwest–southeast trend (numbered 1 to 4 in Figure 24.B).

A peek at hidden layers and terraces

Ancient cratered highlands like Terra Sirenum reveal clues to the character of the crust, which has been drastically altered by impact, tectonic fracturing, erosion, and later deposition. Some of these modifying activities may expose otherwise hidden features, attesting to the origin of highland materials. The inner wall of a small crater (Figure 24.C) reveals continuous, parallel layers – could these be deposits emplaced by volcanism, fluvial activity, or impacts? A ridge near Atlantis Chaos shows a tantalizing, complex array of terraces (Figure 24.D). As on Earth, these terraces may record depositional and erosional processes and conditions at the time they formed – early in the history of Mars.

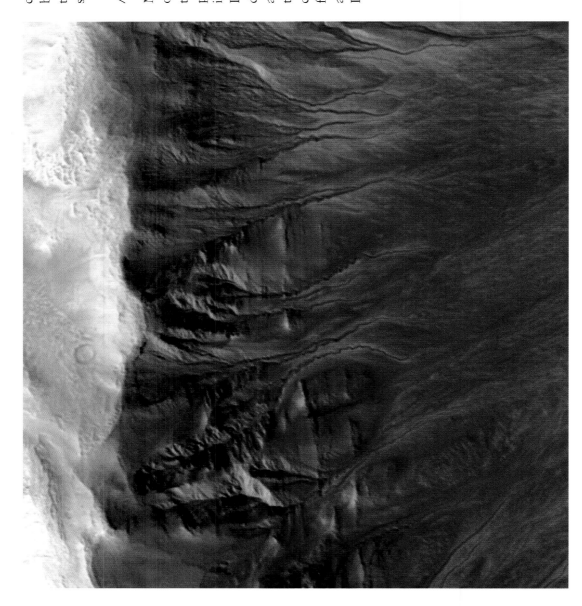

FIGURE 24.C MOC-NA image showing the rim of a small crater in the interior of Newton crater. Contrasts in the layers control the shape of the gullies, several of which contain sinuous channels (image E16–00043, 3 m/pixel, map-projected view 3 km across, north at top, 41.4° S, 202.3° E, NASA/JPL-Caltech/MSSS).

FIGURE 24.D View (4 km by 6 km) of terraces north of Atlantis Chaos, revealed by erosion. The terraces may be original layers in the material, or an artifact of wind erosion (HiRISE red band image ESP_036166_1495, 50 cm/pixel, north at top, 30.0° S, 182.9° E, NASA/ JPL-Caltech/University of Arizona).

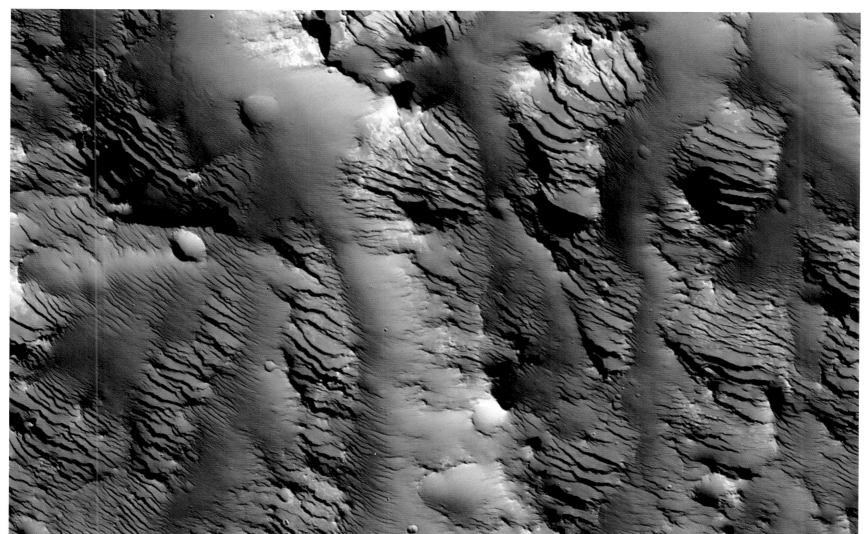

203

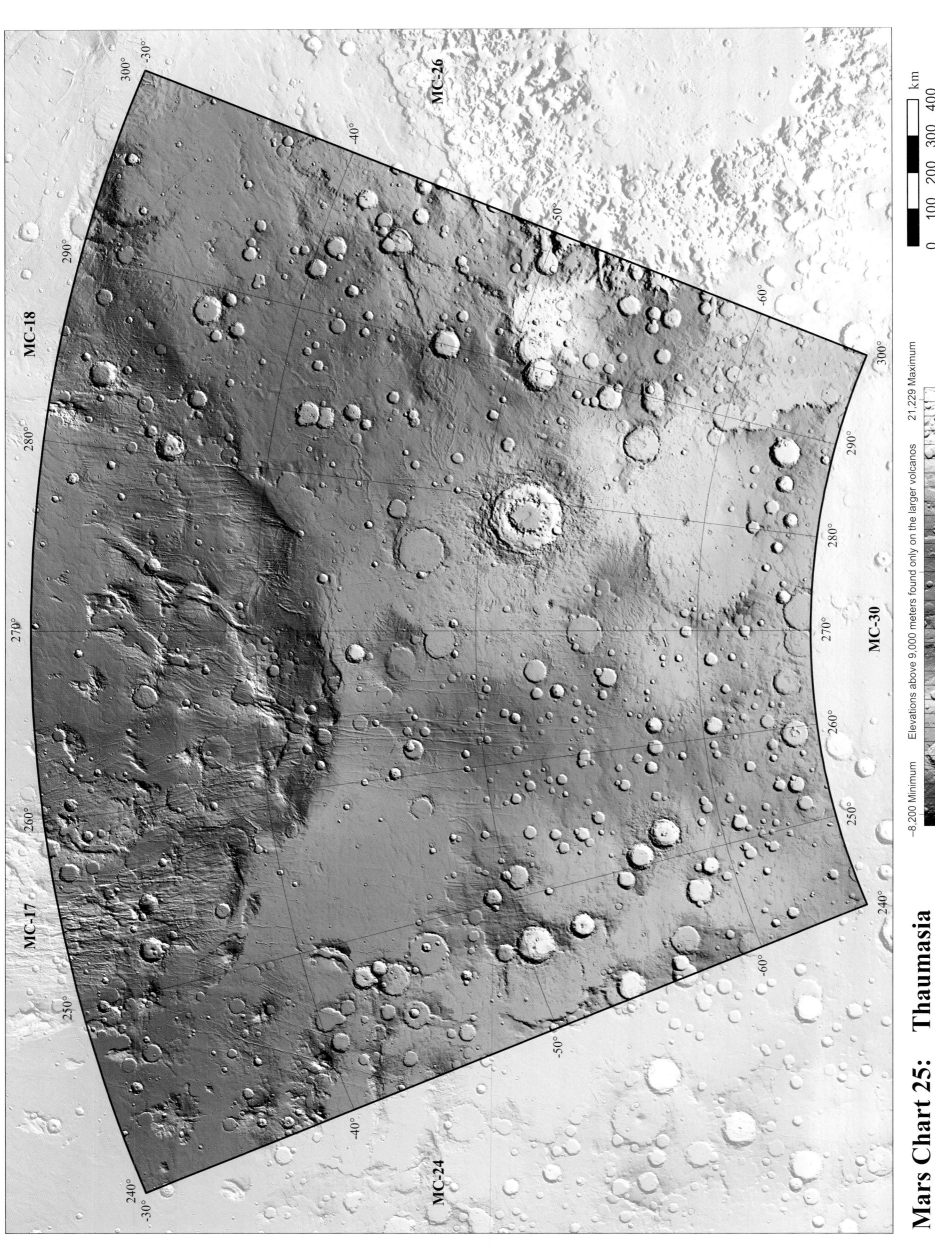

Mars Chart 25:   Thaumasia

km

1:10,000,000
scale at standard parallel

-8,200 Minimum   Elevations above 9,000 meters found only on the larger volcanos   21,229 Maximum

Elevation in meters

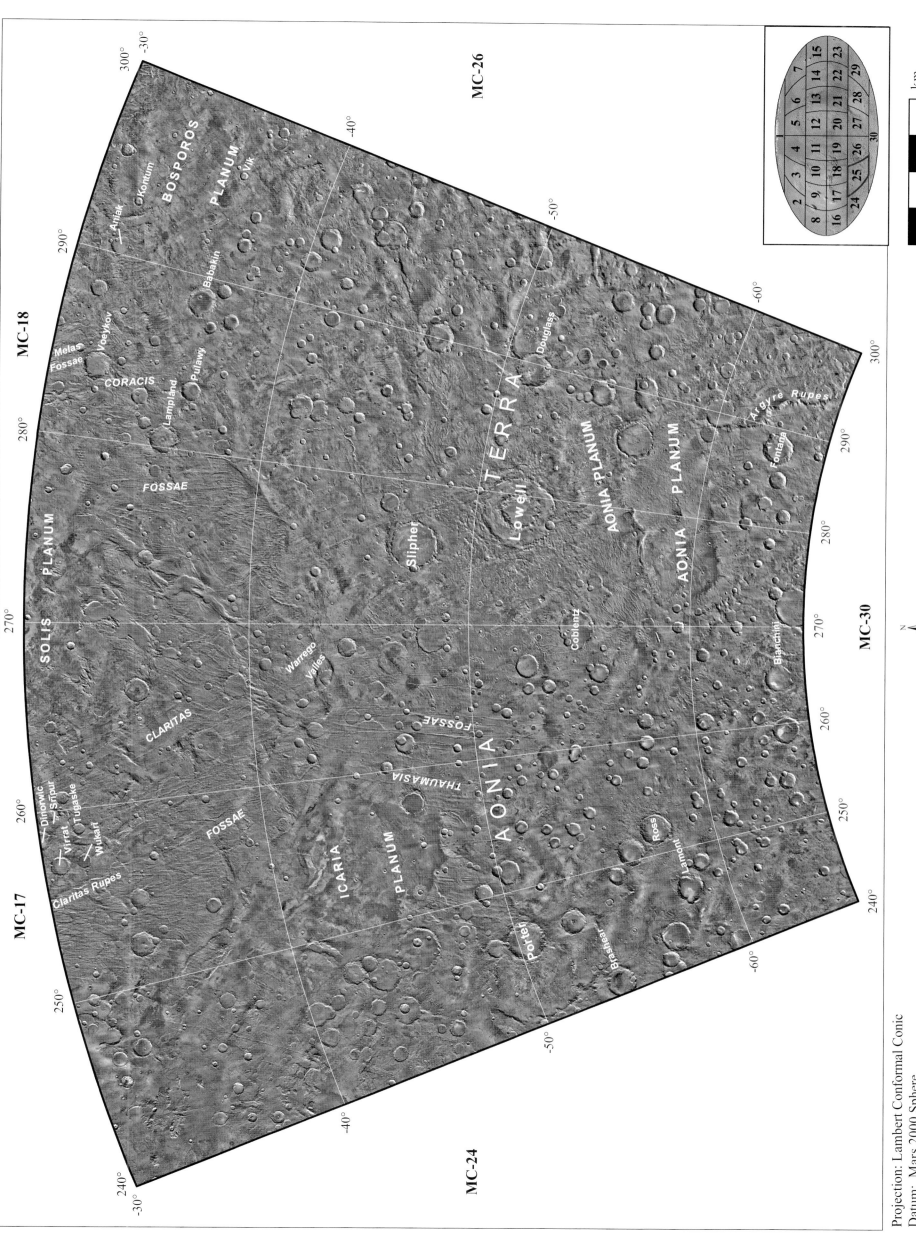

MC-18

MC-17

MC-24

MC-26

MC-30

-30°
300°

BOSPOROS
PLANUM

Aniak

Kontum

Vik

290°

Babakin

Melas
Fossae

Voeykov

CORACIS

Lampland

Pulawy

-40°

FOSSAE

280°

SOLIS
PLANUM

270°

CLARITAS

FOSSAE

Dinorwic
Sripur

Virrat
Tugaske

Wukari

Claritas Rupes

260°

250°

240°

-40°

-50°

-60°

ICARIA

PLANUM

AONIA

PLANUM

Warrego
Valles

THAUMASIA

FOSSAE

Slipher

Lowell

TERRA

Douglass

AONIA PLANUM

AONIA PLANUM

Coblentz

Bianchini

Fontana

Argyre Rupes

Porter

Brashear

Ross

Lamont

-50°

-60°

300°

290°

280°

270°

260°

250°

| 15 |
7	14	23
6	13	22	29
5	12	21	28
4	11	20	27
3	10	19	26
2	9	18	25
8	17	24	30
16			

km

0   100 200 300 400

1:10,000,000

scale at standard parallel

Projection: Lambert Conformal Conic
Datum: Mars 2000 Sphere
Central Meridian: 270.0000
Standard Parallel 1: -36.1516
Standard Parallel 2: -59.4669

Thaumasia (MC-25)

Geography

A mountain range, which has been informally referred to as the Thaumasia highlands, is prominent in the north-central part of the quadrangle; it lies at elevations of 4–7 km above datum, and forms the southern margin of the Thaumasia plateau (informal name; see Figure 4.2). Cratered highlands of Aonia Terra, interspersed with younger plains and basins, ranging from near datum to 4 km elevation, such as Icaria Planum and Aonia Planum, make up the rest of the area. The outer rim of the Argyre basin is just visible along the eastern edge of the quadrangle.

Geology

The initial development of the Thaumasia highlands, south of Solis Planum, began in the Early to Middle Noachian Epochs. The highlands are recognized as the oldest known volcanic and tectonic terrains associated with the Tharsis region. They

are cut by fault and rift systems of varying trends and forms, along which magma ascended to the surface and produced volcanoes. The trends of many of the structures tie them to the development of Tharsis-related centers of deformation, located north of the quadrangle (Anderson *et al.*, 2001). Among the earliest structures are highly degraded promontories, interpreted to be volcanic and tectonic constructs, which dot the cratered highlands, generally to the west and southwest of the mountain range. These features extend into the Terra Sirenum region (MC-24) to the west. South of the Thaumasia highlands, the Noachian cratered terrain shows progressively less Tharsis-related deformation southward. The Argyre impact appears to have produced scarps, ridges, and massifs, southeast of the Thaumasia highlands, which are concentric about the Argyre basin, located east of the quadrangle (MC-26). Icaria and Aonia Plana and other basins and plains within the terrains south of the Thaumasia highlands are filled with mostly locally derived, Late Noachian and Hesperian volcanic rock materials and/or sedimentary deposits, which include local

indications of clay and salt minerals. The southernmost basins also include icy deposits that appear to be outliers of the south polar ice cap. The Early Hesperian, 200-km-diameter Lowell crater forms one of the best-preserved double-ring basins on Mars. Among the youngest geologic activity in the quadrangle was the emplacement of Tharsis-related lava plains (1) near the

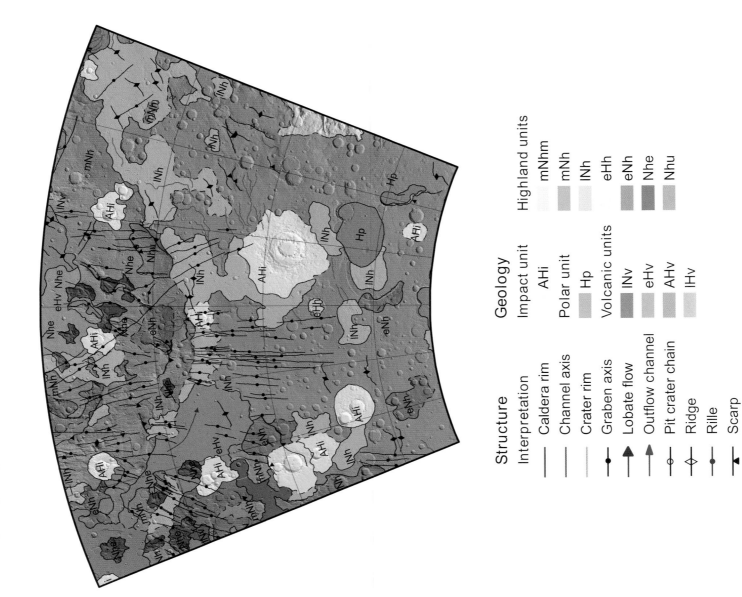

Structure

Interpretation

— Caldera rim
— Channel axis
— Crater rim
⫶▲ Graben axis
▲ Lobate flow
⊶ Outflow channel
⬥ Pit crater chain
◁ Ridge
⬤ Rille
◀ Scarp
⬤ Spiral trough
◆ Wrinkle ridge
⬍ Yardangs

Geology

Impact unit
AHi

Polar unit
Hp

Volcanic units
lNv
eHv
AHv
lHv

Highland units
mNhm
mNh
lNh
eHh
eNh
Nhe
Nhu

groundwater activity (Tanaka *et al.*, 1998). These channels were active in the Late Noachian and Early Hesperian Epochs and dissect older grabens, some of which (Thaumasia Fossae) appear in the southwestern part of Figure 25.A. The northwest-trending grabens (Claritas Fossae) in the northwestern part of Figure 25.A cut Noachian- and Hesperian-age rocks (Dohm *et al.*, 2001b).

All cracked up

The northern and central parts of the quadrangle show the numerous intersecting narrow graben and fracture sets seen in all the areas marginal to Tharsis (Figures 25.A, 25.C). Many, but not all, are radial to central components of the Tharsis rise such as Syria Planum. Note that relatively higher-standing surfaces are generally more densely cratered, fractured, and degraded. Thus, by mapping distinctive surfaces and fracture sets based on their landform characteristics, overlaps, and cross-cutting relationships, geologists have been able to determine that the terrains and fault systems developed in multiple stages. These began in the Noachian Period (higher-relief terrains and features in the southeastern part of Figure 25.C) and extended into the Late Hesperian Epoch (smooth surface at north). Elsewhere in Tharsis, similar faulting extended into the Amazonian Period.

In the terrain cut by Claritas Fossae, the extensional nature of the fractures is evident where they cross old impact craters (Figure 25.D). The east–west extensional strain of 5–10 percent has changed the craters from a circular to an elliptical shape.

Lowell's big impression

Lowell crater, an Early Hesperian impact 200 km across, is an example of a multi-ring basin, with networking channels that partly surround the basin. The channels are thought to result from the impact event, thus indicating volatile-rich target materials (Lias *et al.*, 1997). At this latitude (52° S), the appearance changes with the local seasons, with frost spreading during autumn (Figures 25.E, 25.F).

FIGURE 25.B Dendritic pattern of channels, forming part of Warrego Valles (THEMIS visible image V06907004, 17 m/pixel, view 17 km by 62 km, north at top, center 42° S, 267.5° E, NASA/JPL-Caltech/Arizona State University).

north-central margin of the quadrangle, during the Late Hesperian Epoch, sourced from Syria Planum to the north, and (2) in the northwest part of the map region during the Hesperian to Amazonian Periods, a marginal area of the vast Daedalia Planum flow field.

Doubled up?

The south Tharsis ridge belt is an informally named, regional feature that extends from the Coprates rise (MC-18) through the Thaumasia highlands, across this quadrangle to the Memnonia region (MC-16). Shown here (left center of Figure 25.A; see Table 2, Fig. 11, Fig. 12 of Schultz and Tanaka, 1994) is one east–west trending ridge system, which has a sinuous outline, similar to areas on Earth where rocks of equivalent stratigraphic position have been pushed together and doubled up along shallow ("thrust") faults. This suggests a possible contractional origin, though elsewhere in the south Tharsis ridge belt both contraction and extension have been suggested (MC-16, MC-18, Schultz and Tanaka, 1994; Anderson *et al.*, 2012; Karsozen *et al.*, 2012).

It runs downhill

The rugged Thaumasia highlands between Solis Planum and Aonia Terra include numerous run-off channels. Warrego Valles (Figures 25.A, 25.B) are a system of networking channels, sourcing from one of the most elevated reaches of the highlands. These channels are an example of heavy dissection, interpreted by some geologists to indicate that there was precipitation long ago on Mars (Ansan and Mangold, 2006), along with

FIGURE 25.A THEMIS daytime infrared mosaic of Warrego Valles region, including the fractured Thaumasia highlands that form part of the south Tharsis ridge belt. The dendritic drainage of Warrego Valles lies between the "W" labels. Box shows location of Figure 25.B (100 m/pixel, view about 600 km by 800 km, north at top, 36° S to 46° S, 260° E to 278° E, NASA/JPL-Caltech/Arizona State University).

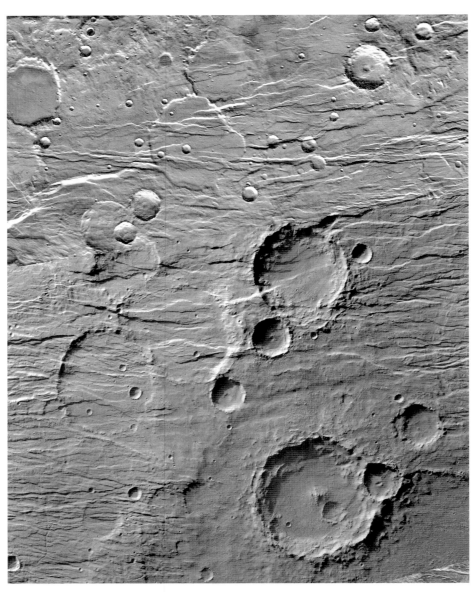

FIGURE 25.C HRSC color image of the margin between Solis Planum (volcanic plains to the north) and the Thaumasia highlands (deformed plateaus and massifs to the south), showing fractures and narrow grabens in at least three sets trending east–northeast, north–northeast, and northwest. Some of the north–northeast-trending grabens occur within, and follow the margins of, a wide trough that may be an ancient, lava-filled crustal rift valley in the western half of the image (press release 082, orbit 0431, 48 m/pixel, view 100 km by 140 km, rotated so that north is at top, centered on 33° S, 271° E, ESA/DLR/FU-Berlin).

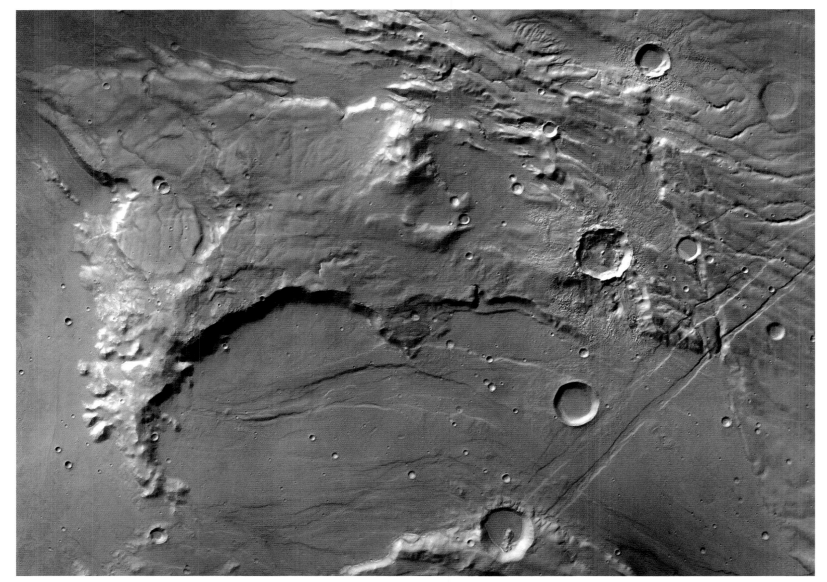

FIGURE 25.D Impact craters sliced up by Claritas Fossae. The north-trending scarps and narrow troughs here are extensional faults and grabens that accommodated east–west stretching of the region (HRSC image, H0563_0000_ND4, 50 m/pixel, view 300 by 250 km, north at top, centered at 33° S, 255° E, ESA/DLR/FU-Berlin).

FIGURES 25.E, 25.F MOC-WA views of 200-km-diameter Lowell crater in local autumn, taken the same Mars year (25.E, left image: M1800922, $L_s = 036°$; 25.F, right image: M2001003, $L_s = 064°$, both red filter with north at top), The images reveal an increase in coverage of the relatively low interior of Lowell by frost as the season progresses (centered at 52° S, 278° E, NASA/JPL-Caltech/MSSS).

The Atlas of Mars

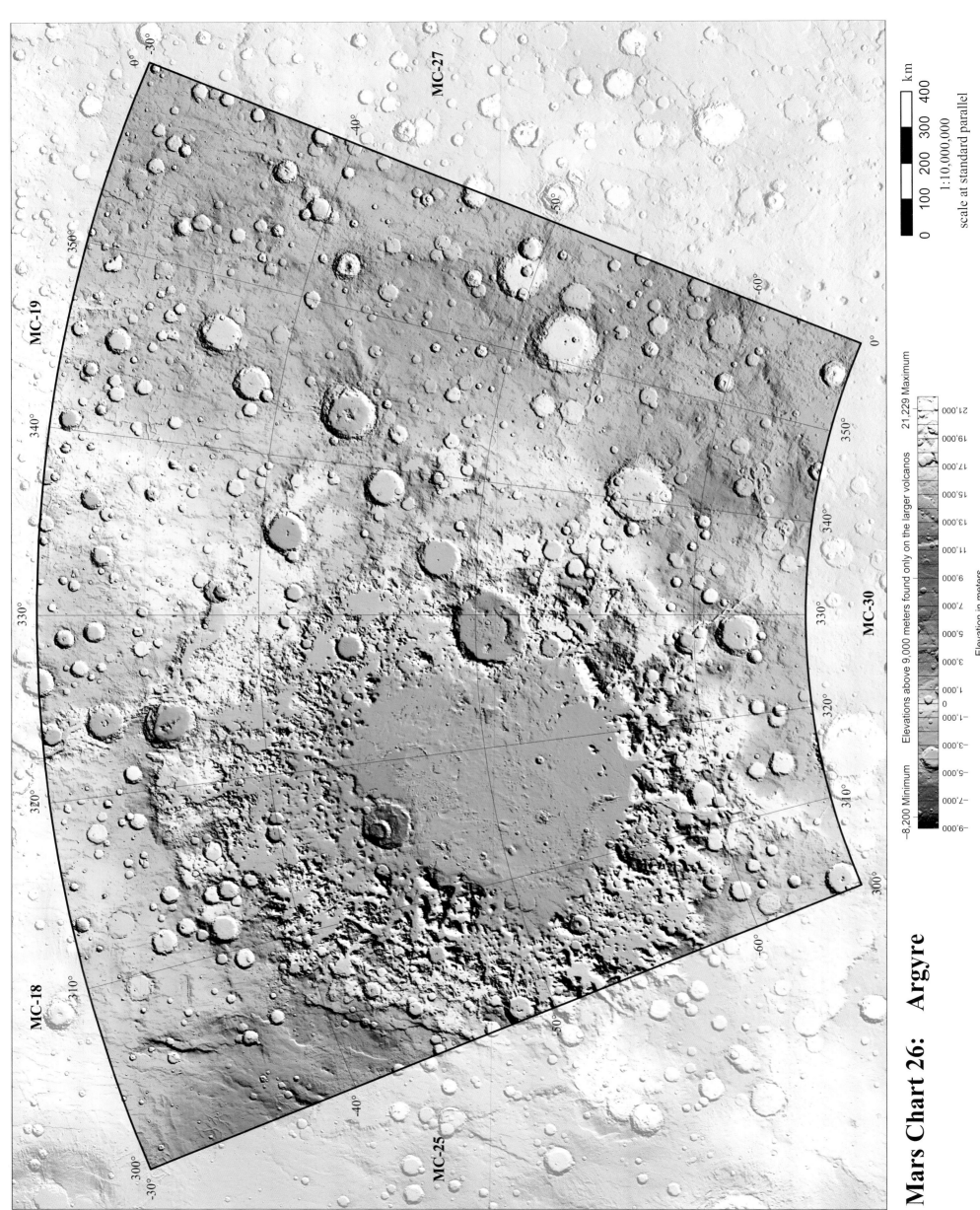

Mars Chart 26: Argyre

MC-19

MC-18

MC-27

MC-25

MC-30

km

1:10,000,000
scale at standard parallel

Elevations above 9,000 meters found only on the larger volcanos

−8,200 Minimum   21,229 Maximum

Elevation in meters

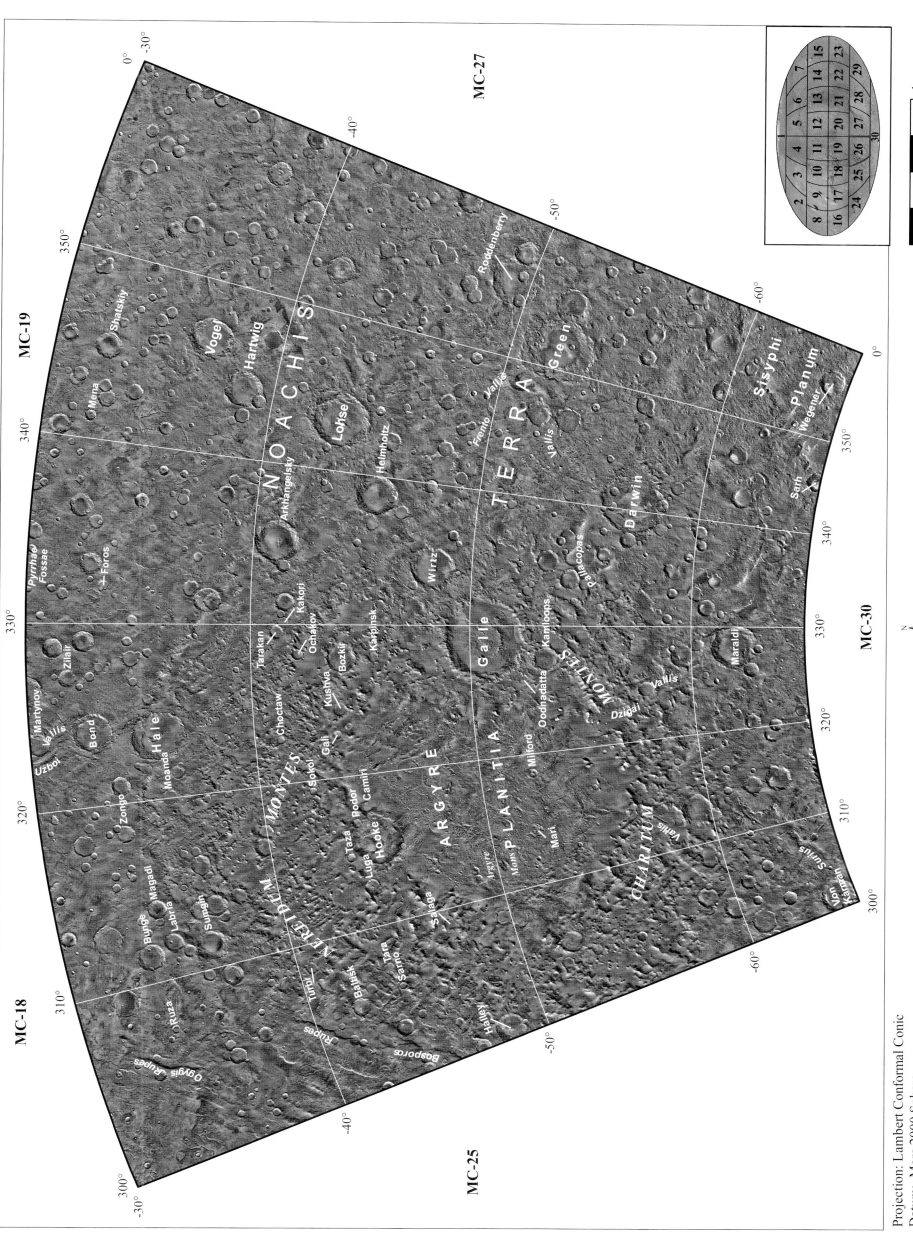

MC-19

MC-18

MC-27

MC-25

MC-30

NOACHIS

TERRA

Sisyphi Planum

ARGYRE PLANITIA

CHARITUM MONTES

NEREIDUM MONTES

Shatskiy
Mena
Vogel
Hartwig
Lohse
Helmholtz
Arkhangelsky
Wirtz
Roddenberry
Green
Frento Vallis
Vallis
Darwin
Palacopas
Sarh
Wegener
Pyrrhae Fossae
Foros
Zilair
Martynov
Bond
Hale
Moanda
Zongo
Uzboi
Vallis
Ruza
Bunge
Labria
Suungin
Magadi
Oxygis Rupes
Rupes
Turbi
Batusk
Tara
Sarno
Saleqa
Halley
Bosporos
Luga
Taza
Podor
Camiri
Gali
Sokol
Choctaw
Tarakan
Ochakov
Kakori
Bozkir
Karpinsk
Kustna
Hooke
Argyre Mons
Mari
Milford
Galle
Oodnadatta
Kamloops
Dzigai
Vallis
Maraldi
Von Kármán
Vallis
Sirius

N
E
S
W

Projection: Lambert Conformal Conic
Datum: Mars 2000 Sphere
Central Meridian: 330.0000
Standard Parallel 1: -36.1516
Standard Parallel 2: -59.4669

km
0 100 200 300 400
1:10,000,000
scale at standard parallel

7
15
6 14 23
5 13 22 29
4 12 21 28
3 11 20 27
10 19 26
2 9 18 25
8 17 24
16 30

Argyre (MC-26)

Geography

The Argyre basin spans the west half of the quadrangle, while part of Noachis Terra, at 0–2 km elevation, lies to the east. Argyre, as deep as –3 km elevation, is the best preserved of the largest multiringed impact basins on Mars, and is comparable in size to the Orientale basin of the Moon. The size and number of rings in the basin, which are generally expressed by discontinuous, concentric ridges and basin-facing scarps, are debated (three to seven rings or more), owing to later modification. The most common diameter assigned to a prominent, inner ring is 800–900 km, while the entire structure may be 1800 km or more across. Valleys drain toward Argyre from the south and east, while large channels may connect Argyre to the Uzboi–Ladon–Morava (see MC-19) system to the north. Drainage into the northwestern flank of the basin from surrounding plains is blocked by concentric, broad ridges. The hummocky floor of Argyre is 3–4 km below the average terrain elevation beyond the rim (Hiesinger and Head, 2002) and includes a variety of landforms. Noachis Terra is typical of the southern cratered highlands of Mars and gives its name to the oldest period of geologic time on Mars (MC-27).

Geology

The Argyre impact on the pre-existing highland surface formed a central basin floor, defined by Argyre Planitia, mountainous rim materials including Nereidum and Charitum Montes, and a complex basement structural fabric that includes radial and concentric escarpments, systems of massifs, and structurally controlled local basins (Dohm *et al.*, 2001c; Nimmo and Tanaka, 2005; Frey 2006). The Argyre impact event apparently took place subsequent to the shutdown of the planetary dynamo, as the basin and surroundings lack, and likely destroyed, remanent magnetic signatures.

Since its formation, the Argyre basin has been a regional catchment for volatiles and sedimentary materials. The basin is also a dominant influence on the formation and routing of surface water and ice (e.g. glaciers) through its topographic features, as well as of groundwater along basement structures. Argyre rock materials have been formed and modified by impact, magmatic, eolian, tectonic, fluvial, lacustrine, glacial, alluvial, and tectonic processes. Ongoing resurfacing since the Noachian Period involves wind transport, gravity-

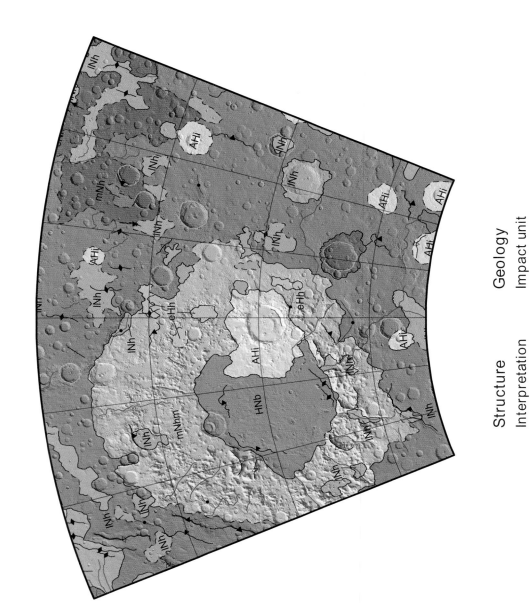

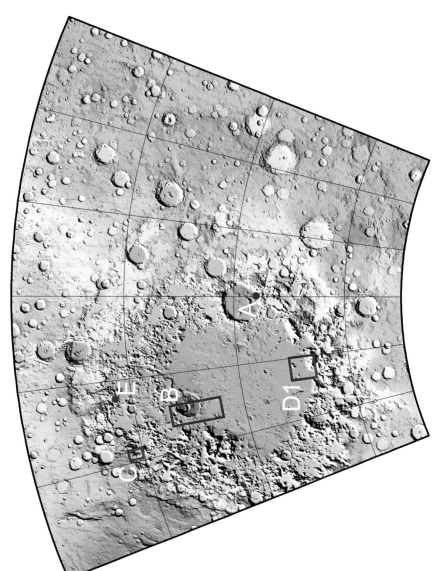

Structure

Interpretation

——	Caldera rim
——	Channel axis
——	Crater rim
●→	Graben axis
▲	Lobate flow
→	Outflow channel
↤	Pit crater chain
◇	Ridge
●—	Rille
▶—	Scarp
●—	Spiral trough
◆—	Wrinkle ridge
↕	Yardangs

Geology

Impact unit
AHi

Polar unit
Hp

Basin unit
HNb

Highland units
HNb
mNhm
mNh
lNh
eHh
eNh
Nhu

FIGURE 26.A MOC-NA mosaic (3 m/pixel; view ~4 km wide, north at top, 52.0° S, 330.1° E, release MOC2–1494) of a spectacular sequence of layered deposits with unconformities in Galle crater. A dust devil is evident at the lower right part of the layered sequence (NASA/JPL-Caltech/MSSS).

driven movement of basin materials, and glacial and periglacial activity (El Maarry *et al*., 2013; Soare *et al*., 2014).

Post-impact adjustment of impact-related concentric structures surrounding Argyre basin is also apparent. In addition, structures expressed within stratigraphic sequences (such as gaps and tilting during accumulation of layers) within and near the basin may mark links to major large-scale geologic activity, centered away from the basin, developed largely through the growth of the Tharsis rise (El Maarry *et al*., 2013).

Great lakes or small?

The association of the Argyre basin and neighboring regions with inflow valleys and outflow channels suggests that it contained one or more lakes in the past. Layered deposits (Figure 26.A), which may or may not be waterlain, show unconformities, or gaps in deposition, marked by a change in angle of the layers. If all the channels and valleys were active at the same time, the earliest high stand of the lake could have been very deep, filling Argyre and creating a lake comparable in extent to the Mediterranean Sea (Parker *et al*., 2000; Clifford and Parker, 2001; Hiesinger and Head, 2002; Dohm *et al*., 2011). Such a lake would have been high enough to flow out through Uzboi Vallis. Uzboi extends from near the degraded rim of the basin all the way to the northern plains, dissecting prominent impact craters such as Ladon, Holden, and Eberswalde (MC-19) along its route. Alternatively, Uzboi Valles may have had their largest flows much later, flowing from and/or into a

smaller lake. Different parts of the outflow channels may have been active at different times (Parker *et al*., 2000; Clifford and Parker, 2001; Hiesinger and Head, 2002).

An icy past

Among the glacial and periglacial landforms in Argyre (Figures 26.B, 26.C) are possible eskers and pingos. Eskers (Figures 26.D1, 26.D2) are ridges of sediment left by rivers of melt water that flowed beneath the ice of a glacier (Hiesinger and Head, 2002). Because they flow under pressure, they can go uphill or downhill. Pingos are ice-cored mounds created by groundwater pressure in permafrost ground. Candidates that are 100 m or more across have been proposed within the northern margin of Argyre basin, though they could have a different origin (Figure 26.E; Burr *et al*., 2009b; Soare *et al*., 2014).

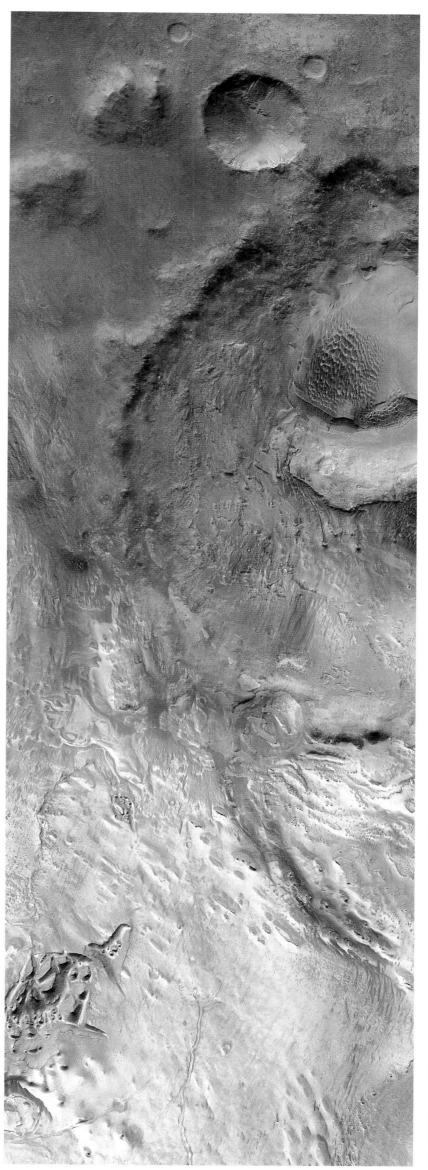

FIGURE 26.B HRSC image of Hooke crater shows examples of modifications that are common in Arygre. Dunes are evident inside the crater, whereas wind-sculpted hills, or yardangs, are evident farther south (to left) on the Argyre floor, where carbon dioxide frost also occurs. The image was taken after the southern hemisphere winter solstice (L_s = 121°; HRSC orbit 10743, press release 570, 22 m/pixel, 100 km by 270 km, north at right, centered at 46° S, 314° E, ESA/DLR/FU-Berlin).

The Atlas of Mars

213

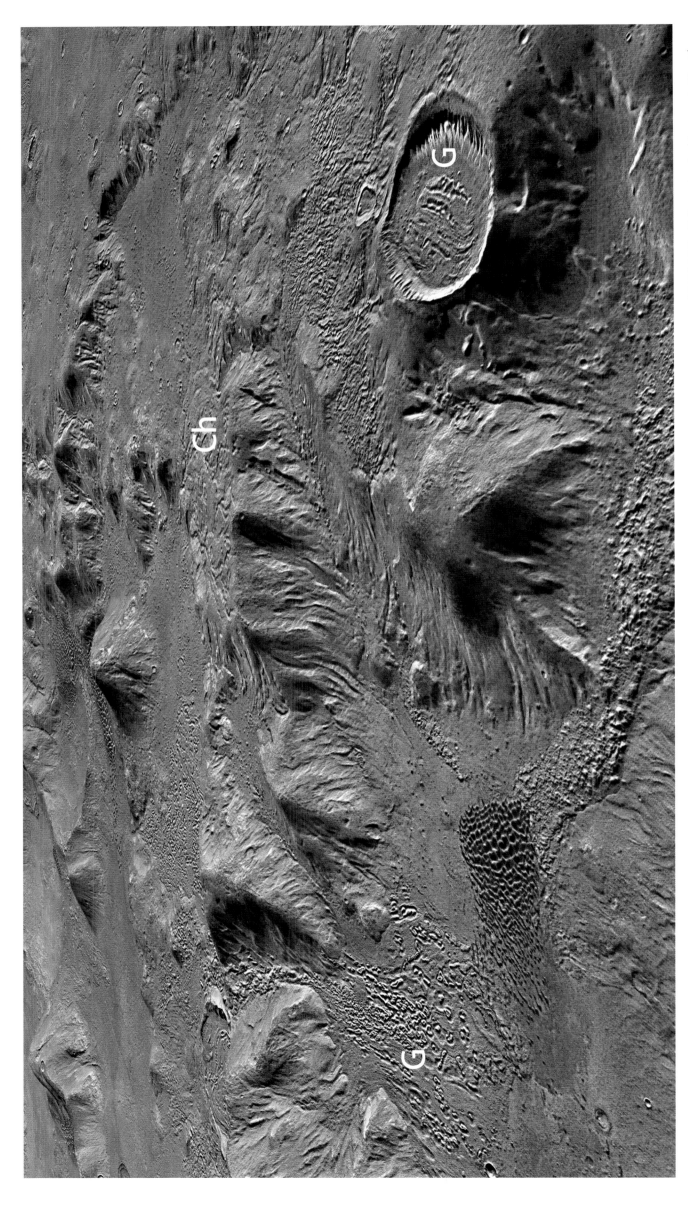

FIGURE 26.C Oblique view looking west in Nereidum Montes along the northwest margin of Argyre Planitia. A surface sloping away from the viewer at the upper center of the image (Ch) has channels in a dendritic pattern. Possible rock-covered glaciers or glacial deposits, coarsely textured by effects of glacial flow and ablation (G), in the right foreground, fill a crater and, at left, flow down a slope toward a field of wind-blown dunes (derived from HRSC orbit 10736, 23 m/pixel, view in foreground about 60 km across, centered near 39° S, 311° E, ESA/DLR/FU-Berlin).

FIGURE 26.E HiRISE image showing possible pingos (mounds aligned near the bottom of image; image ESP_020720_1410, 25 cm/pixel, view 5 km by 7 km, north at top, centered at 38.4° S, 317.7° E, NASA/JPL-Caltech/University of Arizona).

ESP_020720_1410_RED

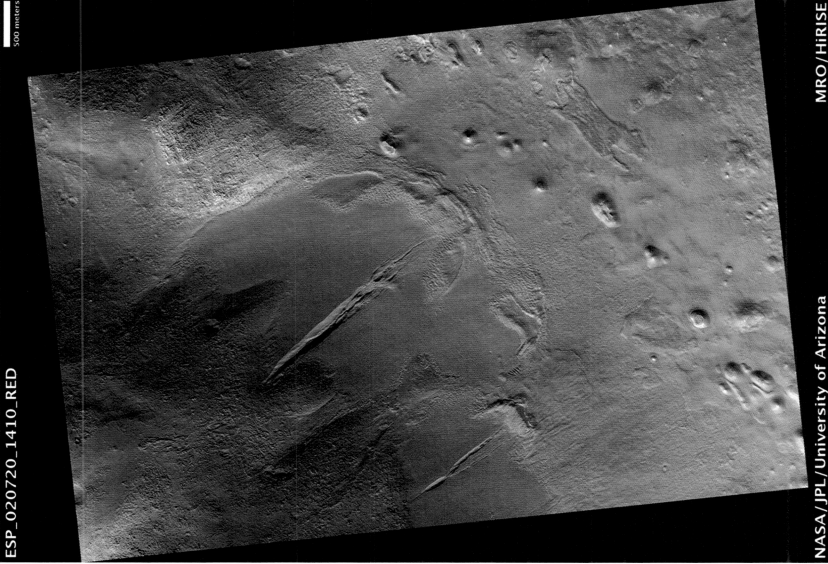

500 meters

MRO/HiRISE

NASA/JPL/University of Arizona

215

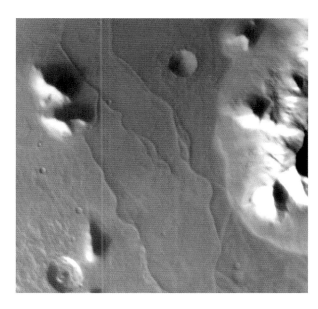

FIGURE 26.D1 MOC image of possible esker ridges (sediment forms deposited by ice-bounded, subglacial rivers) in southern Argyre basin (image M0002982, 230 m/pixel, non-map projected, 110 km by 120 km, north slightly to right of top, centered at 55.7° S, 318.3° E, NASA/JPL-Caltech/MSSS).

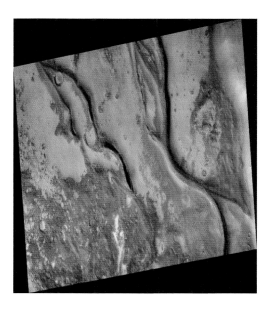

FIGURE 26.D2 CTX image, showing braided forms of possible esker ridges in the center of Figure 26.D1. Auxo Dorsum is near the top while Charis Dorsum crosses the lower half of the image (image P02_001640_1240, 6 m/pixel, view about 40 km across, north at top, centered at 56.1° S, 318.2° E, NASA/JPL-Caltech/MSSS).

The Atlas of Mars

Mars Chart 27: Noachis

MC-28

MC-21

MC-20

MC-30

MC-26

1:10,000,000
scale at standard parallel

km

0   100 200 300 400

−8,200 Minimum   Elevations above 9,000 meters found only on the larger volcanos   21,229 Maximum

Elevation in meters

−9,000 −7,000 −5,000 −3,000 −1,000 0 1,000 3,000 5,000 7,000 9,000 11,000 13,000 15,000 17,000 19,000 21,000

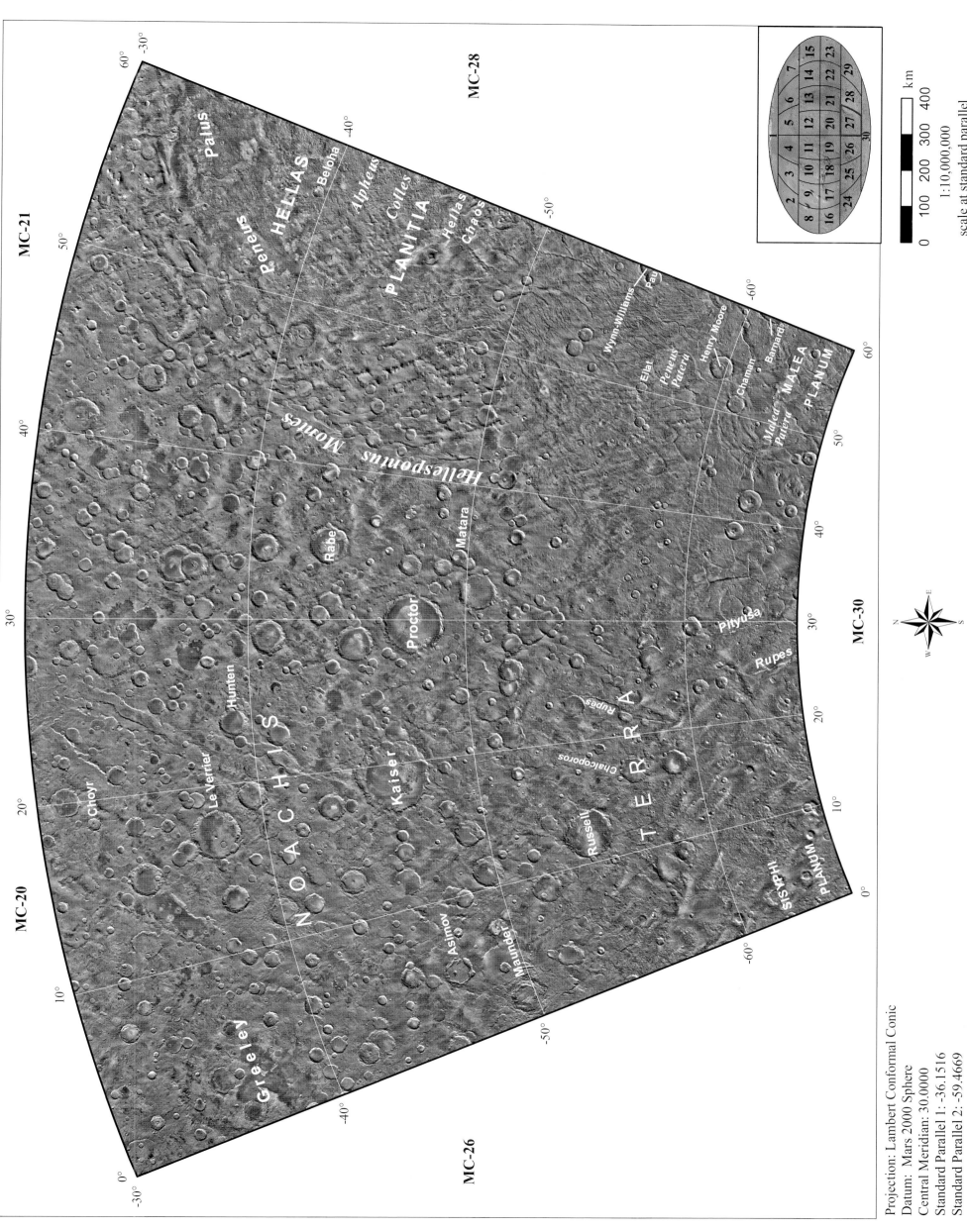

MC-21

MC-28

MC-20

MC-30

MC-26

Palus

HELLAS

Peneus

Beloha

Alpheus

Colles

PLANITIA

Hellas

Chaos

Wynn-Williams

Pau

Eilat

Peneus
Patera

Henry Moore

Chaman

Barnardi

Maleu
Patera

MALEA

PLANUM

Hellespontus Montes

Rabe

Matara

Proctor

Pityusa

Rupes

Hunten

Rupes

Chalcoporos

Le Verrier

NOACHIS

Kaiser

TERRA

Choyr

Russell

Greeley

Asimov

Maunder

SISYPHI

PLANUM

Projection: Lambert Conformal Conic
Datum: Mars 2000 Sphere
Central Meridian: 30.0000
Standard Parallel 1: -36.1516
Standard Parallel 2: -59.4669

km

0   100 200 300 400

1:10,000,000

scale at standard parallel

Noachis (MC-27)

Geography

The western two-thirds of the Noachis quadrangle is dominated by the heavily cratered highlands of Noachis Terra (1–3 km elevation), which are bordered to the east by Hellas Planitia. The latter, as low as −7 km, forms the floor of the ~2,300-km-diameter Hellas basin. The highland–basin margin is marked by the arcuate Hellespontus Montes, composed of basin-ring ridge systems, hundreds of kilometers long. The southern part of the quadrangle consists of the high, ridged plains of Malea and Sisyphi Plana. Broad, lengthy but relatively subtle trough systems occur west of and concentric to Hellas basin.

Geology

Noachis Terra typifies the ancient, Noachian, heavily cratered southern highlands of Mars (Scott

and Carr, 1978). The cratered terrain likely consists mostly of ejecta from the Hellas basin-forming impact (e.g. Smith *et al.*, 1999). Helle-spontus Montes and outer, concentric troughs (Figure 27.A) define basin rings where post-impact deformation occurred (Tanaka and Leonard, 1995). Enhanced erosion of the southern rim of Hellas basin may have resulted from the erosive activity of both flowing water and ice in the Middle to Late Noachian Epochs, perhaps ener-gized by volcanic eruptions from huge centers, including Peneus Patera (Tanaka and Leonard, 1995). Some have suggested that the isolated mountains in the southwest corner of the map area resulted from volcanism beneath an ice sheet (Head and Pratt, 2001), or even from cryovolcan-ism (of volatile-rich, molten slurries) (Tanaka and Kolb, 2001; see Sisyphi Montes in MC-30). Eroded material was deposited in Hellas Planitia, a process that continued on into the Hesperian Period (Leonard and Tanaka, 2001). Ridge systems cut the lower elevation plains in Hellas

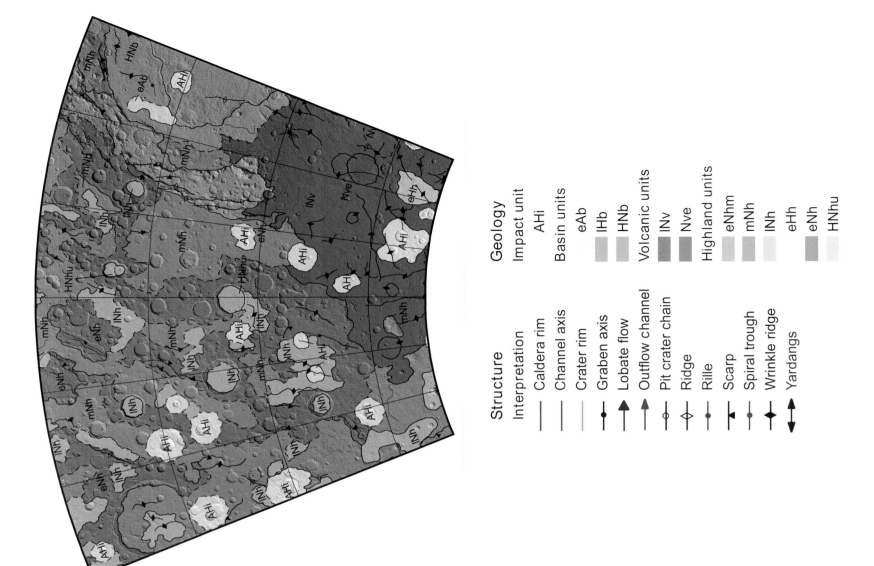

Structure

Interpretation

— Caldera rim
| Channel axis
| Crater rim
•— Graben axis
▲ Lobate flow
↑ Outflow channel
⊶ Pit crater chain
◇ Ridge
• Rille
◄ Scarp
⊢• Spiral trough
◆ Wrinkle ridge
↕ Yardangs

Geology

Impact unit
AHi

Basin units
eAb
lHb
HNb

Volcanic units
lNv
Nve

Highland units
eNhm
mNh
lNh
eHh
eNh
HNhu

as well as in the higher-standing plains, attesting to significant crustal contraction.

Ancestral, celestial gardening

The large impact craters in Noachis Terra (Figure 27.A) may date back as much as 4 billion years, a time when the inner solar system was much denser with asteroids. In fact, astrophysicists have suggested that movements in the orbits of the large, gas-giant planets, farther from the Sun, may have disturbed the asteroid belt between Mars and Jupiter, causing an episode of particularly heavy bombardment at that time (Gomes *et al.*, 2005; Bottke *et al.*, 2010). That same early bombardment formed large circular basins on the Moon, which later became flooded by dark lavas – the dark, smooth "seas" that are clearly visible during a full Moon. Craters saturate the oldest surfaces on Mars, as younger impacts completely obscure older ones (Figure 27.A). In effect, the upper few kilometers of the ancient Martian highlands were broken up by repeated, high-powered impacts of rocky objects flying at many kilometers per second.

Subdued remnants of explosions

Peneus Patera is one of a group of old volcanic centers, including Malea Patera (to the south of Peneus), Amphitrites Patera (MC-28), and Pityusa Patera (MC-30). These lie in Malea Planum, south of the Hellas basin, and record some of the oldest eruptions from central vents on Mars, as opposed to fissure or flood-type eruptions. In Malea

Planum, explosive eruptions of ash created broad, relatively flat mounds that were easily eroded. Later lava flows ran down the flanks and filled the complex summit calderas (Greeley *et al.*, 2007). The volcanic centers are found close to rings and related features of the Hellas impact basin, suggesting that crustal weaknesses that were created by the impact controlled the locations of volcanic activity (Rodriguez and Tanaka, 2006; see also MC-13). Sublimation of ice in near-surface deposits plays a significant role in recent modification of Peneus Patera (Figure 27.B).

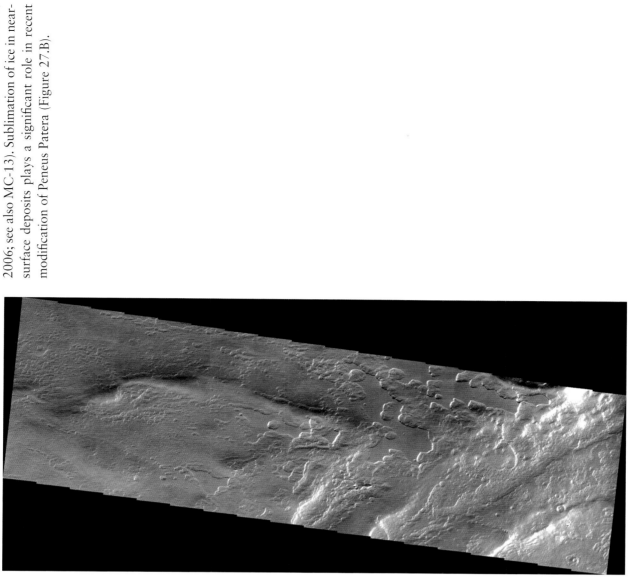

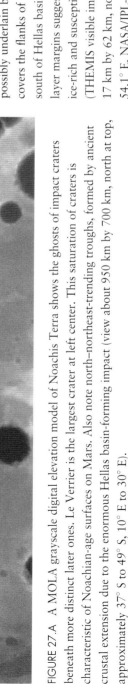

FIGURE 27.B A highly eroded layer of smooth material, possibly underlain by additional, hummocky layers, covers the flanks of the volcano Peneus Patera, just south of Hellas basin. The rounded pits and scalloped layer margins suggest that the surface materials may be ice-rich and susceptible to removal by sublimation (THEMIS visible image V06390008, 17 m/pixel, view 17 km by 62 km, north at top, centered at 57.0° S, 54.1° E, NASA/JPL-Caltech/Arizona State University).

FIGURE 27.A A MOLA grayscale digital elevation model of Noachis Terra shows the ghosts of impact craters beneath more distinct later ones. Le Verrier is the largest crater at left center. This saturation of craters is characteristic of Noachian-age surfaces on Mars. Also note north–northeast-trending troughs, formed by ancient crustal extension due to the enormous Hellas basin-forming impact (view about 950 km by 700 km, north at top, approximately 37° S to 49° S, 10° E to 30° E).

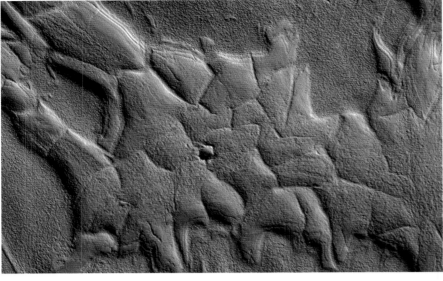

FIGURE 27.F Polygons with ridged margins exposed where overlying layered sequence has retreated, from north end of HiRISE image in Figure 27.C (view 3 km by 5 km, north at top).

A cold and muddy origin

Hellas Planitia displays varied features that may represent interactions of water, ice, and sediment in an ice-covered lake that once filled the Hellas basin. The HiRISE image (Figure 27.C) shows part of the lowest portion of the basin, around 6 km below datum. The oval form (Figure 27.D) resembles an impact crater; subsequent distortion may be due to motion of sediment and associated ice (McEwen, 2012). The curving ridges (Figure 27.E), also called reticulate or honeycomb material, suggest deformation of soft, muddy material by settling ice blocks. Ridged polygons (Figure 27.F) may be sediment-rich casts of cracks, exposed where overlying ice has retreated (Moore and Wilhelms, 2001).

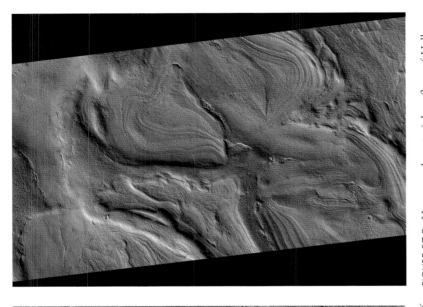

FIGURE 27.E Honeycomb material on floor of Hellas Planitia includes partly truncated sequences of layered materials and large, ovoid, ridged-bounded basin forms, from north-central part of HiRISE image in Figure 27.C (view 5 km by 9 km, north at top).

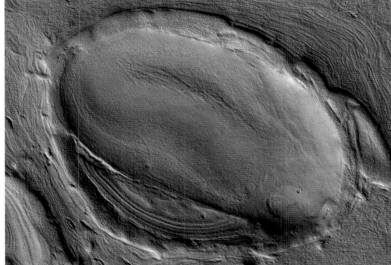

FIGURE 27.D Possible distorted impact crater on floor of Hellas Planitia, from center of HiRISE image in Figure 27.C (view 4.2 km across, north at top).

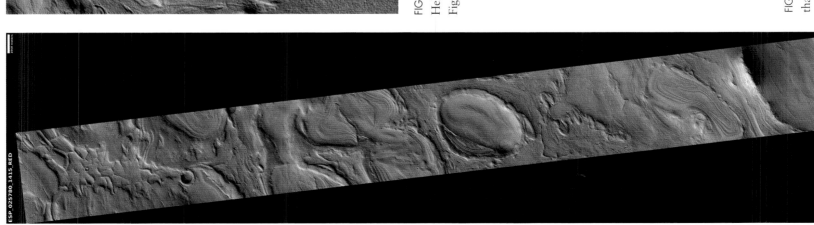

FIGURE 27.C Floor of Hellas Planitia, showing deposits that were possibly deformed by water and ice (HiRISE ESP_025780_1415, red band, 50 cm/pixel, view 5 km by 42 km, north at top, centered at 38.0° S, 53.8° E, NASA/JPL-Caltech/University of Arizona). Figures 27-D, 27-E, and 27-F are details of this image.

What an old basin goes through . . . .

Greeley crater displays a rugged, partly obliterated rim, more than 400 km in diameter, formed during the Early Noachian Epoch (Figure 27.G). The rim and floor are pocked by many craters, tens of kilometers in diameter. The steeper slopes of the crater rim are dissected by Noachian valleys, which would have contributed to infilling of the crater interior. The outer, lighter-colored parts of Greeley's floor are Middle Noachian in age and are relatively more rugged. These areas show evidence for fluvial deposition and exhumation in the form of sinuous ridges, interpreted to be inverted channel systems. Dark, less densely cratered patches in the lowest part of the floor, and within a few large, degraded craters, appear to be made up of relatively resistant lavas, emplaced during the Late Noachian Epoch. Wrinkle ridges and other structurally controlled landforms cross all parts of the floor, especially in the older materials, indicating that planetary contraction as well as local deformation was occurring in the Late Noachian Epoch and into the Hesperian Period.

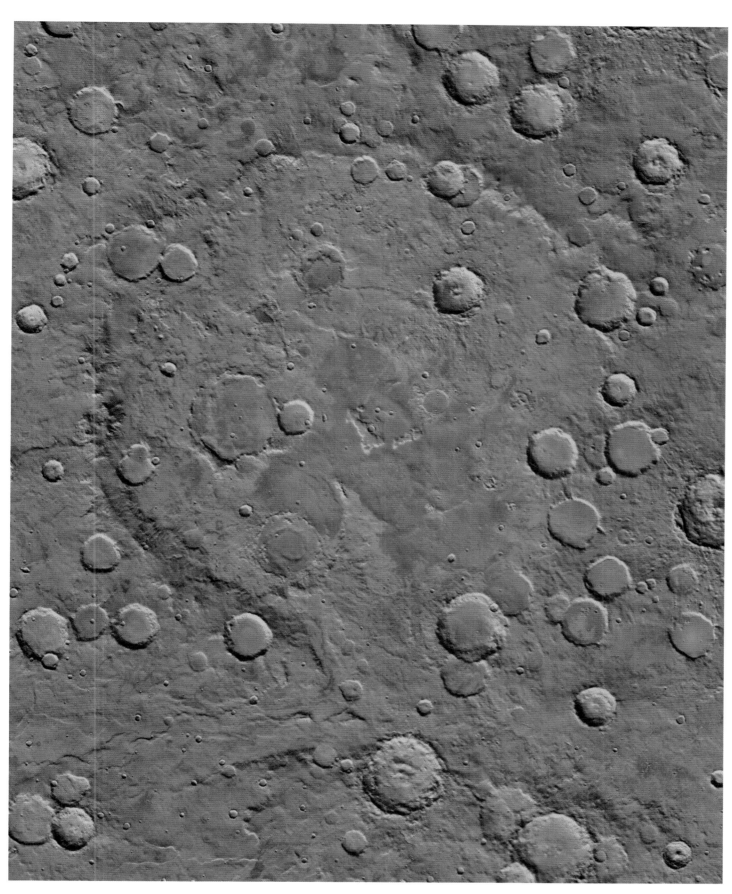

FIGURE 27.G Greeley impact basin on Mars, more than 400 km in diameter, is centered at 37° S, 3° E in Noachis Terra. The rim is densely cratered and dissected. The youngest floor material is dark and likely consists of lava flows (THEMIS daytime infrared mosaic overlying MOLA color shaded-relief DEM, view about 600 km across, north at top, NASA/JPL–Caltech/Arizona State University).

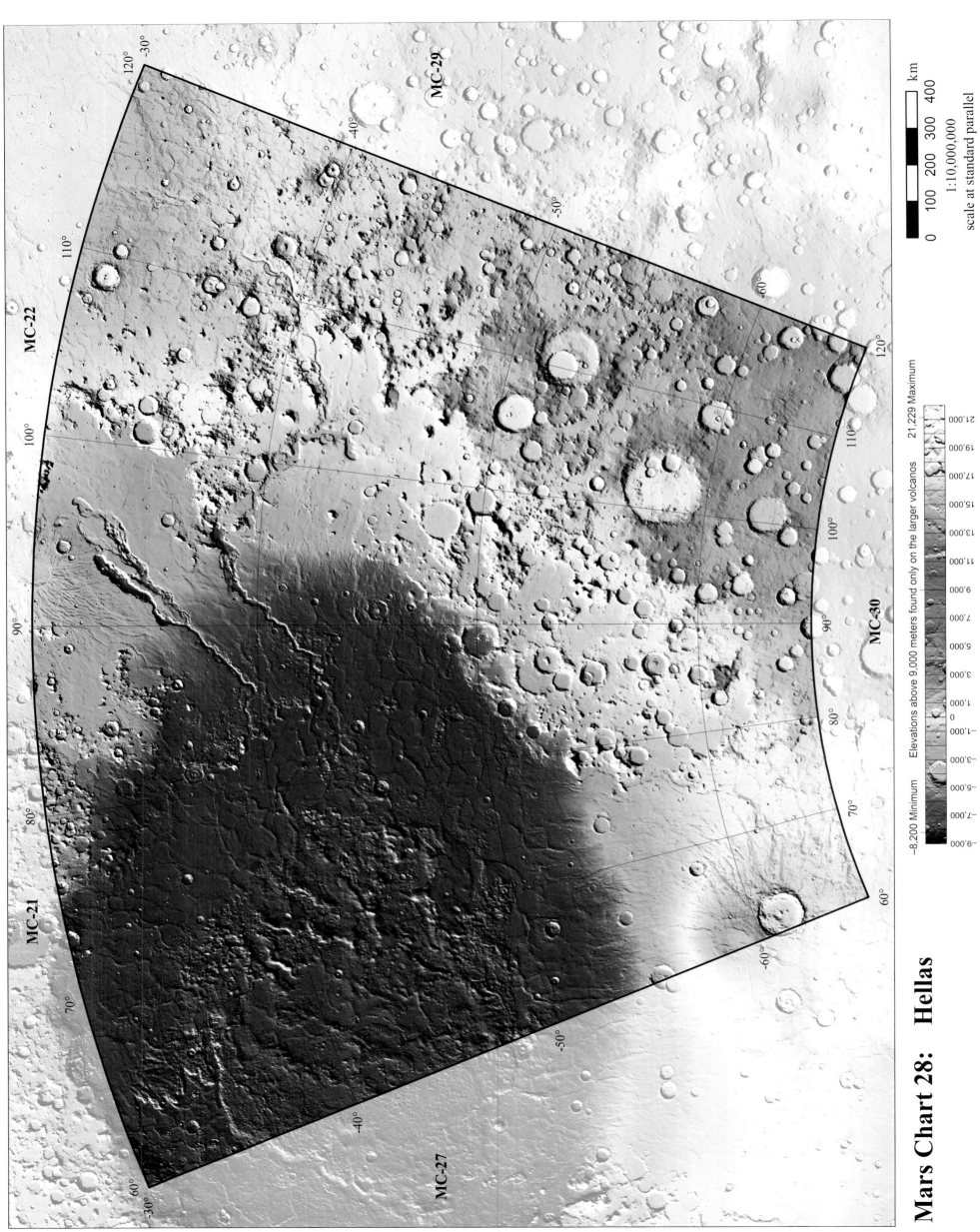

Mars Chart 28: Hellas

MC-29

MC-22

MC-21

MC-27

MC-30

1:10,000,000

scale at standard parallel

km

0   100   200   300   400

-8,200 Minimum   Elevations above 9,000 meters found only on the larger volcanos   21,229 Maximum

Elevation in meters

-9,000   -7,000   -5,000   -3,000   -1,000   0   1,000   3,000   5,000   7,000   9,000   11,000   13,000   15,000   17,000   19,000   21,000

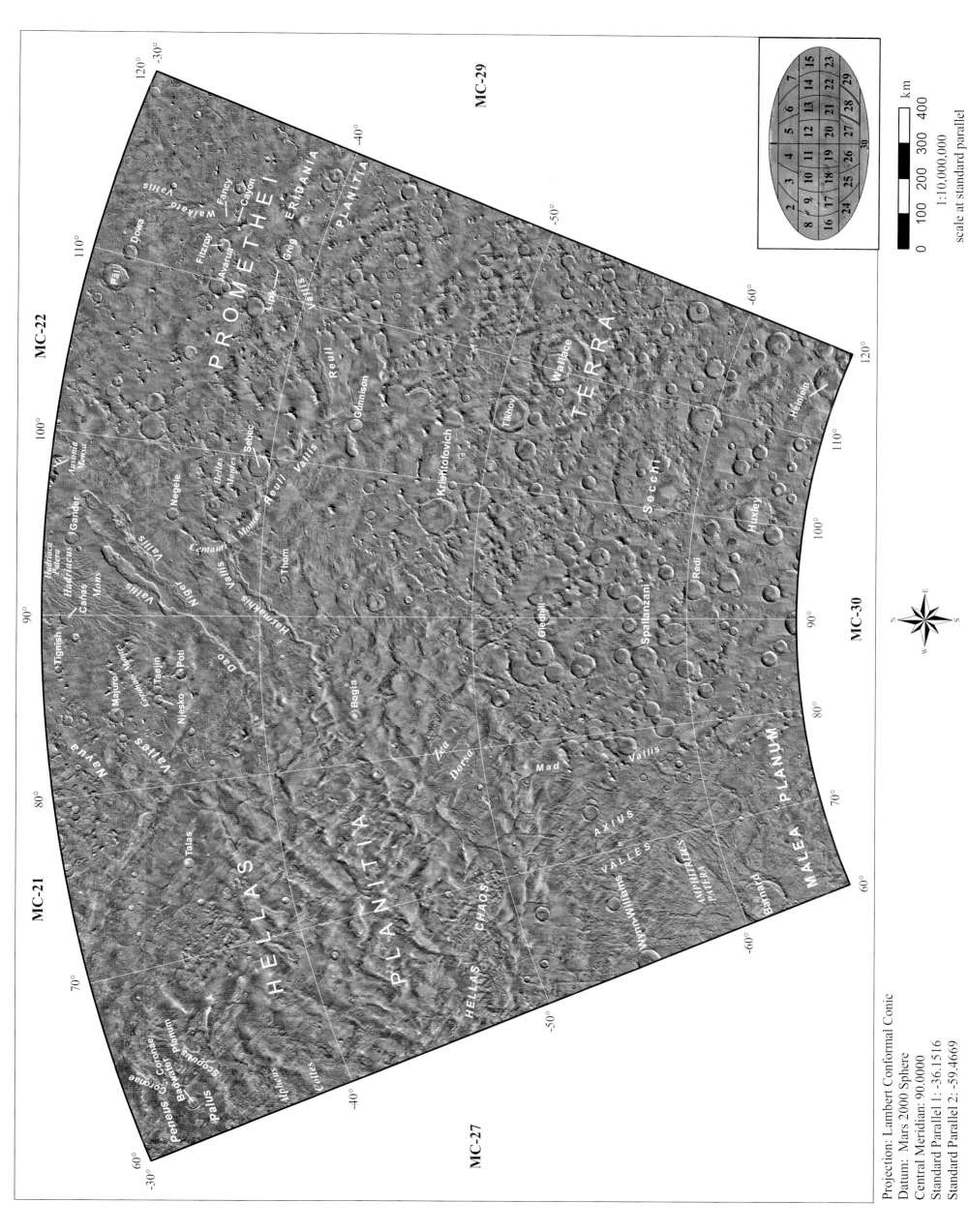

MC-22

MC-21

MC-29

PROMETHEI

ERIDANIA
PLANITIA

Waikato Vallis
Fancy
Cajon
Dowa
Fitzroy
Avarua
Greg
Lipik
Vallis

Reull

Gunnison

Negele

Hadriaca
Patera
Hadriacus
Mons
Canas
Niger
Vallis
Hellas
Montes
Sebec
Vallis
Reull Vallis
Centauri
Montes
Thom

Krishtofovich

Tikhov
Wallace
TERRA

Tignish
Harmakhis
Vallis

Secchi

Heimlein

Gander
Ausonia
Mensa

Huxley

Majuro
Cocytus
Valles
Taelin
Poti
Dao
Vallis

Gledhill

Redi

Spallanzani

Talas
Niesko

Naevus
Valles

Bogia

Zea
Dorsa

Mad

Vallis

AXIUS

HELLAS
PLANITIA

Wynn-Williams

HELLAS
CHAOS

VALLES

AMPHITRITES
PATERA

Barnard

MALEA
PLANUM

Peneus
Palus
Aphens
Corotae
Badwater
Planum
Scopulus
Colles

MC-27

MC-30

1		
2	3	4	5	6	7
8	9	10	11	12	13	14	15
16	17	18	19	20	21	22	23
24	25	26	27	28	29
		30		

km

1:10,000,000

0   100   200   300   400

scale at standard parallel

Projection: Lambert Conformal Conic
Datum: Mars 2000 Sphere
Central Meridian: 90.0000
Standard Parallel 1: -36.1516
Standard Parallel 2: -59.4669

Hellas (MC-28)

Geography

Hellas basin occupies the northwestern part of the quadrangle and makes up the deepest depression on Mars and the largest, well-preserved impact structure, at ~2,300 km inner diameter. The basin floor, Hellas Planitia, is largely a hummocky plain, marked by a network of wrinkle ridges and patches of knobby and troughed terrains, including Hellas Chaos. The lowest elevation on the planet (−8,200 m) occurs within the 33-km-diameter Badwater crater in the northwest corner of the quadrangle in Hellas Planitia. The basin rim is partly dominated by plains and low shield structures to the south (Malea Planum and Amphitrites Patera) and northeast (the southern edge of Hesperia Planum in MC-22, and Hadriacus Mons). The plains lie at 0–3 km elevation. Slopes north of Malea Planum are furrowed by dense, shallow valley systems (including Axius and Mad Valles), whereas the

northeastern slope is deeply dissected by Dao, Niger, Harmakhis, and Reull Valles. Except for Reull, these valleys have sinuous, trough-like rille forms, which narrow in the downstream direction. The Promethei Terra highlands, east of Hellas basin rim, are characterized by particularly rugged terrain, including high-standing massifs and deep crater forms.

Geology

The oldest materials in the quadrangle form large, irregular Early Noachian highland massifs of Promethei Terra, which resulted from uplift of the crust and piling of basin ejecta from the Hellas impact event. This terra forms part of the most extensive massif terrain around Hellas and may have resulted from a low-angle, southeast-directed impact (Tanaka and Leonard, 1995). The volcanic centers on the Hellas rim produced planar deposits during the Late Noachian Epoch, which extended downslope from Hadriacus Mons on the northeast flank and Amphitrites Patera on

the southern flank. Volcanism continued on the northeastern flanks in the Hesperian Period, emplacing volcanic plains and driving groundwater release to form the large, rille-like, sinuous channel systems. These channels deposited materials into Hellas Planitia, which shows evidence for multiple stands of lake level and remnant ice sheets (Moore and Wilhelms, 2001). In addition,

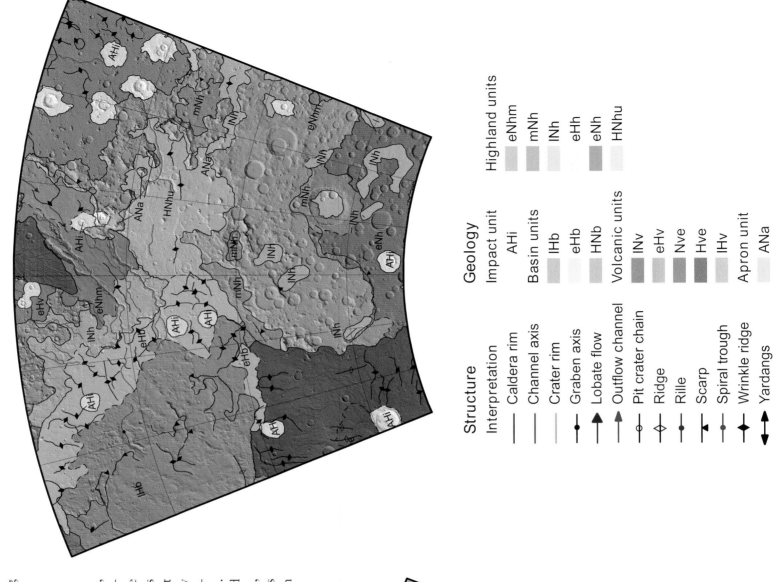

Structure
Interpretation

——	Caldera rim
——	Channel axis
——	Crater rim
●—	Graben axis
▲	Lobate flow
▲	Outflow channel
⊶	Pit crater chain
◇—	Ridge
●—	Rille
◀—	Scarp
●—	Spiral trough
◆—	Wrinkle ridge
↕	Yardangs

Geology

Impact unit
| | AHi |

Basin units
	lHb
	eHb
	HNb

Volcanic units
	lNv
	eHv
	Nve
	Hve
	lHv

Apron unit
| | ANa |

Highland units
	eNhm
	mNh
	lNh
	eHh
	eNh
	HNhu

the basin may have accumulated dust, sand, and ice from the atmosphere by wind transport and precipitation, and possibly volcanism, for more than 3 billion years, from the Late Noachian Epoch into present times.

It is frozen but it moves: debris aprons

Promethei Terra, east of Hellas, has numerous examples of sloping areas in valleys and next to peaks (Figures 28.A1, 28.A2, 28.B, 28.C). These areas are young (they generally lack craters) and show ridges and surface textures that indicate movement of rock debris downhill (e.g. Pierce and Crown, 2003). The shape is also consistent with a flow origin (Li et al., 2006). The debris aprons have been imaged and studied since the Viking mission, and radar images indicate that, beneath the surface, the aprons are made up of water ice. A common interpretation is that the debris aprons represent flowing ice and rock, similar to glaciers on Earth (Plaut et al., 2009). The possibility of remnant glaciers at mid-latitudes indicates that Mars had a different climate, possibly with water ice being more widespread at the

surface than at present. Furthermore, the near absence of craters on the possible glacial material suggests that this episode of glaciation was during the Late Amazonian Epoch, perhaps during a climate excursion in the past few million years (Head et al., 2005; HRSC, 2005; 2006; Dickson and Head, 2008) that followed an earlier, more extensive glaciation (Dickson et al., 2008).

Valleys with an odd shape

On Earth, most valleys have a shape in cross-section that reflects their origin: V-shape if carved by water, and U-shape if made by a valley glacier.

Reull Vallis on Mars has a very broad U-shape, several kilometers wide and several hundred meters deep (HRSC, 2004b; 2013; Figure 28.D). One explanation is that rocks and ice filled the valley floor and created the textures there in a manner similar to rock-covered valley glaciers on Earth. The upper ends of Dao Vallis and Niger Vallis are made up of steep-sided, broad depressions and lack tributary valley networks that are typical of precipitation-fed runoff valley systems. The steep walls and chaotic and rough textures of the valley floors and shallow feeder troughs thus suggest that discharge of water and valley erosion may have been due to sudden melting of

subsurface ice, as proposed in other chaos regions on Mars (HRSC, 2004a; Figure 28.E).

Ancient shoreline?

As noted, one explanation for the forms on Hellas Planitia is that it once held a lake. Figure 28.F shows a possible shoreline at ~5.8 km (Moore and Wilhelms, 2001) where Dao Vallis intersects it. Note how the sharply defined valley upslope (to the right) changes to a much less-defined, substantially narrower form, downslope (west) of the postulated shoreline. This is similar to the forms made on Earth by rivers at shorelines.

FIGURE 28.B Isolated peak in Promethei Terra near Reull Vallis, surrounded by lobate debris aprons (from HRSC image H0451_0000_ND4, 12.5 m/pixel, view 100 km across, north at top, 40.6° S, 103.0° E, ESA/DLR/FU-Berlin).

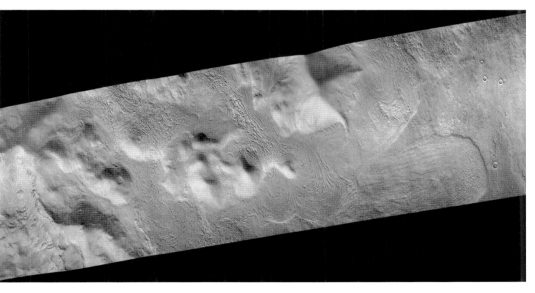

FIGURE 28.A2 CTX image showing detail in center of Figure 28.A1 (image G05_020293_1414, 5.4 m/pixel, view 25 km by 95 km, north at top, centered at 38.4° S, 97.3° E, NASA/JPL-Caltech/MSSS).

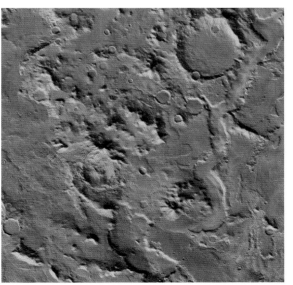

FIGURE 28.A1 Hellas Montes, centered at 97.5° E, 38.5° S, showing lobate forms around peaks, which may be glacial in origin (MOLA color over THEMIS daytime infrared, view 275 km across, north toward top, NASA/JPL-Caltech/Arizona State University).

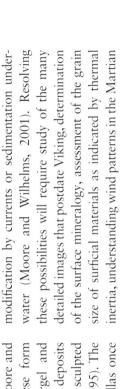

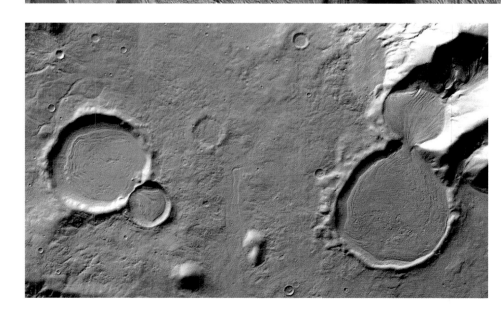

FIGURE 28.C Just east of Hellas Montes are pairs of adjoining craters with unusual fill material displaying ridge forms that are characteristic of debris aprons. In the lower pair, the flow is from right to left (HRSC H0451_0000_ND4, 12.5 m/pixel, approx. 36 km by 60 km view, north at top, near 39° S, 103° E, ESA/DLR/FU-Berlin).

FIGURE 28.D Perspective view toward the northeast of upper Reull Vallis, from a color HRSC image (HRSC orbit 10657, 16 m/pixel, view about 60 km across, view at 41.5° S, 106.5° E, ESA/DLR/FU-Berlin).

Bumps on the floor

Hellas Planitia has varied terrain and landforms that have inspired a great variety of interpretations. For example, elongated, kilometer-scale mounds or knobs (Figure 28.G) that appear in Viking orbiter images have been called fluvial deposits (Greeley and Guest, 1987), dunes or mudflows (Moore and Edgett, 1993), drumlins (on Earth, these form beneath moving glacial ice sheets; Kargel and Strom, 1992), remnants of dust and ice deposits (Tanaka, 2000), and yardangs (landscape sculpted by wind erosion; Tanaka and Leonard, 1995). The interpretation by some scientists that Hellas once held a large lake also suggests the possibility of modification by currents or sedimentation underwater (Moore and Wilhelms, 2001). Resolving these possibilities will require study of the many detailed images that postdate Viking, determination of the surface mineralogy, assessment of the grain size of surficial materials as indicated by thermal inertia, understanding wind patterns in the Martian atmosphere, and other related investigations.

FIGURE 28.F Dao Vallis meets the postulated –5.8 km shoreline of the Hellas basin at the center of the image. The character of the valley changes from higher elevations (incised valley in rugged terrain) in the east to more subdued form, in what may have been a submerged channel, crossing a smoother region to the west (H2664_0001_ND4, HRSC nadir image, 50 m/pixel, 150 km by 200 km, north at top, centered at 39.5° S, 83.5° E, ESA/DLR/FU-Berlin).

FIGURE 28.E Uppermost Dao and Niger Valles, showing steep trough walls and channel floors in various stages of collapse and degradation, including shallower feeder troughs at upper right (HRSC nadir image, orbit 528, 40 m/pixel, view 170 km across, north at top, 34° S, 93° E, ESA/DLR/FU-Berlin).

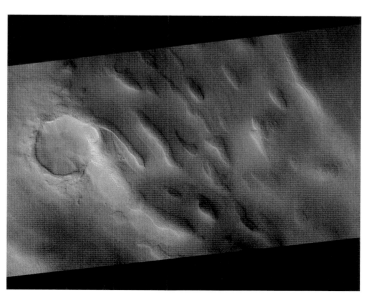

FIGURE 28.G CTX image of enigmatic mounds on the floor of Hellas Planitia (image B20_017472_1338_XN_46S285W, 5 m/pixel, view 25 km by 40 km, north at top, 47.0° S, 74.7° E, NASA/JPL-Caltech/MSSS).

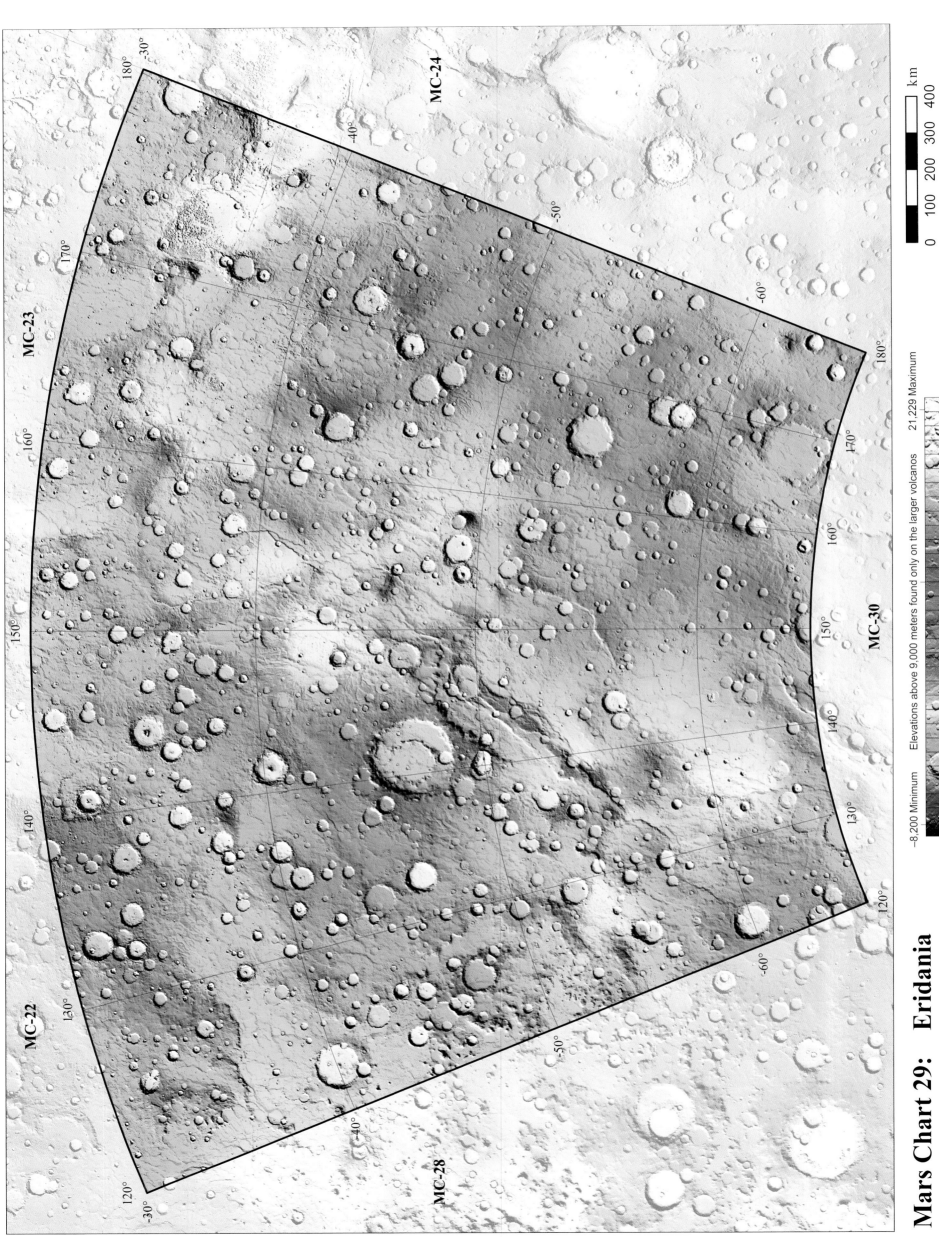

Mars Chart 29:   Eridania

1:10,000,000
scale at standard parallel

km

0   100   200   300   400

-8,200 Minimum   Elevations above 9,000 meters found only on the larger volcanos   21,229 Maximum

Elevation in meters

-9,000 -7,000 -5,000 -3,000 -1,000   0   1,000   3,000   5,000   7,000   9,000   11,000 13,000 15,000 17,000 19,000 21,000

MC-22

MC-23

MC-24

MC-28

MC-30

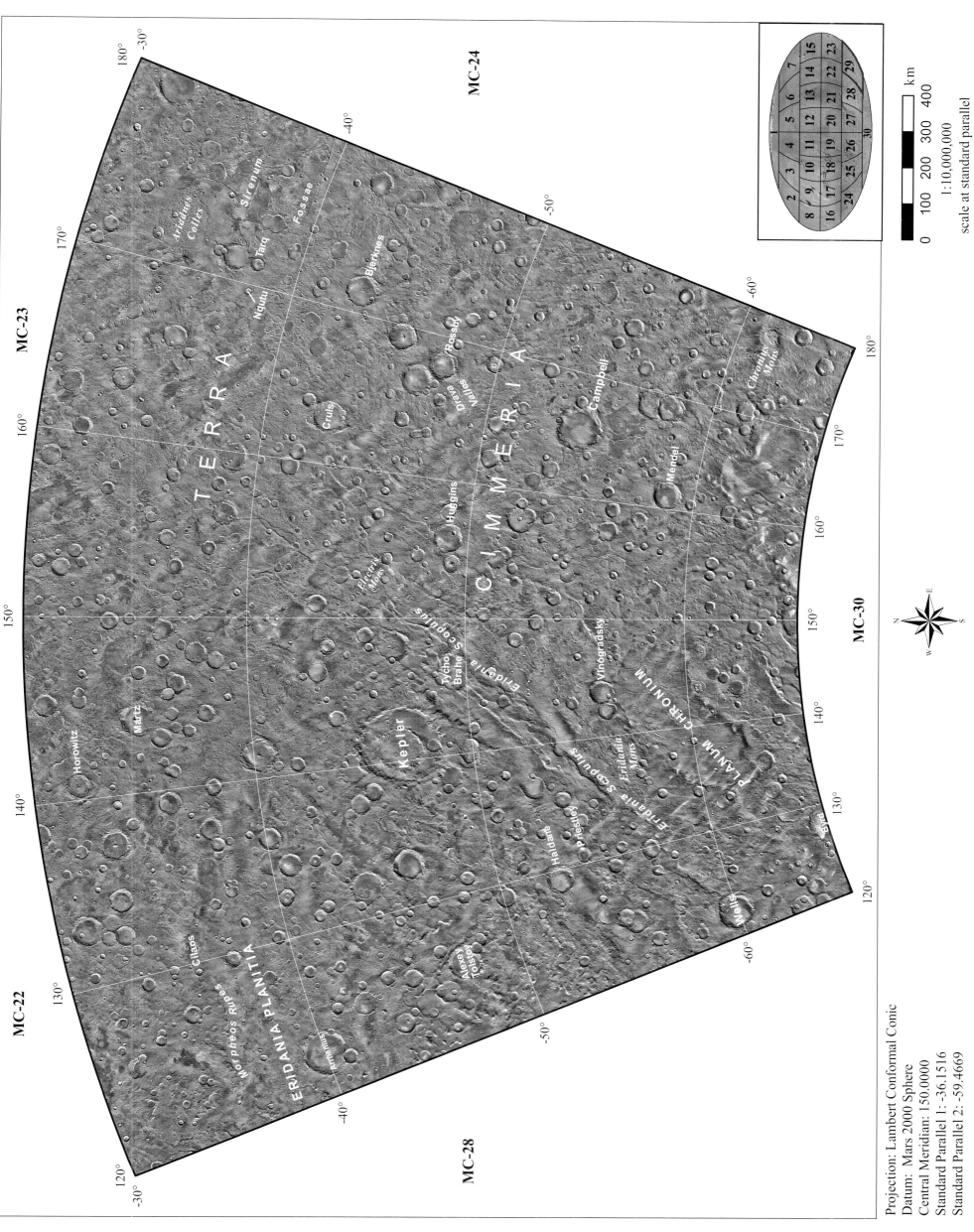

MC-23

MC-24

MC-22

MC-30

MC-28

TERRA

CIMMERIA

ERIDANIA PLANITIA

PLANUM CHRONIUM

Ariadnes Colles

Sirenum Fossae

Tarq

Ngutu

Bjerknes

Cruls

Rossby

Drava Valles

Huggins

Electris Mons

Eridania Scopulus

Campbell

Mendel

Chronius Mons

Tycho Brahe Scopulus

Vinogradsky

Kepler

Martz

Horowitz

Chaos

Morpheos Rupes

Amenmus

Priestley

Haldane

Eridania Mons

Eridania Scopulus

Byrd

Sitek

Alexey Tolstoy

-30°

-40°

-50°

-60°

-30°

-40°

-50°

-60°

180°

170°

160°

150°

140°

130°

120°

180°

170°

160°

150°

140°

130°

120°

N
E
W
S

Projection: Lambert Conformal Conic
Datum: Mars 2000 Sphere
Central Meridian: 150.0000
Standard Parallel 1: -36.1516
Standard Parallel 2: -59.4669

km

0   100 200 300 400

1:10,000,000
scale at standard parallel

1							7
	5	6	13	14	15		
	4	12	13	14	22	23	
2	5	11	12	20	21	29	
3	9	10	19	20	27	28	
8	9	17	18	25	26		
	16	17	24	25			
			30				

Eridania (MC-29)

Geography

Eridania quadrangle is composed almost entirely of the ancient cratered highland terrain of Terra Cimmeria, at 0–2 km elevation. The largest crater, Kepler, is about 230 km in diameter. Less-cratered, relatively low-lying plains are scattered throughout the quadrangle, including Eridania Planitia in the northwest corner and Planum Chronium in the southwest part of the quadrangle. Ridge systems occur throughout the quadrangle, with northeast-trending Eridania Scopulus forming the most prominent ridge.

Geology

This region includes ancient Noachian crust having some of the highest magnetization readings on Mars as recorded by the Mars Global Surveyor spacecraft. The magnetization consists of east–west-trending normal and reversed polarity stripes, each ~100 km wide and exceeding 1,000 km in length in some cases (see map of magnetization in Figure 3.5). These stripes indicate that a Mars dynamo was active early in the planet's history and produced reversing magnetic fields, as the Earth's dynamo has done throughout its history. The buried crust, of pre-Noachian age, has not been subjected to high-temperature demagnetization. Elsewhere on Mars, crustal

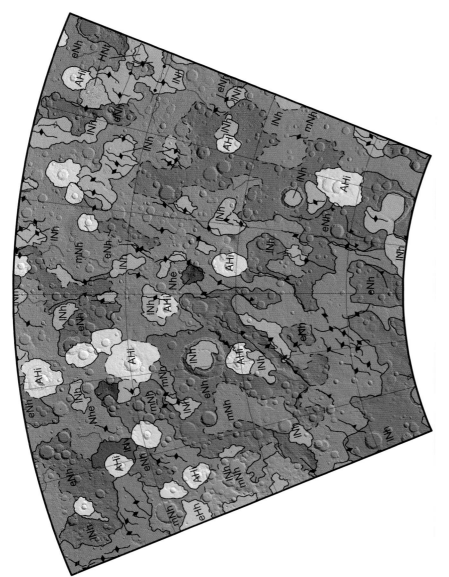

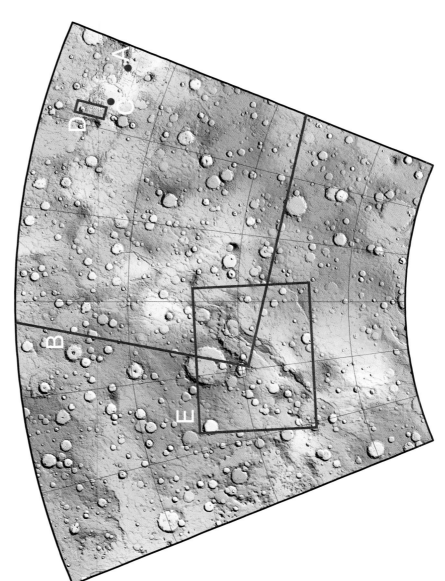

Structure

Interpretation

——	Caldera rim
——	Channel axis
——	Crater rim
•—•	Graben axis
▲	Lobate flow
▲	Outflow channel
o—	Pit crater chain
⬦—	Ridge
•—	Rille
◄—	Scarp
•—	Spiral trough
◆—	Wrinkle ridge
↕	Yardangs

Geology

Impact unit

AHi

Volcanic units

lNv

eHv

Transition unit

HNt

Highland units

eNhm

mNh

lNh

eHh

eNh

Nhe

demagnetization occurred widely due to effects of later basin-forming impacts, such as Hellas and Argyre, magmatic heating in association with volcanic activity as at Tharsis, and formation of the northern plains, possibly by basin-forming impact and/or magmatic activity. Eridania Scopulus is roughly concentric to Hellas basin and may result from crustal deformation due to that impact. Other scarps and broad ridges, including Morpheos Rupes, may be remnant impact or tectonic features. Minor geologic activity in the Hesperian Period formed local plains deposits, of which Ariadnes Colles make up knobby, dissected remnants of an eroded, likely fine-grained material.

Gullies and aprons and possible fountains of youth?

Although the majority of geologic activity in the southern highlands of Mars dates back to the Noachian and Hesperian Periods, more than 3.5 billion years ago, high-resolution image data record erosion and deposition along crater walls as recently as within the past million years. Aprons of meters-thick deposits blanket poleward-facing slopes in middle latitudes, and some of these are dissected by gullies that originate from steep, rocky alcoves. Downslope, the aprons splay into fan deposits, where the surface flattens out (Figure 29.A; Malin

and Edgett, 2000a). Thousands of these features have been observed, and a few appear to be seasonally active, including on dune surfaces. Proposed mechanisms for their formation, and corresponding Earth analogs, are diverse and include discharges of groundwater, snow or ice melt, dry avalanching, and carbon dioxide fluidization. The gullies thus may be indicators of surface dust and volatile accumulation from the atmosphere, crustal thermal and hydrologic conditions and variations, and/or climate change. Those that formed within the current climate appear to have done so during winter, most likely resulting from seasonal carbon dioxide build-ups that drive avalanches. Some of the best examples of these features occur in the Eridania quadrangle (Harrison et al., 2012).

A former, vast upland lake system

North of the Eridania quadrangle lies Ma'adim Vallis (in MC-23), a 900-km-long channel. Unlike many outflow channels on Mars, it does not originate in a region of chaotic ground collapse. Rather, a region of interconnected basins, extending thousands of kilometers across much of the northeastern quarter of the Eridania quadrangle and into adjacent areas (Figure 29.B), is

breached at its northern edge (in MC-23) by the head of Ma'adim Vallis. If the enclosed basins once contained lakes, then an overflow flood could explain the erosion of Ma'adim Vallis. The presence of sustained lakes in the basins during a Late Noachian stage of runoff also could explain why the basins lack the flat crater floors that are seen elsewhere on Mars, and why they lack valleys below 1,100 m in elevation, the postulated highest lake level (Irwin et al., 2002; 2004). Whether the lake water came from precipitation, groundwater, or subsurface ice is unresolved (Baker and Head, 2012).

Knobs of Ariadnes Colles

One of the Eridania basins that may have once been a lake has a floor covered by a field of knobs; these are Ariadnes Colles (Figures 29.C, 29.D). The light-colored knobs give indications in the CRISM data of magnesium- and iron-rich phyllosilicate minerals (possibly water-altered volcanic ash, as most phyllosilicates on Mars are considerably older) and sulfate minerals (possibly evaporites; Moore and Howard, 2003; Wendt et al., 2008; 2012; Annex and Howard, 2011; Golder and Gilmore, 2012). An originally flat-lying deposit may have been deformed by water expulsion, eroded by wind or under a lake to leave the knobs, or they might even represent the rising of salt through shallower layers, a process known as salt tectonics. The knobs are surrounded by younger, dark basaltic rocks.

Crunch time

Along and near Eridania Scopulus, closely spaced wrinkle ridges and high scarps locally cross and deform impact craters (Figure 29.E; Tanaka and Schultz, 1993). The dominant northeast trend of the scarps and wrinkle ridges and adjacent troughs is not consistent with activity in Tharsis to the northeast (Lias et al., 1999), while the relationship to the Hellas impact structure is only approximately concentric.

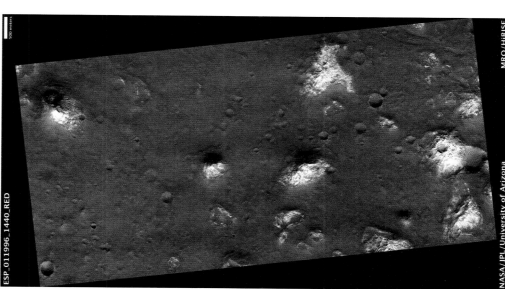

ESP_011996_1440_RED

NASA/JPL/University of Arizona

MRO/HiRISE

FIGURE 29.C HiRISE image showing the fractured, light-colored hills of Ariadnes Colles, rising above darker deposits (ESP_011996_1440 red image, 25 cm/pixel, view 5 km by 10 km, north at top, centered at 35.5° S, 173° E, NASA/JPL-Caltech/University of Arizona).

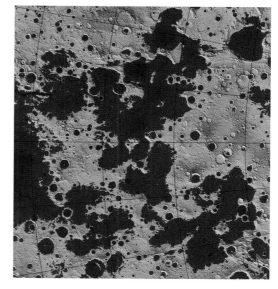

FIGURE 29.B A color elevation map of the connected basins from which Ma'adim Vallis drains northward. Areas lower than the 1,100-m contour, the postulated highest lake level, are shown in black (modified from MOLA color hillshade, shows parts of MC-16, 23, 24, and 29, longitude-latitude grid at 10° spacing, north at top).

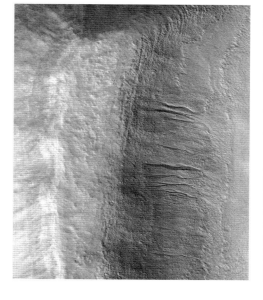

FIGURE 29.A HiRISE image showing gullies on a hill of layered material in Terra Cimmeria. The margins of several aprons of mantling material are visible on this south-facing slope (ESP_024325_1435, 25 cm/pixel, view 1.7 km by 1.4 km, north at top, 36.3° S, 177.4° E, NASA/JPL-Caltech/University of Arizona).

FIGURE 29.D Color overview of Ariadnes Colles (HRSC orbit 4209, 13 m/pixel, view about 150 km by 50 km, north to the right, large crater is 30 km across, 33.0° S to 35.5° S, 171.0° E to 172.3° E, ESA/DLR/ FU-Berlin).

FIGURE 29.E Eridania Scopulus trends northeast–southwest across the image. Ridges crosscut and distort craters. The largest crater at the top is Kepler, 228 km in diameter (Themis daytime infrared mosaic, view about 900 km across, north at top. NASA/JPL–Caltech/Arizona State University).

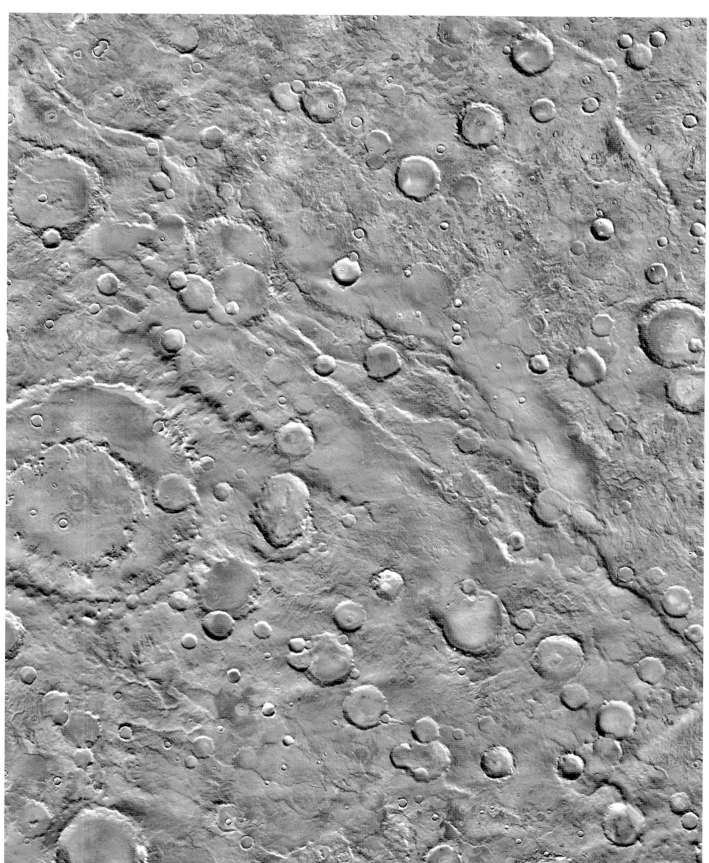

233

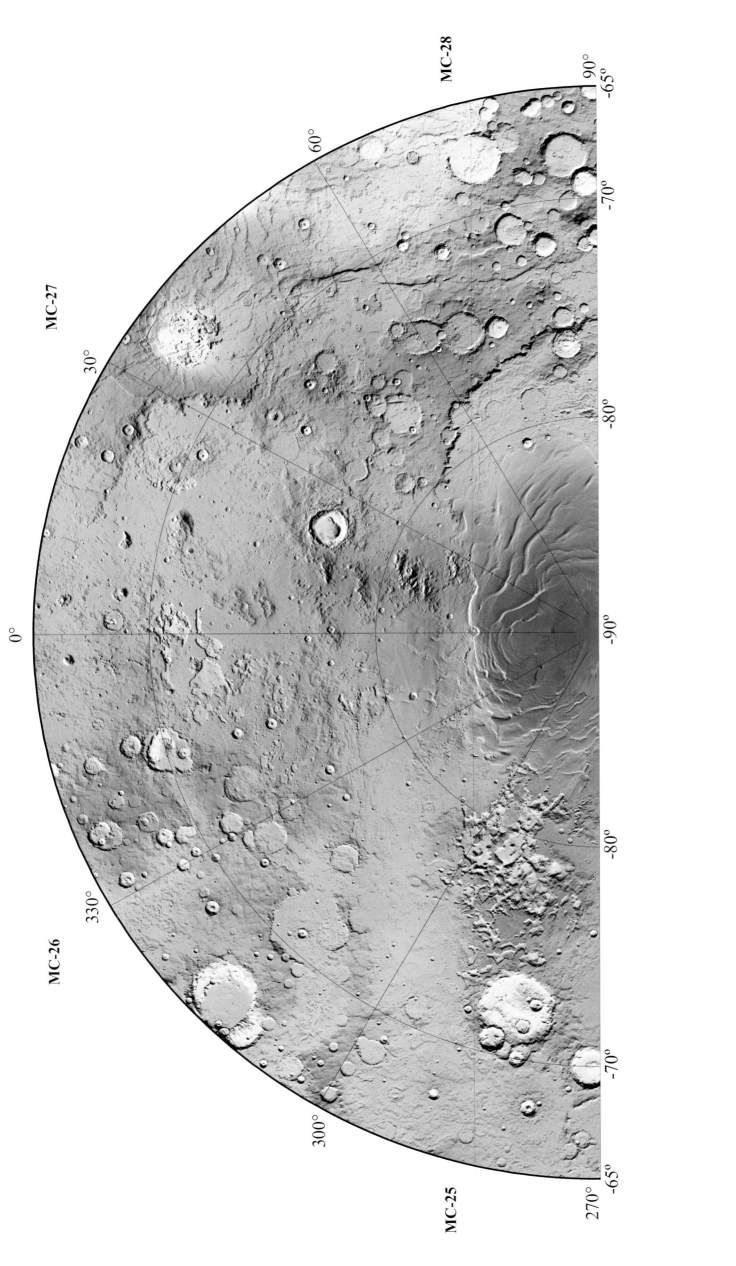

Mars Chart 30: Mare Australe

MC-28

MC-27

MC-26

MC-25

km

400

300

200

100

0

1:10,000,000

scale at standard parallel

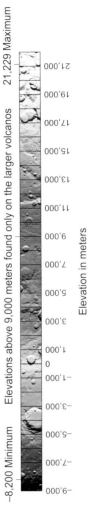

-8,200 Minimum   Elevations above 9,000 meters found only on the larger volcanos   21,229 Maximum

21,000

19,000

17,000

15,000

13,000

11,000

9,000

7,000

5,000

3,000

1,000

0

-1,000

-3,000

-5,000

-7,000

-9,000

Elevation in meters

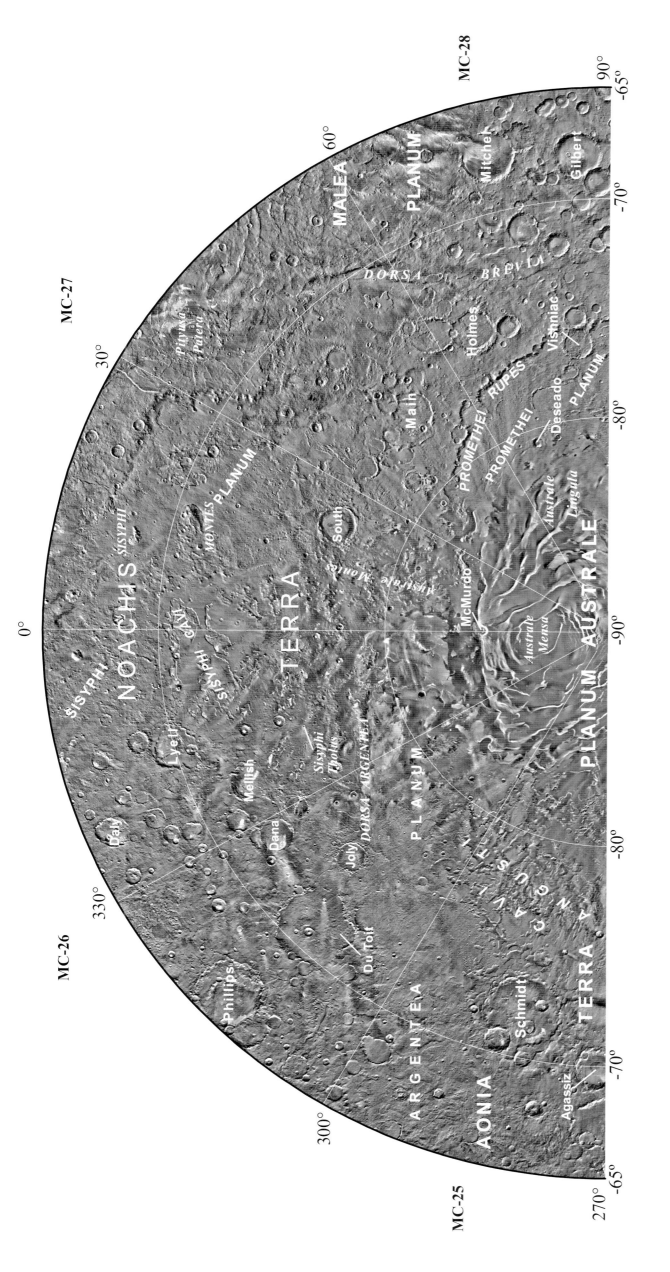

MC-28

MC-27

MC-26

MC-25

MALEA PLANUM

NOACHIS

SISYPHI

SISYPHI

Pityusa Patera

MONTES PLANUM

SISYPHI CAVI

Lyell

Mellish

Daly

Phillips

Dana

Joly

Du Toit

Sisyphi Tholus

DORSA ARGENTEA

Schmidt

ARGENTEA

AONIA

TERRA

Agassiz

TERRA

South

Australe Montes

DORSA BREVIA

Main

Holmes

PROMETHEI RUPES

PROMETHEI

Deseado

Vishniac

Mitchel

Gilbert

McMurdo

Australe

Ungula

Australe Mensa

PLANUM AUSTRALE

PLANUM

60°

30°

0°

330°

300°

270°

90°

-70°

-80°

-90°

-80°

-70°

-65°

-65°

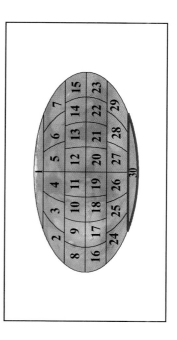

km

1:10,000,000

0 100 200 300 400

scale at standard parallel

Projection: Stereographic South Pole
Datum: Mars 2000 Sphere Polar
Central Meridian: 0.000
Standard Parallel 1: -90.000

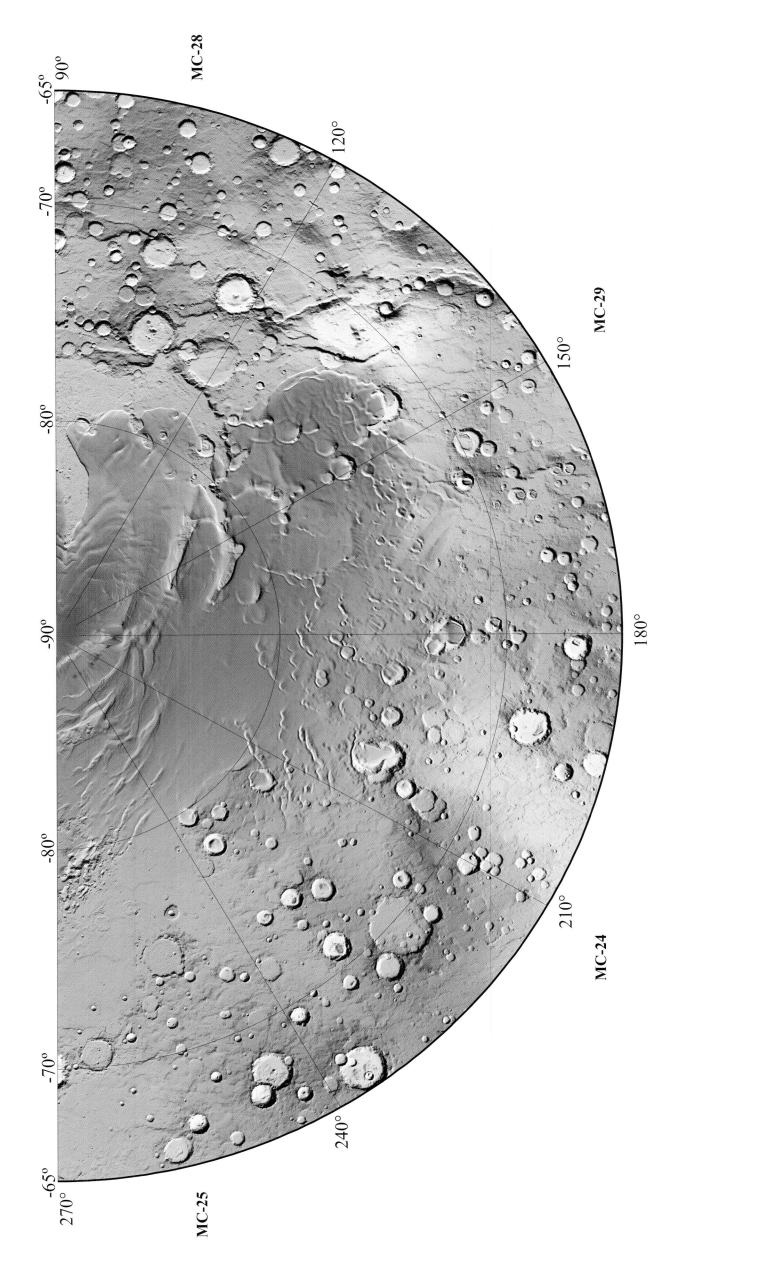

Mars Chart 30: **Mare Australe**

1:10,000,000
scale at standard parallel

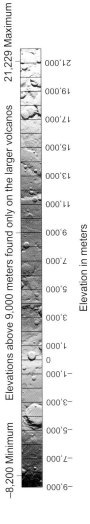

Elevation in meters

−8,200 Minimum   Elevations above 9,000 meters found only on the larger volcanos   21,229 Maximum

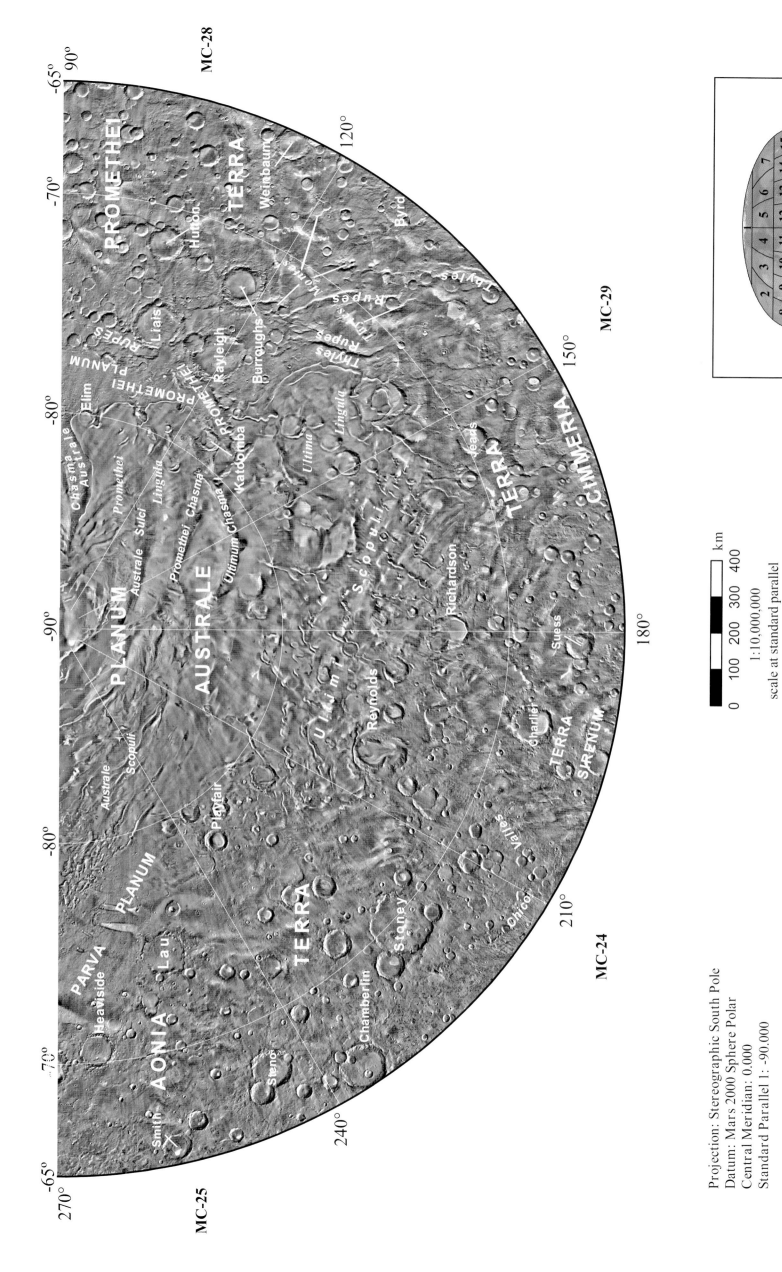

Projection: Stereographic South Pole
Datum: Mars 2000 Sphere Polar
Central Meridian: 0.000
Standard Parallel 1: -90.000

1:10,000,000
scale at standard parallel

km
0   100 200 300 400

Mare Australe (MC-30)

Geography

The Mare Australe quadrangle is dominated by the south polar plateau, Planum Australe (Figure 30.A), rising to about 4,800 m in elevation on its highest part, Australe Mensa (Figure 30.B). Planum Australe is roughly circular, with horizontal dimensions of 1,100 km by 1,400 km. The plateau is dissected by the large troughs Chasma Australe, Promethei Chasma, and Ultimum Chasma, which divide parts of the plateau into the tongue-shaped forms of Australe Lingula and Promethei Lingula. Also present are systems of lower relief,

concentrically and obliquely trending troughs, and asymmetric ridges (Figure 30.A). The surrounding cratered highlands lie at 1–3-km elevations and include some rather exotic terrains, unique to the south polar region. Cavi Angusti form depressions tens of kilometers across and reaching depths of a kilometer. Branching ridges form Dorsa Argentea and other ridge systems that cover Argentea, Promethei, and Parva Plana, surrounding Planum Australe.

Geology

Like its north polar counterpart, a stack of meter-thick layers of Amazonian age forms Planum

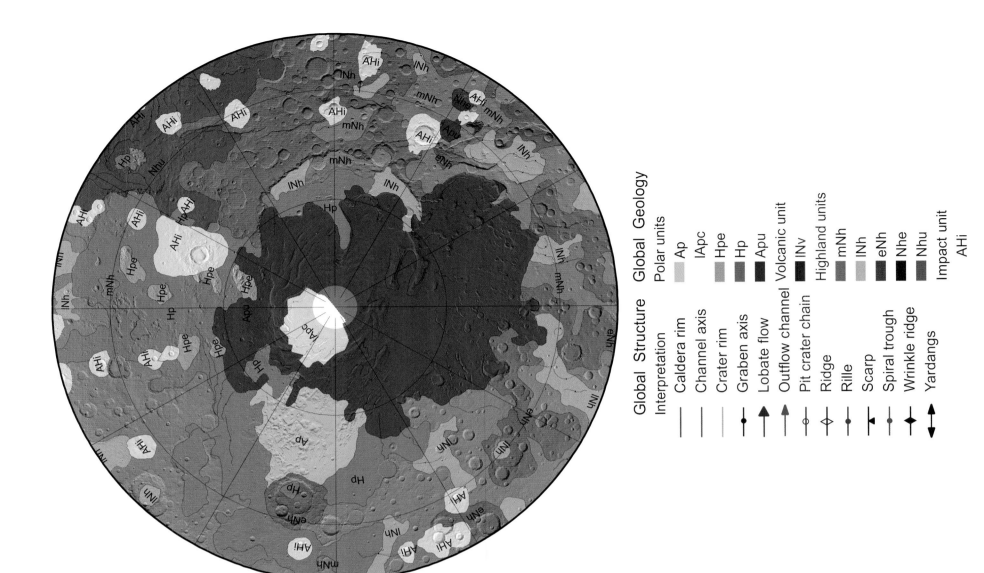

Global Structure
Interpretation

Caldera rim	
Channel axis	
Crater rim	
Graben axis	
Lobate flow	
Outflow channel	
Pit crater chain	
Ridge	
Rille	
Scarp	
Spiral trough	
Wrinkle ridge	
Yardangs	

Global Geology

Polar units
Ap
lApc

Global Geology
Hpe
Hp
Apu

Volcanic unit
lNv

Highland units
mNh
lNh
eNh
Nhe
Nhu

Impact unit
AHi

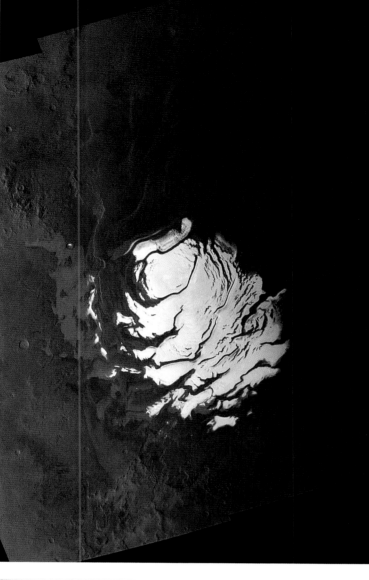

FIGURE 30.A MOLA color hillshade image of Planum Australe and the informally named south polar (or Promethei) basin. The arc of Promethei Rupes is prominent in the right half of the image (south pole at center of image, 0° longitude toward top, view about 1,200 km by 1,800 km).

FIGURE 30.B Mosaic of Viking orbiter 2 images, showing the south polar cap during southern summer (L_s = 341°), when it was about 400 km across (JPL PIA00301, 0° longitude toward top, south pole at lower right edge of polar cap, NASA/JPL-Caltech/USGS).

Australe (Figure 30.C), with a composite thickness exceeding 3,000 m in Australe Mensa. Part of the planum surface is eroded along arcuate scarps that form Australe Sulci, revealing layering and deformational ridges and scarps (Figure 30.D). The layers appear to be mostly water ice, but in part may include a substantial body of carbon dioxide ice (dry ice) in the upper part of Australe Mensa, as revealed in shallow radar and geologic mapping studies (Phillips *et al.*, 2011). Older, Hesperian-age ice sheets, hundreds of meters thick, as indicated by radar sounding data, form the circumpolar plains and may have formed as a result of glaciation or eruption of groundwater. The Dorsa Argentea (Figure 30.E) are in these contexts either eskers, resulting from subglacial river deposits (Scanlon and Head, 2014), or inverted channel systems, produced by river deposition and subsequent removal of adjacent materials. Cavi Angusti (Figure 30.F) and Sisyphi Cavi form extensive, irregular depressions within

the ice sheets and are indicative of ice removal, perhaps stimulated in part by sub-ice volcanism. Surrounding cratered highlands and intercrater plains, of mostly Noachian age, lack evidence for icy compositions at depth.

Solidified air

The carbon dioxide deposit buried within Australe Mensa (Figure 30.G) has a volume of about 10,000 km³, sufficient to nearly double the mass of the Martian atmosphere if unleashed during high obliquity. This might account for increased surface winds, leading to greater dust-storm activity and heightened migration of sand dunes (Phillips *et al.* 2011). On the surface, overlying some of the same locations of the buried deposit, is a much smaller body of dry ice made up of several layers of surficial dry ice and amounting to several meters in thickness. This deposit includes circular and arcuate depressions that resemble Swiss

cheese (Figure 30.H) and are actively enlarging due to solar heating and wall collapse (Buhler *et al.*, 2017; Figures 30.I, 30.J). The flat floors of the depressions indicate that more stable water ice lies immediately below the carbon dioxide layers. Although this dry ice patch is thin, it accounts for the bright residual ice cap that can be seen in southern summer (Figure 30.B).

Spiders on Mars

HiRISE images show spider-like forms consisting of systems of branching, meter-deep channels and troughs, radiating from local, topographically high centers that are associated with dark, fan-shaped surface coatings. The origin of these features is uncertain. One leading hypothesis considers how the south polar surface is cold enough to freeze a meter-thick slab of dry ice in the winter. The slab is potentially transparent enough that springtime solar heating of the material beneath

the frozen carbon dioxide sublimates the base of the dry ice. The trapped carbon dioxide gas may reach sufficiently high pressures to rupture the ice cover, resulting in geysers that spread dark dirt over the surface, downwind. As the gas rushes to the eruption site from under the ice, it may erode channels into the base of the ice slab (Kieffer *et al.* 2006; Hansen *et al.*, 2007a; 2007b; Figures 30.K, 30.L, 30.M, 30.N).

A buried bruise

Promethei Planum is bounded by the semicircular arc of Promethei Rupes. This is the remnant of an impact basin, 850 km in diameter, half of which is now buried beneath Planum Australe (Figure 30.A). This is large enough that multiple rings were probably created when the basin formed, but only Promethei Rupes is visible now. The basin has been informally called "south polar basin" and "Promethei basin."

The Atlas of Mars

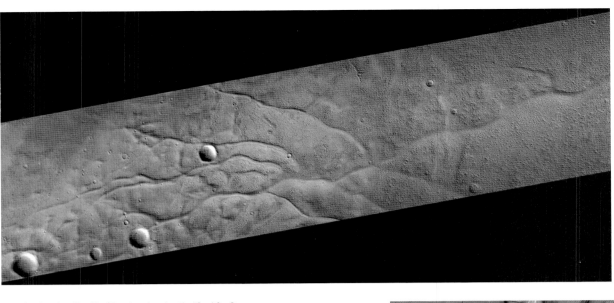

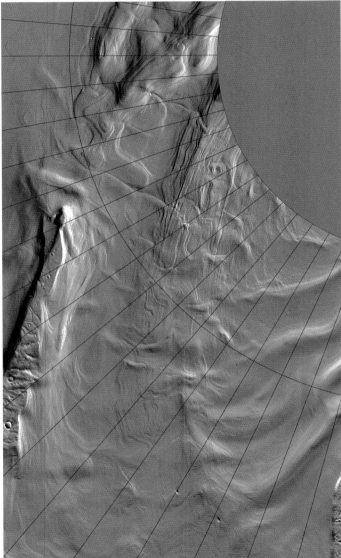

mountains have irregular forms, flat tops, circular moats that may have rims, or summit craters (Figure 30.O). These mountains do not resemble ordinary impact craters, nor are they like volcanoes elsewhere on Mars. The mountains may be the products of volcanic activity beneath an ice sheet, perhaps localized at ancient small impact craters (Ghatan and Head, 2002; Rodriguez and Tanaka, 2006). The location of the mountains in and near the Dorsa Argentea is consistent with a subglacial origin. Sisyphi Montes lie along an arcuate trend that is concentric to the Hellas basin and may represent an impact basin ring and associated fractures that allowed magma to reach the surface.

FIGURE 30.C THEMIS visible mosaic of south polar layered deposits (bottom half of image) overlying Promethei Planum. Crater Elim at margin of ice is 44 km across and lies at 80° S, 96° E (JPL PIA04910, 36 m/pixel, view about 80 km by 250 km, north along the 100° E meridian is toward the top of image, NASA/JPL-Caltech/Arizona State University).

Fire and ice

A chain of isolated, circular mountains, Sisyphi Montes, extends northward from Sisyphi Planum; similar forms occur in the southwest part of the Noachis (MC-27) quadrangle. Some of the

FIGURE 30.D (above) View of layering that is visible in Australe Sulci. Promethei Chasma is at top; part of Chasma Australe is at lower left (MOLA nadir data limit at 87° S; latitude circle at 85° S and longitude every 5°, view 270 km by 470 km, rotated so that illumination appears from the top; south pole to lower right, center at 85° S, 140° E).

FIGURE 30.E This CTX image shows Dorsa Argentea, south of crater Joly, where branching ridges record ancient river channels that may have formed beneath an ice sheet (image P13_006282_1046_XN_75S043W, 5 m/pixel, about 25 km by 100 km, north toward top, 76° S, 317° E, NASA/JPL-Caltech/MSSS).

FIGURE 30.F (left) THEMIS visible image, showing a network of interconnected, ridge-shaped erosional remnants in Cavi Angusti (image V40611004, 17 m/pixel, 17 km by 62 km, north toward top left, 81.5° S, 297° E, NASA/JPL-Caltech/Arizona State University).

FIGURE 30.G SHARAD radargram 596801, showing layering in Australe Mensa, which likely consists of dust and water ice. The relatively transparent regions on top of the layered deposits at center right are carbon dioxide ice (Phillips et al., 2011; radargram processed to show depth, view about 3 km high and 450 km long, looking south; courtesy of J. Holt, NASA/JPL-Caltech/Sapienza University of Rome/Southwest Research Institute).

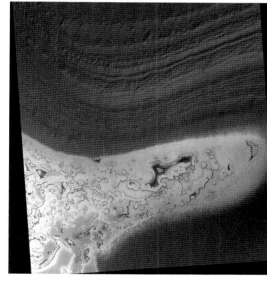

FIGURE 30.H (above) HiRISE view of rounded depressions formed in a bright dry ice layer by seasonal sublimation processes (image PSP_005466_0950, taken in southern summer,

$L_s = 320°$, 25 cm/pixel, view 5 km across, north to upper left, 84.8° S, 298.5° E, NASA/JPL-Caltech/University of Arizona).

241

The Atlas of Mars

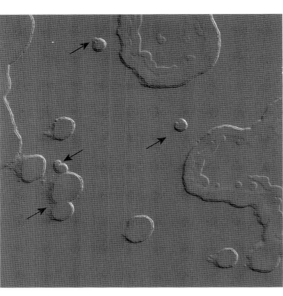

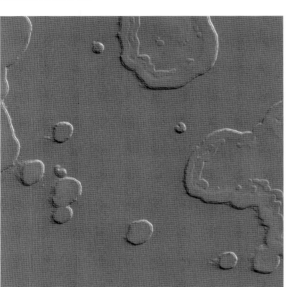

FIGURE 30.J1 AND FIGURE 30.J2 Detailed views of the two images in Figures 30.I1 and 30.I2 reveal, in the depressions indicated by arrows, how much ice was lost (views 1,500 m square).

Spiders:

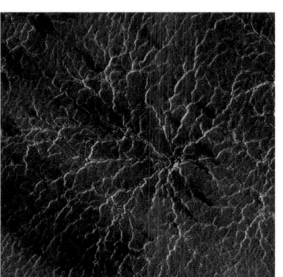

FIGURES 30.K (left), 30.L (middle), 30.M (right)

Series of HiRISE images showing example of seasonal changes from spring into summer in "spiders," and the temporary dark dust fans in the south polar region. View is about 500 m across, north is to the upper right. (Hansen, *et al.*, 2007a; 2007b; L_s is 199.6° [Figure 30.K], 233.1° [Figure 30.L], and 325.4° [Figure 30.M], where $L_s = 270°$ is the southern summer solstice; 86.4° S, 99.1° E; NASA/JPL-Caltech/University of Arizona).

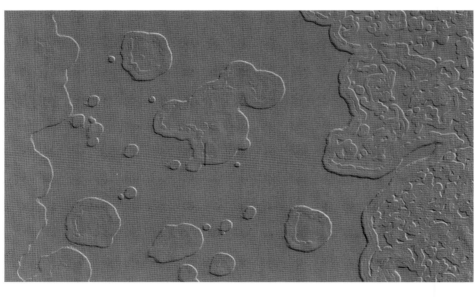

FIGURE 30.I1 AND FIGURE 30.I2 HiRISE images of the same region of "Swiss cheese" terrain, showing how cavities that were created by sublimation of the surficial dry ice layer expanded. Both images taken during southern spring just over 2 Martian years apart (PSP_004000_0945, [left] $L_s = 250°$ on Sol 486 of Mars year 28 vs. ESP_021880_0945, [right] $L_s = 263°$ on Sol 504 of Mars year 30; 25 cm/pixel, view 3 km by 5 km, north to upper right, at 85.6° S, 6.2° E, both images NASA/JPL-Caltech/University of Arizona).

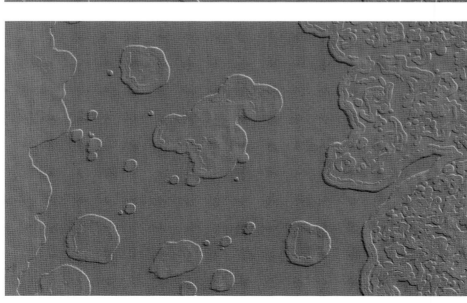

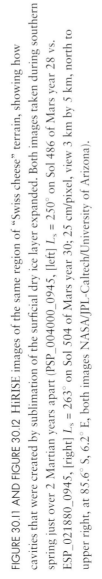

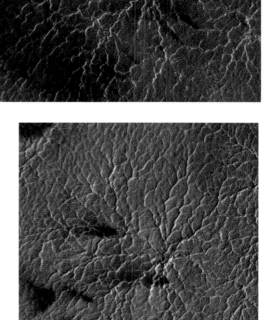

FIGURE 30.N Painting depicting the process that creates the "spiders" and associated dark dust fans near the south pole of Mars (ASU/Ron Miller).

FIGURE 30.O THEMIS daytime infrared image of one of Sisyphi Montes at 66.4° S, 4.0° E (image I26185010, 100 m/pixel, view 30 km by 80 km, north toward top, NASA/JPL-Caltech/Arizona State University).

243

Moons: Phobos and Deimos

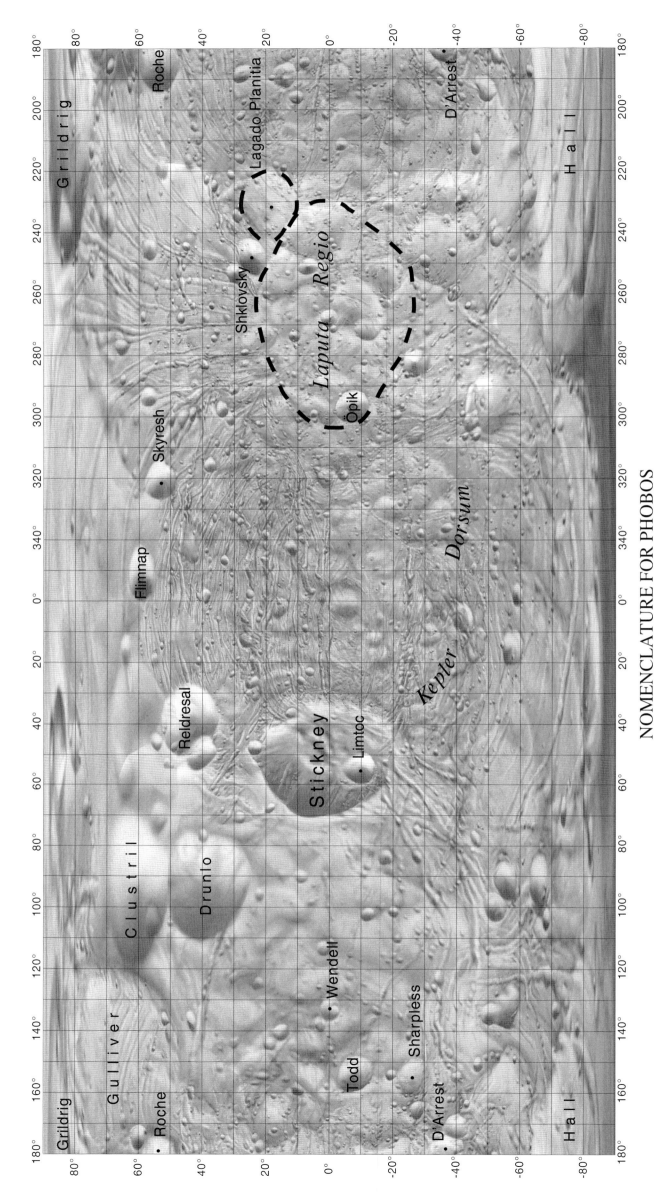

NOMENCLATURE FOR PHOBOS

Nomenclature and shaded relief from US Geological Survey, control by Peter Thomas, and rectification by Phil Stooke.

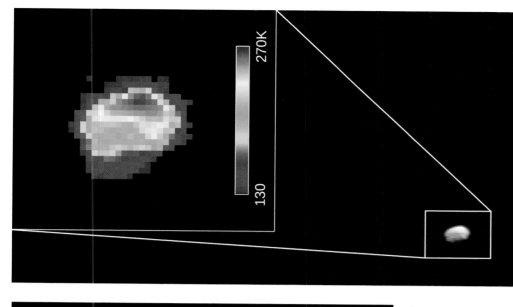

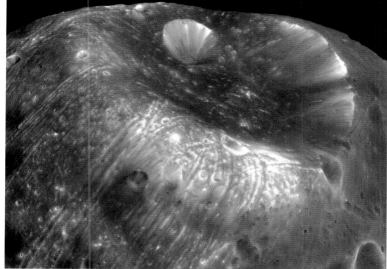

FIGURE M.3 Thermal infrared view of Phobos; the right edge is in morning sunlight and is at the warmest temperature (THEMIS infrared image combining four bands, color coded by temperature in kelvins, NASA/JPL-Caltech/Arizona State University/Space Science Institute).

FIGURE M.2 Part of the image in Figure M.1 with color accentuated. The bluish areas may be more recently exposed than the rest of the surface.

impacts from larger craters, such as Stickney, on Phobos. Study of recent, better-resolution images suggests that the grooves, which are absent on the trailing end of Phobos, may be chains of secondary impacts caused by debris from impacts on Mars (Murray and Heggie, 2014), or debris from impacts on Phobos that orbited the moon before impacting it (Nayak and Asphaug, 2016). Lineations within Stickney (Figure M.2) may be from landslides in the moon's weak gravity. Study of the thermal properties of the surface may determine whether the surface material is loose or relatively coherent, as these would show different rates of heating and cooling (Figure M.3).

FIGURE M.1 HiRISE image of Phobos shows Stickney crater at the right, near the leading end of the moon, and grooves, some of which are clearly lines of small craters (image PSP_007769_9010, false color composite of blue-green, red, and near-infrared channels, 7 m/pixel, NASA/JPL-Caltech/University of Arizona).

Phobos is the larger of the two moons of Mars, with a mean diameter[2] of about 22 km and an orbital radius of 9,376 km. Phobos orbits faster than Mars rotates, so it rises in the west and sets in the east as viewed from the planet's surface.

The low orbit has made it a target for robotic spacecraft orbiting Mars. Images show craters along with numerous grooves (Figure M.1). Explanations proposed for their origin include tidal-stress induced fracturing and secondary

[2] All numbers from NASA-JPL Solar System Dynamics, http://ssd.jpl.nasa.gov.

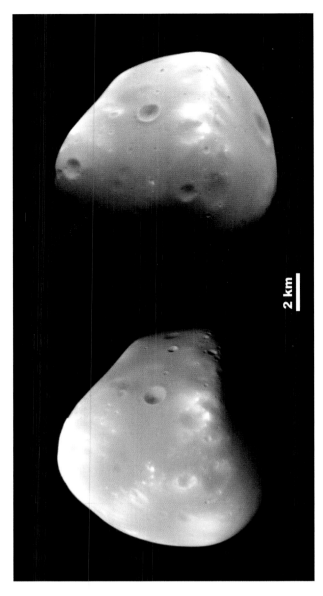

2 km

FIGURE M.4 Two views of Deimos taken by HiRISE, about 5.5 hours apart; the geometry in each is similar but the lighting is different, giving a varied view of the surface (images ESP_012065_9000 and ESP_012068_9000, 20 m/pixel, NASA/JPL-Caltech/University of Arizona).

Deimos is smaller (12 km in diameter) and farther from Mars (orbital radius 23,458 km) than Phobos (Figure M.4). Because it orbits more slowly than Mars rotates, Deimos rises in the east and sets in the west when observed from the surface of Mars.

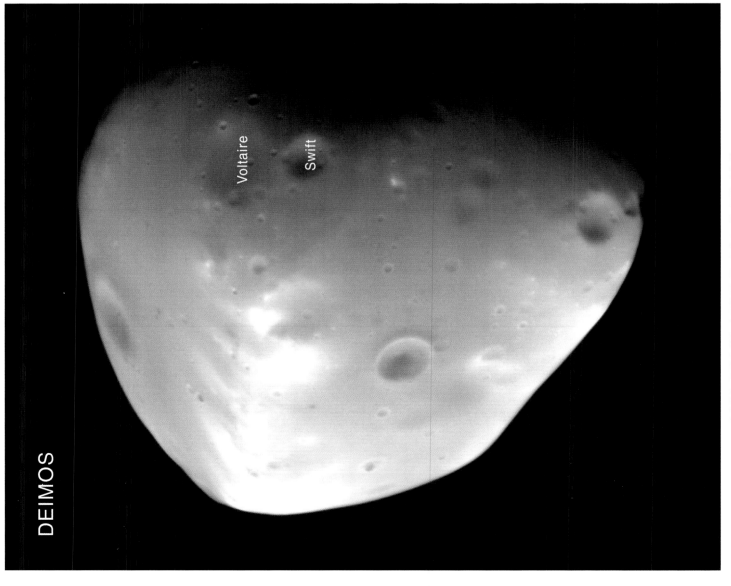

DEIMOS

Voltaire

Swift

NOMENCLATURE FOR DEIMOS

Courtesy US Geological Survey (image NASA/JPL-Caltech/University of Arizona).

Units on Geologic Map of Mars

The geologic map of Mars (Tanaka et al., 2014; available online from the web page for this atlas: www.cambridge.org/atlasofmars) assigned rocks to 44 units in eight unit groups. These are listed in the right column of Table A.1, and detailed descriptions are available in Tanaka et al. (2014). The correlation of the units is given in Figure A.1 (taken from Tanaka et al., 2014). These units are used in the geologic maps presented for each of the 30 map sheets, each of which has a legend listing the units and symbols used on that map sheet.

For the global geologic map presented in Chapter 5 (Figures 5.1a, 5.1b, and 5.1c) the units were combined into 20 map units given in the middle column of Table A.1. Figure 5.1d shows the correlation of these 20 units, while Table 5.1 describes the units.

TABLE A.1 Geologic map units

Unit group type	20 condensed units, global map	44 units, regional maps
Polar	Ap	lApc Late Amazonian polar cap unit lApd Late Amazonian polar dunes unit Apu Amazonian polar undivided unit Ap Amazonian polar unit
	Hp	Hp Hesperian polar unit Hpu Hesperian polar undivided unit Hpe Hesperian polar edifice unit
Impact	AHi	AHi Amazonian and Hesperian impact unit
Volcanic	Av	lAv Late Amazonian volcanic unit lAvf Late Amazonian volcanic field unit Av Amazonian volcanic unit Ave Amazonian volcanic edifice
	Hv	lHv Late Hesperian volcanic unit lHvf Late Hesperian volcanic field unit eHv Early Hesperian volcanic unit Hve Hesperian volcanic edifice unit
	AHv	AHv Amazonian and Hesperian volcanic unit
	Nv	lNv Late Noachian volcanic unit Nve Noachian volcanic edifice unit
Apron	Aa	lAa Late Amazonian apron unit Aa Amazonian apron unit
	ANa	ANa Amazonian and Noachian apron unit
Basin	eAb	eAb Early Amazonian basin unit
	HNb	lHb Late Hesperian basin unit eHb Early Hesperian basin unit HNb Hesperian and Noachian basin unit
Lowland	mAl	mAl Middle Amazonian lowland unit
	lHl	lHl Late Hesperian lowland unit
Highland	Hh	eHh Early Hesperian highland unit HNhu Hesperian and Noachian highland undivided unit
	Nh	Nhu Noachian undivided highland unit lNh Late Noachian highland unit Nhe Noachian highland edifice unit
	mNh	mNh Middle Noachian highland unit mNhm Middle Noachian highland massif unit
	eNh	eNh Early Noachian highland unit eNhm Early Noachian highland massif unit
Transition	AHtu	AHtu Amazonian and Hesperian transitional undivided unit
	Ht	Htu Hesperian transition undivided unit lHt Late Hesperian transition unit eHt Early Hesperian transition unit Ht Hesperian transition unit Hro Hesperian transition outflow unit
	HNt	HNt Hesperian and Noachian transition unit

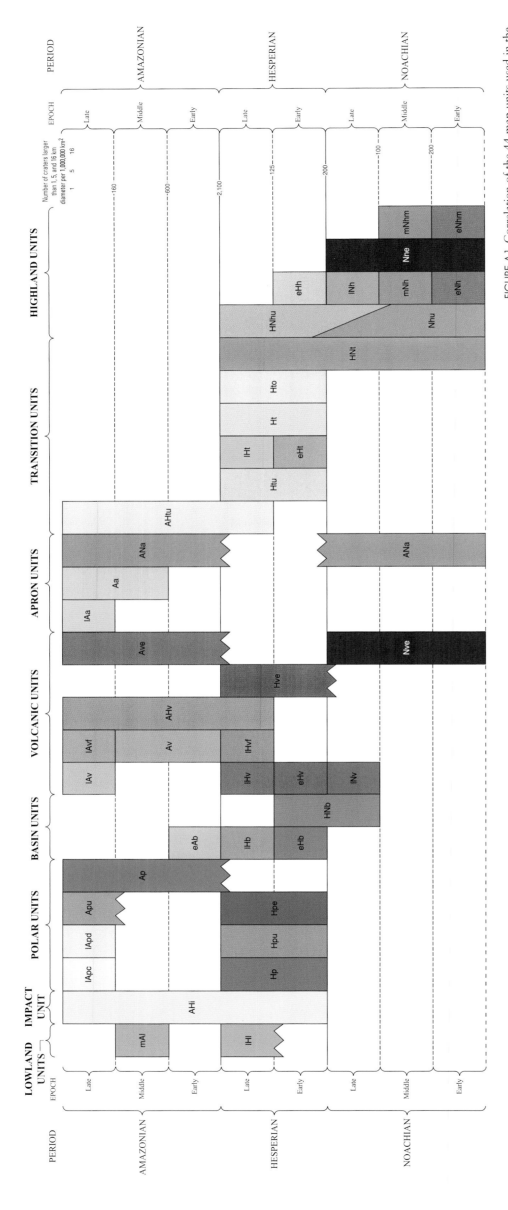

FIGURE A.1 Correlation of the 44 map units used in the geologic maps accompanying each map sheet. Map unit ages are resolved to the nearest epoch; hachured box edges indicate possible extended durations. Cumulative crater densities are shown for epoch boundaries at 1-km, 5-km, or 16-km diameter. See Tanaka *et al.* (2014) for detailed discussions of individual map units, unit groups, and age-dating methodologies.

SI/English Unit Conversions

Length

1 micrometer or micron (µm) = 3.937×10^{-5} inches

1 centimeter (cm) = 0.3937 inches

1 inch = 2.54 cm

1 meter (m) = 100 cm = 39.37 inches

1 kilometer (km) = 1,000 m = 3,280.8 feet = 0.621 miles

Temperature

The temperature in degrees Celsius (°C) is related to degrees Fahrenheit (°F) as follows:

Temperature (°F) = [Temperature (°C) × (1.8)] + 32

The temperature in degrees Celsius (°C) is related to kelvins (K) as follows:

Temperature (K) = Temperature (°C) + 273.15 K, absolute zero is 0 K, and the freezing point of water on Earth is 273.15 K

Pressure

100 kPa (kilopascals) = 1 bar = 1000 millibar (mbar)

1 standard atmosphere = 101.325 kPa

Source: Thompson and Taylor, 2008.

Latin Descriptors

Descriptors (Latin designations) are shown in the table for feature types on Mars. Except for impact craters, the descriptor follows the feature name (definitions from the Gazetteer of Planetary Nomenclature, with additions from Russell *et al.*; 1992).

Feature, plural form	Description
Catena, catenae	Chain of closely spaced depressions or craters
Cavus, cavi	Hollow, irregular steep-sided depression, usually in arrays or clusters
Chaos, chaoses	Distinctive area of broken terrain
Chasma, chasmata	A deep, elongated, steep-sided depression
Collis, colles	Small hill or knob
Dorsum, dorsa	Ridge
Fluctus, fluctūs	Flow terrain
Fossa, fossae	Long, narrow, shallow depression
Labes, labēs	Landslide
Labyrinthus, labyrinthi	Complex of intersecting valleys or ridges
Lingula, lingulae	Extension of plateau having rounded lobate or tongue-like boundaries
Mensa, mensae	A flat-topped prominence with cliff-like edges
Mons, montes	Mountain
Palus, paludes	"Swamp"; small plain
Patera, paterae	Irregular crater, or complex one with scalloped edges
Planitia, planitiae	Low plain
Planum, plana	Plateau or high plain
Rupes, rupēs	Relatively straight cliff or scarp
Scopulus, scopuli	Lobate or irregular cliff or scarp
Serpens, serpentes	Sinuous feature with segments of positive and negative relief along its length
Sulcus, sulci	Subparallel furrow and ridge
Terra, terrae	Extensive land mass; older, heavily cratered terrain
Tholus, tholi	Small domical mountain or hill
Unda, undae	Dune
Vallis, valles	Valley
Vastitas, vastitates	Extensive plain

Glossary of Terms

See also the Latin descriptors of feature types defined in Appendix. Words in italic have their own entries in the Glossary. Sources from which some of the definitions were adapted include NASA, 1978; 1997; 1998; Neuendorf et al., 2005; Snyder, 1987; Tauxe, 2010; Marshak, 2012.

aeolian See *eolian*.

aerobraking A means of slowing a spacecraft in an elliptical orbit by frictional drag in a planetary atmosphere at the low point of the orbit, reducing the high point of the orbit; a strategy to save fuel.

albedo The ratio of the electromagnetic radiation reflected by a body to the amount incident upon it, often expressed as a percentage. The albedo of the Earth is 34%.

alluvial Pertaining to unconsolidated detrital material (e.g. sand, gravel) deposited by running water.

Amazonian Refers to the youngest *period* of time in the geologic *stratigraphy* of Mars; also the *system* of *rocks* originating during that time.

anomaly A departure from an expected or nominal value, as in magnetism, heat flow, or elemental concentration.

aquifer A porous subsurface body of *rock*, having permeability and saturated with water, thus holding and capable of transmitting *groundwater*.

ash Fine (less than 2 mm) *pyroclastic* material, generally loose or unconsolidated.

ash flow A hot mixture of volcanic gases and *pyroclastic* material that flows along the ground as a density current.

ashfall Airborne *ash* that falls from a volcanic eruption cloud, and the resulting deposit.

aureole On Mars, the blocky, lobate deposits that surround Olympus Mons.

barchan An isolated, crescent-shaped sand dune, the terminations of which point in the direction of the prevailing wind.

basalt Dark-colored *mafic igneous rock*, having fine or microscopic crystals and composed primarily of *plagioclase feldspar* and *pyroxene*; other minerals such as *olivine* and *ilmenite* are usually present.

basaltic andesite An *igneous rock* having fine or microscopic crystals and somewhat greater silica (SiO_2) content than *basalt*.

base level The lowest level towards which the erosion of the surface progresses.

basement The *crust* below sedimentary deposits. See also *impact basin*.

basin See also *impact basin*. A low topographic area where materials tend to accumulate; OR an area where a thick sequence of layered *rocks* have accumulated. On Mars, many but not all basins in this sense originated as impact basins.

bedrock *In situ* solid *rock*.

Bouguer gravity anomaly Difference between measured gravity and that expected after corrections are made for elevation and density differences from *datum*.

braided channel A *stream channel* that frequently branches and rejoins after separation by bars or islands.

breccia *Rock* consisting of angular, coarse fragments of preexisting rock, embedded in a fine-grained matrix.

caldera Large, circular to subcircular depression, associated with a volcanic vent. Calderas result from collapse, explosion, or erosion.

calving The breaking away of a mass of ice from a near-vertical ice face.

canyon A long, deep, relatively narrow steep-sided valley in a plateau or mountainous area.

carbonaceous Rich in carbon.

carbonate A mineral having CO_3 in its structure; a sediment formed of carbonates.

channel See also *outflow channel*. The bed or course where surface water or lava flowed or may flow.

chaos An area of broken terrain.

circumpolar Surrounding or circulating around a polar region.

cobble A *rock* fragment, 64–256 mm in diameter; informally, a rock larger than a pebble and smaller than a boulder.

convection Heat transfer involving movement of a fluid (e.g. atmosphere, water, the mantle) due to buoyancy; also, the motion that results.

core The central region of a planet or moon, frequently made of different materials than the surrounding regions (*mantle* and *crust*). Earth and the Moon are thought to have cores of iron and nickel.

corona (pl. coronae) Geologic feature formed by magma rising from deep in the planet, collected below the surface, creating a circular uplift after which the escape of magma to surface volcanism or elsewhere underground leads to collapse and fracturing in radial and circumferential directions. Coronae are common on Venus.

Coriolis force The acceleration which a body in motion experiences when observed in a rotating frame of reference. This force acts at right angles to the direction of the angular velocity. On Earth and Mars, this acceleration is to the right in the northern hemisphere and to the left in the southern hemisphere.

coulee Here used to denote the valleys of the northwest USA that are long, steep-walled, abandoned overflow channels, likely created by meltwater floods.

crater See also *impact crater*. A hole or depression. Most are roughly circular or oval in outline. On Mars most craters are of impact origin.

cross-bedding In stratified *rocks*, individual layers or beds, inclined at an angle to the main planes of stratification.

crust The outermost layer of a planet or moon, above the *mantle*. The crust of Mars is estimated to range from less than 40 km to 60 km or more thick (Neumann et al., 2004; Genova et al., 2016).

cryovolcanism Volcanism at low (icy) temperatures, usually involving aqueous solutions, with or without solids and gases.

cuesta An asymmetrical ridge formed by tilted, layered *rock*, sloping steeply on one side and gentle (parallel to the layers) on the other.

cumulate An *igneous rock*, formed by the accumulation of crystals that settle out from a *magma* by the action of gravity.

datum Here used with respect to elevation: a reference ellipsoid, centered at the center of mass of the planet, and *equipotential* on a large scale with respect to gravity.

deliquescence The absorption of atmospheric water by substances on the surface, sufficient to produce a solution.

delta The low, flat *alluvial* tract at the mouth of a river where it enters a body of water; commonly triangular in shape.

demagnetization A reduction of the magnetic field intensity inside a body (of *rock*) because of changes in its *magnetization*, commonly resulting from heating; may also result from weathering of magnetic minerals.

dendritic Pertaining to a network of valleys having a trunk *channel* that is fed by branching channels, called tributaries; in map view, having the structure of a tree.

detachment A shallowly dipping to subhorizontal *fault* or shear zone, typically having kilometers of displacement and where faults and folds in *rocks* above the detachment do not extend across it.

dichotomy See *global dichotomy*.

dike See also *intrusive*. A tabular body of *intrusive rock* that cuts across the planar structure of the surrounding (and older) rocks.

distributary channel A divergent stream, flowing away from the main stream and not returning to it, as in a *delta* or on an *alluvial* plain.

drumlin A low, smoothly rounded, elongate oval hill of glacially transported material, built under the ice and shaped by its flow so that the long direction is parallel to the movement of the ice.

dry ice The solid form of carbon dioxide, which condenses from gas at 195 K at Earth atmospheric pressure and at 148 K at 0.7 kPa (typical Mars atmospheric pressure; Carr, 2006).

dynamo See *magnetic dynamo*.

eccentricity (of an ellipse) A measure of the elongation of an ellipse, zero for a perfect circle and approaching 1 as the elongation increases.

ejecta The deposit surrounding an *impact crater*, composed of material (*rock* fragments, glass) that are thrown from the crater during its formation.

eolian Pertaining to or transported by wind.

epoch The primary subdivision of a *period*, during which the *rocks* of the corresponding *series* were formed.

equipotential Here used for a surface on which the potential due to the gravitational force is everywhere constant. On Earth, water at rest, and not subject to other forces, has an equipotential surface.

esker A long, narrow, sinuous, steep-sided ridge composed of sand and gravel, deposited by a subglacial or intraglacial stream and left behind when the ice melted.

extrusive Pertaining to volcanic material that is emitted (as liquid *lava* or as solid fragments) onto the surface of a planetary body; also, the *rock* so formed.

fault A fracture or zone of fractures in *rock* along which the sides are displaced relative to one another.

fluvial Of, or pertaining to, rivers.

footwall The material below an inclined *fault*.

fracture General term for any break in a *rock* or rock unit due to mechanical failure by *stress* (includes cracks and joints).

friable Denotes material that crumbles or breaks easily.

Ga Giga-annum, one billion (10^9) years, where *annum* is age in years before present.

global dichotomy The division of the surface of Mars into a southern region, which is heavily cratered, has thicker *crust*, and is high in elevation, and a northern region, with few craters, a thinner *crust*, and a lower elevation.

graben An elongate crustal depression, bounded by *normal faults* on its long sides and formed by the downward displacement of the central block of *crust*.

groundwater Subsurface water, as distinct from surface water.

gypsum A mineral made of calcium sulfate, $CaSO_4 \cdot 2H_2O$.

hanging wall The material above an inclined *fault*.

hematite A mineral made of iron and oxygen, Fe_2O_3; retains a durable *magnetization* below about 685 °C.

Hesperian Refers to the intermediate *period* of time in the geologic *stratigraphy* of Mars; also the *system* of *rocks* originating during that time.

hydrated Water-bearing; for example, hydrated minerals have whole or partial water molecules incorporated into the crystal structure of the mineral.

hydrofracture The propagation of fractures in *rock* by pressure that is exerted by a fluid, as during crystallization of *magma* or by water injection.

hydrothermal Pertaining to or resulting from the interaction of hot water and existing *rocks*; e.g. chemical or mineral products.

igneous rock *Rock* formed by the cooling of *magma* to a solid form.

ilmenite A mineral of iron, titanium, and oxygen, $FeTiO_3$; it occurs in *igneous rocks*.

impact In planetology, the collision of objects, ranging in size from tiny micrometeoroids to planetesimals. The forceful striking of one body, such as an asteroid, against another body, such as a moon or planet.

impact basin A large circular area, typically 300 km or more in diameter, surrounded by one or more mountainous rings; may or

may not be lower in elevation than the surrounding terrain, considered to be the result of *impact*.

impact crater A *crater* formed by *impact*.

inlier An area or complex of *rocks* surrounded by younger rocks.

insolation Incoming (electromagnetic) solar radiation.

intrusive Pertaining to the process of emplacing *magma* into preexisting *rock*; also, the rock so formed (for example, a *dike*).

inverted channel Former path of water, capped by material that is more resistant than adjacent material to erosion after water flow ends, leaving an elevated landform.

jarosite A mineral of potassium, iron, and sulfate; $KFe_3(SO_4)_2(OH)_6$.

katabatic wind A wind that moves downslope, driven by density differences, especially by cooling, as on the surface of ice.

lacustrine Pertaining to, produced by, or formed in a lake.

lahar A mudflow that is composed chiefly of *volcaniclastic* materials on the flank of a volcano.

Lambert conformal projection A *map projection* on a cone in contact with the surface along two standard parallels of latitude.

lava *Magma* that reaches the surface of a planet or moon; erupted from a volcano, fissure, or vent. It typically includes the *crust* and the upper part of the *mantle*.

lava tube A tube within a body of partly or completely solidified *lava* through which liquid *lava* has flowed. If near the surface, *rocks* above the tube may collapse, resulting in a *rille*, a channel-like depression on the upper surface of the lava body.

lobate Having a rounded, tongue-like projection or convex apron.

L_s Aerocentric longitude: The angle, centered on the Sun, between the position of Mars at northern spring equinox and the position in question.

lithosphere The stiff upper layer of a planetary body; the solid outer part of a planet. It typically includes the *crust* and the upper part of the *mantle*.

Ma Mega-annum, one million (10^6) years, where *annum* is age in years before present.

mafic rock *Igneous rock*, composed chiefly of dark, iron- and magnesium-rich minerals and having 45–52 weight percent silicon dioxide (SiO_2).

magma Molten *rock* in the interior of a planet or moon. When it reaches the surface, magma is called *lava*. Magma cools to form *igneous* rocks.

magmatic Of or pertaining to *magma*.

magnetic dynamo Circulating, electrically conducting fluid, such as liquid iron, in the *core* or interior of a planet; gives rise to a planetary magnetic field.

magnetite A mineral of iron and oxygen, Fe_3O_4; noteworthy for retaining a durable *magnetization* below about 580 °C.

magnetization The property of *rock* or other material of having a magnetic field with a defined direction.

mantle A mostly solid layer, lying beneath the *crust* and above the *core*; on Mars, as on Earth, it consists primarily of iron, magnesium, silicon, and oxygen.

mantle plume A vertical, cylindrical segregation within the *mantle*, hotter than its surroundings, within which larger-than-normal amounts of heat are conducted upward to form a "hot spot" at the surface.

map projection A systematic representation of all or part of the surface of a round, convex body, such as a planet, on a plane. Some distortion of shape or area is inherent in any map projection.

mass wasting The downslope movement of *rock* material, solely under the influence of gravity; includes slow displacement such as creep and rapid displacements such as earth flows, rock slides, and avalanches.

massif As used here, a discrete mountain mass; commonly composes part of the uplifted mountainous ring(s) around circular basins.

meandering channel A *channel* having a series of regular sinuous curves, bends, or loops.

Mercator projection A *map projection* on a cylinder in contact with the surface along the equator.

mesa An isolated, nearly level-topped landmass, standing distinctly above the surrounding country and bounded by steep slopes.

mineral Naturally occurring inorganic solid with a specific chemical composition and crystal structure.

moraine A mound, ridge, or other accumulation of glacially transported material.

Noachian Refers to the oldest *period* of time in the geologic *stratigraphy* of Mars; also

the *system* of *rocks* originating during that time.

normal fault A *fault* in which the *hanging wall* has moved down relative to the *footwall* during extensional *strain*.

obliquity The angle of a planet's rotational axis relative to the plane of its orbit. An axis at a right angle to the plane of orbit has an obliquity of 0°.

olivine *Mineral* found in *basalt*; ranges from Mg_2SiO_4 to Fe_2SiO_4.

opposition This occurs when Earth is directly between Mars and the Sun.

outflow channel On Mars, a broad valley system, tens to hundreds of kilometers wide and hundreds to thousands of kilometers long. Commonly, these connect canyons or depressions that contain chaotic assemblages of knobs and mesas to the northern lowlands.

pedestal crater Crater upon a platform that is at an elevation well above the surrounding surface.

perchlorate A compound of the ClO_4^- ion.

periglacial Processes, areas, and climates at or typical of the immediate margins of former and existing glaciers and ice sheets and influenced by the cold temperature of the ice.

period A fundamental unit of the geologic timescale, during which *rocks* of the corresponding *system* were formed.

permafrost Soil or surficial *rock* that is consistently below freezing in temperature; in the case of Mars, commonly implies some content of frozen water.

phyllosian era Earliest of three informal, mineralogical eras proposed by Bibring *et al.* (2006); corresponds to much, but not the last part, of the Noachian Period.

phyllosilicate *Minerals* in which the SiO_4 units (which are tetrahedral) are arranged in flat sheets; includes clay minerals.

pingo A conical mound of soil-covered ice, raised by the pressure of water within or below the *permafrost*. The crest may have collapsed due to melting of the ice.

pit crater Used in two different senses: (1) A collapse feature (the sense used in this atlas). (2) An *impact crater* with a smaller pit in the center of the floor or on a central peak of the crater; also known as central pit craters.

pixel In a digital image, the area represented by a single digital value.

plagioclase Common *mineral*; ranges from NaAlSi$_3$O$_8$ to CaAl$_2$Si$_2$O$_8$. Also known as plagioclase feldspar, as it is one of the feldspar group of minerals.

planetocentric Denotes latitude that is measured as the angle between the equatorial plane and a vector directed at a point of interest, measured at the center of mass of the planet.

planetographic Denotes latitude that is measured as the angle between a vector perpendicular to a (non-spherical) reference surface at a point of interest and the equatorial plane.

plate tectonics The theory of planetary dynamics in which the *lithosphere* (the *crust* and upper *mantle*) is broken into individual plates that are moved by *convection* of the deeper *mantle*. The formation on Earth of mountains and volcanoes, and the occurrence of earthquakes, have been explained using this theory; on Mars, some ancient terrains and crustal *remanent magnetization* signatures suggest possible early plate tectonics.

polar stereographic projection A *map projection* on a plane in contact with the surface at the north or the south pole.

pre-Noachian Refers to the interval of geologic time on Mars from the origin of the planet to the beginning of the *Noachian Period*; also *rocks* formed during this interval.

pyroclastic eruption Explosive eruption of *lava*, producing and ejecting hot fragments of *rock* and *lava*, including fine-grained *ash*.

pyroxene Dark *mineral* that is common in *igneous rocks* having calcium, sodium, magnesium, iron, or other cations with silicon and oxygen.

quadrangle A map showing a region as a four-sided polygon; the opposite sides are not necessarily parallel. In casual use, refers to any map of part of a planet, even those having a different shape.

reach A straight, continuous, or extended part of a stream, commonly where the *channel* size or form is consistent.

regolith Loose, unconsolidated *rock*, *mineral*, and glass fragments. On the Moon and Mars, this debris is produced primarily by *impacts* and blankets the surface.

remanent magnetization That component of a *rock's magnetization* that has a fixed direction relative to the rock and is independent of moderate applied magnetic fields, including the planetary field.

reverse fault A *fault* in which the *hanging wall* has moved up relative to the *footwall*; commonly indicative of contractional motion.

rift A long, narrow trough, bounded by *normal faults*, a *graben* of regional extent.

rille Trench or crack-like valleys; on Mars, up to several hundred kilometers long and 1–2 km wide. May be sinuous in form; probably formed either as an open channel in a *lava* flow, or as an underground *lava tube* which collapsed as the lava flowed out.

rock A naturally formed solid that is an aggregate of one or more *minerals*.

scarp Cliff produced by *tectonic*, *impact*, or erosion processes.

sediment See also *regolith*. Solid *rock* or *mineral* fragments that are transported and deposited by wind, water, gravity, or ice; precipitated by chemical reactions; or (on Earth) secreted by organisms; accumulated as layers in loose, unconsolidated form.

sedimentary rock Rock formed of sediment that is cemented into a solid.

series The *rocks* formed during an *epoch* of geologic time, also the primary subdivision of a *system*.

serpentine A group of secondary *minerals* formed by the alteration of magnesium-rich *silicate* minerals such as *olivine*.

shear Deformation of a body by sliding along parallel planes.

shield A volcanic mountain in the shape of a broad, flattened dome.

sidereal Measured relative to the stars, as for the rotation or revolution of a planet.

siderikian era Most recent of three informal, mineralogical eras proposed by Bibring *et al.* (2006); corresponds to later part of the Hesperian and all of the Amazonian Periods.

silica Dioxide of silicon (SiO$_2$), occurs as quartz and other minerals.

silicate A compound whose crystal structure contains SiO$_4$ tetrahedra; the most common type of mineral on the surface of Earth.

sol One solar day; here specific to Mars (24 hours and 40 minutes) as contrasted with the 24-hour solar day of Earth.

solifluction The slow, viscous, downslope flow of saturated or waterlogged soil or material, especially when underlain by frozen material.

spectrum (spectral) Visible light or other electromagnetic radiation, arrayed according to its wavelengths (colors).

strain Change in size or shape of a material in response to an applied *stress*.

stratigraphy Science of *rock* strata; concerned with the original succession and age relations of rock strata as well as their form, distribution, and composition.

stress Force that *strains* or deforms a material, such as *rock*.

structure The general disposition (attitude, arrangement, or position) of the *rock* masses of a region or area. The term "structure" is also applied to individual features that modify rock masses, such as *grabens*, *faults*, or *basins*.

structurally controlled Influenced by structural features, such as folds or faults.

subduction (subducting) The process in which one region or plate of the *lithosphere* descends beneath another. On Earth, this process is responsible for much earthquake and volcanic activity and the recycling of lithosphere into the Earth's interior.

sublimate To convert from solid directly to gas.

sulfate Any *mineral* having SO$_4$ in its structure.

synodic period The interval between successive *oppositions* of Mars (averages 780 days).

system The *rocks* formed during a *period* of geologic time.

talus Loose fragmental *rock* material derived from a cliff or slope and lying at or near its base.

tectonic Refers to the deformation of planetary materials, as in faulting of the *crust*.

tectonism Process involving movement of the *lithosphere*, displacing large masses, whether by uplift, subsidence, large-scale folding and faulting, or large-scale impact events.

terrace Part of a former *channel* surface left when the channel cuts down to a new, lower level.

theiikian era Second of three informal, mineralogical eras proposed by Bibring *et al.* (2006); corresponds to the latest Noachian and much of Hesperian Periods.

thermal inertia Measure of the response of a material to temperature changes. A material that heats or cools very slowly has high thermal inertia.

thermokarst Features formed by the removal of interstitial ground ice or permafrost, including steep-sided, flat-floored valleys and curvilinear depressions as well as other collapse or chaotic landforms.

thrust fault A *reverse fault* having a relatively flat-lying orientation, along which one *rock* mass has moved upward and over another by contractional *strain*.

topography The three-dimensional configuration of the solid surface of a planetary body, including its relief and the position of its features. The term also refers to its graphical description, usually on maps or charts.

transform fault A *fault* in which the *rocks* have been shifted horizontally past each other along the fault.

tsunami A sea wave produced by any large-scale, short-duration disturbance of the ocean floor, such as a submarine earthquake, slump, volcanic eruption, or impact. The effects at a shoreline can be severe.

tuff Volcanic *ash* and related material, consolidated into *rock*.

ultramafic rock *Igneous rock* composed of iron- and magnesium-rich minerals, and lower in silica (less than 45 weight percent SiO$_2$) than *mafic rock*.

unconformity A break in the depositional record, typically followed by a change in the angle of the beds.

volatile A substance that can be readily vaporized; on Mars, includes but is not necessarily limited to water and carbon dioxide.

volcaniclastic Pertaining to volcanic material formed by any process of fragmentation.

volcanotectonic Controlled by both volcanic and tectonic processes.

wrinkle ridge A long, sinuous, irregular, segmented, crenulated ridge on planetary surfaces. They probably form by *tectonic* folding and *thrust* faulting of surface materials.

yardang A long, irregular, sharp-crested ridge between round-bottomed troughs, carved on a plain by wind erosion, typically in soft but coherent material.

GAZETTEER

Features shown on atlas sheets are in bold. MC-1 and MC-30 are divided into two maps at longitudes of 90° E and 270° E.

Regional Features

Large features shown on Figure 4.1

Acidalia Planitia	Chryse Planitia	Olympus Mons	Tharsis Montes
Amazonis Planitia	Elysium Planitia	Promethei Terra	Tyrrhena Terra
Aonia Terra	Hellas Planitia	Tempe Terra	Utopia Planitia
Arabia Terra	Lunae Planum	Terra Cimmeria	Valles Marineris
Arcadia Planitia	Margaritifer Terra	Terra Sabaea	Vastitas Borealis
Argyre Planitia	Noachis Terra	Terra Sirenum	Xanthe Terra

Features on Map Sheets

Feature name	Diameter (km)	Center latitude	Center longitude	Feature type	Quadrangle	Origin
Abalos Colles	235.83	76.83	288.35	Collis, colles	MC-1	Named for classical albedo feature at 72° N, 70° W
Abalos Mensa	129.18	81.17	284.4	Mensa, mensae	MC-1	Classical albedo feature name
Abalos Scopuli	109.16	80.72	283.44	Scopulus, scopuli	MC-1	Classical albedo feature name
Abalos Undae	**442.74**	**78.52**	**272.5**	**Unda, undae**	**MC-1**	**Classical albedo feature at 72° N, 70° W**
Aban	4.28	15.91	111.1	Crater, craters	MC-14	Town in Russia
Abus Vallis	60.99	−5.49	212.8	Vallis, valles	MC-16	Classical name for Humber River in England
Achar	5.36	45.43	123.16	Crater, craters	MC-7	Town in Uruguay
Acheron Catena	**421.77**	**37.47**	**259.2**	**Catena, catenae**	**MC-3**	**Named for classical albedo feature at 35° N, 140 W**
Acheron Fossae	**703.11**	**38.27**	**224.98**	**Fossa, fossae**	**MC-2**	**From classical albedo feature at 35° N, 140° W**
Acidalia Colles	**356.3**	**50.34**	**336.91**	**Collis, colles**	**MC-4**	**From classical albedo feature name**
Acidalia Mensa	**226.86**	**46.69**	**334.66**	**Mensa, mensae**	**MC-4**	**From classical albedo feature name**
Acidalia Planitia	**3505**	**49.76**	**339.26**	**Planitia, planitiae**	**MC-4, MC-5, Figure 4.1**	**From classical albedo feature name**
Ada	2.09	−3.06	356.78	Crater, craters	MC-19	Town in Oklahoma, USA
Adamas Labyrinthus	**853**	**35.7**	**105.12**	**Labyrinthus, labyrinthi**	**MC-6**	**Classical albedo feature name; "A River of Diamonds"; today's River Sarbarnarekha in India**
Adams	**90.22**	**30.91**	**163.1**	**Crater, craters**	**MC-7**	**Walter S.; American astronomer (1876–1956)**
Aeolis Dorsa	**459.17**	**−5.05**	**152.63**	**Dorsum, dorsa**	**MC-23**	**Classical albedo feature name**
Aeolis Mensae	**785.09**	**−3.25**	**140.63**	**Mensa, mensae**	**MC-23**	**Classical albedo feature name**

Feature name	Diameter (km)	Center latitude	Center longitude	Feature type	Quadrangle	Origin
Aeolis Mons	**88.99**	**−5.08**	**137.85**	**Mons, montes**	**MC-23**	**Classical albedo feature name**
Aeolis Palus	111.63	−4.47	137.42	Palus, paludes	MC-23	Classical albedo feature name
Aeolis Planum	**852.81**	**−1.14**	**144.76**	**Planum, plana**	**MC-15, MC-23**	**Classical albedo feature name**
Aeolis Serpens	538.55	−1.38	149.57	Serpens, serpentes	MC-15, MC-23	Classical albedo feature name
Aesacus Dorsum	**276.69**	**36.82**	**153.15**	**Dorsum, dorsa**	**MC-7**	**From albedo feature at 45° N, 205° W**
Aganippe Fossa	**537.16**	**−8.49**	**234**	**Fossa, fossae**	**MC-17**	**Classical albedo feature name**
Agassiz	**108.77**	**−69.88**	**271.11**	**Crater, craters**	**MC-30**	**Jean L.; American naturalist (1807–1873)**
Airy	**43.05**	**−5.14**	**0.05**	**Crater, craters**	**MC-19, MC-20**	**George B.; British astronomer (1801–1892)**
Airy-0	0.79	−5.07	0	Crater, craters	MC-20	Small crater within crater Airy that defines zero degrees longitude on Mars
Ajon	8.08	16.49	103.14	Crater, craters	MC-14	Town in Russia
Aki	7.87	−35.46	299.76	Crater, craters	MC-25	Town in Japan
Aktaj	5.01	20.41	313.51	Crater, craters	MC-10	Town in Russia
Al-Qahira Vallis	**600**	**−18.23**	**162.41**	**Vallis, valles**	**MC-23**	**Word for "Mars" in Arabic, Indonesian, Malay**
Alamos	6.44	23.48	322.88	Crater, craters	MC-11	Town in Mexico
Alba Catena	**144.86**	**35.04**	**245.42**	**Catena, catenae**	**MC-3**	**Classical albedo name**
Alba Fossae	**2072.02**	**49.39**	**253.18**	**Fossa, fossae**	**MC-3**	**Classical albedo name**
Alba Mons	**548.02**	**41.08**	**249.29**	**Mons, montes**	**MC-3**	**Classical albedo feature name**
Alba Patera	**65.98**	**39.53**	**250.82**	**Patera, paterae**	**MC-3**	**Classical albedo name**
Albany	2.15	22.96	310.98	Crater, craters	MC-10	American colonial town (New York)
Albi	9.07	−41.47	324.99	Crater, craters	MC-26	Town in France

Feature name	Diameter (km)	Center latitude	Center longitude	Feature type	Quadrangle	Origin
Albor Fossae	155	18.09	150.78	Fossa, fossae	MC-15	From albedo feature at 20° N, 205° W
Albor Tholus	158.38	18.87	150.47	Tholus, tholi	MC-15	Classical albedo feature name
Alexey Tolstoy	93.04	-47.44	125.34	Crater, craters	MC-29	Aleksey; Soviet writer (1882–1945)
Alga	18.72	-24.34	333.32	Crater, craters	MC-19	Town in Kazakhstan
Alitus	50	-34.91	321.86	Crater, craters	MC-26	(Alytus), town in Lithuania
Allegheny Vallis	171.08	-9.01	306.1	Vallis, valles	MC-18	River in Pennsylvania, USA
Alpheus Colles	633.03	-39.38	61.53	Collis, colles	MC-27, MC-28	From albedo feature at 45° S, 292° W
Amazonis Mensa	414.04	-1.98	213.1	Mensa, mensae	MC-16	Classical albedo feature name
Amazonis Planitia	2809.04	25.75	197.09	Planitia, planitiae	MC-2, MC-8, Figure 4.1	Classical albedo feature name; home of the Amazons
Amazonis Sulci	250.59	-2.15	216.29	Sulcus, sulci	MC16	Albedo feature name; home of the Amazons
Amenthes Cavi	1330.5	16.23	114.52	Cavus, cavi	MC-14	Classical albedo feature name
Amenthes Fossae	850	9.07	102.68	Fossa, fossae	MC-14	Classical albedo feature name
Amenthes Planum	0	3.4	105.92	Planum, plana	MC-14, MC-22	Classical albedo feature name
Amenthes Rupes	335.25	1.51	110.68	Rupes, rupes	MC14	Classical albedo feature name
Amphitrites Patera	129.8	-58.7	60.87	Patera, paterae	MC-28	Mare Amphitrites; classical albedo feature name
Amsterdam	1.66	23	313	Crater, craters	MC-10	Dutch port
Andapa	11	-5.33	355.27	Crater, craters	MC-19	Town in Madagascar
Angu	2.08	20.01	105.64	Crater, craters	MC-14	Town in Zaire
Angustus Labyrinthus	67.52	-81.62	296.61	Labyrinthus, labyrinthi	MC-30	Classical albedo feature name
Aniak	50.97	-31.84	290.44	Crater, craters	MC-25	Town in Alaska, USA
Anio Vallis	54	37.75	55.89	Vallis, valles	MC-5	Classical river in Italy; modern Aniene and Teverone rivers
Annapolis	1.11	23.16	312.27	Crater, craters	MC-10	American colonial town (Maryland)
Anseris Mons	52.51	-29.81	86.65	Mons, montes	MC-21	From albedo feature Anseris Fons
Antoniadi	400.95	21.38	60.83	Crater, craters	MC-13	Eugène Michael; Turkish-born French astronomer (1870–1944)
Aonia Mons	27.07	-53.33	272.08	Mons, montes	MC-25	Classical albedo feature name.
Aonia Planum	563.45	-57.9	281.33	Planum, plana	MC-25	Classical albedo feature name
Aonia Terra	3873.48	-60.2	262.95	Terra, terrae	MC-25, MC-30, Figure 4.1	Classical albedo feature name
Aonia Tholus	53.69	-59.04	279.96	Tholus, tholi	MC-25	Classical albedo feature name
Apia	10.06	-37.28	89.02	Crater, craters	MC-28	Town in Samoa
Apollinaris Mons	275.4	-9.17	174.79	Mons, montes	MC-23	Classical albedo feature name
Apollinaris Patera	89.6	-8.57	174.18	Patera, paterae	MC-23	Classical albedo feature name
Apollinaris Sulci	188.64	-11.06	177.47	Sulcus, sulci	MC-23	Albedo feature at 5° S, 187° W
Apollinaris Tholus	32.39	-17.64	175.75	Tholus, tholi	MC-23	Albedo feature name

Feature name	Diameter (km)	Center latitude	Center longitude	Feature type	Quadrangle	Origin
Apsus Vallis	121.56	34.91	134.99	Vallis, valles	MC-7	Classical river in ancient Macedonia, present-day Greece
Apt	9.57	39.88	350.53	Crater, craters	MC4	Town in France
Arabia Terra	4851.74	21.25	5.72	Terra, terrae	MC-4, MC-5, MC-11, MC-12, MC-19, Figure 4.1	Classical albedo feature name
Arago	152.35	10.22	29.93	Crater, craters	MC-12	Dominique F.; French astronomer (1786–1853)
Aram Chaos	283.81	2.52	337.61	Chaos, chaoses	MC-11	Classical albedo feature name
Aram Dorsum	83.31	7.8	348.76	Dorsum, dorsa	MC-11	
Arandas	24.76	42.41	344.97	Crater, craters	MC-4	Town in Mexico
Arcadia Dorsa	1952.65	55.9	222.44	Dorsum, dorsa	MC-2	Named for classical albedo feature at 45° N, 120° W
Arcadia Planitia	1871.97	47.19	184.31	Planitia, planitiae	MC-2, MC-7, Figure 4.1	From classical albedo feature at 45° N, 120° W
Arda Vallis	173.67	-20.4	327.69	Vallis, valles	MC-19	Ancient European river (Bulgaria)
Arena Colles	580.12	24.63	82.93	Collis, colles	MC-13	Classical albedo feature at 13° N, 294° W
Arena Dorsum	372.03	12.71	68.94	Dorsum, dorsa	MC-13	Word for "Mars" in Greek
Ares Vallis	1757.67	10.29	334.39	Vallis, valles	MC-11	Town in Russia
Argas	3.55	23.33	309.83	Crater, craters	MC-10	
Argentea Planum	1370.64	-72.49	298.33	Planum, plana	MC-30	Classical albedo feature name
Argyre Cavi	72.33	-48.31	319.88	Cavus, cavi	MC-26	Albedo name
Argyre Mons	60.58	-50.37	311.91	Mons, montes	MC-26	Classical albedo feature name
Argyre Planitia	892.93	-49.84	316.69	Planitia, planitiae	MC-26, Figure 4.1	Classical albedo feature name
Argyre Rupes	335.42	-62.15	291.25	Rupes, rupes	MC-25	Classical albedo feature name
Ariadnes Colles	188.01	-34.5	172.78	Collis, colles	MC-29	Classical albedo feature name
Arica	15.77	-23.8	110.24	Crater, craters	MC-22	Town in Colombia
Arima	53.59	-15.84	296.32	Crater, craters	MC-18	Town in Trinidad and Tobago
Arimanes Rupes	192.67	-9.84	212.3	Rupes, rupes	MC-16	Albedo feature; classical Persian deity of wickedness
Arkhangelsky	116.83	-41.09	335.21	Crater, craters	MC-26	A. D.; Russian geologist
Arnus Vallis	311.61	13.97	70.61	Vallis, valles	MC-13	Classical and present day Arno River in Tuscany, Italy (previously named Arena Rupes)
Aromatum Chaos	72.8	-1.03	317.03	Chaos, chaoses	MC-19	Classical albedo feature name
Arrhenius	122.72	-40.04	122.71	Crater, craters	MC-29	Svante; Swedish physical chemist (1859–1927)
Arsia Chasmata	97.06	-7.47	240.65	Chasma, chasmata	MC-17	Albedo name
Arsia Mons	470	-8.26	239.91	Mons, montes	MC-17	Arsia Silva, classical albedo feature name
Arsia Sulci	500	-6.29	230.19	Sulcus, sulci	MC-17	Albedo name
Arsinoes Chaos	200.08	-7.66	332.08	Chaos, chaoses	MC-19	Daughter of Ptolemy Lagun and Bernice

255

Feature name	Diameter (km)	Center latitude	Center longitude	Feature type	Quadrangle	Origin
Arta	3.96	21.38	305.71	Crater, craters	MC-10	Town in Russia
Artik	5.36	-34.8	130.98	Crater, craters	MC-29	Town in Armenia
Artynia Catena	279.28	47.69	240.55	Catena, catenae	MC-3	Named for classical albedo feature at 54° N, 137° W
Asau	25.05	-3.63	154.68	Crater, craters	MC-23	Village in Tuvalu
Ascraeus Chasmata	105.2	8.77	254.37	Chasma, chasmata	MC-9	Classical albedo name
Ascraeus Mensa	34.86	11.72	252.11	Mensa, mensae	MC-9	Classical albedo name
Ascraeus Mons	456.4	11.92	255.92	Mons, montes	MC-9	Ascraeus Lacus, classical albedo feature name
Ascraeus Sulci	138.7	12.06	251.25	Sulcus, sulci	MC-9	Classical albedo name
Ascuris Planum	617.66	40.59	279.22	Planum, plana	MC-3	Classical albedo name
Asimov	80.82	-46.97	4.93	Crater, craters	MC-27	Isaac; American biochemist and writer (1920-1992)
Asopus Vallis	40.82	-4.29	210.39	Vallis, valles	MC-16	Classical name for modern Hagios River, Greece
Aspen	18.48	-21.39	336.85	Crater, craters	MC-19	Town in Colorado, USA
Aspledon Undae	215.2	73.06	309.65	Unda, undae	MC-1	Classical albedo feature name
Astapus Colles	597	35.46	88.08	Collis, colles	MC-6	From albedo feature at 35° N, 269° W
Athabasca Valles	270	8.54	155.01	Vallis, valles	MC-15	River in Canada (changed from Athabasca Vallis)
Atlantis Chaos	181.37	-34.28	182.69	Chaos, chaoses	MC-24	From albedo feature at 30° N, 173° W
Atrax Fossa	34.42	38.19	271.02	Fossa, fossae	MC-3	Classical town
Auce	37.01	-27.17	80.14	Crater, craters	MC-21	Town in Latvia
Auki	40	-15.76	96.87	Crater, craters	MC-22	Town in the Solomon Islands
Auqakuh Vallis	347	30.25	60.41	Vallis, valles	MC-6, MC-13	Word for "Mars" in Quechua (Inca)
Aureum Chaos	351.03	-3.89	333.04	Chaos, chaoses	MC-19	Classical albedo feature name
Aurorae Chaos	713.92	-8.47	325.19	Chaos, chaoses	MC-19	Classical albedo name
Aurorae Planum	564.49	-10.41	311.38	Planum, plana	MC-18	Classical albedo feature name
Ausonia Cavus	49.5	-31.92	96.55	Cavus, cavi	MC-28	Albedo feature Ausonia
Ausonia Mensa	102.51	-30.02	97.72	Mensa, mensae	MC-28	Albedo feature name
Ausonia Montes	333.13	-25.42	99.04	Mons, montes	MC-22	Albedo feature name
Australe Lingula	436.33	-84.05	68.56	Lingula, lingulae	MC-30	Classical albedo feature name
Australe Mensa	172	-86.88	357.24	Mensa, mensae	MC-30	Classical albedo feature name
Australe Montes	411.67	-80.19	14.05	Mons, montes	MC-30	Classical albedo feature name
Australe Scopuli	504.58	-83.48	247.06	Scopulus, scopuli	MC-30	Classical albedo feature name
Australe Sulci	357.91	-84.99	133.06	Sulcus, sulci	MC-30	Classical albedo feature name
Auxo Dorsum	82.05	-55.72	318.24	Dorsum, dorsa	MC-26	One of the Graces
Avan	3.3	-10.97	290.23	Crater, craters	MC-18	Village in Armenia
Avarua	49.99	-35.93	109.66	Crater, craters	MC-28	Town in the Cook Islands
Aveiro	9.11	21.28	281.03	Crater, craters	MC-10	Town in Portugal

Feature name	Diameter (km)	Center latitude	Center longitude	Feature type	Quadrangle	Origin
Avernus Cavi	115	-3.72	172.52	Cavus, cavi	MC-23	Named for albedo feature at 10° S, 195° W
Avernus Colles	238.7	-1.73	171.02	Collis, colles	MC-23	Named for albedo feature at 10° S, 195° W
Avernus Dorsa	296.59	-6.03	170.9	Dorsum, dorsa	MC-23	From albedo feature at 4° S, 190° W
Avernus Rupes	223.32	-9.2	172.8	Rupes, rupes	MC-23	From albedo feature Avernus at 10° S, 195° W
Avire	6.85	-40.82	200.24	Crater, craters	MC-24	Town in the Republic of Vanuatu
Axius Valles	435.95	-54.53	70.72	Vallis, valles	MC-28	Ancient European River (Vardar River of Greece)
Ayacucho	2.59	38.18	267.97	Crater, craters	MC-3	Town in Bolivia
Ayr	12.74	-38.98	91.58	Crater, craters	MC-28	Town in Queensland, Australia
Azul	19.53	-42.07	317.49	Crater, craters	MC-26	Town in Argentina
Azusa	39.25	-5.48	319.68	Crater, craters	MC-19	Town in California, USA
Babakin	76.66	-36	288.56	Crater, craters	MC-25	Soviet builder of unmanned space stations (1914-1970)
Bacht	7.86	18.66	102.7	Crater, craters	MC-14	Town in Uzbekistan
Bacolor	21.58	32.99	118.6	Crater, craters	MC-6	Town in the Philippines
Bada	2.52	20.35	309.31	Crater, craters	MC-10	Town in Russia
Badwater	33.14	-32.79	62.14	Crater, craters	MC-28	Town in California, USA
Baetis Chaos	66.66	-0.17	299.6	Chaos, chaoses	MC-18	Classical albedo feature name
Baetis Chasma	92.21	-4.29	295.13	Chasma, chasmata	MC-18	From classical albedo feature at 5° S, 60° W; changed from Iamunae Chasma
Baetis Labes	88	-3.67	288.54	Labes, labes	MC-18	Classical albedo feature, Baetis
Baetis Mensa	181.1	-5.17	287.55	Mensa, mensae	MC-18	From albedo feature at 7° S, 63° W
Bahn	11.93	-3.5	316.68	Crater, craters	MC-19	Town in Liberia
Bahram Vallis	269.68	20.42	302.86	Vallis, valles	MC-10	Word for "Mars" in Persian
Bak	3.13	18.05	103.78	Crater, craters	MC-14	Town in Hungary
Bakhuysen	152.9	-22.97	15.73	Crater, craters	MC-20	Hendrik G.; Dutch astronomer (1838-1923)
Balboa	21.95	-3.82	326.12	Crater, craters	MC-19	Town in the Republic of Panama, formerly within the US Panama Canal Zone
Baldet	181.31	22.76	65.48	Crater, craters	MC-13	Fernand; French astronomer (1885-1964)
Balta	17.28	-23.82	333.45	Crater, craters	MC-19	Town in Ukraine
Baltisk	50.75	-42.27	305.34	Crater, craters	MC-26	Town in Russia
Balvicar	20.35	16.2	306.76	Crater, craters	MC-10	Town in Scotland
Bam	6.8	-25.79	115.67	Crater, craters	MC-22	Town in Kermán province in Iran
Bamba	22.57	-3.36	318.41	Crater, craters	MC-19	Town in Zaire
Bamberg	55.7	39.71	356.9	Crater, craters	MC-4	Town in Germany
Banff	5	17.51	329.29	Crater, craters	MC-11	Town in Alberta, Canada
Banh	14.4	19.42	304.5	Crater, craters	MC-10	Town in Burkina Faso (formerly Upper Volta)
Baphyras Catena	95.52	38.83	275.84	Catena, catenae	MC-3	Classical river
Bar	2.06	-25.25	340.5	Crater, craters	MC-19	Town in Ukraine
Barabashov	120.67	47.33	291.25	Crater, craters	MC-3	Nikolay P.; Russian astronomer (1894-1971)
Barnard	121.11	-61.06	61.59	Crater, craters	MC-27, MC-28	Edward E.; American astronomer (1857-1923)
Baro	16.93	-24.8	110.7	Crater, craters	MC-22	Town in Nigeria

Feature name	Diameter (km)	Center latitude	Center longitude	Feature type	Quadrangle	Origin
Bok	7.34	20.58	328.4	Crater, craters	MC-11	Town in New Guinea
Bole	8.54	25.35	306.01	Crater, craters	MC-10	Town in Ghana
Bombala	37.14	-27.6	106.13	Crater, craters	MC-22	Town in New South Wales, Australia
Bond	104.69	-32.79	324.06	Crater, craters	MC-26	George P.; American astronomer (1825–1865)
Bonestell	40.67	42	329.61	Crater, craters	MC-4	Chesley; American space artist (1888–1986)
Boola	17.25	81.26	254.81	Crater, craters	MC-1	Town in Guinea
Bopolu	18.89	-2.96	353.7	Crater, craters	MC-19	Town in Liberia
Bor	4.28	18.17	326.32	Crater, craters	MC-11	Town in Russia
Bordeaux	2.01	23.13	311.11	Crater, craters	MC-10	French port
Boreales Scopuli	1.13	88.88	269.84	Scopulus, scopuli	MC-1	Classical albedo feature name
Boreum Cavus	62.13	84.64	339.85	Cavus, cavi	MC-1	Classical albedo feature name
Boru	10.87	-24.34	332.13	Crater, craters	MC-19	Town in Russia
Bosporos Planum	729.58	-33.87	295.51	Planum, plana	MC-25	Classical albedo feature name
Bosporos Rupes	531.42	-42.74	302.45	Rupes, rupes	MC-26	Classical albedo feature name
Bouguer	107.78	-18.46	27.27	Crater, craters	MC-20	Pierre; French physicist-hydrographer (1698–1758)
Boulia	9.99	-22.89	111.32	Crater, craters	MC-22	Town in Queensland, Australia
Bozkir	79.89	-44.14	327.82	Crater, craters	MC-26	Town in Turkey
Bradbury	63.2	2.58	85.8	Crater, craters	MC-13	Raymond Douglas "Ray"; American author (1920–2012)
Brashear	77.45	-53.81	240.97	Crater, craters	MC-25	John A.; American astronomer (1840–1920)
Brazos Valles	387.51	-6.08	18.7	Vallis, valles	MC-20	River in Texas, USA
Bree	28.79	37.64	149.63	Crater, craters	MC-7	Town in Belgium
Bremerhaven	2.69	23.7	311.36	Crater, craters	MC-10	German port
Briault	93.06	-9.98	89.68	Crater, craters	MC-21, MC-22	P.; French astronomer (d. 1922)
Bridgetown	1.57	21.9	312.92	Crater, craters	MC-10	Port of Barbados
Bristol	3	22.09	313.07	Crater, craters	MC-10	English port
Broach	11.31	23.51	303.11	Crater, craters	MC-10	Town in India
Bronkhorst	17.75	-10.7	304.79	Crater, craters	MC-18	Town in the Netherlands
Brush	6.49	21.7	111.34	Crater, craters	MC-14	Town in Colorado, USA
Bulhar	18.24	50.36	134.52	Crater, craters	MC-7	Town in Somalia.
Bunge	70.83	-33.82	311.41	Crater, craters	MC-26	Andrey Aleksandrovich; Russian zoologist, permafrost investigator (late nineteenth century), and Nicolai A.; Russian chemist
Bunnik	29	-38.07	217.9	Crater, craters	MC-24	Village in the Netherlands
Burroughs	112.69	-72.29	117.1	Crater, craters	MC-30	Edgar R.; American novelist (1875–1950)
Burton	119.26	-13.88	203.67	Crater, craters	MC-16	Charles E.; British astronomer (1846–1882)
Buta	11.27	-23.25	327.59	Crater, craters	MC-19	Town in Zaire
Butte	12.55	-5.08	321.09	Crater, craters	MC-19	Town in Montana, USA
Buvinda Vallis	134.17	33.17	151.96	Vallis, valles	MC-7	Classical river in Hibernia; present Boyne River, Ireland

Feature name	Diameter (km)	Center latitude	Center longitude	Feature type	Quadrangle	Origin
Barsukov	68.45	7.97	330.98	Crater, craters	MC-11	Valery L.; Soviet geochemist and planetologist (1928–1992)
Bashkaus Valles	246.93	-25.68	356.74	Vallis, valles	MC-19	River in the Altai Republic, Russia
Basin	15.53	17.82	107	Crater, craters	MC-14	Town in Wyoming, USA
Batos	17.33	21.5	330.5	Crater, craters	MC-11	Town in Romania
Batoka	14.96	-7.55	323.35	Crater, craters	MC-19	Town in Zambia
Baucau	17.94	-28.37	304.9	Crater, craters	MC-18	Town in Timor-Leste
Baum	62	-24.72	28.3	Crater, craters	MC-20	William Alvin; American astronomer (1924–2012)
Baykonyr	3.9	46.41	132.68	Crater, craters	MC-7	Soviet launch site
Bazas	16.43	-27.78	93.38	Crater, craters	MC-22	Town in France
Becquerel	165.23	21.89	352.06	Crater, craters	MC-11	Antoine H.; French physicist (1852–1908)
Beer	85.5	-14.47	351.83	Crater, craters	MC-19	Wilhelm; German astronomer (1797–1850)
Beloha	31.74	-39.58	56.71	Crater, craters	MC-27	Town in Madagascar
Beltra	7.17	18.01	102.4	Crater, craters	MC-14	Town in Ireland
Belyov	0.2	-45.02	201.99	Crater, craters	MC-24	(Belev) Town in Tula region, Russia
Belz	10.21	21.57	316.77	Crater, craters	MC-11	Town in Ukraine
Bend	3.7	-22.4	332.26	Crater, craters	MC-19	Town in Oregon, USA
Bentham	11.36	-55.78	319.45	Crater, craters	MC-26	Town in England
Bentong	10.32	-22.31	340.96	Crater, craters	MC-19	Town in Malaysia
Bernard	128.1	-23.24	205.79	Crater, craters	MC-16	P.; French atmospheric scientist
Berseba	36.83	-4.4	322.4	Crater, craters	MC-19	Town in Namibia
Beruri	45.12	5.27	81.24	Crater, craters	MC-13	Town in Brazil
Betio	32.44	-23.13	281.35	Crater, craters	MC-18	Village in Kiribati
Bhor	5.73	41.75	134.53	Crater, craters	MC-7	Town in India
Bianchini	70.71	-63.85	264.71	Crater, craters	MC-25	Francesco; Italian astronomer (1662–1729)
Biblis Patera	53.65	2.36	236.18	Patera, paterae	MC-9	Classical albedo feature name
Biblis Tholus	168.6	2.52	235.62	Tholus, tholi	MC-9	Classical albedo feature name
Bigbee	20.86	-24.78	325.25	Crater, craters	MC-19	Town in Mississippi, USA
Bira	2.7	25.1	314.46	Crater, craters	MC-10	Town in Russia
Bise	9.21	20.22	303.17	Crater, craters	MC-10	Town in Okinawa
Bison	15.28	-26.31	330.85	Crater, craters	MC-19	Town in Kansas, USA
Bjerknes	88.64	-43.01	171.48	Crater, craters	MC-29	Vilhelm F.; Norwegian physicist (1862–1951)
Bland	6.63	18.3	108.76	Crater, craters	MC-14	Town in Missouri, USA
Bled	7.69	21.58	328.54	Crater, craters	MC-11	Town in Slovenia
Blitta	12.95	-25.9	339.04	Crater, craters	MC-19	Town in Togo
Blois	11.58	23.6	304.15	Crater, craters	MC-10	Town in France
Bluff	6.75	23.47	110.03	Crater, craters	MC-14	Town in New Zealand
Blunck	66.49	-27.23	323.1	Crater, craters	MC-19	Jürgen; German historian (1935–2008)
Boeddicker	107.12	-14.82	162.49	Crater, craters	MC-23	Otto; German astronomer (1853–1937)
Bogia	37.68	-44.31	83.27	Crater, craters	MC-28	Town in Papua New Guinea
Bogra	21.31	-24.16	331.2	Crater, craters	MC-19	Town in Bangladesh

Feature name	Diameter (km)	Center latitude	Center longitude	Feature type	Quadrangle	Origin
Byala	26.23	-25.73	293.53	Crater, craters	MC-18	Town in Bulgaria
Byrd	123.27	-65.22	127.83	Crater, craters	MC-29, MC-30	Richard Edwin; American aviator–explorer (1888–1975)
Byske	12.56	-4.97	326.05	Crater, craters	MC-19	Town in Sweden
Cádiz	1.38	23.15	310.97	Crater, craters	MC-10	Spanish port
Cairns	8.73	23.56	312.54	Crater, craters	MC-10	Town in Australia
Calahorra	34.22	26.46	321.35	Crater, craters	MC-11	Town in Spain
Calamar	7.21	18.27	305.13	Crater, craters	MC-10	Town in Colombia
Calbe	13.14	-25.14	331.13	Crater, craters	MC-19	Town in Germany
Calydon Fossa	351.25	-7.43	272.02	Fossa, fossae	MC-18	The son of Ares and Astynome
Camargo	4.77	17.7	109.64	Crater, craters	MC14	Town in Bolivia
Camichel	65.26	2.26	308.39	Crater, craters	MC-10	Henri; French astronomer (1907–2003)
Camiling	21.91	-0.71	322	Crater, craters	MC-19	Town in the Philippines
Camiri	31.36	-44.65	317.83	Crater, craters	MC-26	Town in Bolivia
Campbell	125.26	-54.25	165.58	Crater, craters	MC-29	John W.; Canadian physicist (1889–1955); William W.; American astronomer (1862–1938)
Campos	8.25	-21.8	332.19	Crater, craters	MC-19	Town in Brazil
Can	8.62	48.21	345.41	Crater, craters	MC-4	Town in Turkey
Canala	12	24.35	279.92	Crater, craters	MC-10	Town in New Caledonia
Cañas	41.52	-31.19	89.86	Crater, craters	MC-28	Town in Puerto Rico
Canaveral	3.18	46.83	135.83	Crater, craters	MC7	American launch site
Canberra	3.1	47.2	132.66	Crater, craters	MC7	Australian tracking site
Candor Chaos	0	-6.94	287.42	Chaos, chaoses	MC-18	From albedo feature at 5 N, 75 W
Candor Chasma	810.61	-6.53	289.22	Chasma, chasmata	MC-18	Classical albedo feature name
Candor Colles	37.38	-6.63	284.43	Collis, colles	MC-18	Classical albedo feature name
Candor Labes	134.94	-4.79	284.01	Labes, labes	MC-18	From classical albedo feature at 5 N, 75 W
Candor Mensa	116.56	-6.26	286.48	Mensa, mensae	MC-18	Classical albedo feature
Candor Sulci	73.36	-4.92	283.15	Sulcus, sulci	MC-18	Classical albedo feature name
Cangwu	13.64	41.85	270.41	Crater, craters	MC-3	Town in China
Canillo	33.93	10.23	116.48	Crater, craters	MC-14	Town in Andorra
Cankuzo	48.45	-19.42	52.03	Crater, craters	MC-21	Town in Burundi
Canso	26.42	21.36	299.38	Crater, craters	MC-10	Town in Nova Scotia, Canada
Cantoura	51.59	14.84	308.28	Crater, craters	MC-10	Town in Venezuela
Capen	68.99	6.58	14.31	Crater, craters	MC-12	Charles F., Jr.; American astronomer (1926–1986)
Capri Chasma	1471.56	-8.27	317.93	Chasma, chasmata	MC-18, MC-19	Classical albedo feature name
Capri Mensa	282.35	-13.73	312.81	Mensa, mensae	MC-18	Classical albedo feature name
Caralis Chaos	103.35	-37.2	178.6	Chaos, chaoses	MC29	After Caralis Fons, classical albedo feature named after Caralis Lacus, now Beyşehir Gölü, in Isauria (Asia Minor)
Cardona	13.7	-19.65	328.02	Crater, craters	MC-19	Town in Uruguay
Cartago	36.57	-23.25	342.03	Crater, craters	MC-19	Town in Costa Rica
Cassini	408.23	23.35	32.11	Crater, craters	MC-12	Giovanni; Italian astronomer (1625–1712)
Castril	2.19	-14.7	175.3	Crater, craters	MC-23	Town in Spain
Catota	1.3	51.67	333.02	Crater, craters	MC-4	Village in Angola
Cave	8.2	21.61	324.36	Crater, craters	MC-11	Town in New Zealand
Cavi Angusti	640.04	-78.16	285.25	Cavus, cavi	MC-30	Classical albedo feature name
Caxias	25.88	-28.95	259.32	Crater, craters	MC-17	Town in Brazil
Cayon	27.31	-35.93	113.62	Crater, craters	MC-28	Town in Saint Kitts and Nevis
Cefalù	5.53	23.64	321.11	Crater, craters	MC11	Town in Sicily
Centauri Montes	271	-38.67	95.52	Mons, montes	MC-28	Albedo feature Centauri Lacus
Ceraunius Catena	50.49	37.1	251.91	Catena, catenae	MC-3	Named for classical albedo feature at 35 N, 96 W
Ceraunius Fossae	1166.63	27	249.85	Fossa, fossae	MC-3, MC-9	Classical albedo feature name
Ceraunius Tholus	128.58	24	262.75	Tholus, tholi	MC-9	Classical albedo feature name
Cerberus Dorsa	623.05	-13.74	105.29	Dorsum, dorsa	MC-22	Classical albedo feature name
Cerberus Fossae	1235	11.28	166.37	Fossa, fossae	MC-15	From albedo feature at 10 N, 212 W; changed from Cerberus Rupes
Cerberus Palus	466.68	5.78	148.15	Palus, paludes	MC-15	Classical albedo feature name
Cerberus Tholi	698	4.48	164.41	Tholus, tholi	MC-15	Classical albedo feature name
Cerulli	114.28	32.2	22.12	Crater, craters	MC-5	Vicenzo; Italian astronomer (1859–1927)
Ceti Chasma	49.77	-5.03	291.63	Chasma, chasmata	MC-18	From albedo feature at 10 S, 74 W
Ceti Labes	11.05	-6.78	284.27	Labes, labes	MC-18	Classical albedo feature name
Ceti Mensa	133.95	-5.89	283.98	Mensa, mensae	MC-18	Albedo feature Ceti Lacus
Chafe	4.67	15.1	102.41	Crater, craters	MC-14	Town in Nigeria
Chalce Fossa	33.97	-51.67	320.41	Fossa, fossae	MC-26	Albedo name
Chalce Montes	100	-53.72	322.35	Mons, montes	MC-26	Albedo name
Chalcoporos Rupes	404.98	-55.64	20.57	Rupes, rupes	MC-27	From albedo feature at 50 S, 6 W
Chaman	47.92	-60.86	50.96	Crater, craters	MC-27	Town in Pakistan
Chamberlin	120.25	-65.84	235.71	Crater, craters	MC-30	Thomas C.; American geologist (1843–1928)
Changsŏng	33.54	23.47	302.66	Crater, craters	MC-10	Town in the Democratic People's Republic of Korea
Chapais	36.67	-22.35	339.45	Crater, craters	MC-19	Town in Quebec, Canada
Charis Dorsum	2.51	-55.86	318.53	Dorsum, dorsa	MC-26	One of the Graces
Charitum Montes	933.54	-58.1	319.71	Mons, montes	MC-26	Classical albedo feature name
Charleston	1.96	22.63	312.2	Crater, craters	MC-10	American colonial town (South Carolina)
Charlier	106.28	-68.56	191.53	Crater, craters	MC-30	Carl V.; Swedish astronomer (1862–1934)
Charlieu	18.63	38.15	276.01	Crater, craters	MC3	Town in France

Feature name	Diameter (km)	Center latitude	Center longitude	Feature type	Quadrangle	Origin
Coblentz	101.75	-54.9	269.69	Crater, craters	MC-25	William W.; American physicist (1873–1962)
Cobres	93.76	-11.7	206.4	Crater, craters	MC-16	Village in Argentina
Coimbra	34.53	4.18	354.69	Crater, craters	MC-11	Town in Portugal
Colón	1.36	22.75	312.93	Crater, craters	MC10	Port of Panama
Colles Nili	653.67	38.72	62.88	Colles, colles	MC-5, MC-6	From classical albedo feature Portus Nili, at 38 N, 295 W
Coloe Fossae	575.89	36.65	56.78	Fossa, fossae	MC-5, MC-6	Classical albedo feature name
Columbia Valles	84.82	-9.44	317.1	Vallis, valles	MC-19	River in Washington, USA
Columbus	112.6	-29.29	194.02	Crater, craters	MC-16	Christopher; Italian explorer (1451–1506)
Comas Sola	120.24	-19.59	201.49	Crater, craters	MC-16	Jose; Spanish astronomer (1868–1937)
Conches	20.89	-4.22	325.8	Crater, craters	MC-19	Town in France
Concord	20.46	16.53	325.98	Crater, craters	MC-11	Town in Massachusetts, USA
Coogoon Valles	300	17.19	338.26	Vallis, valles	MC-11	River in Australia
Cooma	17.85	-23.69	251.65	Crater, craters	MC-17	Town in New South Wales, Australia
Copernicus	301.83	-48.84	191.17	Crater, craters	MC-24	Nicolaus; Polish astronomer (1473–1543)
Coprates Catena	302.06	-15	297.91	Catena, catenae	MC-18	Classical albedo feature name
Coprates Chasma	958.31	-13.37	299.26	Chasma, chasmata	MC-18	Classical albedo feature name
Coprates Labes	61.97	-11.82	292.21	Labes, labes	MC-18	From albedo feature at 14 S, 65 W
Coprates Mensa	240	-12.2	288.84	Mensa, mensae	MC-18	Classical martian albedo feature, Coprates
Coprates Montes	350	-13	294.61	Mons, montes	MC-18	Classical martian albedo feature, Coprates
Caracis Fossae	748.96	-35.82	279.14	Fossa, fossae	MC-25	From albedo feature at 46 S, 87 W
Corby	6.62	42.88	137.56	Crater, craters	MC-7	Town in England
Corinto	13.69	16.95	141.71	Crater, craters	MC-15	Town in El Salvador
Coronae Montes	247.4	-34.31	86.11	Mons, montes	MC-28	Albedo feature name
Coronae Planum	265	-32.68	65.41	Planum, plana	MC-28	Albedo feature name
Coronae Scopulus	245.24	-33.26	64.94	Scopulus, scopuli	MC-28	From albedo feature at 26 S, 276 W
Corozal	8.33	-38.79	159.42	Crater, craters	MC-29	Town in Belize
Cost	11.07	14.98	104.02	Crater, craters	MC-14	Town in Texas, USA
Cray	6.98	44.1	343.88	Crater, craters	MC-4	Town in England
Creel	9.19	-6.05	321.15	Crater, craters	MC-19	Town in Mexico
Crewe	3.68	-24.84	340.47	Crater, craters	MC-19	Town in England
Crivitz	6.19	-14.55	174.79	Crater, craters	MC-23	Town in Germany
Crommelin	110.08	5.08	349.86	Crater, craters	MC-11	Andrew C.; British astronomer (1865–1939)
Cross	66.57	-30.2	202.31	Crater, craters	MC-24	Charles Arthur; British astronomer and cartographer (1920–1980)
Crotone	6.28	82.21	290.69	Crater, craters	MC-1	Town in Italy
Cruls	87.89	-42.91	163.03	Crater, craters	MC-29	Luiz; Brazilian astronomer (1848–1908)
Cruz	5.33	38.46	358.03	Crater, craters	MC-4	Town in Venezuela

259

Feature name	Diameter (km)	Center latitude	Center longitude	Feature type	Quadrangle	Origin
Charybdis Scopulus	551.26	-24.14	20.08	Scopulus, scopuli	MC-20	From albedo feature at 19 S, 320 W
Chasma Australe	352.61	-82.35	95.03	Chasma, chasmata	MC-30	Classical albedo feature name
Chasma Boreale	459.88	82.54	312.36	Chasma, chasmata	MC-1	Classical albedo feature name
Chaturat	7.84	35.38	265.06	Crater, craters	MC-3	Town in Thailand
Chauk	9.97	23.35	304.09	Crater, craters	MC-10	Town in Burma
Cheb	8.28	-24.2	340.56	Crater, craters	MC-19	Town in Czechoslovakia
Chefu	11.27	-22.91	112.24	Crater, craters	MC-22	Town in Mozambique
Chekalin	87.78	-24.28	333.19	Crater, craters	MC-19	Town in Russia
Chia	91.91	1.57	300.35	Crater, craters	MC-10	Town in Spain
Chico Valles	446.35	-66.77	207.77	Vallis, valles	MC-30	River in Argentina
Chimbote	62.86	-1.42	320.32	Crater, craters	MC-19	Town in Peru
Chincoteague	34.03	41.2	124.12	Crater, craters	MC-7	Town in Virginia, USA
Chinju	65.71	-4.52	317.86	Crater, craters	MC-19	Town in the Republic of Korea
Chinook	18.1	22.5	304.54	Crater, craters	MC-10	Town in Alberta, Canada
Chive	9.1	21.68	303.99	Crater, craters	MC-10	Town in Bolivia
Choctaw	23.96	-41.19	322.76	Crater, craters	MC-26	Town in Ohio, USA
Chom	5.58	38.57	357.48	Crater, craters	MC-4	Town in China (Tibet)
Choyr	36.42	-32.43	18.72	Crater, craters	MC-27	Town in Mongolia
Chronius Mons	56.14	-61.49	178.01	Mons, montes	MC-29	Classical albedo feature name
Chrysas Mensa	167	-6.77	289.9	Mensa, mensae	MC18	Classical albedo feature name, Chrysas
Chryse Chaos	658.89	9.86	322.81	Chaos, chaoses	MC-11	Classical albedo feature name
Chryse Colles	48.66	8.15	318.14	Collis, colles	MC11	Classical albedo feature name
Chryse Planitia	1542.44	28.43	319.69	Planitia, planitiae	MC-4, MC-10, MC-11, Figure 4.1	Classical albedo feature name
Chupadero	8.04	6.13	83.43	Crater, craters	MC-13	Village in New Mexico, USA
Chur	4.39	16.93	330.69	Crater, craters	MC-11	Town in Russia
Cilaos	21.4	-35.71	129.48	Crater, craters	MC-29	Town on Réunion Island
Circle	11.8	-22.17	334.47	Crater, craters	MC-19	Town in Montana, USA
Clanis Valles	58	33.24	58.47	Vallis, valles	MC-5	Classical river in Etruria; present Chiana River, Italy
Claritas Fossae	2030.64	-27.89	255.76	Fossa, fossae	MC-17, MC-25	Classical albedo feature name
Claritas Rupes	952.87	-25.04	254.74	Rupes, rupes	MC-17, MC25	From albedo feature at 25 S, 110 W
Clark	97.5	-55.14	226.8	Crater, craters	MC-24	Alvan; American optician-astronomer (1804–1887)
Clasia Vallis	147	33.77	57.04	Vallis, valles	MC-5	Classical river in Umbria, Italy
Cleia Dorsum	131.69	-54.86	314.01	Dorsum, dorsa	MC-26	One of the Graces
Clogh	11.2	20.56	312.33	Crater, craters	MC-10	Town in Ireland
Clota Vallis	114.36	-25.59	339.5	Vallis, valles	MC-19	Ancient name for present River Clyde, Scotland
Clova	7.75	21.47	307.95	Crater, craters	MC-10	Town in Quebec, Canada
Cluny	14.84	-23.86	332.7	Crater, craters	MC-19	Town in France
Cobalt	10.53	-25.79	332.97	Crater, craters	MC19	Town in Connecticut, USA

Feature name	Diameter (km)	Center latitude	Center longitude	Feature type	Quadrangle	Origin
Cue	10.54	−35.84	93.23	Crater, craters	MC-28	Town in Western Australia
Culter	4.87	−8.84	306.07	Crater, craters	MC-18	Village near Aberdeen, Scotland, also called Peterculter
Curie	111.11	28.78	355.25	Crater, craters	MC-11	Pierre; French physicist-chemist (1859–1906)
Cusus Valles	250.24	14.05	50.37	Vallis, valles	MC-13	Classical name for modern Hron River in Slovakia
Cyane Catena	204.06	36.25	241.7	Catena, catenae	MC-3	Classical albedo feature
Cyane Fossae	913.17	31.25	238.83	Fossa, fossae	MC-2, MC-3, MC-9	Cyane Fons Classical albedo feature name
Cyane Sulci	335.94	25.4	231.34	Sulcus, sulci	MC-9	Classical albedo feature name
Cydnus Rupes	1550.81	52.53	112.21	Rupes, rupes	MC-6, MC-7	From albedo feature at 70 N, 248 W
Cydonia Colles	362.78	39.07	347.78	Collis, colles	MC-4	Named for classical albedo feature at 50° N, 355° W
Cydonia Labyrinthus	344.05	41.29	347.94	Labyrinthus, labyrinthi	MC-4	Named for classical albedo feature at 50° N, 355° W
Cydonia Mensae	764.96	34.56	347.67	Mensa, mensae	MC-4	From albedo feature at 50° N, 355° W
Cypress	14.67	−47.28	312.65	Crater, craters	MC-26	Town in Illinois, USA
Da Vinci	96.3	1.47	320.74	Crater, craters	MC-11	Leonardo; Italian artist-scientist (1452–1519)
Daan	12.22	−40.48	91.58	Crater, craters	MC-28	Town in China
Daedalia Planum	1922.02	−18.35	234.05	Planum, plana	MC-16, MC-17	Classical albedo feature name
Daet	10.58	−7.29	318.2	Crater, craters	MC-19	Town in the Philippines
Daga Vallis	49.86	−12.07	317.58	Vallis, valles	MC-19	River in Burma
Daly	79.72	−66.29	336.88	Crater, craters	MC-30	Reginald A.; Canadian-born American geologist (1871–1957)
Dana	88.49	−72.49	327.21	Crater, craters	MC-30	James D.; American geologist (1813–1895)
Danielson	64.3	7.97	352.95	Crater, craters	MC-11	G. Edward Danielson, Jr.; American Engineer (1939–2005)
Dank	8.29	21.96	107	Crater, craters	MC-14	Town in Oman
Dao Vallis	794	−37.61	88.89	Vallis, valles	M-28	Word for "star" in Thai
Darvel	22.36	17.78	308.99	Crater, craters	MC-10	Town in Scotland
Darwin	176.38	−56.97	340.85	Crater, craters	MC-26	Charles R.; British naturalist (1809–1882) and George H.; British astronomer (1845–1912)
Davies	48.06	45.96	0.09	Crater, craters	MC-4, MC-5	Merton Edward; American engineer, planetary geodesist (1917–2001)
Dawes	185.32	−9.11	38.06	Crater, craters	MC-20	William R.; British astronomer (1799–1868)
de Vaucouleurs	302.27	−13.31	171	Crater, craters	MC-23	Gerard Henri; American astronomer (1918–1995)
Deba	8.8	−23.95	342.7	Crater, craters	MC-19	Town in Nigeria
Degana	57	−23.72	314.5	Crater, craters	MC-18	Town in India
Dein	25.52	38.21	357.49	Crater, craters	MC-4	Town in New Guinea
Dejnev	152.09	−25.14	195.36	Crater, craters	MC-16	Semen Ivanovich; Russian geographer, explorer, and navigator (1605–1673)

Feature name	Diameter (km)	Center latitude	Center longitude	Feature type	Quadrangle	Origin
Delta	7.89	−45.96	320.83	Crater, craters	MC-26	Town in Louisiana, USA
Denning	159.71	−17.43	33.52	Crater, craters	MC-20	William F.; British astronomer (1848–1931)
Dersu	5.88	22.64	308.11	Crater, craters	MC-10	Town in Russia
Dese	13.14	−45.41	329.39	Crater, craters	MC-26	Town in Ethiopa
Deseado	27	−80.62	70.29	Crater, craters	MC-30	Town in Argentina
Dessau	9.95	−42.76	306.87	Crater, craters	MC-26	Town in Germany
Deuteronilus Colles	59.01	41.95	21.7	Collis, colles	MC-5	Classical albedo feature name
Deuteronilus Mensae	919.17	45.11	23.92	Mensa, mensae	MC-5	From albedo feature at 35° N, 355° W
Deva Vallis	53.65	−7.67	203.13	Vallis, valles	MC-16	Classical name for Dee River in Scotland; new position and coordinates
Dia-Cau	29.47	−0.36	317.34	Crater, craters	MC-19	Town in the Socialist Republic of Vietnam
Dilly	2.13	13.27	157.22	Crater, craters	MC-15	Town in Mali
Dingo	15.56	−23.71	342.51	Crater, craters	MC-19	Town in Australia
Dinorwic	51.33	−30.03	258.54	Crater, craters	MC-25	Town in Ontario, Canada
Dison	20.67	−25.03	343.5	Crater, craters	MC-19	Town in Belgium
Dittaino Valles	157.42	−1.43	293.14	Vallis, valles	MC-18	Modern river in Italy
Dixie	28.44	17.78	304.09	Crater, craters	MC-10	Town in Georgia, USA
Doanus Vallis	139.91	−63.02	334.41	Vallis, valles	MC-26	Classical river shown in Ptolemy's map; may be modern Mekong River of Burma
Doba	25.89	10.92	119.62	Crater, craters	MC-14	Town in the Republic of Chad
Dogana	41.2	−10.01	306.33	Crater, craters	MC-18	Town in the Republic of San Marino
Dokka	51.1	77.17	214.24	Crater, craters	MC-1	Town in Norway
Dokuchaev	74.74	−60.62	232.92	Crater, craters	MC-24	Vasily Vasil'evich; Russian soil scientist; founded modern genetical soil science (1840–1903)
Dollfus	363.08	−21.59	355.74	Crater, craters	MC-19	Audouin Charles; French astronomer (1924–2010)
Domoni	13.82	51.38	234.39	Crater, craters	MC-2	Town in the Union of the Comoros
Doon	3.79	23.53	109.51	Crater, craters	MC-14	Town in Ontario, Canada
Dorsa Argentea	339.26	−77.63	326.61	Dorsum, dorsa	MC-30	Classical albedo feature name
Dorsa Brevia	650.99	−71.05	63.18	Dorsum, dorsa	MC-30	Classical albedo feature name
Douglass	92.95	−51.34	289.46	Crater, craters	MC-25	Andrew E.; American astronomer (1867–1962)
Dowa	40.84	−31.66	110.24	Crater, craters	MC-28	Town in Malawi
Downe	28.13	−15.98	175.78	Crater, craters	MC-23	Town in England
Drava Valles	159.03	−48.86	165.99	Vallis, valles	MC-29	Modern river in Yugoslavia
Drilon Vallis	118.53	7.17	307.66	Vallis, valles	MC-10	Classical name for present Drin River, Albania
Dromore	14.75	19.88	310.42	Crater, craters	MC-10	Town in Northern Ireland
Du Martheray	96.12	5.45	93.58	Crater, craters	MC-14	Maurice; Swiss astronomer (1892–1955).
Du Toit	81.82	−71.62	310.4	Crater, craters	MC-30	Alexander L.; South African geologist (1878–1948)

Feature name	Diameter (km)	Center latitude	Center longitude	Feature type	Quadrangle	Origin
Dubis Vallis	45.42	−5.16	211.87	Vallis, valles	MC-16	Classical name for modern Doubs River, France
Dubki	9.19	−34.97	304.8	Crater, craters	MC-26	Town in Russia
Dukhan	**34.04**	**7.76**	**320.86**	**Crater, craters**	**MC-11**	**Town in Qatar**
Dulce Vallis	32.46	−4.82	136.54	Vallis, valles	MC-23	River in Argentina
Dulovo	17.38	3.62	84.56	Crater, craters	MC-13	Town in Bulgaria
Dunhuang	11.73	−80.84	311.47	Crater, craters	MC-30	Town in China
Dunkassa	8	−37.49	222.94	Crater, craters	MC-24	Town in Benin
Durius Vallis	**240**	**−17.3**	**171.98**	**Vallis, valles**	**MC-23**	**Classical name for modern Douro River, Portugal**
Dush	2.39	22.49	305.98	Crater, craters	MC-10	Town in Egypt
Dzeng	10.88	−80.51	289.53	Crater, craters	MC-30	Town in Cameroon
Dzigai Vallis	**327.31**	**−58.1**	**323.41**	**Vallis, valles**	**MC-26**	**Word for "valley" in Navajo**
E. Mareotis Tholus	4.6	35.92	274.87	Tholus, tholi	MC-3	Classical albedo name
Eads	2.74	−28.48	330.09	Crater, craters	MC-19	Town in Colorado, USA
Eagle	12.5	43.81	351.83	Crater, craters	MC-4	Town in Idaho, USA
Eberswalde	**62.19**	**−23.98**	**326.7**	**Crater, craters**	**MC-19**	**Town in Germany**
Echt	2.15	−21.97	331.81	Crater, craters	MC-19	Town in Scotland
Echus Chaos	**480.51**	**10.79**	**285.28**	**Chaos, chaoses**	**MC-10**	**From albedo feature Echus Lacus at 1 N, 90 W**
Echus Chasma	**391.1**	**2.47**	**280.04**	**Chasma, chasmata**	**MC-10, MC-18**	**Classical albedo feature name**
Echus Fossae	**421.03**	**2.61**	**283.25**	**Fossa, fossae**	**MC-10**	**Classical albedo feature name**
Echus Montes	**397.06**	**7.81**	**282.05**	**Mons, montes**	**MC-10**	**Albedo feature name**
Echus Palus	**715**	**12.29**	**282.73**	**Palus, paludes**	**MC-10**	**Classical albedo feature name, Echus Lacus**
Edam	19.49	−26.28	339.96	Crater, craters	MC-19	Town in the Netherlands
Eddie	**86.38**	**12.32**	**142.2**	**Crater, craters**	**MC-15**	**Lindsay A.; South African astronomer (1845–1913)**
Eden Patera	**80**	**33.77**	**348.94**	**Patera, paterae**	**MC-4**	**Classical albedo feature name**
Eger	12.25	−48.29	308.13	Crater, craters	MC-26	Town in Hungary
Ehden	**57.4**	**8.23**	**119.01**	**Crater, craters**	**MC-14**	**Town in Lebanon**
Eil	5.56	41.73	350.26	Crater, craters	MC-4	Town in Somalia
Eilat	**29.7**	**−56.53**	**50.2**	**Crater, craters**	**MC-27**	**Town in Israel**
Eijriksson	**46.63**	**−19.2**	**186.17**	**Crater, craters**	**MC-16**	**Leif; Norse explorer (c. 1000)**
Elath	13.23	45.87	346.4	Crater, craters	MC-4	Town in Israel
Elaver Vallis	**178.92**	**−9.38**	**310.48**	**Vallis, valles**	**MC-18**	**Classical name for modern Allier River, France**
Electris Mons	**104.47**	**−45.67**	**152.73**	**Mons, montes**	**MC-29**	**Classical albedo feature name**
Elim	**43.63**	**−80.17**	**96.8**	**Crater, craters**	**MC-30**	**Town in South Africa**
Ellsley	10.94	36.29	276.7	Crater, craters	MC-3	Town in England
Elorza	**45.16**	**−8.76**	**304.79**	**Crater, craters**	**MC-18**	**Town in Venezuela**
Ely	10.43	−23.62	332.7	Crater, craters	MC-19	Town in Nevada, USA
Elysium Catena	**48.5**	**17.69**	**149.73**	**Catena, catenae**	**MC-15**	**Albedo feature name**
Elysium Chasma	**130**	**22.39**	**141.51**	**Chasma, chasmata**	**MC-15**	**Albedo feature name**

Feature name	Diameter (km)	Center latitude	Center longitude	Feature type	Quadrangle	Origin
Elysium Fossae	**1044**	**24.08**	**146.14**	**Fossa, fossae**	**MC-15**	**Classical albedo feature name**
Elysium Mons	**401**	**25.02**	**147.21**	**Mons, montes**	**MC-15**	**Classical albedo feature name**
Elysium Planitia	**3000.79**	**2.98**	**154.74**	**Planitia, planitiae**	**MC-14, MC-15, MC-23, Figure 4.1**	**Classical albedo feature name**
Elysium Rupes	140.5	25.24	148.04	Rupes, rupes	MC-15	Albedo feature name
Endeavour	**21.78**	**−2.28**	**354.8**	**Crater, craters**	**MC-19**	**Town in Canada**
Enipeus Vallis	**527.07**	**36.8**	**267.2**	**Vallis, valles**	**MC-3**	**Classical river**
Eos Chaos	**497.85**	**−16.82**	**313.48**	**Chaos, chaoses**	**MC-18**	**Greek name of Aurora; albedo feature**
Eos Chasma	**1305.69**	**−12.15**	**320.83**	**Chasma, chasmata**	**MC-18, MC-19**	**Classical albedo feature name**
Eos Mensa	**346.67**	**−11.01**	**317.84**	**Mensa, mensae**	**MC-19**	**Classical albedo feature name**
Erebus Montes	**811.67**	**35.66**	**185.02**	**Mons, montes**	**MC-2**	**From albedo feature at 26 N, 182 W**
Eridania Mons	**143.29**	**−57.02**	**137.86**	**Mons, montes**	**MC-29**	**Classical albedo feature name**
Eridania Planitia	**1062.13**	**−38.15**	**122.21**	**Planitia, planitiae**	**MC-28, MC-29**	**Classical albedo feature name**
Eridania Scopulus	**1017.66**	**−52.61**	**141.79**	**Scopulus, scopuli**	**MC-29**	**Classical albedo feature name**
Erythraea Fossa	**155.19**	**−27.27**	**329.06**	**Fossa, fossae**	**MC-19**	**Classical albedo feature name**
Erythraeum Chaos	**147.63**	**−21.84**	**347.62**	**Chaos, chaoses**	**MC-19**	**Classical albedo feature name**
Escalante	**75.26**	**0.19**	**115.39**	**Crater, craters**	**MC-14**	**F.; Mexican astronomer (c. 1930)**
Escorial	**22.24**	**76.89**	**304.96**	**Crater, craters**	**MC-1**	**Town in Spain**
Esira	16.26	8.96	313.39	Crater, craters	MC-10	Town in Madagascar
Esk	3.67	45.21	352.98	Crater, craters	MC-4	Town in Australia
Espino	12.05	−19.69	110.34	Crater, craters	MC-22	Town in Venezuela
Eudoxus	**98.51**	**−44.52**	**212.78**	**Crater, craters**	**MC-24**	**Greek astronomer (c. 408–355 BC)**
Eumenides Dorsum	**569.26**	**4.79**	**203.6**	**Dorsum, dorsa**	**MC-8**	**Classical albedo feature name**
Euphrates Patera	20.27	38.43	10.26	Patera, paterae	MC-5	Classical albedo feature name
Euripus Mons	88.91	−44.82	105.18	Mons, montes	MC28	Albedo feature name
Evpatoriya	1.04	46.95	134.36	Crater, craters	MC-7	Soviet tracking site (Soviet spelling used)
Evros Vallis	358.01	−12.65	13.83	Vallis, valles	MC-20	River in Greece
Faith	5.3	42.92	348.17	Crater, craters	MC-4	Town in South Dakota, USA
Falun	10.01	−23.96	335.33	Crater, craters	MC-19	Town in Sweden
Fancy	**49.44**	**−35.45**	**113.6**	**Crater, craters**	**MC-28**	**Town in Saint Vincent and the Grenadines**
Faqu	12.25	−24.54	106.34	Crater, craters	MC-22	Town in Jordan
Farah Vallis	76.37	−6.03	136.8	Vallis, valles	MC-23	River in Afghanistan
Farim	3.92	−44.31	139.28	Crater, craters	MC-29	Town in the Republic of Guinea-Bissau
Fastov	11.12	−25.04	339.63	Crater, craters	MC-19	Town in Ukraine
Felis Dorsa	**244**	**−21.87**	**294.1**	**Dorsum, dorsa**	**MC-18**	**Classical albedo feature name**
Fenagh	6.22	34.29	144.37	Crater, craters	MC-7	Town in Ireland

Feature name	Diameter (km)	Center latitude	Center longitude	Feature type	Quadrangle	Origin
Fesenkov	87.38	21.66	273.47	Crater, craters	MC-10	Vasilii G.; Russian astrophysicist (1889–1972)
Firsoff	90	2.73	350.63	Crater, craters	MC-11	Valdemar Axel; English astronomer (1910–1981)
Fitzroy	38.17	-35.69	112.06	Crater, craters	MC-28	Town in the Falkland Islands
Flammarion	173.7	25.22	48.28	Crater, craters	MC-13	Camille; French astronomer (1842–1925)
Flat	3.04	-25.42	340.45	Crater, craters	MC-19	Town in Alaska, USA
Flateyri	9.5	-35.86	330.92	Crater, craters	MC-26	Village in Iceland
Flaugergues	236.06	-16.8	19.22	Crater, craters	MC-20	Honoré; French astronomer (1755–1835)
Floq	2.55	14.94	107.16	Crater, craters	MC-14	Town in Albania
Flora	18.31	-44.67	308.55	Crater, craters	MC-26	Town in Mississippi, USA
Focas	72.02	33.56	12.75	Crater, craters	MC-5	Jean H.; Greco-French astronomer (1909–1969)
Fontana	80.06	-62.91	287.88	Crater, craters	MC-25	Francesco; Italian astronomer (1585–1646)
Foros	24.54	-33.4	332.13	Crater, craters	MC-26	Town in Ukraine
Fortuna Fossae	324.28	4.64	267.31	Fossa, fossae	MC-9	Classical albedo feature name
Fournier	114.28	-4.3	72.64	Crater, craters	MC-21	Georges; French astronomer (1881–1954)
Freedom	12.74	43.36	351.02	Crater, craters	MC-4	Town in Oklahoma, USA
Frento Vallis	251.09	-50.03	345.16	Vallis, valles	MC-26	Classical name for river in Italy
Funchal	1.62	22.98	310.56	Crater, craters	MC-10	Port of Madeira Islands
Gaan	3.01	38.66	356.6	Crater, craters	MC-4	Town in Somalia
Gagra	13.41	-20.64	337.9	Crater, craters	MC-19	Town in the Republic of Georgia
Gah	2.78	-44.69	327.36	Crater, craters	MC-26	Town in Indonesia
Galap	5.99	-37.66	192.93	Crater, craters	MC-24	Town in the Republic of Palau
Galaxias Chaos	234.48	33.83	146.52	Chaos, chaoses	MC-7	Albedo feature name
Galaxias Colles	610.34	36.8	147.48	Collis, colles	MC-7	Albedo feature name
Galaxias Fluctus	607.03	30.96	143.03	Fluctus, fluctis	MC-7, MC-15	Albedo feature name
Galaxias Fossae	552	36.63	142	Fossa, fossae	MC-7	Albedo feature name
Galaxius Mons	22.23	34.76	142.31	Mons, montes	MC-7	Classical albedo name
Galdakao	33.44	-13.34	176.63	Crater, craters	MC-23	Euskadi (Basque) town in Spain
Gale	154.08	-5.37	137.81	Crater, craters	MC-23	Walter F.; Australian astronomer (1865–1945)
Gali	25.86	-43.75	322.81	Crater, craters	MC-26	Town in the Republic of Georgia
Galilaei	137.17	5.72	333.09	Crater, craters	MC-11	Galileo;Italianastronomer andphysicist(1564–1642)
Galle	223.53	-50.63	329	Crater, craters	MC-26	Johann G.; German astronomer (1812–1910)
Galu	13.57	-22.08	338.33	Crater, craters	MC-19	Town in Democratic Republic of Congo
Gamboa	30.82	40.77	315.64	Crater, craters	MC-4	Town in Panama
Gan	20.63	61.72	229	Crater, craters	MC-2	Town in the Republic of the Maldives
Gander	36.08	-31.26	94.22	Crater, craters	MC-28	Town in Newfoundland, Canada
Gandu	9.49	-45.38	312.72	Crater, craters	MC-26	Town in Brazil
Gandzani	51.91	34.24	269.17	Crater, craters	MC-3	Town in the Republic of Georgia

Feature name	Diameter (km)	Center latitude	Center longitude	Feature type	Quadrangle	Origin
Ganges Catena	81.27	-2.7	291.22	Catena, catenae	MC-18	Classical albedo feature name
Ganges Cavus	43.11	-10.09	308.55	Cavus, cavi	MC-18	Classical albedo feature name
Ganges Chaos	113.73	-9.76	313.96	Chaos, chaoses	MC-18	Classical albedo feature name
Ganges Chasma	574.08	-7.96	312.11	Chasma, chasmata	MC-18, MC19	Classical albedo feature name
Ganges Mensa	135.9	-7.23	311.25	Mensa, mensae	MC-18	Classical albedo feature name
Gardo	15.53	-26.67	335.23	Crater, craters	MC-19	Town in Somalia
Gari	9.43	-35.88	288.77	Crater, craters	MC-25	Town in Russia
Garm	4.8	48.25	350.94	Crater, craters	MC-4	Town in Tadzhikistan
Garni	2.57	-11.52	290.31	Crater, craters	MC-18	Village in Armenia
Gasa	7.03	-35.72	129.4	Crater, craters	MC-29	Town in Bhutan
Gastre	7.1	24.61	112.53	Crater, craters	MC-14	Town in Argentina
Gediz Vallis	8.3	-4.85	137.44	Vallis, valles	MC-23	River in Turkey
Gemina Lingula	772.89	81.87	2.59	Lingula, lingulae	MC-1	Classical albedo feature name
Gemini Scopuli	1000.36	80.39	26.1	Scopulus, scopuli	MC-1	Classical albedo feature name
Geryon Montes	377.83	-7.72	278.38	Mons, montes	MC-18	Classical albedo feature
Gigas Fossae	190	3.55	230.44	Fossa, fossae	MC-9	Albedo feature name
Gigas Sulci	418.56	10.02	232.27	Sulcus, sulci	MC-9	Classical albedo feature name
Gilbert	121.34	-68	86.08	Crater, craters	MC-30	Grove K.; American geologist (1843–1918)
Gill	83.17	15.76	5.55	Crater, craters	MC-12	David; British astronomer (1843–1914)
Glazov	22.02	-20.62	333.41	Crater, craters	MC-19	Town in Russia
Gledhill	78.47	-53.17	87.1	Crater, craters	MC-28	Joseph; British astronomer (1836–1906)
Glendore	8.23	18.34	308.33	Crater, craters	MC-10	Town in Ireland
Glide	9.85	-8.13	316.82	Crater, craters	MC-19	Town in Oregon, USA
Globe	50.74	-23.68	332.65	Crater, craters	MC-19	Town in Arizona, USA
Goba	10.9	-23.22	338.99	Crater, craters	MC-19	Town in Ethiopia
Goff	7.95	23.26	104.86	Crater, craters	MC-14	Town in Somalia
Gol	9.53	47.15	349.36	Crater, craters	MC-4	Town in Norway
Gold	8.91	20.03	328.76	Crater, craters	MC-11	Town in Pennsylvania, USA
Golden	19.59	-22.01	326.51	Crater, craters	MC-19	Town in Illinois, USA
Goldstone	1.03	47.77	134.58	Crater, craters	MC-7	American tracking site
Gonnus Mons	49.38	41.21	269.12	Mons, montes	MC-3	Classical town
Gordii Dorsum	481.56	4.11	215.86	Dorsum, dorsa	MC-8	Classical albedo feature name
Gordii Fossae	369	14.83	232.4	Fossa, fossae	MC-9	From classical albedo feature name
Gorgonum Chaos	150.71	-37.26	189.1	Chaos, chaoses	MC-24	From albedo feature at 24° S, 154° W
Gori	6.43	-22.95	331.17	Crater, craters	MC-19	Town in the Republic of Georgia
Grjótá Valles	343.77	15.38	166.38	Vallis, valles	MC-15	River in Iceland
Grójec	37.31	-21.47	329.16	Crater, craters	MC-19	Town in Poland
Graff	154.49	-21.18	153.81	Crater, craters	MC-23	Kasimir; German astronomer (1878–1950)
Granicus Valles	777.78	30.58	129.97	Vallis, valles	MC-7, MC-14, MC-15	Ancient name for river in Turkey
Gratteri	7.56	-7.71	199.94	Crater, craters	MC-16	Town on the island of Sicily, Italy

Feature name	Diameter (km)	Center latitude	Center longitude	Feature type	Quadrangle	Origin
Greeley	457.45	-36.79	3.92	Crater, craters	MC-27	Ronald; American geologist (1939–2011)
Green	182.07	-52.3	351.46	Crater, craters	MC-26	Nathan E.; British astronomer (1823–1899)
Greg	68.12	-38.59	112.89	Crater, craters	MC-28	Percy; English writer (1836–1889)
Grindavik	11.71	25.4	321.01	Crater, craters	MC-11	Town in Iceland
Gringauz	71.02	-20.67	324.3	Crater, craters	MC-19	Konstantin Iosifovich; Russian cosmophysicist (1918–1993)
Groves	10.27	-4.06	315.45	Crater, craters	MC-19	Town in Texas, USA
Guaymas	20.12	25.66	314.97	Crater, craters	MC-10	Town in Mexico
Guir	18.2	-21.54	339.5	Crater, craters	MC-19	Town in Mali
Gulch	8.32	15.85	109.02	Crater, craters	MC-14	Town in Ethiopia
Gunjur	26.85	-0.17	146.66	Crater, craters	MC-23	Town in the Republic of Gambia
Gunnison	39.57	-43.67	102.92	Crater, craters	MC-28	Town in Colorado, USA
Gusev	158.12	-14.53	175.52	Crater, craters	MC-23	Matvei M.; Russian astronomer (1826–1866)
Gwash	4.75	38.96	356.84	Crater, craters	MC-4	Town in Pakistan
Hadley	115.46	-19.26	156.97	Crater, craters	MC-23	George; British meteorologist (1685–1768)
Hadriaca Patera	66.04	-30.2	92.79	Patera, paterae	MC-28	Classical albedo feature name
Hadriacus Cavi	59.09	-27.25	78.05	Cavus, cavi	MC-21	Classical albedo feature name
Hadriacus Mons	450	-31.29	91.86	Mons, montes	MC-22, MC-28	Classical albedo feature name
Hadriacus Palus	176.33	-27.25	77.3	Palus, paludes	MC-21	Classical albedo feature name
Halba	31.41	-26.01	303.86	Crater, craters	MC-18	Town in Lebanon
Haldane	76.75	-52.75	129.26	Crater, craters	MC-29	John B.; British physiologist-geneticist (1892–1964)
Hale	137.31	-35.69	323.64	Crater, craters	MC-26	George E.; American astronomer (1868–1938)
Halex Fossae	147.25	27.35	233.96	Fossa, fossae	MC-9	From albedo feature at 40° N, 110° W
Halley	83.72	-48.34	300.73	Crater, craters	MC-26	Edmund; British astronomer (1656–1742)
Ham	1.59	-44.67	327.5	Crater, craters	MC-26	Town in France
Hamoguir	0.82	48.68	132.51	Crater, craters	MC-7	Algerian launch site
Hamelin	9.74	20.25	327.25	Crater, craters	MC-11	Old German town referred to in the Pied Piper fairy tale
Handlová	4.39	37.69	271.41	Crater, craters	MC-3	Town in Slovakia
Harad	8.06	-27.46	331.99	Crater, craters	MC-19	Town in Saudi Arabia
Hargraves	60.28	20.74	75.74	Crater, craters	MC-13	Robert B.; American geoscientist (1928–2003)
Harmakhis Vallis	526.66	-40.98	90.06	Vallis, valles	MC-28	Ancient Egyptian word for "Mars"
Harris	81.56	-21.9	66.81	Crater, craters	MC-21	Daniel Lester III; American astronomer (1919–1962)
Hartwig	99.33	-38.66	344.14	Crater, craters	MC-26	Ernst; German astronomer (1851–1923)
Hashir	16.15	3.19	85.01	Crater, craters	MC-13	Town in Turkey
Havel Vallis	240.19	0.77	302.54	Vallis, valles	MC-10	River in Germany

Feature name	Diameter (km)	Center latitude	Center longitude	Feature type	Quadrangle	Origin
Heaviside	83.28	-70.5	264.78	Crater, craters	MC-30	Oliver; British physicist (1850–1925)
Hebes Chasma	316.74	-1.07	283.94	Chasma, chasmata	MC-18	Classical albedo feature name
Hebes Mensa	112.46	-1.02	283.22	Mensa, mensae	MC-18	Classical albedo feature; name of goddess of youth
Hebrus Valles	325	19.88	126.74	Vallis, valles	MC-14	Ancient river in Greece
Hecates Tholus	181.57	32.12	150.24	Tholus, tholi	MC-7	Classical albedo feature name
Hegemone Dorsum	143.63	-54.72	315.1	Dorsum, dorsa	MC-26	One of the Graces
Heimdal	10.49	68.33	235.44	Crater, craters	MC-1	Town in Norway
Heinlein	85.34	-64.48	116.31	Crater, craters	MC-28	Robert A.; American author (1907–1988)
Hellas Chaos	590.62	-47.12	64.41	Chaos, chaoses	MC-27, MC-28	Named for albedo feature Hellas
Hellas Chasma	148	-34.64	65.47	Chasma, chasmata	MC-28	Classical albedo feature name
Hellas Montes	159.65	-37.63	97.61	Mons, montes	MC-28	Albedo feature Hellas
Hellas Planitia	2299.16	-42.43	70.5	Planitia, planitiae	MC-27, MC-28, Figure 4.1	Classical albedo feature name
Hellespontus Montes	711.46	-44.37	42.76	Mons, montes	MC-27	Classical albedo feature name
Helmholtz	111.26	-45.4	338.73	Crater, craters	MC-26	Hermann von; German physicist (1821–1894)
Henbury	25.36	-63.49	212.27	Crater, craters	MC-24	Town in Australia
Henry	167.57	10.79	23.45	Crater, craters	MC-12	Paul; French astronomer (1848–1905). Prosper; French astronomer (1849–1903)
Henry Moore	65.47	-59.72	53.9	Crater, craters	MC-27	Henry J. Moore; American astrogeologist (1928–1998)
Hephaestus Fossae	633.32	20.84	122.85	Fossa, fossae	MC-14	Classical albedo feature name
Hephaestus Rupes	1707.44	23.54	114.9	Rupes, rupes	MC-6, MC-14	Named for classical albedo feature at 20° N, 240° W
Her Desher Vallis	117.29	-25.08	312.07	Vallis, valles	MC-18	Egyptian name for Mars
Herculaneum	34.71	19.31	301.35	Crater, craters	MC-10	Town in Italy
Hermus Vallis	53.32	-5.32	212.19	Vallis, valles	MC-16	Classical name for river in ancient Lydia (modern Turkey)
Herschel	297.92	-14.48	129.89	Crater, craters	MC-22	John F.; British astronomer (1792–1871); William H.; British astronomer (1738–1822)
Hesperia Dorsa	818.25	-22.8	113.16	Dorsum, dorsa	MC-22	Albedo feature name
Hesperia Planum	1601.73	-21.42	109.89	Planum, plana	MC-22	Classical albedo feature name
Hibes Montes	140	3.79	171.34	Mons, montes	MC-15	From albedo feature at 17° N, 186° W
Hiddekel Cavus	23.3	29.43	16.24	Cavus, cavi	MC-12	Classical albedo feature name, Hiddekel
Himera Vallis	175	-21.54	337.34	Vallis, valles	MC-19	Ancient name for Italian river

Feature name	Diameter (km)	Center latitude	Center longitude	Feature type	Quadrangle	Origin
Hipparchus	94.81	-44.45	208.8	Crater, craters	MC-24	Greek astronomer (c. 160–125 BC)
Hit	7.09	47.06	138.35	Crater, craters	MC-7	Town in Iraq
Holden	152.66	-26.04	325.98	Crater, craters	MC-19	Edward S.; American astronomer (1846–1914)
Holmes	114.06	-74.86	66.55	Crater, craters	MC-30	Arthur; British geologist (1890–1965)
Honda	9.26	-22.4	343.6	Crater, craters	MC-19	Town in Colombia
Hooke	137.65	-44.92	315.6	Crater, craters	MC-26	Robert; British physicist-astronomer (1635–1703)
Hope	7.26	44.84	349.7	Crater, craters	MC-4	Town in British Columbia, Canada
Horarum Mons	20.5	-51.05	323.44	Mons, montes	MC26	Albedo name
Horowitz	64.9	-32.06	140.75	Crater, craters	MC-29	Norman H.; American biologist and geneticist (1915–2005)
Houston	1.98	48.23	135.95	Crater, craters	MC7	American mission control site
Hrad Vallis	974.4	38.17	135.91	Vallis, valles	MC-7	Word for "Mars" in Armenian
Hsüanch'eng	1.99	46.72	132.69	Crater, craters	MC7	Chinese launch site
Huallaga Vallis	92.5	-26.67	79.07	Vallis, valles	MC-21	River in Peru
Huancayo	24.34	-3.64	320.23	Crater, craters	MC-19	Town in Peru
Huggins	82.64	-49.04	155.84	Crater, craters	MC-29	William; British astronomer (1824–1910)
Hunten	82.44	-39.18	23.69	Crater, craters	MC-27	Donald M.; American atmospheric physicist (1925–2010)
Huo Hsing Vallis	332.3	30.19	66.61	Vallis, valles	MC-6, MC-13	Word for "Mars" in Chinese
Hussey	99.71	-53.32	233.41	Crater, craters	MC-24	William J.; American astronomer (1862–1926)
Hutton	91.74	-71.63	104.6	Crater, craters	MC-30	James; British geologist (1726–1797)
Huxley	106.52	-62.67	100.77	Crater, craters	MC-28	Thomas H.; British biologist (1825–1895)
Huygens	467.25	-13.88	55.58	Crater, craters	MC-21	Christiaan; Dutch physicist-astronomer (1629–1695)
Hyblaeus Catena	10.49	21.6	140.62	Catena, catenae	MC15	Albedo feature name
Hyblaeus Chasma	56.61	21.98	141.26	Chasma, chasmata	MC15	Albedo feature name
Hyblaeus Dorsa	887.53	13.16	130.32	Dorsum, dorsa	MC-14	Named for albedo feature
Hyblaeus Fossae	375	21.44	137.06	Fossa, fossae	MC-14, MC-15	Albedo feature name
Hydaspis Chaos	336.04	3.09	333.07	Chaos, chaoses	MC-11	Classical albedo feature name
Hydrae Cavus	64.5	-7.93	298.69	Cavus, cavi	MC-18	Classical albedo feature name
Hydrae Chaos	66	-5.9	300.03	Chaos, chaoses	MC-18	Classical albedo feature name
Hydrae Chasma	55.18	-6.75	297.99	Chasma, chasmata	MC-18	Classical albedo feature name
Hydraotes Chaos	419.04	1.12	324.71	Chaos, chaoses	MC-11, MC-19	Classical albedo feature name
Hydraotes Colles	47.5	-0.02	326.32	Collis, colles	MC-19	Classical albedo feature name
Hypanis Valles	220	9.46	313.58	Vallis, valles	MC-10	Classical name for river in Scythia; present Kuban River in Russia

Feature name	Diameter (km)	Center latitude	Center longitude	Feature type	Quadrangle	Origin
Hyperborea Lingula	124.8	80.32	306.46	Lingula, lingulae	MC-1	Classical albedo feature name
Hyperboreae Undae	463.65	79.96	310.51	Unda, undae	MC-1	Classical albedo feature name
Hyperborei Cavi	92.81	79.91	310.3	Cavus, cavi	MC-1	Classical albedo feature name
Hyperboreus Labyrinthus	111.97	80.28	300.25	Labyrinthus, labyrinthi	MC-1	Classical albedo feature name
Hypsas Vallis	36.47	33.63	57.99	Vallis, valles	MC-5	Classical name for river in Sicily
Iamuna Chaos	21.72	-0.28	319.39	Chaos, chaoses	MC-19	Classical albedo feature name
Iamuna Dorsa	38.18	20.97	309.6	Dorsum, dorsa	MC-10	From classical albedo feature Iamunae Sinus
Iani Chaos	450.51	-2.19	342.96	Chaos, chaoses	MC-19	Classical albedo feature name
Iazu	6.83	-2.71	354.82	Crater, craters	MC-19	Town in Romania
Iberus Vallis	87.26	21.25	152.07	Vallis, valles	MC-15	Classical name for present Ebro River in NE Spain
Ibragimov	86.77	-25.43	300.43	Crater, craters	MC-18	Nadir Baba Ogly; Soviet astronomer (1932–1977)
Icaria Fossae	2115.45	-48.09	234.84	Fossa, fossae	MC-24	From albedo feature at 44° S, 130° W
Icaria Planum	566.59	-43.27	253.96	Planum, plana	MC-25	Classical albedo feature name
Idaeus Fossae	202.01	37.33	308.8	Fossa, fossae	MC-4	Classical albedo feature name
Igal	8.83	-20.09	110.9	Crater, craters	MC-22	Town in Hungary; name changed from Igol (incorrect spelling)
Ikej	4.51	20.96	112.5	Crater, craters	MC-14	Town in Russia
Imgr	3.42	19.12	111.18	Crater, craters	MC-14	Town in Russia
Indus Vallis	342	18.95	38.88	Vallis, valles	MC-12	Ancient and modern name for river in Pakistan
Innsbruck	59	-6.39	320.04	Crater, craters	MC-19	Town in Austria
Ins	2.78	24.49	108.9	Crater, craters	MC-14	Town in Switzerland
Inta	16.12	-24.36	334.9	Crater, craters	MC-19	Town in Russia
Inuvik	20.52	78.59	331.68	Crater, craters	MC-1	Town in Northwest Territories, Canada
Irbit	12.73	-24.34	335.09	Crater, craters	MC-19	Town in Russia
Irharen	6.48	34.49	140.82	Crater, craters	MC-7	Town in Algeria
Isara Valles	5.36	-5.31	213.58	Vallis, valles	MC16	Classical name for modern Oise River in France
Isidis Dorsa	1074.7	12.92	88.21	Dorsum, dorsa	MC-13, MC-14	Named for classical albedo feature at 25° N, 270° W
Isidis Planitia	1224.58	13.94	88.38	Planitia, planitiae	MC-13, MC-14	Classical albedo feature name
Isil	77.12	-27.02	87.93	Crater, craters	MC-21	Town in Spain
Ismenia Patera	82	38.55	1.8	Patera, paterae	MC-5	Classical albedo feature name
Ismeniae Fossae	286.91	41.31	38.35	Fossa, fossae	MC-5	From albedo feature at 40° N, 333° W
Ismenius Cavus	90.61	33.9	17.08	Cavus, cavi	MC-5	Classical albedo feature name
Issedon Paterae	5.31	38.13	269.75	Patera, paterae	MC3	Classical albedo name
Issedon Tholus	54.53	36.05	265.17	Tholus, tholi	MC3	Classical albedo name
Ister Chaos	109.1	12.95	303.44	Chaos, chaoses	MC-10	From classical albedo feature at 10° N, 56° W
Istok	4.82	-45.1	274.18	Crater, craters	MC-25	Town in Kosovo

Feature name	Diameter (km)	Center latitude	Center longitude	Feature type	Quadrangle	Origin
Iluxi Vallis	123.07	25.45	153.32	Vallis, valles	MC-15	River in Brazil
Ius Chasma	**839.91**	**-7.29**	**275.61**	**Chasma, chasmata**	**MC-18**	**Classical albedo feature name**
Ius Labes	61.18	-7.47	281.54	Labes, labes	MC-18	Classical albedo feature name
Ius Mensa	160	-8.6	284.68	Mensa, mensae	MC-18	Classical Martian albedo feature, Ius
Izendy	**22.26**	**-28.88**	**258.56**	**Crater, craters**	**MC-17**	**Town in Russia**
Jal	4.81	-26.25	331.24	Crater, craters	MC-19	Town in New Mexico, USA
Jama	2.9	21.39	306.82	Crater, craters	MC-10	Town in Tunisia
Jampur	**27.9**	**38.71**	**278.45**	**Crater, craters**	**MC-3**	**Town in Pakistan**
Janssen	**153.63**	**2.69**	**37.61**	**Crater, craters**	**MC-12**	**Pierre Jules César; French astronomer (1824–1907)**
Jarry-Desloges	**93.36**	**-9.37**	**83.85**	**Crater, craters**	**MC-21**	**Rene; French astronomer (1868–1951)**
Jeans	**73.6**	**-69.64**	**154.18**	**Crater, craters**	**MC-30**	**James H.; British physicist, astronomer (1877–1946)**
Jen	8.88	39.88	349.43	Crater, craters	MC-4	Town in Nigeria
Jezza	9.22	-48.42	322.08	Crater, craters	MC-26	Town in Russia
Jezero	**47.52**	**18.41**	**77.69**	**Crater, craters**	**MC-13**	**Town in Bosnia-Herzegovina**
Jijiga	16.16	25.11	306.05	Crater, craters	MC-10	Town in Ethiopia
Jodrell	3.02	47.47	132.3	Crater, craters	MC-7	United Kingdom tracking site.
Johannesburg	1.22	47.92	133.19	Crater, craters	MC-7	Republic of South Africa tracking site
Johnstown	3.36	-9.8	308.93	Crater, craters	MC-18	Town in Pennsylvania, USA
Jojutla	19.32	81.59	190.2	Crater, craters	MC-1	Town in Mexico
Joly	**76.99**	**-74.5**	**317.31**	**Crater, craters**	**MC-30**	**John; Irish geologist (1857–1933)**
Jones	**90.11**	**-18.88**	**340.17**	**Crater, craters**	**MC-19**	**Harold S.; British astronomer (1890–1960)**
Jörn	**20.47**	**-27.19**	**76.43**	**Crater, craters**	**MC-21**	**Town in Sweden**
Jovis Fossae	**348.63**	**19.77**	**244.17**	**Fossa, fossae**	**MC-9**	**From albedo feature at 16° N, 111° W**
Jovis Tholus	**58.07**	**18.2**	**242.59**	**Tholus, tholi**	**MC-9**	**Classical albedo feature name**
Jumla	**49.23**	**-21.29**	**86.44**	**Crater, craters**	**MC-21**	**Town in Nepal**
Juventae Cavi	94.4	-3.91	301.84	Cavus, cavi	MC-18	Classical albedo feature name
Juventae Chasma	**304.99**	**-3.37**	**298.61**	**Chasma, chasmata**	**MC-18**	**Classical albedo feature name**
Juventae Dorsa	**481.41**	**0.39**	**288.98**	**Dorsum, dorsa**	**MC-10, MC18**	**From albedo feature at 4° S, 63° W**
Juventae Mensa	116	-7.93	294.37	Mensa, mensae	MC-18	Classical albedo feature name, Juventae Fons
Kachug	4.86	18.15	107.59	Crater, craters	MC-14	Town in Russia
Kagoshima	1.32	47.32	135.73	Crater, craters	MC-7	Japanese launch site
Kagul	9.13	-23.73	340.97	Crater, craters	MC-19	Town in Moldova
Käid	7.67	-4.46	315.3	Crater, craters	MC-19	Town in Iraq
Kaiser	**201.67**	**-46.19**	**19.11**	**Crater, craters**	**MC-27**	**Frederick; Dutch astronomer (1808–1872)**
Kaj	1.83	-27.05	330.61	Crater, craters	MC-19	Town in Russia
Kakori	**28.09**	**-41.49**	**330.15**	**Crater, craters**	**MC-26**	**Town in India**
Kalba	14.15	-5.89	154.83	Crater, craters	MC-23	Town in the United Arab Emirates

Feature name	Diameter (km)	Center latitude	Center longitude	Feature type	Quadrangle	Origin
Kaliningrad	1.47	48.47	134.96	Crater, craters	MC-7	Soviet mission control site
Kalocsa	**34.15**	**6.92**	**353.05**	**Crater, craters**	**MC-11**	**Town in Hungary**
Kamativi	**58.81**	**-20.5**	**99.99**	**Crater, craters**	**MC-22**	**Town in Zimbabwe**
Kamloops	**63.96**	**-53.45**	**327.4**	**Crater, craters**	**MC-26**	**Town in Canada**
Kamnik	10.37	-37.21	198.21	Crater, craters	MC-24	Town in Slovenia
Kampot	13.23	-41.78	314.41	Crater, craters	MC-26	Town in Democratic Kampuchea (Cambodia)
Kanab	14.55	-27.2	341	Crater, craters	MC-19	Town in Utah, USA
Kandi	8.24	-32.75	122.1	Crater, craters	MC-29	Town in Benin
Kankossa	16.5	-11.83	304.46	Crater, craters	MC-18	Town in Mauritania
Kansk	**33.35**	**-20.52**	**342.73**	**Crater, craters**	**MC-19**	**Town in Russia**
Kantang	**52.44**	**-24.44**	**342.42**	**Crater, craters**	**MC-19**	**Town in Thailand**
Karpinsk	**28.84**	**-45.57**	**327.85**	**Crater, craters**	**MC-26**	**Town in Russia**
Karshi	21.52	-23.28	340.68	Crater, craters	MC-19	Town in Uzbekistan
Kartabo	19.41	-40.85	307.54	Crater, craters	MC-26	Town in Guyana
Kārūn Valles	64	-35.94	174.1	Vallis, valles	MC-29	River in Iran
Karzok	15.29	18.4	228.26	Crater, craters	MC-9	Village in Kashmir
Kasabi	**41.09**	**-27.77**	**89.06**	**Crater, craters**	**MC-21**	**Town in Zambia**
Kasei Valles	**1580**	**25.14**	**297.12**	**Vallis, valles**	**MC-10**	**Word for "Mars" in Japanese**
Kashira	**65.8**	**-27.09**	**341.69**	**Crater, craters**	**MC-19**	**Town in Russia**
Kasimov	**87.18**	**-24.63**	**337.06**	**Crater, craters**	**MC-19**	**Town in Russia**
Kasra	3.46	21.98	103.63	Crater, craters	MC-14	Town in Tunisia
Katoomba	**51.24**	**-79.01**	**127.81**	**Crater, craters**	**MC-30**	**Town in Australia**
Kaup	3.21	22.63	326.84	Crater, craters	MC-11	Town in New Guinea
Kaw	10.72	16.4	104.28	Crater, craters	MC-14	Town in French Guiana
Kayne	**33.82**	**-15.5**	**173.56**	**Crater, craters**	**MC-23**	**Town in Botswana**
Keeler	**90.19**	**-60.69**	**208.76**	**Crater, craters**	**MC-24**	**James E.; American astronomer (1857–1900)**
Kem'	3.62	-44.94	327.03	Crater, craters	MC-26	Town in Russia
Kepler	**228.24**	**-46.69**	**140.98**	**Crater, craters**	**MC-29**	**Johannes; German astronomer (1571–1630)**
Keren	28.63	20.98	337.5	Crater, craters	MC-11	Town in Eritrea
Keul'	5.81	45.99	122.23	Crater, craters	MC-7	Town in Russia
Khanpur	2.68	20.73	102	Crater, craters	MC-14	Town in Pakistan
Kholm	11.08	-7.21	318	Crater, craters	MC-19	Town in Russia
Khurli	8.78	-20.94	112.96	Crater, craters	MC-22	Town in Pakistan
Kibuye	7.14	-29.13	181.82	Crater, craters	MC-16	Town in Rwanda
Kifri	13.89	-45.64	305.69	Crater, craters	MC-26	Town in Iraq
Kimry	20.64	-20.14	343.68	Crater, craters	MC-19	Town in Russia
Kin	8.1	20.2	326.62	Crater, craters	MC-11	Town in Japan
Kinda	14.04	-25.69	254.85	Crater, craters	MC-17	Town in Democratic Republic of Congo
Kingston	1.52	22.11	312.96	Crater, craters	MC-10	Jamaican port
Kinkora	**51.09**	**-24.95**	**112.88**	**Crater, craters**	**MC-22**	**Town in Prince Edward Island, Canada**
Kipini	**67.26**	**25.86**	**328.44**	**Crater, craters**	**MC-11**	**Town in Kenya**
Kirs	3.46	-26.31	340.56	Crater, craters	MC-19	Town in Russia

Feature name	Diameter (km)	Center latitude	Center longitude	Feature type	Quadrangle	Origin
Kirsanov	15.08	-22.2	334.88	Crater, craters	MC-19	Town in Russia
Kisambo	15.22	34.07	271.08	Crater, craters	MC-3	Town in Democratic Republic of Congo
Kita	10.72	-22.78	342.82	Crater, craters	MC-19	Town in Mali
Knobel	123.31	-6.57	133.31	Crater, craters	MC-22	Edward B.; British astronomer (1841–1930)
Koga	19.17	-28.96	256.24	Crater, craters	MC-17	Town in Tanzania
Kok	6.13	15.65	331.93	Crater, craters	MC-11	Town in Malaysia (Sarawak)
Kolonga	41.09	8.32	305.06	Crater, craters	MC-10	Town in the Kingdom of Tonga
Kong	11.66	-5.36	321.43	Crater, craters	MC-19	Town in Ivory Coast
Kontum	22.26	-32.04	292.93	Crater, craters	MC-25	Town in Vietnam
Korolev	81.37	72.77	164.58	Crater, craters	MC-1	Sergey Pavlovich; Russian engineer (1906–1966)
Korph	7.33	19.34	105.45	Crater, craters	MC-14	Town in Russia
Koshoba	10.33	22.93	77	Crater, craters	MC-13	Village in Turkmenistan
Kotka	39.45	19.25	169.88	Crater, craters	MC-15	Town in Finland
Kourou	1.84	46.73	132.78	Crater, craters	MC-7	French Guianan launch site
Koval'sky	296.67	-29.56	218.46	Crater, craters	MC-16, MC-24	M. A.; Russian astronomer (1821–1884)
Koy	7.12	21.47	309.59	Crater, craters	MC-10	Town in Russia
Krasnoye	6.55	35.85	143.84	Crater, craters	MC-7	Town in Russia
Kribi	13.18	-43	316.49	Crater, craters	MC-26	Town in the United Republic of Cameroon
Krishtofovich	111.09	-48.09	97.34	Crater, craters	MC-28	Afrikan Nikolaevich; Soviet paleobotanist (1885–1953)
Krupac	10	-7.79	86.01	Crater, craters	MC-21	Town in the municipality of Pirot, Serbia
Kuba	26.59	-25.31	340.36	Crater, craters	MC-19	Town in Azerbaijan
Kufra	37.48	40.36	120.3	Crater, craters	MC-7	Town in Libya
Kuiper	81.78	-56.99	202.87	Crater, craters	MC24	Gerard P.; American astronomer (1905–1973)
Kulal	8.48	16.39	108.14	Crater, craters	MC-14	Town in Russia
Kumak	13.5	-35.47	291.93	Crater, craters	MC-25	Town in Russia
Kumara	11.87	43.03	128.56	Crater, craters	MC-7	Town in New Zealand
Kunes	15.13	-25.24	107.94	Crater, craters	MC-22	Town in Norway
Kunowsky	66.29	56.82	350.36	Crater, craters	MC-4	George K.; German astronomer (1786–1846)
Kushva	37.55	-43.96	324.49	Crater, craters	MC-26	Town in Russia
La Paz	1.39	21.05	310.97	Crater, craters	MC-10	Mexican port
Labeatis Catenae	220.6	19.49	266.83	Catena, catenae	M-9	Classical albedo feature name
Labeatis Fossae	1496.36	24.58	275.47	Fossa, fossae	MC-3, MC-9, MC-10	Previously named feature at 30° N, 75° W; expanded coordinates
Labeatis Mensa	124.67	25.5	285.53	Mensa, mensae	MC-10	Classical albedo feature name
Labeatis Mons	42.78	37.48	284.14	Mons, montes	MC-3	Named for albedo feature Labeatis Lacus
Labou Vallis	257.79	-8.63	205.58	Vallis, valles	MC-16	Origin unknown
Labria	52.64	-34.94	311.93	Crater, craters	MC-26	Town in Brazil
Lachute	15.15	-4.27	320.24	Crater, craters	MC-19	Town in Canada
Ladon Valles	244.59	-22.43	331.39	Vallis, valles	MC-19	Ancient name for Greek river

Feature name	Diameter (km)	Center latitude	Center longitude	Feature type	Quadrangle	Origin
Laf	2.86	48.01	354.1	Crater, craters	MC-4	Town in the United Republic of Cameroon
Lagarto	19.79	49.86	351.71	Crater, craters	MC-4	Town in Brazil
Lamas	22.99	-26.99	339.36	Crater, craters	MC-19	Town in Peru
Lambert	92.53	-19.97	25.39	Crater, craters	MC-20	Johann H.; German physicist (1728–1777)
Lamont	76.62	-58.17	246.46	Crater, craters	MC-25	Johann von; German astronomer (1805–1879)
Lampland	76.78	-35.54	280.48	Crater, craters	MC-25	Carl O.; American astronomer (1873–1951)
Land	5.2	48.26	351.28	Crater, craters	MC-4	Town in Alabama, USA
Langtang	9.8	-38.12	224.04	Crater, craters	MC-24	Village in Nepal
Lapri	3.01	20.32	107.49	Crater, craters	MC-14	Town in Russia
Lar	6.85	-25.83	330.9	Crater, craters	MC-19	Town in Iran
Lassell	85.6	-20.61	297.54	Crater, craters	MC-18	William; British astronomer (1799–1880)
Lasswitz	108.04	-9.31	138.31	Crater, craters	MC-23	Kurd; German author (1848–1910)
Lau	106.92	-74.3	252.52	Crater, craters	MC-30	Hans E.; Danish astronomer (1879–1918)
Layla	19.36	-61.11	107.12	Crater, craters	MC-28	Town in Saudi Arabia
Le Verrier	137.55	-37.71	17.1	Crater, craters	MC-27	Urbain J.; French astronomer (1811–1877)
Lebu	19.34	-20.29	340.53	Crater, craters	MC-19	Town in Chile
Lederberg	87.25	13.01	314.08	Crater, craters	MC-10	Joshua; American molecular biologist (1925–2008)
Leighton	65.94	3.08	57.75	Crater, craters	MC-13	Robert B.; American physicist (1919–1997)
Leleque	8.43	36.46	138.17	Crater, craters	MC-7	Town in Argentina
Lemgo	15.73	-42.5	325.21	Crater, craters	MC-26	Town in Germany
Lenya	14.96	-26.72	253.2	Crater, craters	MC-17	Town in Burma
Lethe Vallis	236.65	3.16	154.97	Vallis, valles	MC-15	River in Katmai National Monument, Alaska, USA
Leuk	3.44	23.91	304.99	Crater, craters	MC-10	Town in Switzerland
Lexington	5.17	21.81	311.37	Crater, craters	MC-10	American colonial town (Massachusetts)
Li Fan	105.58	-46.88	206.94	Crater, craters	MC-24	Chinese astronomer (c. AD 85)
Liais	122.78	-75.3	106.93	Crater, craters	MC-30	Emmanuel; French astronomer (1826–1900)
Liberta	25.1	35.23	304.55	Crater, craters	MC-4	Towns in Antigua and Barbuda
Libertad	31.19	23.06	330.59	Crater, craters	MC-11	Town in Venezuela
Libya Montes	1043.63	1.44	88.23	Mons, montes	MC-13, MC-14	Classical albedo feature name
Licus Vallis	240	-3.05	126.35	Vallis, valles	MC-22	Ancient name for modern Lech River, France
Linpu	18.16	18.14	113.21	Crater, craters	MC-14	Town in China (Chekiang)
Lins	6.17	15.76	330.2	Crater, craters	MC-11	Town in Brazil
Lipany	50.1	-0.22	79.67	Crater, craters	MC-21	Town in Slovakia
Lipik	48.95	-38.41	111.61	Crater, craters	MC-28	Town in Croatia
Liris Valles	596.24	-10.5	58.25	Vallis, valles	MC-21	Ancient name for modern Liri River, Italy
Lisboa	1.17	21.24	312.41	Crater, craters	MC-10	Portuguese port
Lismore	9.34	27.04	318.35	Crater, craters	MC-11	Town in Ireland
Littleton	7.35	15.7	107.14	Crater, craters	MC-14	Town in Maine, USA

Feature name	Diameter (km)	Center latitude	Center longitude	Feature type	Quadrangle	Origin
Liu Hsin	134.51	-53.2	188.45	Crater, craters	MC-24	Chinese astronomer (d. AD 22)
Livny	9.29	-27.16	330.88	Crater, craters	MC-19	Town in Russia
Llanesco	29.4	-28.18	258.89	Crater, craters	MC-17	Probably named for Llanes, a town in Spain
Lobo Vallis	80	26.82	298.83	Vallis, valles	MC-10	Modern river in Ivory Coast
Locana	6.64	-3.39	321.91	Crater, craters	MC-19	Town in Italy
Lockyer	71.35	27.84	160.51	Crater, craters	MC-15	Joseph N.; British astronomer (1836-1920)
Locras Valles	351.32	8.84	48.26	Vallis, valles	MC-13	Ancient name for river on Corsica
Lod	7.6	20.98	328.46	Crater, craters	MC-11	Town in Israel
Lodwar	15.01	-55.09	316.68	Crater, craters	MC-26	Town in Kenya
Lohse	151.01	-43.24	343.31	Crater, craters	MC-26	Oswald; German astronomer (1845-1915)
Loire Valles	790	-17.69	342.97	Vallis, valles	MC-19	Modern river in France
Loja	9.9	41.22	136.21	Crater, craters	MC7	Town in Ecuador
Lomela	11.16	-81.65	303.81	Crater, craters	MC-30	Town in Democratic Republic of Congo
Lomonosov	130.53	65.04	350.76	Crater, craters	MC-1, MC-4	Mikhail Vasilievich; Russian poet, scientist, and grammarian (1711-1765)
Lonar	11.07	72.99	38.29	Crater, craters	MC-1	Town in India
Longa	10.97	-20.67	334.06	Crater, craters	MC-19	Town in Angola
Loon	7.71	-18.84	113.45	Crater, craters	MC-22	Town in Ontario, Canada
López	85	-14.57	98.04	Crater, craters	MC-22	Epidio López; Mexican astronomer and author (1879-1965)
Lorica	58.49	-19.83	331.67	Crater, craters	MC-19	Town in Colombia
Los	8.05	-35.08	283.77	Crater, craters	MC-25	Town in Sweden
Lota	14.68	46.32	348.2	Crater, craters	MC-4	Town in Chile
Loto	22.14	-21.88	337.56	Crater, craters	MC-19	Town in Democratic Republic of Congo
Louros Valles	516.14	-8.41	278.23	Vallis, valles	MC-18	Modern river in Greece
Louth	36.29	70.19	103.24	Crater, craters	MC-1	Town in Ireland
Lowbury	17.18	42.41	267.08	Crater, craters	MC3	Named for a town in New Zealand, probably Lowburn, incorrectly recorded as Lowbury
Lowell	202.22	-51.96	278.5	Crater, craters	MC-25	Percival; American astronomer (1855-1916)
Luba	38.33	-18.26	323	Crater, craters	MC-19	Town in the Republic of Equatorial Guinea
Lucaya	34.21	-11.55	51.91	Crater, craters	MC-21	Town in the Commonwealth of the Bahamas
Luck	7.75	17.26	323.09	Crater, craters	MC-11	Town in Wisconsin, USA
Lucus Planum	899.87	-4.99	182.83	Planum, plana	MC-16, MC-23	Albedo feature name
Luga	44.56	-44.25	312.58	Crater, craters	MC-26	Town in Russia
Luki	20.8	-29.53	322.63	Crater, craters	MC-19	Town in Ukraine
Lunae Mensa	114.75	23.91	297.5	Mensa, mensae	MC-10	Albedo feature name

Feature name	Diameter (km)	Center latitude	Center longitude	Feature type	Quadrangle	Origin
Lunae Planum	1817.66	10.79	294.49	Planum, plana	MC-10, MC-18, Figure 4.1	Classical albedo feature name
Luqa	17.14	-18.23	131.82	Crater, craters	MC-22	Town in Malta
Lutsk	4.85	38.7	356.91	Crater, craters	MC-4	Town in Ukraine
Luzin	101.04	27.06	31.28	Crater, craters	MC-12	N. N.; Russian mathematician (1883-1950)
Lycus Sulci	1350.61	28.14	215.53	Sulcus, sulci	MC-2, MC-8, MC-9	Classical albedo feature name
Lydda	33.83	24.42	328.05	Crater, craters	MC-11	Town in Israel
Lyell	121.83	-69.91	344.53	Crater, craters	MC-30	Charles; British geologist (1797-1875)
Lyot	221.53	50.47	29.34	Crater, craters	MC-5	Bernard; French astronomer (1897-1952)
Ma'adim Vallis	913.11	-21.98	177.5	Vallis, valles	MC-23	Word for "Mars" in Hebrew
Mädler	124.16	-10.65	2.77	Crater, craters	MC-20	Johann H. von; German astronomer (1794-1874)
Mad Vallis	537.37	-56.27	76.47	Vallis, valles	MC-28	Modern river, Vermont, USA
Madrid	3.8	48.45	135.44	Crater, craters	MC-7	Spanish tracking site
Mafra	13.35	-44.02	306.85	Crater, craters	MC-26	Town in Brazil
Magadi	50.79	-34.52	313.93	Crater, craters	MC-26	Town in Kenya
Mago	2.74	15.92	105.36	Crater, craters	MC-14	Town in Russia
Magong	46.56	11.89	313.31	Crater, craters	MC-10	Town in Taiwan (also spelled Makung)
Magelhaens	103.8	-32.36	185.42	Crater, craters	MC-24	Fernao de; Portuguese navigator (1480-1521)
Maidstone	9.39	-41.56	305.78	Crater, craters	MC-26	Town in England
Main	110.99	-76.54	49.01	Crater, craters	MC-30	Robert; British astronomer (1808-1878)
Maggini	139.06	27.78	9.5	Crater, craters	MC-12	Mentore; Italian astronomer (1890-1941)
Maja Valles	1515	10.23	301.62	Vallis, valles	MC-10	Nepali word for "Mars"
Majuro	43.43	-33.26	84.33	Crater, craters	MC-28	Capital of the Republic of the Marshall Islands
Makhambet	15.85	28.43	319.53	Crater, craters	MC-11	Town in Kazakhstan
Malea Patera	241.61	-63.54	51.59	Patera, paterae	MC-27	Classical albedo feature name
Malea Planum	872.47	-65.82	62.94	Planum, plana	MC-27, MC-28, MC-30	From albedo feature at 60 S, 290 W
Mambali	31	-23.51	27.03	Crater, craters	MC-20	Town in Tanzania
Mamers Valles	1020	40.65	17.94	Vallis, valles	MC-5	Word for "Mars" in Oscan
Manah	9.9	-4.66	326.39	Crater, craters	MC-19	Town in Oman
Mandora	55.94	12.22	306.37	Crater, craters	MC-10	Town in Australia
Mangala Fossa	695	-17.27	214.12	Fossa, fossae	MC-16	Named for nearby valles
Mangala Valles	900	-11.32	208.61	Vallis, valles	MC-16	Word for "Mars" in Sanskrit
Manti	15.64	-3.58	322.43	Crater, craters	MC-19	Town in Utah, USA
Manzi	7.52	-22.15	332.53	Crater, craters	MC-19	Town in Burma
Maraldi	118.24	-61.92	328.04	Crater, craters	MC-26	Giacomo F.; French astronomer (1665-1729)

Feature name	Diameter (km)	Center latitude	Center longitude	Feature type	Quadrangle	Origin
Marbach	24.74	17.65	111.03	Crater, craters	MC-14	Town in Switzerland
Marca	78.35	-9.98	201.85	Crater, craters	MC-16	Village in Peru
Mareotis Fossae	1907.94	44.34	283.88	Fossa, fossae	MC-3, MC-4	From albedo feature at 32 N, 9° W
Margaritifer Chaos	383.67	-9.3	338.3	Chaos, chaoses	MC-19	Classical albedo feature name
Margaritifer Terra	2733.22	-1.85	335.08	Terra, terrae	MC-11, MC-19, Figure 4.1	Classical albedo feature name
Mari	37.05	-52.01	314.12	Crater, craters	MC-26	Ruined city in Syria
Maricourt	9.9	53.34	288.83	Crater, craters	MC-3	Town in Canada
Marikh Vallis	1147.22	-19.16	4.32	Vallis, valles	MC-20	Malaysian word for "Mars"
Mariner	156.58	-34.68	195.76	Crater, craters	MC-24	Named for Mariner IV spacecraft
Marte Vallis	231.43	14.08	182.9	Vallis, valles	MC-8, MC-15	Spanish word for "Mars"
Marth	96.69	12.94	356.55	Crater, craters	MC-11	Albert; German astronomer (1828–1897)
Martin	61.11	-21.34	290.75	Crater, craters	MC-18	James S., Jr.; American engineer (1920–2002)
Martynov	61.13	-30.36	323.59	Crater, craters	MC-26	Dmitry Yakovlevich; Russian astronomer (1906–1989)
Martz	92.74	-34.91	144.18	Crater, craters	MC-29	Edwin P.; American physicist (1916–1967)
Masursky	115.34	12.07	327.7	Crater, craters	MC-11	Harold; American astrogeologist (1922–1990)
Matara	48.84	-49.61	34.59	Crater, craters	MC-27	Town in Sri Lanka
Matrona Vallis	61.28	-7.66	176.19	Vallis, valles	MC-23	Classical name for present Marne River, France
Maumee Valles	390	19.51	307.15	Vallis, valles	MC-10	North American river (Indiana, Ohio)
Maunder	90.84	-49.6	1.75	Crater, craters	MC-27	Edward W.; British astronomer (1851–1928)
Mawrth Vallis	634.63	22.43	343.03	Vallis, valles	MC-11	Welsh word for "Mars"
Mazamba	52.29	-27.53	290.33	Crater, craters	MC-18	Town in Mozambique
McLaughlin	90.92	21.9	337.63	Crater, craters	MC-11	Dean B.; American astronomer (1901–1965)
McMurdo	26.9	-84.38	0.59	Crater, craters	MC-30	American station in Antarctica
Medissa	19.52	18.64	303.43	Crater, craters	MC-10	Town in Algeria
Medusae Fossae	278.52	-2.17	195.8	Fossa, fossae	MC-16	Classical albedo feature name
Medusae Sulci	191.52	-5.04	200.3	Sulcus, sulci	MC-16	Albedo feature name
Mega	16.95	-1.43	323.1	Crater, craters	MC-19	Town in Ethiopia
Meget	4.59	18.86	107.31	Crater, craters	MC-14	Town in Russia
Melas Chasma	563.52	-10.52	287.46	Chasma, chasmata	MC-18	Classical albedo feature name
Melas Dorsa	486.81	-18.92	287.9	Dorsum, dorsa	MC-18	Classical albedo feature name
Melas Fossae	568.18	-26.28	288.48	Fossa, fossae	MC-18, MC-25	Classical albedo feature name
Melas Labes	107.24	-8.53	288.3	Labes, labes	MC-18	From albedo feature at 10 S, 74 W

Feature name	Diameter (km)	Center latitude	Center longitude	Feature type	Quadrangle	Origin
Melas Mensa	245	-10.72	285.85	Mensa, mensae	MC18	Classical Martian albedo feature, Melas
Mellish	104.95	-72.63	336.26	Crater, craters	MC-30	John E.; American amateur astronomer (1886–1970)
Mellit	22.53	7.12	358.27	Crater, craters	MC-11	Town in Sudan
Memnonia Fossae	1585.28	-23.63	206.18	Fossa, fossae	MC-16	Classical albedo feature name
Memnonia Sulci	452.66	-7.16	184.17	Sulcus, sulci	MC-16	Albedo feature name.
Mena	29.91	-32.11	341.24	Crater, craters	MC-26	Town in Russia
Mendel	77.32	-58.78	161.25	Crater, craters	MC-29	Gregor J.; Austrian biologist (1822–1884)
Mendota	8.86	35.83	138.33	Crater, craters	MC7	Town in Illinois, USA
Meridiani Planum	1058.53	-0.04	356.86	Planum, plana	MC-11, MC-12, MC-19	Classical albedo feature name
Meroe Patera	52.6	6.98	68.77	Patera, paterae	MC-13	Classical albedo feature name
Micoud	51.85	50.56	16.34	Crater, craters	MC-5	Town in Saint Lucia
Mie	100.91	48.16	139.65	Crater, craters	MC-7	Gustav; German physicist (1868–1957)
Mila	10.87	-27.16	339.25	Crater, craters	MC-19	Town in Algeria
Milankovič	113.51	54.46	213.42	Crater, craters	MC-2	Milutin; Yugoslav geophysicist, astrophysicist (1879–1958)
Milford	24.97	-52.41	318.51	Crater, craters	MC-26	Town in Utah, USA
Millman	73.84	-53.95	210.36	Crater, craters	MC-24	Peter; Canadian astronomer (1906–1990)
Millochau	112.89	-21.19	85.1	Crater, craters	MC-21	Gaston; French astronomer (b. 1866)
Milna	27.48	-23.46	347.76	Crater, craters	MC-19	Town in the Republic of Croatia
Minio Vallis	90	-4.38	208.33	Vallis, valles	MC-16	Classical name for river in Italy
Mirtos	6.38	22.12	308.24	Crater, craters	MC-10	Town in Greece (Crete)
Mistretta	16.56	-24.68	250.87	Crater, craters	MC-17	Town in Sicily
Mitchel	135.9	-67.53	76.01	Crater, craters	MC-30	Ormsby M.; American astronomer (1809–1862)
Miyamoto	145.21	-2.87	353.05	Crater, craters	MC-19	Shotaro; Japanese astronomer (1912–1992)
Mliba	11.85	-39.61	87.98	Crater, craters	MC-28	Town in Swaziland
Moa Valles	265	35.62	305.3	Vallis, valles	MC-4	River in Sierra Leone
Moanda	38.88	-35.93	320.05	Crater, craters	MC-26	Town in Gabon
Mohawk	17.49	42.89	354.65	Crater, craters	MC-4	Town in New York, USA
Mojave	57.97	7.48	327.01	Crater, craters	MC-11	Town in California, USA
Molesworth	168.87	-27.5	149.27	Crater, craters	MC-23	Percy B.; British astronomer (1867–1908)
Moni	5.44	-47.01	18.77	Crater, craters	MC-27	Village in Cyprus
Montevallo	50.42	15.25	305.73	Crater, craters	MC-10	Town in Alabama, USA
Morava Valles	364.1	-13.57	335.8	Vallis, valles	MC-19	River in the Czech Republic
Morella	76.97	-9.58	308.61	Crater, craters	MC-18	Town in Spain
Moreux	131.55	41.79	44.54	Crater, craters	MC-5	Theophile; French astronomer and meteorologist (1867–1954)

Feature name	Diameter (km)	Center latitude	Center longitude	Feature type	Quadrangle	Origin
Moroz	116.3	-23.77	339.43	Crater, craters	MC-19	Vasily Ivanovich; Russian planetary scientist (1931–2004)
Morpheos Rupes	404.15	-36	125.58	Rupes, rupes	MC-29	Classical albedo feature name
Mosa Vallis	191.6	-15.09	22.2	Vallis, valles	MC-20	Modern river in Western Europe
Moss	9.09	19.23	109.49	Crater, craters	MC-14	Town in Norway
Muara	3.83	24.32	340.69	Crater, craters	MC-11	Town in Brunei
Müller	120.51	-25.74	127.89	Crater, craters	MC-22	Hermann J.; American geneticist (1890–1967), and Carl H.; German astronomer (1851–1925)
Munda Vallis	9.13	-5.37	213.83	Vallis, valles	MC-16	Classical name for river in ancient Lusitania, (modern Mondega River in Portugal)
Murgoo	22.64	-23.64	337.55	Crater, craters	MC-19	Town in Australia
Murray	92	-23.29	28.06	Crater, craters	MC-20	Bruce Churchill; American planetary scientist (1931–2013)
Mut	6.97	22.36	324.24	Crater, craters	MC-11	Town in Turkey
Mutch	198.81	0.6	304.79	Crater, craters	MC-10, MC-18	Dr. Thomas A.; American geologist, Viking Lander Imaging Team leader (1931–1980)
N. Mareotis Tholus	3.58	36.38	273.79	Tholus, tholi	MC-3	Classical albedo name
Naar	11.34	22.91	317.87	Crater, craters	MC-11	Town in Egypt
Naic	8.68	24.45	107.44	Crater, craters	MC-14	Town in the Philippines
Nain	6.89	41.47	126.84	Crater, craters	MC-7	Town in Newfoundland, Canada
Naju	8.03	44.99	122.86	Crater, craters	MC-7	Town in the Republic of Korea
Naktong Vallis	669.63	4.89	33.39	Vallis, valles	MC-12	Modern name for river in the Republic of Korea
Nakusp	7.26	24.73	324.55	Crater, craters	MC-11	Town in British Columbia
Nan	2.29	-26.69	340.06	Crater, craters	MC-19	Town in Thailand
Nanedi Valles	550	5.05	311.38	Vallis, valles	MC-10	Word for "planet" in Sesotho, national language of Lesotho, Africa
Nansen	74.63	-49.92	219.58	Crater, craters	MC-24	Fridtjof; Norwegian explorer (1861–1930)
Napo Vallis	87.5	-25.97	78.03	Vallis, valles	MC-21	River in Ecuador
Nardo	25.1	-27.51	327.16	Crater, craters	MC-19	Town in Italy
Naro Vallis	442.72	-4	60.71	Vallis, valles	MC-21	Ancient name for modern Neretva River, Bosnia and Herzegovina
Naruko	4.17	-36.24	198.3	Crater, craters	MC-24	Former town in Japan
Naryn	3.94	14.89	123.3	Crater, craters	MC-14	Town in Kyrgyzstan
Naukan	7.47	21.25	329.42	Crater, craters	MC-11	Town in Russia
Navan	24.86	-25.89	336.5	Crater, craters	MC-19	Town in Ireland
Navua Valles	500	-33.94	82.68	Vallis, valles	MC-28	River in Fiji
Nazca	15.01	-31.63	93.67	Crater, craters	MC-28	Town in Peru
Nectaris Fossae	623.07	-23.09	302.84	Fossa, fossae	MC-18	Classical albedo feature name
Nectaris Montes	220	-14.64	305.35	Mons, montes	MC-18	Classical Martian albedo feature, Nectaris

Feature name	Diameter (km)	Center latitude	Center longitude	Feature type	Quadrangle	Origin
Negele	36.93	-35.8	96	Crater, craters	MC-28	Town in Ethiopia
Negril	52	20.19	69.43	Crater, craters	MC-13	Town in Jamaica
Neive	2.79	23.18	107.07	Crater, craters	MC-14	Town in Italy
Nema	14.54	20.7	307.87	Crater, craters	MC-10	Town in Russia
Nepa	16.16	-24.97	340.33	Crater, craters	MC-19	Town in Russia
Nepenthes Mensae	2176.23	9.19	119.42	Mensa, mensae	MC-14	Classical albedo feature name
Nepenthes Planum	1650.14	14.01	113.79	Planum, plana	MC-14	Classical albedo feature name
Nereidum Montes	1142.58	-37.57	316.79	Mons, montes	MC-26	Classical albedo feature name
Nestus Valles	38.25	-7.03	201.52	Vallis, valles	MC-16	Classical name for river in Macedonia (Greece)
Never	2.75	23.5	105.77	Crater, craters	MC-14	Town in Russia
Neves	22.13	-3.39	151.31	Crater, craters	MC-23	Town in the Republic of Sao Tome and Principe
New Bern	1.73	21.53	310.85	Crater, craters	MC-10	American colonial town (North Carolina)
New Haven	1.51	22.08	310.74	Crater, craters	MC-10	American colonial town (Connecticut)
New Plymouth	31.54	-15.78	175.87	Crater, craters	MC-23	Town in Idaho, USA
Newcomb	254.13	-24.27	1.04	Crater, craters	MC-19, MC-20	Simon; American astronomer (1835–1909)
Newport	1.97	22.24	311.04	Crater, craters	MC-10	American colonial town (Rhode Island)
Newton	299.94	-40.5	201.97	Crater, craters	MC-24	Isaac; British physicist (1643–1727)
Nhill	23.7	-28.68	256.67	Crater, craters	MC-17	Town in Victoria, Australia
Nia Chaos	48	-6.74	292.62	Chaos, chaoses	MC-18	Classical albedo feature name, Nia
Nia Fossae	379.56	-14.73	288.23	Fossa, fossae	MC-18	Classical albedo feature name
Nia Mensa	95	-7.72	292.68	Mensa, mensae	MC-18	Classical albedo feature name, Nia
Nia Tholus	34.01	-6.59	285.05	Tholus, tholi	MC-18	Classical albedo feature name, Nia
Nia Vallis	140	-53.53	325.19	Vallis, valles	MC-26	Lowell canal name; also classical river name
Nicer Vallis	22.6	-6.96	201.81	Vallis, valles	MC-16	Classical name for present Neckar River, Germany
Nicholson	102.45	0.21	195.57	Crater, craters	MC-8, MC-16	Seth Barnes; American astronomer (1891–1963)
Nier	46.3	42.79	106.11	Crater, craters	MC-6	Alfred O.C.; American physicist (1911–1994)
Niesten	114.81	-28	57.75	Crater, craters	MC-21	Louis; Belgian astronomer (1844–1920)
Nif	8.48	19.91	303.76	Crater, craters	MC-10	Town in the Caroline Islands (Yap)
Niger Vallis	360	-34.96	92.57	Vallis, valles	MC-28	River in Africa
Nili Fossae	727.91	22.02	76.69	Fossa, fossae	MC-13	Classical albedo feature name
Nili Patera	67.51	8.97	67.17	Patera, paterae	MC-13	Classical albedo feature name
Nili Tholus	7	9.15	67.35	Tholus, tholi	MC-13	Classical albedo feature name
Nilokeras Fossa	267	24.59	302.17	Fossa, fossae	MC-10	Classical albedo feature name

Feature name	Diameter (km)	Center latitude	Center longitude	Feature type	Quadrangle	Origin
Oglala	17.59	-3.11	321.89	Crater, craters	MC-19	Town in South Dakota, USA
Ogygis Rupes	**184.23**	**-33.03**	**305.47**	**Rupes, rupes**	**MC-26**	**Classical albedo feature name**
Ogygis Undae	87.7	-49.66	293.79	Unda, undae	MC-25	Classical albedo feature name: Ogygis Regio
Ohara	9.37	4.92	82.48	Crater, craters	MC-13	Town in Japan
Okavango Valles	285.12	38.1	8.97	Vallis, valles	MC-5	River in Botswana
Okhotsk	1.67	22.97	312.67	Crater, craters	MC-10	Russian port
Okotoks	**21.78**	**-21.21**	**84.41**	**Crater, craters**	**MC-21**	**Town in Alberta, Canada**
Olenek	3.06	19.87	305.78	Crater, craters	MC-10	Town in Russia
Olom	5.89	22.96	302.34	Crater, craters	MC-10	Town in Russia
Ollis Vallis	169.3	-23.5	338.35	Vallis, valles	MC-19	Ancient name for modern Lot river, France
Olympia Cavi	**342.78**	**85.06**	**182.23**	**Cavus, cavi**	**MC-1**	**Classical albedo feature name**
Olympia Mensae	**335.42**	**78**	**119.98**	**Mensa, mensae**	**MC-1**	**Classical albedo feature name**
Olympia Planum	**804.39**	**82.18**	**188.81**	**Planum, plana**	**MC-1**	**Classical albedo feature name**
Olympia Rupes	**1197.04**	**86.04**	**174.16**	**Rupes, rupes**	**MC-1**	**Classical albedo feature name**
Olympia Undae	**1507.96**	**81.16**	**178.48**	**Unda, undae**	**MC-1**	**Classical albedo feature name**
Olympica Fossae	**420**	**24.85**	**246.08**	**Fossa, fossae**	**MC-9**	**From albedo feature at 17° N, 134° W**
Olympus Mons	**610.13**	**18.65**	**226.2**	**Mons, montes**	**MC-8, MC-9, Figure 4.1**	**Classical albedo feature name**
Olympus Rupes	**1914.77**	**18.4**	**226.44**	**Rupes, rupes**	**MC-8, MC-9**	**Classical albedo feature name**
Ome	2.85	20.6	104.02	Crater, craters	MC-14	Town in Japan
Ōmura	8.47	-25.36	334.79	Crater, craters	MC-19	Town in Japan
Onon	3.42	16.13	102.48	Crater, craters	MC-14	Town in Mongolia
Oodnadatta	**25.44**	**-52.43**	**325.83**	**Crater, craters**	**MC-26**	**Town in Australia**
Ophir Catenae	509	-9.46	300.6	Catena, catenae	MC-18	Classical albedo feature name
Ophir Cavus	36.72	-9.89	304.96	Cavus, cavi	MC-18	Classical albedo feature name
Ophir Chasma	**314.71**	**-4**	**287.65**	**Chasma, chasmata**	**MC-18**	**Classical albedo feature name**
Ophir Labes	**92.53**	**-11.01**	**291.72**	**Labes, labes**	**MC-18**	**From albedo feature at 10° S, 65° W**
Ophir Mensa	103.33	-3.99	286.51	Mensa, mensae	MC-18	Classical albedo feature name
Ophir Planum	**642.24**	**-8.45**	**302.18**	**Planum, plana**	**MC-18**	**Classical albedo feature name**
Oraibi	**32.37**	**17.22**	**327.66**	**Crater, craters**	**MC-11**	**Town in Arizona, USA**
Orcus Patera	**387.64**	**14.13**	**178.35**	**Patera, paterae**	**MC-15**	**Classical albedo feature name**
Ore	7.15	16.78	326.07	Crater, craters	MC-11	Town in Nigeria
Orinda	9.03	45.37	126.98	Crater, craters	MC-7	Town in California, USA
Orson Welles	**115.99**	**-0.19**	**314.1**	**Crater, craters**	**MC-10, MC-18**	**George Orson; American radio and motion picture actor and director (1915-1985)**
Ortygia Colles	**255.62**	**53.9**	**350.7**	**Collis, colles**	**MC-4**	**Named for classical albedo feature at 65° N, 350° W**
Ostrov	**72.98**	**-26.55**	**331.89**	**Crater, craters**	**MC-19**	**Town in Russia**

Feature name	Diameter (km)	Center latitude	Center longitude	Feature type	Quadrangle	Origin
Nilokeras Mensae	**450.88**	**30.48**	**308.05**	**Mensa, mensae**	**MC-4, MC-10**	**Albedo feature name**
Nilokeras Scopulus	**901.48**	**31.72**	**304.15**	**Scopulus, scopuli**	**MC-3, MC-4**	**From albedo feature at 30° N, 55° W**
Nilosyrtis Mensae	**676.03**	**34.77**	**68.47**	**Mensa, mensae**	**MC-6**	**Classical albedo feature name**
Nilus Chaos	**283**	**25.39**	**283.05**	**Chaos, chaoses**	**MC-10**	**Named for albedo feature at 20° N, 65° W**
Nilus Dorsa	**292.9**	**20.68**	**280.94**	**Dorsum, dorsa**	**MC-10**	**Named for albedo feature at 20° N, 65° W**
Nilus Mensae	**206.84**	**22.2**	**287.77**	**Mensa, mensae**	**MC-10**	**Named for albedo feature at 20° N, 65° W**
Nipigon	8.89	33.76	278.16	Crater, craters	MC-3	Town in Canada
Niquero	10.7	-38.79	194.03	Crater, craters	MC-24	Town in the Republic of Cuba
Nirgal Vallis	**610**	**-28.16**	**318.32**	**Vallis, valles**	**MC-18, MC-19**	**Word for "Mars" in Babylonian**
Nitro	**29.34**	**-21.26**	**336**	**Crater, craters**	**MC-19**	**Town in West Virginia, USA**
Njesko	**27.77**	**-35.25**	**85.11**	**Crater, craters**	**MC-28**	**Town in the Czech Republic**
Noachis Terra	**5519.45**	**-50.41**	**354.84**	**Terra, terrae**	**MC-18, MC-19, MC-20, MC-26, MC-27, MC-30, Figure 4.1**	**Classical albedo feature name**
Noctis Fossae	**712.73**	**-2.69**	**261.15**	**Fossa, fossae**	**MC-17**	**Classical albedo feature at 10° S, 96° W**
Noctis Labyrinthus	**1190.31**	**-6.36**	**258.81**	**Labyrinthus, labyrinthi**	**MC-17**	**Classical albedo feature name**
Noma	**40.49**	**-25.43**	**335.69**	**Crater, craters**	**MC-19**	**Town in Namibia**
Noord	7.8	-19.27	348.73	Crater, craters	MC-19	Town in Aruba
Nordenskiöld	**85.6**	**-52.37**	**201.24**	**Crater, craters**	**MC-24**	**Nils Adolf Erik; Swedish geologist and geographer, Arctic researcher (1832–1901)**
Northport	18.44	18.52	305.52	Crater, craters	MC-10	Town in Alabama, USA
Novara	**86.98**	**-24.9**	**349.31**	**Crater, craters**	**MC-19**	**Town in Italy**
Nqutu	**21**	**-38.04**	**169.55**	**Crater, craters**	**MC-29**	**Nquthu, town in South Africa**
Nune	8.47	17.55	321.24	Crater, craters	MC-11	Town in Mozambique
Nutak	11.26	17.41	329.74	Crater, craters	MC-11	Town in Newfoundland, Canada
Nybyen	6.2	-37.02	343.34	Crater, craters	MC-26	Town in Norway
Obock	14.45	-2.01	150.53	Crater, craters	MC-23	Town in the Republic of Djibouti
Ocampo	7.16	32.66	138.3	Crater, craters	MC-7	Town in Mexico
Oceanidum Fossa	167.16	-61.58	330.49	Fossa, fossae	MC-26	Classical albedo feature name
Oceanidum Mons	33.39	-54.93	318.77	Mons, montes	MC-26	Name change from Charitum Tholus
Ochakov	**31.05**	**-42.11**	**328.14**	**Crater, craters**	**MC-26**	**Town in Ukraine**
Ochus Valles	127	7.07	314.96	Vallis, valles	MC-10	Classical name for present Hari-Rud River in Turkmenistan
Octantis Cavi	71.21	-52.57	314.03	Cavus, cavi	MC-26	Albedo name
Octantis Mons	19.09	-55.26	317.15	Mons, montes	MC-26	Albedo name
Oenotria Planum	**61.25**	**-8.14**	**76.64**	**Planum, plana**	**MC-21**	**Classical albedo feature name**
Oenotria Scopuli	**1425**	**-6.62**	**77.11**	**Scopulus, scopuli**	**MC-21**	**Classical albedo feature name**

Feature name	Diameter (km)	Center latitude	Center longitude	Feature type	Quadrangle	Origin
Osuga Valles	164	-15.31	321.41	Vallis, valles	MC-19	River in Russia
Oti Fossae	373.59	-9.63	242.91	Fossa, fossae	MC-17	Classical albedo feature
Ottumwa	51.62	24.58	304.25	Crater, craters	MC-10	Town in Iowa, USA
Oudemans	124.16	-9.84	268.23	Crater, craters	MC-17	Jean A.; Dutch astronomer (1827–1906)
Oxia Chaos	24.12	0.22	320.13	Chaos, chaoses	MC-11	Classical albedo feature name
Oxia Colles	595.24	21.24	333.73	Collis, colles	MC-11	From albedo feature at 25° N, 24° W
Oxus Cavus	37.87	37.41	359.48	Cavus, cavi	MC-4	Classical albedo feature name
Oxus Patera	33.42	38.97	359.66	Patera, paterae	MC-4	Classical albedo feature name
Oyama	100.78	23.57	339.89	Crater, craters	MC-11	Vance I.; American biochemist (1922–1998)
Pabo	9.18	-26.9	336.92	Crater, craters	MC-19	Town in Uganda
Padus Vallis	57.42	-4.52	210.02	Vallis, valles	MC-16	Classical name for modern Po River in Italy
Paks	6.9	-7.66	317.96	Crater, craters	MC-19	Town in Hungary
Pál	71.21	-31.31	108.7	Crater, craters	MC-1, MC-7	George; American–Hungarian film producer (1908–1980)
Palana	4.53	21.04	102.02	Crater, craters	MC-14	Town in Kamchatka, Russia
Palikir	15.57	-41.57	202.14	Crater, craters	MC-24	Capital of the Federated States of Micronesia
Pallacopas Vallis	134.77	-54.73	339.52	Vallis, valles	MC-26	Lowell canal name; also classical river name
Palos	54.82	-2.69	110.9	Crater, craters	MC-22	Town in Spain
Panchaia Rupes	1113.4	64.37	129.83	Rupes, rupes	MC-1, MC-7	Named for classical albedo feature at 62° N, 220° W
Pangboche	10.16	17.28	226.6	Crater, craters	MC-9	Village in Nepal
Paraná Valles	329.13	-23.19	350.2	Vallis, valles	MC-19	Ancient and modern name for South American river (Brazil, Argentina)
Paros	34.61	21.99	261.87	Crater, craters	MC-9	Famous in antiquity for its marble quarries
Parva Planum	1027.32	-73.67	264.93	Planum, plana	MC-30	Classical albedo feature name
Pasithea Dorsum	282.2	-55.14	318.42	Dorsum, dorsa	MC-26	One of the Graces
Pasteur	116.15	19.31	24.62	Crater, craters	MC-12	Louis; French chemist (1822–1895)
Patapsco Vallis	172.87	23.7	152.51	Vallis, valles	MC-15	Modern river in Maryland, USA
Pau	42.2	-55.4	59.3	Crater, craters	MC-27	Town in France
Pavonis Chasma	45.94	2.73	248.98	Chasma, chasmata	MC-9	Albedo name
Pavonis Fossae	156.08	4.15	248.71	Fossa, fossae	MC-9	Albedo name
Pavonis Mons	366.53	1.48	247.04	Mons, montes	MC-9, MC-17	Classical albedo feature name
Pavonis Sulci	425.8	4.01	242.63	Sulcus, sulci	MC-9	Albedo name
Peace Vallis	35.24	-4.21	137.23	Vallis, valles	MC-23	River in British Columbia and Alberta, Canada

Feature name	Diameter (km)	Center latitude	Center longitude	Feature type	Quadrangle	Origin
Pebas	5.43	-2.6	359.04	Crater, craters	MC-19	Town in Peru
Peixe	9.35	20.33	312.4	Crater, craters	MC-10	Town in Brazil
Peneus Palus	870	-35.06	56.71	Palus, paludes	MC-27, MC-28	From classical albedo feature at 48° S, 290° W
Peneus Patera	128.5	-57.82	52.65	Patera, paterae	MC-27	From albedo feature at 48° S, 290° W
Penticton	8.19	-38.37	96.76	Crater, craters	MC-28	Town in British Columbia, Canada
Peraea Cavus	56.24	-29.61	95.43	Cavus, cavi	MC-22	Albedo feature name
Peraea Mons	14.94	-31.08	86.11	Mons, montes	MC-28	Albedo feature name
Perepelkin	77.46	52.44	295.17	Crater, craters	MC-3	Evgenii J.; Russian astronomer (1906–1938)
Peridier	94.21	25.51	83.91	Crater, craters	MC-13	Julien; French astronomer (1882–1967)
Perrotin	82.82	-2.82	282.06	Crater, craters	MC-18	Henri A.; French astronomer, studied dark lineations on Mars (1845–1904)
Persbo	19.49	8.57	156.88	Crater, craters	MC-15	Town in Sweden
Peta	75.75	-21.26	350.9	Crater, craters	MC-19	Town in Greece
Pettit	92.49	12.25	186.13	Crater, craters	MC-8	Edison; American astronomer (1890–1962)
Phaenna Dorsum	164.16	-53.79	316.71	Dorsum, dorsa	MC-26	One of the Graces
Phedra	20.31	13.84	123.88	Crater, craters	MC-14	Town in Suriname
Philadelphia	1.65	21.76	312.02	Crater, craters	MC-10	American colonial town (Pennsylvania)
Phillips	185.45	-66.34	315.11	Crater, craters	MC-30	John; British geologist (1800–1874), Theodore E.; British astronomer (1868–1942)
Phison Rupes	203.07	26.7	50.35	Rupes, rupes	MC-13	Classical albedo feature name
Phlegethon Catena	399.69	38.83	256.72	Catena, catenae	MC-3	From albedo feature at 38° N, 125° W
Phlegra Dorsa	2818.61	25.08	170.37	Dorsum, dorsa	MC-7, MC-15	Named for classical albedo feature at 35° N, 195° W
Phlegra Montes	1350.65	40.4	163.71	Mons, montes	MC-7, MC-15	Classical albedo feature name
Phon	10.02	15.53	102.79	Crater, craters	MC-14	Town in Thailand
Pica	2.4	19.82	306.77	Crater, craters	MC-10	Town in Chile
Pickering	115.2	-33.48	227.39	Crater, craters	MC-24	Edward Charles; American astronomer (1846–1919), William Henry; American astronomer (1858–1938), Sir William Hayward; New Zealand-American engineer (1910–2004)
Piña	5.05	18.37	111.74	Crater, craters	MC-14	Town in Panama
Pindus Mons	16.3	39.47	271.48	Mons, montes	MC-3	Mountains near Vale of Tempe
Pinglo	15.9	-2.92	323.24	Crater, craters	MC-19	Town in China (Ningsia)
Pital	41.7	-9.27	297.72	Crater, craters	MC-18	Town in Costa Rica
Pityusa Patera	196.51	-66.88	36.86	Patera, paterae	MC-30	Classical albedo feature name

Feature name	Diameter (km)	Center latitude	Center longitude	Feature type	Quadrangle	Origin
Pityusa Rupes	430.14	-63.96	28.32	Rupes, rupes	MC-27	From albedo feature at 58° S, 319° W
Piyi	11.63	-22.88	106.63	Crater, craters	MC-22	Town in Cyprus
Planum Angustum	206.41	-79.8	276.8	Planum, plana	MC-30	Classical albedo feature name
Planum Australe	1429.87	-83.35	157.7	Planum, plana	MC-30	Classical albedo feature
Planum Boreum	354.63	87.32	54.96	Planum, plana	MC-1	Classical albedo feature name
Planum Chronium	576.38	-59.14	139.5	Planum, plana	MC-29	From albedo feature at 58° S, 90° W
Platte	3.42	16.03	113.18	Crater, craters	MC-14	Town in South Dakota, USA
Playfair	62.21	-77.91	234.22	Crater, craters	MC-30	John; British geologist and mathematician (1748-1819)
Plum	2.76	-26.07	340.93	Crater, craters	MC-19	Town in Wisconsin, USA
Podor	25.08	-44.11	316.86	Crater, craters	MC-26	Town in Senegal
Pollack	96.35	-7.79	25.26	Crater, craters	MC-20	James B.; American physicist (1938-1994)
Polotsk	30.12	-19.89	333.66	Crater, craters	MC-19	Town in Belarus
Pompeii	31.13	18.98	300.9	Crater, craters	MC-10	Ruined town in Italy
Poona	19.87	23.76	307.68	Crater, craters	MC-10	Town in India
Port-Au-Prince	1.52	21.1	311.82	Crater, craters	MC-10	Port of Hispaniola Island, Haiti
Porter	103.99	-50.36	246.24	Crater, craters	MC-25	Russell W.; American astronomer (1871-1949)
Porth	9.52	21.19	104.21	Crater, craters	MC-14	Town in Wales
Portsmouth	1.5	22.55	310.93	Crater, craters	MC-10	American colonial town (New Hampshire)
Porvoo	9.85	-43.3	319.19	Crater, craters	MC-26	Town in Finland
Poti	30.55	-36.31	86.56	Crater, craters	MC-28	Town in Georgia
Poynting	69.7	8.42	247.25	Crater, craters	MC-9	J.H.; English astrophysicist (1852-1914)
Prao	20	-11.16	56.61	Crater, craters	MC-21	Town in Vietnam
Priestley	42.26	-54.12	130.7	Crater, craters	MC-29	Joseph; British chemist (1733-1804)
Princeton	2.16	21.69	310.89	Crater, craters	MC-10	American colonial town (New Jersey)
Proctor	172.56	-47.63	29.72	Crater, craters	MC-27	Richard A.; British astronomer (1837-1888)
Promethei Chasma	295.28	-82.66	141.39	Chasma, chasmata	MC-30	Classical albedo feature name
Promethei Lingula	571.63	-82.8	119.89	Lingula, lingulae	MC-30	Classical albedo feature name
Promethei Mons	65.17	-70.57	87.44	Mons, montes	MC-30	Classical albedo feature name
Promethei Planum	831.28	-79.18	88.36	Planum, plana	MC-30	Classical albedo feature name
Promethei Rupes	1379.21	-75.54	90.24	Rupes, rupes	MC-30	Classical albedo feature name
Promethei Terra	3244.3	-64.37	97	Terra, terrae	MC-28, MC-30, Figure 4.1	Classical albedo feature name
Protonilus Mensae	1033.97	43.87	48.86	Mensa, mensae	MC-5	From albedo feature at 42° N, 315° W
Protva Valles	259.71	-29.11	299.42	Vallis, valles	MC-18	River in Russia
Ptolemaeus	165.18	-45.88	202.4	Crater, craters	MC-24	Claudius; Greco-Egyptian astronomer (c. AD 90-160)
Pulawy	51.84	-36.41	283.38	Crater, craters	MC-25	Town in Poland

Feature name	Diameter (km)	Center latitude	Center longitude	Feature type	Quadrangle	Origin
Púnsk	11.24	20.62	318.87	Crater, craters	MC-11	Town in Poland
Pursat	17.55	-37.36	130.76	Crater, craters	MC-29	Town in Cambodia
Puyo	9.94	83.93	137.26	Crater, craters	MC-1	Town in Ecuador
Pylos	18.94	16.79	329.92	Crater, craters	MC-11	Town in Greece
Pyramus Fossae	298.18	50.39	66.31	Fossa, fossae	MC-6	From albedo feature at 65° N, 300° W
Pyrrhae Chaos	162.35	-10.46	331.6	Chaos, chaoses	MC-19	Albedo feature name
Pyrrhae Fossae	430	-29.2	336.2	Fossa, fossae	MC-19, MC-26	Classical Martian albedo feature, Pyrrhae
Qibā	4.08	17.13	103.09	Crater, craters	MC-14	Town in Saudi Arabia
Quenisset	136.66	34.27	40.67	Crater, craters	MC-5	Ferdinand J.; French astronomer (1872-1951)
Quick	13.31	18.19	310.75	Crater, craters	MC-10	Town in British Columbia, Canada
Quines	10.75	-41.86	89.25	Crater, craters	MC-28	Town in Argentina
Quorn	6.33	-5.56	326.38	Crater, craters	MC-19	Town in Australia
Quthing	15.59	0.4	149.29	Crater, craters	MC-15	Town in Lesotho
Rabe	106.95	-43.61	34.91	Crater, craters	MC-27	Wilhelm F.; German astronomer (1893-1958)
Radau	109.96	16.95	355.29	Crater, craters	MC-11	Rodolphe; French astronomer (1835-1911)
Raga	3.43	-48.1	242.42	Crater, craters	MC-25	Town in South Sudan
Rahe	34.44	25.05	262.52	Crater, craters	MC-9	Jurgen; American astronomer and NASA program director (1940-1997)
Rahway Valles	346.19	8.46	173.58	Vallis, valles	MC-15	River in New Jersey, USA
Rakke	18.47	-4.57	316.64	Crater, craters	MC-19	Town in Estonia
Rana	12.33	-25.59	338.2	Crater, craters	MC-19	Town in Norway
Raub	6.95	42.38	135.11	Crater, craters	MC-7	Town in Malaysia
Rauch	32.9	21.56	301.87	Crater, craters	MC-10	Town in Argentina
Rauna	2.53	35.26	327.92	Crater, craters	MC-4	Village in Latvia
Ravi Vallis	148.78	-0.42	319.52	Vallis, valles	MC-19	Ancient Pakistani River
Ravius Valles	388.18	46.12	249.83	Vallis, valles	MC-3	Classical name for river in northwest Ireland
Rayadurg	21.38	-18.45	102.43	Crater, craters	MC-22	Town in India
Rayleigh	125.66	-75.57	118.94	Crater, craters	MC-30	Strutt, John W., third Baron Rayleigh; British physicist (1842-1919)
Redi	60.31	-60.33	92.8	Crater, craters	MC-28	Francesco; Italian physicist (1626-1697)
Renaudot	63.74	42.04	62.68	Crater, craters	MC-6	Gabrielle; French astronomer (1877-1962)
Rengo	13.7	-43.45	316.37	Crater, craters	MC-26	Town in Chile
Resen	7.4	-27.94	108.87	Crater, craters	MC-22	Town in Macedonia
Reull Vallis	1051.94	-42.14	104.95	Vallis, valles	MC-28	Word for "planet" in Gaelic
Reutov	18.02	-45.07	202.29	Crater, craters	MC-24	Town in Moscow region, Russia
Reuyl	84.27	-9.63	166.93	Crater, craters	MC-23	Dirk; American physicist (1906-1972)
Revda	26.6	-24.28	331.5	Crater, craters	MC-19	Town in Russia
Reykholt	52.17	40.48	273.86	Crater, craters	MC-3	Town in Iceland
Reynolds	90.69	-74.99	202.41	Crater, craters	MC-30	Osborne; British physicist (1842-1912)
Rhabon Valles	245	21.21	268.73	Vallis, valles	MC-9	Classical river in Dacia (Romania)
Ribe	11.14	16.49	330.84	Crater, craters	MC-11	Town in Denmark

Feature name	Diameter (km)	Center latitude	Center longitude	Feature type	Quadrangle	Origin
Richardson	89	-72.47	180.14	Crater, craters	MC-30	Lewis F.; British meteorologist, chemist (1881–1953)
Rimac	7.29	44.97	136.06	Crater, craters	MC-7	Town in Peru
Rincon	13.36	-8	316.99	Crater, craters	MC-19	Town in the Netherlands; Antilles (Bonaire)
Ritchey	77.23	-28.42	309.01	Crater, craters	MC-18	George W.; American astronomer (1864–1945)
Robert Sharp	152.08	-4.17	133.42	Crater, craters	MC-22	Robert Phillip; American geologist (1911–2004)
Roddenberry	139.15	-49.37	355.57	Crater, craters	MC-26	Gene; American engineer, television producer (1921–1991)
Roddy	85.82	-21.65	320.61	Crater, craters	MC-19	David John; astrogeologist (1932–2002)
Romny	5.39	-25.4	341.83	Crater, craters	MC-19	Town in Russia
Rong	8.92	22.46	314.65	Crater, craters	MC-10	Town in China (Tibet)
Rongxar	21.63	26.33	304.56	Crater, craters	MC10	Small village in Tibet, near Mt. Everest
Roseau	6.49	-41.69	150.57	Crater, craters	MC-29	Town in Dominica
Ross	82.51	-57.39	252.16	Crater, craters	MC-25	Frank E.; American astronomer (1874–1966)
Rossby	80.42	-47.52	167.92	Crater, craters	MC-29	Carl G.; Swedish-American meteorologist (1898–1957)
Rubicon Valles	308.21	44.41	242.48	Vallis, valles	MC-3	Ancient river in Italy
Ruby	26.43	-25.24	342.93	Crater, craters	MC-19	Town in South Carolina, USA
Rudaux	107.18	38.03	50.96	Crater, craters	MC-5	Lucien; French astronomer (1874–1947)
Ruhea	9.5	-43.26	173.08	Crater, craters	MC-29	Town in Bangladesh
Runa Vallis	36	-28.34	323.29	Vallis, valles	MC-19	Name proposed by Soviets
Runanga	41.36	-26.64	75.96	Crater, craters	MC-21	Town in New Zealand
Rupes Tenuis	669.03	81.6	274.53	Rupes, rupes	MC-1	Classical albedo feature name
Russell	135.08	-54.5	12.43	Crater, craters	MC-27	Henry N.; American astronomer (1877–1957)
Rutherford	107.08	19.03	349.41	Crater, craters	MC-11	Ernest; British physicist (1871–1937)
Ruza	22.25	-34	307.28	Crater, craters	MC-26	Town in Russia
Rynok	8.49	44.13	121.76	Crater, craters	MC-7	Town in Russia
Rypin	18.18	-1.28	319.11	Crater, craters	MC-19	Town in Poland
Sabis Vallis	212.91	-5.01	207.49	Vallis, valles	MC-16	Classical name for present Sambre River in France and Belgium
Sabo	4.39	25.17	311.06	Crater, craters	MC-10	Town in Russia
Sabrina Vallis	280	10.99	310.96	Vallis, valles	MC-10	Classical name for present Severn River, England
Sacra Dorsa	1416	11.21	293.91	Dorsum, dorsa	MC-10	From albedo feature at 20 N, 67 W
Sacra Fossae	950	20.36	290	Fossa, fossae	MC-10	Classical albedo feature name
Sacra Mensa	577	24.64	291.78	Mensa, mensae	MC-10	Albedo feature name
Sacra Sulci	1009.05	22.16	285.3	Sulcus, sulci	MC-10	From albedo feature at 20 N, 67 W
Sagan	90.26	10.72	329.4	Crater, craters	MC-11	Carl E.; American astronomer (1934–1996)

Feature name	Diameter (km)	Center latitude	Center longitude	Feature type	Quadrangle	Origin
Saheki	82.44	-21.74	73.14	Crater, craters	MC-21	Tsuneo; Japanese amateur astronomer (1916–1996)
Salaga	28.03	-47.19	308.89	Crater, craters	MC-26	Town in Ghana
Samara Valles	661.84	-24.17	341.27	Vallis, valles	MC-19	Ancient name for modern Somme River, France
San Juan	1.23	22.87	311.96	Crater, craters	MC-10	Puerto Rican port
Sandila	13.32	-25.56	329.65	Crater, craters	MC-19	Town in India
Sangar	30.33	-27.53	335.66	Crater, craters	MC-19	Town in Russia
Santa Cruz	1.35	21.25	312.74	Crater, craters	MC-10	Port of Canary Islands
Santa Fe	20.3	19.28	312.05	Crater, craters	MC-10	Town in New Mexico, USA
Santaca	15.85	-41.06	87.37	Crater, craters	MC-28	Town in Mozambique
Saravan	46.89	-16.93	305.98	Crater, craters	MC-18	Town in Laos
Sarh	50.27	-64.85	345.42	Crater, craters	MC-26	Town in the Republic of Chad
Sarn	11.37	-77.34	305.28	Crater, craters	MC-30	Town in Wales
Sarno	20.29	-44.37	305.85	Crater, craters	MC-26	Town in Italy
Satka	18.8	-42.68	323.06	Crater, craters	MC-26	Town in Russia
Sauk	3.08	-44.67	327.44	Crater, craters	MC-26	Town in Wisconsin, USA
Savannah	1.34	22.02	312.22	Crater, craters	MC-10	American colonial town (Georgia)
Savich	179.06	-27.49	96.12	Crater, craters	MC-22	Aleksey N. Savich; Russian astronomer (1811–1883)
Say	13.59	-28.07	330.33	Crater, craters	MC-19	Town in Niger
Scamander Vallis	269	15.89	28.53	Vallis, valles	MC-12	Ancient name of river at Troy (modern Turkey)
Scandia Cavi	663.8	77.55	209.65	Cavus, cavi	MC-1	Named for classical albedo feature at 65° N, 150° W
Scandia Colles	1521.68	65.47	220.87	Collis, colles	MC-1, MC-2	From albedo feature name
Scandia Tholi	398.27	73.91	201.28	Tholus, tholi	MC-1	Named for classical albedo feature at 65° N, 150° W
Schaeberle	158.67	-24.37	50.23	Crater, craters	MC-21	John M.; American astronomer (1853–1924)
Schiaparelli	458.52	-2.71	16.77	Crater, craters	MC-12, MC-20	Giovanni V.; Italian astronomer (1835–1910)
Schmidt	201.35	-72.07	282.1	Crater, craters	MC-30	Johann F.; German astronomer (1825–1884); Otto Y.; Russian geophysicist (1891–1956)
Schöner	198.96	19.93	50.7	Crater, craters	MC-13	Johannes; German geographer (1477–1547)
Schroeter	291.59	-1.9	55.99	Crater, craters	MC-21	Johann H.; German astronomer (1745–1816)
Scylla Scopulus	476.91	-25.22	18.34	Scopulus, scopuli	MC-20	From albedo feature at 19° S, 320° W
Sebec	63.54	-39.5	99.41	Crater, craters	MC-28	Town in Maine, USA
Secchi	223.41	-57.84	102.15	Crater, craters	MC-28	Angelo; Italian astronomer (1818–1878)
Sefadu	10.84	28.74	325.03	Crater, craters	MC-11	Town in Sierra Leone

Feature name	Diameter (km)	Center latitude	Center longitude	Feature type	Quadrangle	Origin
Selevac	7.3	−37.39	228.93	Crater, craters	MC24	Village in Serbia
Semeykin	73.51	41.51	8.75	Crater, craters	MC-5	Boris Evgen'evich; Soviet astronomer (1900–1937)
Seminole	20.64	−24.18	340.89	Crater, craters	MC-19	Town in Florida, USA
Senus Vallis	22.18	−5.23	213.04	Vallis, valles	MC-16	Classical river in Ireland
Sepik Vallis	59	−1.01	294.27	Vallis, valles	MC-18	River in New Guinea
Sevel	7.39	79.21	323.78	Crater, craters	MC-1	Town in Denmark
Sevi	3.2	18.89	103.03	Crater, craters	MC-14	Town in Russia
Sfax	6.7	−7.67	316.58	Crater, craters	MC-19	Town in Tunisia
Shalbatana Vallis	1029	7.33	317.91	Vallis, valles	MC-11	Word for "Mars" in Akkadian
Shambe	35.58	−20.58	329.31	Crater, craters	MC-19	Town in Sudan
Shardi	16.71	10.05	344.68	Crater, craters	MC-11	Town in Pakistan
Sharonov	99.92	27	301.47	Crater, craters	MC-10	Vsevolod V.; Russian astronomer (1901–1964)
Shatskiy	69.46	−32.36	345.11	Crater, craters	MC-26	N. S.; Russian geologist
Shawnee	16.71	22.49	328.49	Crater, craters	MC-11	Town in Ohio, USA
Sian	4.06	19.96	312	Crater, craters	MC-10	Town in Russia
Sibiti	33	−12.37	294.74	Crater, craters	MC-18	Town in the Republic of Congo (Congo-Brazzaville)
Sibu	17.63	−23.02	340.28	Crater, craters	MC-19	Town in Malaysia
Sibut	22.18	9.68	310.65	Crater, craters	MC-10	Town in Central African Republic
Sigli	30.3	−20.31	329.19	Crater, craters	MC-19	Town in Indonesia
Silinka Vallis	150.93	9.13	331.95	Vallis, valles	MC-11	River in Russia
Siloe Patera	39.08	35.3	6.55	Patera, paterae	MC-5	Classical albedo feature name
Simois Colles	86.94	−37.72	183.41	Collis, colles	MC-24	Classical albedo feature name of Greek origin
Simud Vallis	987.99	19.09	321.99	Vallis, valles	MC-11	Word for "Mars" in Sumerian
Sinai Dorsa	456.54	−12.77	281.08	Dorsum, dorsa	MC-18	Classical albedo feature name
Sinai Fossae	589.15	−14.08	281.33	Fossa, fossae	MC-18	Classical albedo feature name
Sinai Planum	901.44	−13.72	272.24	Planum, plana	MC-17, MC-18	Classical albedo feature name
Sinda	6.67	15.75	111.28	Crater, craters	MC-14	Town in Russia
Singa	13.14	−22.43	342.67	Crater, craters	MC-19	Town in Sudan
Sinop	14.72	−23.28	110.62	Crater, craters	MC-22	Town in Turkey
Sinton	62.8	40.75	31.73	Crater, craters	MC-5	William M.; American astronomer (1925–2004)
Sirenum Fossae	2731.21	−35.57	197.26	Fossa, fossae	MC-16, MC-24, MC-29	Classical albedo feature name
Sirenum Mons	122.86	−38.22	212.15	Mons, montes	MC-24	Classical albedo feature name
Sirenum Tholus	53.9	−34.64	215.21	Tholus, tholi	MC-24	Classical albedo feature name
Sisyphi Cavi	423.63	−72.2	353.7	Cavus, cavi	MC-30	Classical albedo feature name
Sisyphi Montes	200	−69.65	13.08	Mons, montes	MC-30	From albedo feature at 67° S, 348° W
Sisyphi Planum	1032.87	−69.64	6.41	Planum, plana	MC-26, MC-27, MC-30	Classical albedo feature name
Sisyphi Tholus	27.52	−75.68	341.47	Tholus, tholi	MC-30	Classical albedo feature name

Feature name	Diameter (km)	Center latitude	Center longitude	Feature type	Quadrangle	Origin
Sitka	16.89	−4.28	320.77	Crater, craters	MC-19	Town in Alaska, USA
Siton Undae	222.97	75.55	297.28	Unda, undae	MC-1	Classical albedo feature name
Sitrah	33.09	−59.09	217.72	Crater, craters	MC-24	Village in Bahrain
Sklodowska	109.72	33.52	357.05	Crater, craters	MC-4	Marie; Polish-born French chemist (Mme P. Curie) (1867–1934)
Slipher	127.14	−47.34	275.54	Crater, craters	MC-25	Vesto M.; American astronomer (1875–1969), Earl C.; American astronomer (1883–1964)
Smith	74.33	−65.76	257.27	Crater, craters	MC-30	William; British geologist-engineer (1769–1839)
Soffen	58.31	−23.73	140.86	Crater, craters	MC-23	Gerald A.; American astrobiologist (1926–2000)
Sögel	28.45	21.43	304.85	Crater, craters	MC-10	Town in Germany
Sokol	22.18	−42.37	319.32	Crater, craters	MC-26	Town in Russia
Solano	9	−26.74	108.95	Crater, craters	MC-22	Town in Philippines
Solis Dorsa	779.5	−22.88	280.26	Dorsum, dorsa	MC-18	Classical albedo feature name
Solis Planum	1811.23	−26.4	270.33	Planum, plana	MC-17, MC-18, MC-25	Classical albedo feature name
Somerset	3.33	−9.73	308.74	Crater, craters	MC-18	Town in Pennsylvania, USA
Soochow	30.06	16.73	331.18	Crater, craters	MC-11	Town in China (Kiangsu)
Souris	2.93	19.47	113.31	Crater, craters	MC-14	Town in Manitoba, Canada
South	101.84	−76.94	21.91	Crater, craters	MC-30	James; British astronomer (1785–1867)
Spallanzani	71.69	−58.01	86.38	Crater, craters	MC-28	Lazzaro; Italian biologist (1729–1799)
Spry	7.67	−3.7	321.57	Crater, craters	MC-19	Town in Utah, USA
Spur	8.09	22.02	307.74	Crater, craters	MC-10	Town in Texas, USA
Sripur	22.99	−30.74	259.29	Crater, craters	MC-25	Town in Bangladesh
Stege	76.44	3.75	300.5	Crater, craters	MC-10	Town in Denmark
Steinheim	11.28	54.57	190.65	Crater, craters	MC2	Town in Baden-Württemberg, Germany
Steno	103.54	−67.75	244.63	Crater, craters	MC-30	Nicolaus; Danish geologist (1638–1686)
Stobs	12.06	−4.96	321.7	Crater, craters	MC19	Town in Scotland
Stokes	62.74	55.63	171.29	Crater, craters	MC-7	George G.; British physicist (1819–1903)
Ston	6.49	46.87	122.55	Crater, craters	MC7	Town in Croatia
Stoney	161.37	−69.61	221.49	Crater, craters	MC-30	George J.; Irish physicist (1826–1911)
Stura Vallis	75	22.71	142.47	Vallis, valles	MC15	Classical river east of Rome, Italy
Stygis Catena	65.38	23.25	150.57	Catena, catenae	MC15	From albedo feature at 30° N, 200° W
Stygis Fossae	385	26.92	149.83	Fossa, fossae	MC-15	Albedo feature name
Styx Dorsum	90.67	30.81	151.86	Dorsum, dorsa	MC7	Albedo feature name
Suata	23.9	−18.91	106.67	Crater, craters	MC-22	Town in Venezuela
Subur Vallis	26.2	11.63	306.85	Vallis, valles	MC-10	Classical river in Mauritania
Sucre	13.56	23.69	305.41	Crater, craters	MC-10	Town in Colombia

Feature name	Diameter (km)	Center latitude	Center longitude	Feature type	Quadrangle	Origin
Suess	71.9	−66.88	181.51	Crater, craters	MC-30	Eduard; Austrian geologist, engineer (1831–1914)
Süf	9.41	16.34	321.78	Crater, craters	MC-11	Town in Jordan
Sulak	25	18.17	281.39	Crater, craters	MC-10	Town in Russia
Sulci Gordii	400	19.02	234.27	Sulcus, sulci	MC-9	Classical albedo feature name
Sumgin	78.6	−36.53	311.33	Crater, craters	MC-26	M. I.; Russian cryopedologist
Sungari Vallis	344.82	−40.33	88.5	Vallis, valles	MC-28	River in China, also known as Songhua Jiang
Surinda Valles	80.07	−28.8	324.89	Vallis, valles	MC-19	Name proposed by Soviets; found on Mars-5 Map
Surius Vallis	30.76	−61.2	311.28	Vallis, valles	MC-26	Lowell canal name
Surt	9.85	16.85	329.36	Crater, craters	MC-11	Town in Libya
Suzhi	24.63	−27.41	86.1	Crater, craters	MC-21	Town in China
Swanage	18.68	26.45	326.33	Crater, craters	MC-11	Town in England
Syria Colles	630	−13.46	259.27	Collis, colles	MC-17	Classical albedo feature name, Syria
Syria Mons	73.47	−13.88	255.73	Mons, montes	MC-17	Classical albedo feature name
Syria Planum	735.74	−12.09	256.1	Planum, plana	MC-17	Classical albedo feature name
Syrtis Major Planum	1214.86	9.2	67.1	Planum, plana	MC-13	Albedo feature name; changed from Planitia to Planum
Sytinskaya	89.16	42.42	306.94	Crater, craters	MC-4	Nadezhda Nikolaevna; Soviet astronomer (1906–1974)
Tábor	19.11	−35.5	301.67	Crater, craters	MC-26	Town in the Czech Republic
Tabou	7.68	−45.1	324.96	Crater, craters	MC-26	Town in Ivory Coast
Tader Valles	200	−48.78	207.7	Vallis, valles	MC-24	Ancient name for present Segura River, Spain
Taejin	28.06	−35.2	85.66	Crater, craters	MC-28	Town in the Republic of Korea
Tagus Valles	144.58	−6.68	114.54	Vallis, valles	MC-22	Ancient and modern river in Spain, Portugal
Tak	5.21	−26.02	331.35	Crater, craters	MC-19	Town in Thailand
Tola	8.51	−20.34	112.79	Crater, craters	MC-22	Town in Tunisia
Talas	30	−35.67	75.37	Crater, craters	MC-28	Town in Kyrgyzstan
Talsi	9.59	−41.53	310.63	Crater, craters	MC-26	Town in Latvia
Taltal	10	−39.51	234.21	Crater, craters	MC-24	Town in Chile
Talu	10	−40.35	20.09	Crater, craters	MC-27	Town in Indonesia
Tame	13.8	−22.73	252.01	Crater, craters	MC-17	Town in Colombia
Tana Vallis	56.51	4.78	332.11	Vallis, valles	MC-11	River in Kenya
Tanaica Montes	178.55	39.55	269.17	Mons, montes	MC-3	Classical albedo name
Tanais Fossae	172.95	38.74	273.51	Fossa, fossae	MC-3	Classical albedo feature name
Tantalus Fluctus	794.34	35.93	264.32	Fluctus, fluctūs	MC-3	Albedo feature name
Tantalus Fossae	2361.86	49.83	263.91	Fossa, fossae	MC-3	From albedo feature at 35 N, 110 W
Tara	32.94	−44.01	307.16	Crater, craters	MC-26	Town in Ireland
Tarakan	39.31	−41.21	329.56	Crater, craters	MC-26	Town in Indonesia (Borneo)

Feature name	Diameter (km)	Center latitude	Center longitude	Feature type	Quadrangle	Origin
Tarata	12.27	−3.78	318.78	Crater, craters	MC-19	Town in Bolivia
Tarma	9	16.54	109.85	Crater, craters	MC-14	Town in Peru
Tarq	35.36	−38.1	171.22	Crater, craters	MC-29	Town in Iran
Tarrafal	4.89	24.26	340.82	Crater, craters	MC-11	Town in Cape Verde
Tarsus	18.55	23.12	319.74	Crater, craters	MC-11	Town in Turkey
Tartarus Colles	1672.69	21.24	175.19	Collis, colles	MC-8, MC-15	From albedo feature at 2 N, 183 W
Tartarus Montes	1086.46	15.46	167.54	Mons, montes	MC-15	Classical albedo feature name
Tartarus Rupes	97.46	−6.5	175.71	Rupes, rupes	MC-23	Named for albedo feature at 2 N, 183 W
Tartarus Scopulus	251.31	−4.23	177.25	Scopulus, scopuli	MC-23	Named for albedo feature at 12 S, 182 W
Taus Vallis	10.4	−4.85	211.68	Vallis, valles	MC-16	Classical river in Caledonia (Scotland)
Tavua	31.56	15.62	117.61	Crater, craters	MC-14	Town in Fiji
Taxco	17.36	20.67	319.87	Crater, craters	MC-11	Town in Mexico
Taytay	18.17	7.39	340.4	Crater, craters	MC-11	Town in the Philippines
Taza	24.28	−43.57	314.7	Crater, craters	MC-26	Town in Morocco
Tecolote	47.93	−24.55	253.16	Crater, craters	MC-17	Town in New Mexico, USA
Teisserenc de Bort	114.89	0.43	45.07	Crater, craters	MC-12, MC13	Leon P.; French meteorologist (1855–1913)
Tejn	3.77	15.39	106.42	Crater, craters	MC-14	Town in Denmark
Telz	3.19	21.16	111.12	Crater, craters	MC-14	Town in Germany
Tem'	5.88	41.91	350.55	Crater, craters	MC-4	Town in Russia
Tempe Colles	34.49	33.75	277.44	Collis, colles	MC-3	Classical albedo name
Tempe Fossae	2116.24	40.42	288.6	Fossa, fossae	MC-3, MC-4	From albedo feature at 40 N, 70 W
Tempe Mensa	55.39	27.94	288.41	Mensa, mensae	MC-10	From albedo feature at 40 N, 70 W
Tempe Terra	1954.94	38.69	289.39	Terra, terrae	MC-3, MC-4, Figure 4.1	From albedo feature at 40 N, 70 W
Tenuis Cavus	51.53	84.76	1.39	Cavus, cavi	MC-1	Classical albedo feature name
Tenuis Mensa	120.92	81.13	267.04	Mensa, mensae	MC-1	Classical albedo feature name
Tepko	3.98	15.21	103.51	Crater, craters	MC-14	Town in Australia
Terby	171.5	−27.96	74.14	Crater, craters	MC-21	Francois J.; Belgian astronomer (1846–1911)
Termes Vallis	55.6	−11.11	202.99	Vallis, valles	MC-16	Classical river in ancient Lusitania, present Tormes River, Spain
Terra Cimmeria	5855.87	−32.68	147.75	Terra, terrae	MC-14, MC-22, MC-23, MC-29, MC-30, Figure 4.1	Classical albedo feature name
Terra Sabaea	4688.44	2.72	51.3	Terra, terrae	MC-5, MC-6, MC-12, MC-13, MC-20, MC-21, Figure 4.1	Classical albedo feature name

Feature name	Diameter (km)	Center latitude	Center longitude	Feature type	Quadrangle	Origin
Terra Sirenum	3635.18	-39.49	205.85	Terra, terrae	MC-16, MC-24, MC-30	Classical albedo feature name
Teviot Vallis	143.89	-43.37	102.26	Vallis, valles	MC-28	River in Scotland
Tharsis Montes	2058.91	1.57	247.42	Mons, montes	MC-9, MC-17, Figure 4.1	Classical albedo feature name
Tharsis Tholus	149.3	13.25	269.31	Tholus, tholi	MC-9, MC-10	Classical albedo feature name
Thaumasia Fossae	996.18	-47.75	268.95	Fossa, fossae	MC-25	Classical albedo feature name
Thaumasia Planum	799.6	-21.66	294.78	Planum, plana	MC-18	Albedo feature at 30 S, 75° W
Thermia	2.76	19.67	109.17	Crater, craters	MC-14	Town in Greece
Thila	5.37	18.11	155.52	Crater, craters	MC-15	Town in Yemen
Thira	21.84	-14.47	175.98	Crater, craters	MC-23	Town on Santorini Island in the Aegean Sea
Thom	22.06	-41.11	92.35	Crater, craters	MC-28	Town in Thailand
Thule	13.07	-23.37	334.28	Crater, craters	MC-19	Town in Greenland
Thyles Montes	380	-69.88	126.54	Mons, montes	MC-30	Classical albedo feature name
Thyles Rupes	548.75	-69.32	132.28	Rupes, rupes	MC-30	Name and feature changes in 1984, from Ultimi Cavi and Thyles Chasma
Tibrikot	59.08	12.56	305.13	Crater, craters	MC-10	Town in Nepal
Tignish	20.98	-30.74	87.04	Crater, craters	MC-28	Town in Prince Edward Island, Canada
Tigre Valles	102.82	-12.02	322.91	Vallis, valles	MC-19	River in Peru
Tikhonravov	343.7	13.28	35.93	Crater, craters	MC-12	M. K.; Russian rocket scientist (1900–1974)
Tikhov	110.07	-50.68	105.8	Crater, craters	MC-28	Gavril A.; Russian astronomer (1875–1960)
Tile	8.47	17.73	331.38	Crater, craters	MC-11	Town in Somalia
Timaru	18.4	-25.27	337.66	Crater, craters	MC-19	Town in New Zealand
Timbuktu	65.68	-5.56	322.48	Crater, craters	MC-19	Town in Mali
Timoshenko	86.11	41.76	296	Crater, craters	MC-3	Ivan Fedorovich; Soviet astronomer (1918–1941)
Tinia Valles	17.83	-4.61	211.12	Vallis, valles	MC-16	Classical river in Italy
Tinjar Valles	400.58	37.54	124.27	Vallis, valles	MC-7	Modern river in Sarawak, Malaysia
Tinto Vallis	191.97	-3.97	111.5	Vallis, valles	MC-22	River in Spain
Tisia Valles	384.05	-10.75	46.72	Vallis, valles	MC-21	Ancient name for modern Tisza River, Ukraine
Tithoniae Catena	562	-5.5	278.18	Catena, catenae	MC-18	Classical albedo feature name
Tithoniae Fossae	838	-4.32	276.96	Fossa, fossae	MC-18	Classical albedo name
Tithonium Chasma	802.78	-4.6	275.71	Chasma, chasmata	MC-18	Classical albedo feature name
Tiu Valles	1720	16.23	325.14	Vallis, valles	MC-11	Word for "Mars" in old English (West Germanic)
Tivat	3.62	-45.93	9.53	Crater, craters	MC-27	Town in Montenegro
Tivoli	32.8	-14.33	100.91	Crater, craters	MC-22	Town in Grenada
Tiwi	21.15	-27.56	335.24	Crater, craters	MC-19	Town in Oman
Toconao	17.16	-20.85	285.31	Crater, craters	MC-18	Town in Chile
Tokko	2.7	22.55	109.52	Crater, craters	MC-14	Town in Russia
Tokma	3.28	21.31	108.57	Crater, craters	MC-14	Town in Russia
Tolon	2.64	18.23	104.98	Crater, craters	MC-14	Town in Russia
Tomari	5.29	19.98	113.78	Crater, craters	MC-14	Town in Russia
Tombaugh	59.84	3.56	161.92	Crater, craters	MC-15	Clyde William; American astronomer (1906–1997)
Tombe	5.99	-42.4	315.45	Crater, craters	MC-26	Town in Sudan
Tomini	7.77	16.26	125.88	Crater, craters	MC-14	Town in Indonesia
Tooting	27.86	23.21	207.76	Crater, craters	MC-8	Town in England
Torbay	6.33	17.87	114.08	Crater, craters	MC-14	Town in Australia
Toro	41.4	17.04	71.82	Crater, craters	MC-13	Town in Spain
Torsö	15.3	-44.29	308.82	Crater, craters	MC-26	Town in Sweden
Torup	42.72	-27.89	97.81	Crater, craters	MC-22	Town in Sweden
Tractus Catena	910.57	27	257.21	Catena, catenae	MC-3, MC-9	Classical albedo feature name
Tractus Fossae	403.06	25.89	258.72	Fossa, fossae	MC-9	Classical albedo feature name
Trebia Valles	179.8	32.08	150.12	Vallis, valles	MC-7	Classical name for modern Trebbia River, Italy
Trinidad	27.91	-23.38	109.05	Crater, craters	MC-22	Town in Peru
Triolet	12.14	-37.09	191.98	Crater, craters	MC-24	Town in the Republic of Mauritius
Troika	13.43	16.83	105.14	Crater, craters	MC-14	Town in Russia
Trouvelot	148.77	16.09	347.02	Crater, craters	MC-11	Étienne Léopold; French astronomer (1827–1895)
Troy	9.59	23.17	307.38	Crater, craters	MC-10	Town in Idaho, USA
Trud	2.44	17.68	328.41	Crater, craters	MC-11	Town in Russia
Trumpler	75.35	-61.43	209.29	Crater, craters	MC-24	Robert J.; American astronomer (1886–1956)
Tsau	6.61	49.49	121.06	Crater, craters	MC-7	Town in Botswana
Tsukuba	1.86	48.58	134	Crater, craters	MC-7	Japanese mission control site
Tuapi	4.48	16.98	104.34	Crater, craters	MC-14	Town in Nicaragua
Tugaske	30.89	-31.78	258.89	Crater, craters	MC-25	Town in Saskatchewan
Tumul	8.68	14.71	104.61	Crater, craters	MC-14	Town in Nicaragua
Tungla	16.61	-40.77	89.64	Crater, craters	MC-28	Town in Nicaragua
Tura	14.81	-26.63	338.02	Crater, craters	MC-19	Town in Russia
Turbi	30.59	-40.62	308.55	Crater, craters	MC-26	Town in Kenya
Turma	6.68	17.31	108.11	Crater, craters	MC-14	Town in Russia
Tuscaloosa	59.66	-0.02	28.73	Crater, craters	MC-12, MC-20	Town in Alabama, USA
Tuskegee	62.88	-2.8	323.91	Crater, craters	MC-19	Town in Alabama, USA
Tycho Brahe	105.27	-49.41	146.12	Crater, craters	MC-29	Danish astronomer (1546–1601)
Tyndall	83.05	39.73	169.97	Crater, craters	MC-7	John; British physicist (1820–1893)
Tyras Vallis	99.13	8.33	309.85	Vallis, valles	MC-10	Classical name for present Dniester River, Ukraine
Tyrrhena Dorsa	779.4	-24.2	115.72	Dorsum, dorsa	MC-22	Albedo feature name
Tyrrhena Fossae	305.55	-22.23	105.8	Fossa, fossae	MC-22	Classical albedo name
Tyrrhena Patera	12.64	-21.39	106.63	Patera, paterae	MC-22	Classical albedo feature name
Tyrrhena Terra	2470.14	-11.9	88.84	Terra, terrae	MC-14, MC-21, MC-22, Figure 4.1	Classical albedo feature name

Feature name	Diameter (km)	Center latitude	Center longitude	Feature type	Quadrangle	Origin
Tyrrhenus Labyrinthus	102.68	−16.18	101.12	Labyrinthus, labyrinthi	MC-22	Classical albedo feature name
Tyrrhenus Mons	269.77	−21.63	105.88	Mons, montes	MC-22	Classical albedo feature name
Tyuratam	0.3	−45.04	202.04	Crater, craters	MC-24	[Töretam] Township in Kazakhstan
Ubud	27	−10.63	341.68	Crater, craters	MC-19	Town in Indonesia
Udzha	42.87	81.92	77.35	Crater, craters	MC-1	Village in northern Russia
Ultima Lingula	551.28	−76.32	142.56	Lingula, lingulae	MC-30	Classical albedo feature name
Ultimi Scopuli	560.47	−77.88	179.04	Scopulus, scopuli	MC-30	Classical albedo feature name
Ultimum Chasma	322.09	−81.1	151.37	Chasma, chasmata	MC-30	Classical albedo feature name
Ulu	3.43	22.49	107.32	Crater, craters	MC-14	Town in Russia
Ulya	8.02	−17.9	111.68	Crater, craters	MC-22	Town in Russia
Ulysses Colles	84.81	6.14	236.91	Collis, colles	MC-9	From albedo feature name
Ulysses Fossae	849.94	9.95	236.93	Fossa, fossae	MC-9	Classical albedo feature name
Ulysses Patera	57.86	2.95	238.58	Patera, paterae	MC-9	Classical albedo feature name
Ulysses Tholus	102.47	2.96	238.5	Tholus, tholi	MC-9	Classical albedo feature name
Ulyxis Rupes	383.09	−68.78	160.02	Rupes, rupes	MC-30	Classical albedo feature name
Umatac	17.16	42.52	137.26	Crater, craters	MC-7	Town in Guam, USA
Uranius Dorsum	542.08	23.79	284.96	Dorsum, dorsa	MC-10	Named for albedo feature
Uranius Fossae	394.11	25.29	269.87	Fossa, fossae	MC-9, MC-10	Classical albedo feature name
Uranius Mons	265.17	26.9	267.85	Mons, montes	MC-9	Classical albedo feature name
Uranius Patera	114	26.32	267.2	Patera, paterae	MC-9	Classical albedo feature name
Uranius Tholus	61.39	26.25	262.43	Tholus, tholi	MC-9	Classical albedo feature name
Urk	2.89	23.11	111.42	Crater, craters	MC-14	Town in Netherlands
Ulan	4.73	24.24	113.81	Crater, craters	MC-14	Town in Russia
Utopia Planitia	3560.45	46.74	117.52	Planitia, planitiae	MC-6, MC-14, Figure 4.1	Classical albedo feature name
Utopia Rupes	2492.68	43.53	86.03	Rupes, rupes	MC-17, MC-18, MC-19, Figure 4.1	Named for classical albedo feature at 55° N, 260° W
Uzboi Vallis	353.53	−29.46	323.02	Vallis, valles	MC-19, MC-26	Dry riverbed in Russia
Uzer	9.24	−1.22	358.25	Crater, craters	MC-19	Town in France
Vaals	10.85	−3.96	327.03	Crater, craters	MC-19	Town in the Netherlands
Vaduz	2	38.24	15.79	Crater, craters	MC-5	Capital of the Principality of Liechtenstein
Valga	15.7	−44.32	323.36	Crater, craters	MC-26	Town in Estonia
Valles Marineris	3761.28	−14.01	301.41	Vallis, valles	MC-17, MC-18, MC-19, Figure 4.1	General name of the system of canyons honoring the scientific team of the Mariner 9 program
Valverde	34.92	20.1	304.24	Crater, craters	MC-10	Town in the Dominican Republic

Feature name	Diameter (km)	Center latitude	Center longitude	Feature type	Quadrangle	Origin
Varus Vallis	90.12	−8.57	204.01	Vallis, valles	MC-16	Classical name for present Var River, France
Vastitas Borealis	2002.91	87.73	32.53	Vastitas, vastitates	MC-1, Figure 4.1	Classical albedo feature name
Väto	17.24	−43.61	306.31	Crater, craters	MC-26	Town in Sweden
Vaux	5.99	17.96	327.21	Crater, craters	MC-11	Town in France
Vedra Valles	118	19.12	304.52	Vallis, valles	MC-10	Ancient European river (Great Britain)
Verde Vallis	133	−0.5	29.88	Vallis, valles	MC-20	River in Arizona, USA
Verlaine	38.84	−9.22	64.12	Crater, craters	MC-21	Town in France
Vernal	55.51	5.9	355.55	Crater, craters	MC-11	Town in Utah, USA
Very	114.81	−49.17	182.97	Crater, craters	MC-24	Frank W.; American astronomer (1852–1927)
Viana	29.03	19.18	104.81	Crater, craters	MC-14	Town in Brazil
Vichada Valles	438.31	−19.87	88.13	Vallis, valles	MC-21	River in Colombia
Victoria	0.88	−2.05	354.5	Crater, craters	MC-19	Town in the Republic of Seychelles
Vik	22.32	−36.09	296.06	Crater, craters	MC-25	Town in Iceland
Vils	6.68	39.04	348.32	Crater, craters	MC-4	Town in Austria
Vinogradov	209.66	−19.83	322.26	Crater, craters	MC-19	Aleksander P.; Soviet geochemist (1895–1975)
Vinogradsky	66.26	−56.13	143.85	Crater, craters	MC-29	Sergei N.; Russian microbiologist (1856–1953)
Virrat	50.67	−30.73	257.12	Crater, craters	MC-25	Town in Finland
Vishniac	80.47	−76.52	84.12	Crater, craters	MC-30	Wolf V.; American microbiologist (1922–1974)
Vistula Valles	193	13.41	308.03	Vallis, valles	MC10	Classical name for modern Wisła River in Poland
Vivero	27.13	48.97	118.83	Crater, craters	MC-6	Town in Spain
Voeykov	75.45	−32.11	283.86	Crater, craters	MC-25	A. I.; Russian climatologist and geographer (1842–1916)
Vogel	120.69	−36.77	346.72	Crater, craters	MC-26	Hermann Carl; German astronomer (1841–1907)
Vol'sk	8.46	23	308.76	Crater, craters	MC-10	Town in Russia
Volgograd	1.59	48.1	135.03	Crater, craters	MC-7	Soviet launch site
Von Kármán	90.29	−64.27	301.3	Crater, craters	MC-26	Theodore; Hugarian-American aeronautical engineer (1881–1963)
Voo	2.13	−26.94	340.01	Crater, craters	MC-19	Town in Kenya
Voza	2.71	23.34	306.47	Crater, craters	MC-10	Town in Solomon Islands
W. Mareotis Tholus	13.19	35.56	272.04	Tholus, tholi	MC-3	Classical albedo name
Wabash	40.71	21.36	326.36	Crater, craters	MC-11	Town in Indiana, USA
Wafra	30.19	4.25	148.54	Crater, craters	MC-15	Town in Kuwait
Wahoo	63.07	23.23	326.32	Crater, craters	MC-11	Town in Nebraska, USA
Waikato Vallis	228.03	−33.33	113.78	Vallis, valles	MC-28	River in New Zealand
Wajir	11.85	−27.02	105.54	Crater, craters	MC-22	Town in Kenya
Walla Walla Vallis	22.99	−9.88	305.54	Vallis, valles	MC-18	River in Washington, USA

Feature name	Diameter (km)	Center latitude	Center longitude	Feature type	Quadrangle	Origin
Wallace	170.78	-52.48	110.9	Crater, craters	MC-28	Alfred R.; British biologist (1823–1913)
Wallops	1.84	46.59	132.72	Crater, craters	MC-7	American launch site
Wallula	12.13	-9.92	305.6	Crater, craters	MC-18	Town in Washington, USA
Warra	10.12	20.75	322.37	Crater, craters	MC-11	Town in Australia
Warrego Valles	205.08	-41.84	267.85	Vallis, valles	MC-25	Modern Australian River
Waspam	41.6	20.45	303.37	Crater, craters	MC-10	Town in Nicaragua
Wassamu	16.3	25.57	306.79	Crater, craters	MC-10	Town in Japan
Wau	6.79	-44.86	317.39	Crater, craters	MC-26	Town in New Guinea
Weert	9.49	19.71	308.31	Crater, craters	MC-10	Town in the Netherlands (spelling corrected from Weer)
Wegener	68.51	-64.3	355.93	Crater, craters	MC-26	Alfred L.; German geophysicist (1880–1930)
Weinbaum	82.01	-65.53	114.57	Crater, craters	MC-30	Stanley G.; American novelist (1902–1935)
Wells	98.28	-59.94	122.4	Crater, craters	MC-29	Herbert G.; British novelist (1866–1946)
Wer	3.21	45.67	353.81	Crater, craters	MC-4	Town in India
Wicklow	21.5	-2.01	319.47	Crater, craters	MC-19	Town in Ireland
Wien	115.14	-10.57	139.75	Crater, craters	MC-23	Wilhelm; German physicist (1864–1928)
Williams	123.2	-18.39	195.86	Crater, craters	MC-16	Arthur S.; British astronomer (1861–1938)
Wilmington	1.35	21.6	312.53	Crater, craters	MC-10	American colonial town (Delaware)
Wiltz	1.26	15.54	159.21	Crater, craters	MC-15	Town in Luxembourg
Windfall	17.55	-2.09	316.67	Crater, craters	MC-19	Town in Alberta, Canada
Wink	10.16	-6.51	318.66	Crater, craters	MC-19	Town in Texas, USA
Winslow	1.08	-3.74	59.16	Crater, craters	MC-21	Town in Arizona
Wirtz	120.26	-48.24	334.14	Crater, craters	MC-26	Carl Wilhelm; German astronomer (1876–1939)
Wislicenus	140.15	-18.17	11.39	Crater, craters	MC-20	Walter; German astronomer (1859–1905)
Woking	9.53	5.12	82.99	Crater, craters	MC-13	Town in England
Woolgar	15.31	34.66	274.55	Crater, craters	MC-3	Town in Australia
Woomera	2.26	48.07	132.62	Crater, craters	MC-7	Australian launch site
Worcester	24.05	26.61	309.63	Crater, craters	MC-10	Town in New York, USA
Wright	113.78	-58.51	208.99	Crater, craters	MC-24	William H.; American astronomer (1871–1959)
Wukari	38.21	-31.81	257.2	Crater, craters	MC-25	Town in Nigeria
Wynn-Williams	66.31	-55.1	60.21	Crater, craters	MC-27, MC-28	David D.; English astrobiologist (1946–2002)
Xainza	23.96	0.78	356.06	Crater, craters	MC-11	Town in China
Xanthe Chaos	34.37	11.87	317.78	Chaos, chaoses	MC-11	Classical albedo feature name
Xanthe Dorsa	0	35.9	325.96	Dorsum, dorsa	MC-4, MC-10, MC-11	Classical albedo feature name
Xanthe Montes	499.32	18.13	305.08	Mons, montes	MC-10	Classical albedo feature name
Xanthe Scopulus	59.48	19.38	307.49	Scopulus, scopuli	MC-10	Classical albedo feature name
Xanthe Terra	1867.65	1.6	311.95	Terra, terrae	MC-10, MC-11, MC-18, MC-19, Figure 4.1	Classical albedo feature name

Feature name	Diameter (km)	Center latitude	Center longitude	Feature type	Quadrangle	Origin
Xui	3.15	15.09	112.63	Crater, craters	MC-14	Town in Brazil
Yakima	12.53	43.03	356.85	Crater, craters	MC-4	Town in Washington, USA
Yala	19.65	17.37	321.42	Crater, craters	MC-11	Town in Thailand
Yalata	4.74	21.81	106.17	Crater, craters	MC-14	Town in Australia
Yalgoo	17.38	4.93	84.23	Crater, craters	MC-13	Town in Australia
Yar	6.18	22.27	320.85	Crater, craters	MC-11	Town in Russia
Yaren	9.19	-43.88	222.55	Crater, craters	MC-24	Town in Nauru
Yat	7.46	18.13	330.97	Crater, craters	MC-11	Town in Niger
Yebra	4.85	20.79	105.69	Crater, craters	MC-14	Town in Spain
Yegros	13.98	-22.3	336.34	Crater, craters	MC-19	Town in Paraguay
Yellowknife	0.12	-4.58	137.44	Crater, craters	MC-23	Town in the Northwest Territories, Canada
Yorktown	8.01	22.88	311.35	Crater, craters	MC-10	American colonial town (Virginia)
Yoro	9.61	22.8	331.96	Crater, craters	MC-11	Town in Honduras
Yungay	19.69	-43.87	315.25	Crater, craters	MC-26	Town in Peru
Yuty	19.06	22.16	325.91	Crater, craters	MC-11	Town in Paraguay
Zarand	2.78	-3.41	358.5	Crater, craters	MC-19	Town in Iran
Zaranj	27.41	12.09	113.05	Crater, craters	MC-14	Town in Afghanistan
Zarqa Valles	21.37	0.32	80.59	Vallis, valles	MC-13, MC-21	River in Jordan
Zea Dorsa	249.02	-48.87	80.54	Dorsum, dorsa	MC-28	Classical albedo feature name
Zephyria Fluctus	42	0.72	155.53	Fluctus, fluctus	MC-15	Classical albedo feature name
Zephyria Mensae	333.61	-11.62	171.98	Mensa, mensae	MC-23	Albedo feature name
Zephyria Planum	575.11	-1.08	153.73	Planum, plana	MC-15, MC-23	Classical albedo feature name
Zephyria Tholus	35.95	-19.75	172.92	Tholus, tholi	MC-23	Albedo feature name
Zephyrus Fossae	306.19	23.93	144.19	Fossa, fossae	MC-15	Albedo feature name
Zhigou	21.86	-29.1	257.41	Crater, craters	MC-17	Town in China
Zilair	46.91	-31.81	327.06	Crater, craters	MC-26	Town in Russia
Zir	6.16	18.54	323.46	Crater, craters	MC-11	Town in Turkey
Zongo	46.83	-33.76	318.31	Crater, craters	MC-26	Town in Democratic Republic of Congo
Žulanka	43.12	-2.27	317.84	Crater, craters	MC-19	Town in Russia
Zumba	2.93	-28.67	226.93	Crater, craters	MC-17	Town in Ecuador
Zuni	24.28	19.22	330.42	Crater, craters	MC-11	Town in New Mexico, USA
Zunil	10.26	7.7	166.19	Crater, craters	MC-15	Mayan village in Guatemala
Zutphen	38.29	-13.85	174.32	Crater, craters	MC-23	Town in The Netherlands

Features on Moons

Features shown on atlas sheets are in bold.

Feature name	Target	Diameter (km)	Center latitude	Center longitude	Feature type	Origin
Stickney	**Phobos**	9	1	49	Crater, craters	Angeline; wife of American astronomer A. Hall (1830–1892)
Todd	**Phobos**	2.6	–9	153	Crater, craters	David; American astronomer (1855–1939)
Wendell	**Phobos**	1.7	–1	132	Crater, craters	Oliver C.; American astronomer (1845–1912)

Feature name	Target	Diameter (km)	Center latitude	Center longitude	Feature type	Origin
Swift	**Deimos**	1	12.5	358.2	Crater, craters	Jonathan; British writer (1667–1745).
Voltaire	**Deimos**	1.9	22	3.5	Crater, craters	Francios-Marie Arouet; French writer (1694–1778)

Feature name	Target	Diameter (km)	Center latitude	Center longitude	Feature type	Origin
Clustril	**Phobos**	3.4	60	91	Crater, craters	Character in Lilliput who (along with Drunlo) informed Flimnap that his wife had visited Gulliver privately, in Jonathan Swift's novel, *Gulliver's Travels*
D'Arrest	**Phobos**	2.1	–39	179	Crater, craters	Heinrich L.; German/Danish astronomer (1822–1875)
Drunlo	**Phobos**	4.2	36.5	92	Crater, craters	Character in Lilliput who (along with Clustril) informed Flimnap that his wife had visited Gulliver privately, in Jonathan Swift's novel, *Gulliver's Travels*
Flimnap	**Phobos**	1.5	60	350	Crater, craters	Treasurer of Lilliput in Jonathan Swift's novel, *Gulliver's Travels*
Grildrig	**Phobos**	2.6	81	195	Crater, craters	Name given to Gulliver by the farmer's daughter in the giants' country, Brobdingnag, in Jonathan Swift's novel, *Gulliver's Travels*
Gulliver	**Phobos**	5.5	62	163	Crater, craters	Lemuel Gulliver, surgeon, captain, and voyager, in Jonathan Swift's novel, *Gulliver's Travels*
Hall	**Phobos**	5.4	–80	210	Crater, craters	Asaph; American astronomer, discoverer of Phobos and Deimos (1829–1907)
Kepler Dorsum	**Phobos**	15	–45	356	Dorsum, dorsa	Johannes; German astronomer (1571–1630)
Lagado Planitia	**Phobos**	4	19	231	Planitia, planitiae	Fictional city in Jonathan Swift's *Gulliver's Travels*, ruled by a tyrannical king from the flying island Laputa
Laputa Regio	**Phobos**	14	0	265	Regio, regiones	Fictional flying island in Jonathan Swift's *Gulliver's Travels*
Limtoc	**Phobos**	2	–11	54	Crater, craters	General in Lilliput who prepared articles of impeachment against Gulliver in Jonathan Swift's novel, *Gulliver's Travels*
Öpik	**Phobos**	2	–7	297	Crater, craters	Ernst J.; Estonian astronomer (1893–1985)
Reldresal	**Phobos**	2.9	41	39	Crater, craters	Secretary for Private Affairs in Lilliput, Gulliver's friend, in Jonathan Swift's novel, *Gulliver's Travels*
Roche	**Phobos**	2.3	53	183	Crater, craters	Edouard; French astronomer (1820–1883)
Sharpless	**Phobos**	1.8	–27.5	154	Crater, craters	Bevan P.; American astronomer (1904–1950)
Shklovsky	**Phobos**	2	24	248	Crater, craters	Iosif S., Soviet astronomer (1916–1985)
Skyresh	**Phobos**	1.5	52.5	320	Crater, craters	Skyresh Bolgolam, High Admiral of the Lilliput council, who opposed Gulliver's plea for freedom and accused him of being a traitor, in Jonathan Swift's novel, *Gulliver's Travels*

References

Adeli, S., Hauber, E., Le Deit, L., and Jaumann, R. (2012). Sedimentary evolution of the Eridania paleolake in the Atlantis Chaos basin, Terra Sirenum. *Third Conf. Early Mars*, abs. 7047.

Agee, C. B., Wilson, N. V., McCubbin, F. M., *et al.* (2013). Unique meteorite from Early Amazonian Mars: Water-rich basaltic breccia northwest Africa 7034. *Science*, 339, 780–785.

Aharonson, O., Schorghofer, N., and Gerstell, M. F. (2003). Slope streak formation and dust deposition rates on Mars. *J. Geophys. Res.*, 108, 5138, doi:10.1029/2003JE002123.

Anderson, R. C. and Dohm, J. M. (2011). Unraveling the complex history of faulting for the Terra Sirenum region, Mars. *Lunar Planet. Sci. Conf.*, 42, abs. 2221.

Anderson, R. C., Dohm, J. M., Golombek, M. P., *et al.* (2001). Primary centers and secondary concentrations of tectonic activity through time for the western hemisphere of Mars. *J. Geophys. Res.*, 106, 20,563–20,585, doi:10.1029/2000JE001278.

Anderson, R. C., Dohm, J. M., Robbins, S., Hynek, S. B., and Andrews-Hanna, J. (2012). Terra Sirenum: Window into pre-Tharsis and Tharsis phases of Mars evolution. *Lunar Planet. Sci. Conf.*, 43, abs. 2803.

Andrews-Hanna, J. C. (2011). The formation of Valles Marineris, Mars. *Lunar Planet. Sci. Conf.*, 42, abs. 2182.

Andrews-Hanna, J. C., Zuber, M. T., and Banerdt, W. B. (2008). The Borealis basin and the origin of the Martian crustal dichotomy. *Nature*, 453, 1212–1215, doi:10.1038/nature07011.

Annex. A. M. and Howard, A. D. (2011). Phyllosilicates related to exposed knobs in Sirenum Fossae, Ariadnes Colles. *Lunar Planet. Sci. Conf.*, 42, abs. 1577.

Ansan, V. and Mangold, N. (2006). New observations of Warrego Valles, Mars: Evidence for precipitation and surface runoff. *Planet. Space Sci.*, 54, 219–242.

Ansan V., Loizeau, D., Mangold, N., *et al.* (2011). Stratigraphy, mineralogy, and origin of layered deposits inside Terby crater, Mars. *Icarus*, 211, 273–304.

Arvidson, R. E., Squyres, S. W., Anderson, R. C., *et al.* (2006). Overview of the Spirit Mars Exploration Rover Mission to Gusev crater: Landing site to Backstay Rock in the Columbia Hills. *J. Geophys. Res.*, 111, E02S01, doi:10.1029/2005JE002499.

Baker, D. M. H. and Head, J. W. (2012). The Noachian to Hesperian hydrologic evolution of the Ma'adim Vallis–Eridania basin region, Mars. *Third Conf. Early Mars*, abs. 7058.

Baker, V. R. and Milton, D. J. (1974). Erosion by catastrophic floods on Mars and Earth. *Icarus*, 23, 27–41, doi:10.1016/0019-1035(74)90101-8.

Baker, V. R., Carr, M. H., Gulick, V. C., Williams, C. R., and Marley, M. S. (1992). Channels and valley networks. In Kieffer, H. H., Jakosky, B. M., Snyder, C. W., and Matthews, M. S., eds., *Mars*. Tucson, AZ: University of Arizona Press, pp. 493–522.

Baker, V. R., Bjornstad, B. N., Gaylord, D. R., *et al.* (2016). Pleistocene megaflood landscapes of the Channeled Scabland. In Lewis, R. S., and Schmidt, K. L., eds., *Exploring the Geology of the Inland Northwest*. Geological Society of America Field Guide 41, pp. 1–73, doi:10.1130/2016.0041(01).

Bandfield, J. L. (2002). Global mineral distributions on Mars. *J. Geophys. Res.*, 107, doi:10.1029/2001JE001510.

Baptista, A. R. and Craddock, R. A. (2010). The Galapagos and Hawaii volcanoes: Two analogs of Syria Planum on Mars. *Lunar Planet. Sci. Conf.*, 41, abs. 1768.

Baptista, A. R., Mangold, N., Ansan, V., *et al.*, and HRSC team (2007). Coalesced small shield volcanoes on Syria Planum, Mars, detected by Mars Express-HRSC images. *Seventh Int. Mars Conf.*, abs. 3128.

Bargar, K. E. and Jackson, E. D. (1974). Calculated volumes of individual shield volcanoes along the Hawaiian–Emperor Chain. *J. Res. US Geol. Surv.*, 2, 545–550.

Barlow, N.G. (2008). *Mars: An Introduction to its Interior, Surface, and Atmosphere*. Cambridge: Cambridge University Press.

Barlow, N. G. and T. L. Bradley (1990). Martian impact craters: Correlations of ejecta and interior morphologies with diameter, latitude, and terrain. *Icarus*, 87, 156–179.

Basilevsky, A. T., Neukum, G., Werner, S. C., *et al.* (2009). Episodes of floods in Mangala Valles, Mars, from the analysis of HRSC, MOC and THEMIS images. *Planet. Space Sci.*, 57, 917–943.

Basilevsky, A. T., Lorenz, C. A., Shingareva, T. V., *et al.* (2014). The surface geology and geomorphology of Phobos. *Planet. Space Sci.*, 102, 95–118, doi:10.1016/j.pss.2014.04.013.

Batson, R. M., Bridges, P. M., and Inge, J. L. (1979). *Atlas of Mars*. NASA Special Publication SP-438, Washington, DC: Government Printing Office.

Bell, J. F., ed. (2008). *The Martian Surface: Composition, Mineralogy, and Physical Properties*. Cambridge: Cambridge University Press.

Bell, J. F., Malin, M. C., Caplinger, M. A., *et al.* (2013). Calibration and performance of the Mars Reconnaissance Orbiter Context Camera (CTX). *MARS*, 8, 1–14, doi:10.1555/mars.2013.0001.

Bergonio, J. R., Rottas, K. M., and Schorghofer, N. (2013). Properties of Martian slope streak populations. *Icarus*, 225, 194–199, doi:10.1016/j.icarus.2013.03.023.

Beyer, R. A., Stack, K. M., Griffes, J. L., *et al.* (2012). An atlas of Mars sedimentary rocks as seen by HiRISE. In Grotzinger, J. P., and Milliken, R. E., eds., *Sedimentary Geology of Mars*. SEPM Special Publication 102, pp. 49–95.

Bibring, J.-P., Langevin, Y., Mustard, J. F., *et al.* (2006). Global mineralogical and aqueous Mars history derived from OMEGA/Mars Express data. *Science*, 312, 400–404, doi:10.1126/science.1122659.

Bottke, W. F., Vokrouhlický, D., Nesvorný, D., *et al.* (2010). The E-belt: A possible missing link in the Late Heavy Bombardment. *Lunar Planet. Sci. Conf.*, 41, abs. 1269.

Boynton, W. V., Feldman, W. C., Mitrofanov, I. G., *et al.* (2004). The Mars Odyssey Gamma-Ray Spectrometer Instrument Suite. *Space Sci. Rev.*, 110, 37–83.

Boynton, W. V., Taylor, G. J., Evans, L. G., *et al.* (2007). Concentration of H, Si, Cl, Fe, and Th in the low- and mid-latitude regions of Mars. *J. Geophys. Res.*, 112, E12S99, doi:10.1029/2007JE002887.

Brothers, T. C., Holt, J. W., and Spiga, A. (2015). Planum Boreum basal unit topography, Mars: Irregularities and insights from SHARAD. *J. Geophys. Res.*, 120, 1357–1375, doi:10.1002/2015JE004830.

Broz, P. and Hauber, E. (2013). Hydrovolcanic tuff rings and cones as indicators for phreatomagmatic explosive eruptions on Mars. *J. Geophys. Res.*, 118, 1656–1675, doi:10.1002/jgre.20120.

Buhler, P. B., Ingersoll, A. P., Ehlmann, B. L., Fassett, C. I., and Head, J. W. (2017). How the Martian residual south polar cap develops quasi-circular and heart-shaped pits, troughs, and moats. *Icarus*, 286, 69–93, doi:10.1016/j.icarus.2017.01.012.

Burns, J. A. (1992). Contradictory clues as to the origin of the Martian moons. In Kieffer, H. H., Jakosky, B. M., Snyder, C. W., and Matthews, M. S., eds., *Mars*. Tucson, AZ: University of Arizona Press, pp. 1283–1301.

Burr, D. M., Nega, M.-T., Williams, R. M. E., *et al.* (2009a). Pervasive aqueous paleoflow features in the Aeolis/Zephyria Plana region, Mars. *Icarus*, 200, 52–76, doi:10.1016/j.icarus.2008.10.014.

Burr, D. M., Tanaka, K. L., and Yoshikawa, K. (2009b). Pingos on Earth and Mars. *Planet. Space Sci.*, 57, 541–555, doi:10.1016/j.pss.2008.11.003.

Caprarelli, G., Pondrelli, M., DiLorenzo, S., *et al.* (2007). A description of surface features in north Tyrrhena Terra, Mars: Evidence for extension and lava flooding. *Icarus*, 191, 524–544. doi:10.1016/j.icarus.2007.05.009.

Carr, M. H. (1979). Formation of Martian flood features by release of water from confined aquifers. *J. Geophysical Res.*, 84, 2995–3007.

Carr, M. H. (2006). *The Surface of Mars*. Cambridge: Cambridge University Press.

Carr, M. H. and Evans, N. (1980). *Images of Mars: The Viking Extended Mission*. NASA Special Publication SP-444. Washington, DC: Government Printing Office.

Carr, M. H. and Head, J. W., III (2010). Geologic history of Mars. *Earth and Planet. Sci. Letters*, 294, 185–203, doi:10.1016/j.epsl.2009.06.042.

Carr, M. H., Greeley, R., Blasius, K. R., Guest, J. E., and Murray, J. B. (1977). Some Martian volcanic features as viewed from the Viking orbiters. *J. Geophys. Res.*, 82, 3985–4015, doi:10.1029/JS082i028p03985.

Carter, J., Poulet, F., Bibring, J.-P., Mangold, N., and Murchie, S. (2013). Hydrous minerals on Mars as seen by the CRISM and OMEGA imaging spectrometers: Updated global view. *J. Geophys. Res.*, 118, doi:10.1029/2012JE004145.

Chadwick, J. and McGovern, P. (2011). Modelling subsidence due to the Olympus Mons load using paleo-slope indicators. *Lunar Planet. Sci. Conf.*, 42, abs. 2688.

Chapman, M. G. and Tanaka, K. L. (1993). Geologic map of the MTM-05152 and -10152 quadrangles, Mangala Valles region of Mars. *USGS Misc. Inv. Ser. Map I-2294*, scale 1:500,000.

Chapman, M. G., Neukum, G., Dumke, A., *et al.* (2010a). Amazonian geologic history of the Echus Chasma and Kasei Valles system on Mars: New data and interpretations. *Earth Planet. Sci. Lett.*, 294, 238–255.

Chapman, M. G., Neukum, G., Dumke, A., *et al.* (2010b). Noachian–Hesperian geologic history of the Echus Chasma and Kasei Valles system on Mars: New data and interpretations. *Earth Planet. Sci. Lett.*, 294, 256–271.

Christiansen, E. H. (1989). Lahars in the Elysium region of Mars. *Geology*, 17, 203–206.

Christensen, P. R. and Ruff, S. W. (2004). The formation of the hematite-bearing unit in Meridiani Planum: Evidence for deposition in standing water. *J. Geophys. Res.*, 109, E08003, doi:10.1029/2003JE002233.

Christensen, P. R., Bandfield, J. L., Hamilton, V. E., *et al.* (2001). The Mars Global Surveyor Thermal Emission Spectrometer experiment: Investigation description and surface science results. *J. Geophys. Res.*, 106, 23,823–23,871, doi:10.1029/2000JE001370.

Christensen, P. R., Bandfield, J. L., Bell, J. F., III, *et al.* (2003). Morphology and composition of the surface of Mars: Mars Odyssey THEMIS results. *Science*, 300, 2056–2061.

Christensen, P. R., Jakosky, B. M., Kieffer, H. H., *et al.* (2004). The Thermal Emission Imaging System

(THEMIS) for the Mars 2001 Odyssey mission. *Space Science Reviews*, **110**, 85–130.

Christensen, P. R., Ruff, S. W., Fergason, R. L., *et al.* (2005). Mars Exploration Rover candidate landing sites as viewed by THEMIS. *Icarus*, **187**, 12–43.

Chuang, F. C., Beyer, R. A., McEwen, A. S., and Thomson, B. J. (2007). HiRISE observations of slope streaks on Mars. *Geophys. Res. Letters*, **34**, L20204.

Chuang, F. C., Crown, D. A., Berman, D. C., Skinner, J. A., and Tanaka, K. L. (2011). Martian lobate debris aprons: Compilation of a new GIS-based global map. *Lunar Planet. Sci. Conf.*, **42**, abs. 2294.

Citron, R. I., Genda, H., and Ida, S. (2015). Formation of Phobos and Deimos via a giant impact. *Icarus*, **252**, 334–338, doi:10.1016/j.icarus.2015.02.011.

Clifford, S. M. and Parker, T. J. (2001). The evolution of the Martian hydrosphere: Implications for the fate of a primordial ocean and the current state of the Northern Plains. *Icarus*, **154**, 40–79.

Collins, S. A. (1971). *The Mariner 6 and 7 Pictures of Mars*. NASA Special Publication SP-263, Washington, DC: Government Printing Office.

Connerney, J. E. P., Acuña, M. H., Ness, N. F., *et al.* (2005). Tectonic implications of Mars crustal magnetism. *Proc. Natl. Acad. Sci.*, **102**, 14970–14975, doi:10.1073/pnas.0507469102.

Costard, F., Séjourné, A., Kelfoun, K., *et al.* (2017). Modeling tsunami propagation and the emplacement of thumbprint terrain in an early Mars ocean. *J. Geophys. Res.*, **122**, 633–649, doi:10.1002/2016JE005230.

Craddock, R. A. (2011). Are Phobos and Deimos the result of a giant impact? *Icarus*, **211**, 1150–1161, doi:10.1016/j.icarus.2010.10.023.

Craddock, R. A. and Greeley, R. (1994). Geologic map of the MTM-20147 quadrangle, Mangala Valles region of Mars. *USGS Misc. Inv. Ser. Map 1-2310*, scale 1:500,000.

Craddock, R. A. and Howard, A. D. (2002). The case for rainfall on a warm, wet early Mars. *J. Geophys. Res.*, **107**, 5111, doi:10.1029/2001JE001505.

Crumpler, L. S., Head, J. W., and Aubele, J. C. (1996). Calderas on Mars: Characteristics, structure, and associated flank deformation. *Geol. Soc. London, Special Pub.*, **110**, 307–348, doi:10.1144/GSL.SP.1996.110.01.24.

Cushing, G. E. (2011). Visible evidence of cave-entrance candidates in Martian fresh-looking pit craters. *Lunar Planet. Sci. Conf.*, **42**, abs. 2494.

Cushing, G. E. (2012). Candidate cave entrances on Mars. *J. Cave Karst Studies*, **74**, 33–47.

Davila, A. F., Fairén, A. G., Stokes, C. R., *et al.* (2013). Evidence for Hesperian glaciation along the Martian dichotomy boundary. *Geology*, **41**, 755–758, doi:10.1130/G34201.1.

de Pablo, M. A. and Centeno, J. D. (2012). Geomorphological map of the lower NW flank of Hecates Tholus volcano, Mars. *Lunar Planet. Sci. Conf.*, **43**, abs. 1098.

de Pablo, M. A., Michael, G. G., and Centeno, J. D. (2013). Age and evolution of the lower NW flank of the Hecates Tholus volcano, Mars, based on crater size-frequency distribution on CTX images. *Icarus*, **226**, 455–469, doi:10.1016/j.icarus.2013.05.012.

DiBiase, R. A., Limaye, A. B., Scheingross, J. S., Fischer, W. W., and Lamb, M. P. (2013). Deltaic deposits at Aeolis Dorsa: Sedimentary evidence for a standing body of water on the northern plains of Mars. *J. Geophys. Res.*, **118**, 1285–1302, doi:10.1002/jgre.20100.

Dickson, J. L. and Head, J. W. (2008). Amazonian glaciation in eastern Hellas, Mars: Evidence for high-altitude atmospheric deposition as the source for the hourglass and related deposits. *Lunar Planet. Sci. Conf.*, **39**, abs. 1660.

Dickson, J. L., Head, J. W., and Marchant, D. R. (2008). Late Amazonian glaciation at the dichotomy boundary on Mars: Evidence for glacial thickness maxima and multiple glacial phases. *Geology*, **36**, 411–414, doi:10.1130/G24382A.1.

Dickson, J. L., Head, J. W., and Fassett, C. I. (2011). Ice accumulation and flow on Mars: Orientation trends and implications for climate in the Late Amazonian. *Lunar Planet. Sci. Conf.*, **42**, abs. 1324.

Diniega, S., Hansen, C. J., McElwaine, J. N., and 4 others (2013). A new dry hypothesis for the formation of Martian linear gullies. *Icarus*, **225**, 526–537, doi:10.1016/j.icarus.2013.04.006.

Diniega, S., McEwen, A. S., Dundas, C. M., and Ojha, L. (2014). Signs of water? A review of recent Martian slope features. *Eighth Int. Mars Conf.*, abs. 1423.

Diniega, S., Hansen, C. J., Allen, A., *et al.* (2017). Dune–slope activity due to frost and wind throughout the north polar erg, Mars. In Conway S. J., Carrivick, J. L., Carling, P. A., de Haas, T., and Harrison, T. N., eds., *Martian Gullies and their Earth Analogues*. Geological Society, London, Special Publications, 467, doi:10.1144/SP467.6.

Di Pietro, I., Ori, G. G., Pondrelli, M., and Salese, F. (2018). Geology of Aeolis Dorsa alluvial sedimentary basin, Mars. *J. Maps*, **14**, 212–218, doi:10.1080/17445647.2018.1454350.

Dohm, J. M. and Tanaka, K. L. (1999). Geology of the Thaumasia region, Mars: Plateau development, valley, and magmatic evolution. *Planet. Space Sci.*, **47**, 411–431, doi:10.1016/S0032-0633(98)00141-X.

Dohm, J. M., Anderson, R. C., Baker, V. R., *et al.* (2001a). Latent outflow activity for western Tharsis, Mars: Significant flood record exposed. *J. Geophys. Res.*, **103**, 2000JE00135, 12,301–12,314.

Dohm, J. M., Ferris, J. C., Baker, V. R., *et al.* (2001b). Ancient drainage basin of the Tharsis region, Mars: Potential source for outflow channel systems and putative oceans or paleolakes. *J. Geophys. Res.*, **106**, 32,943–32,958, doi:10.1029/2000JE001468.

Dohm, J. M., Tanaka, K. L., and Hare, T. M. (2001c). Geologic map of the Thaumasia region of Mars. *USGS Misc. Invest. Ser. Map 1-2650*, scale 1: 5,000,000.

Dohm, J. M., Anderson, R. C., Barlow, N. G., *et al.* (2008). Recent geological and hydrological activity on Mars: The Tharsis/Elysium Corridor. *Planet. Space Sci.*, **56**, 985–1013, doi:10.1016/j.pss.2008.01.001.

Dohm, J. M., Anderson, R. C., Williams, J.-P., *et al.* (2009a). Claritas rise: Pre-Tharsis magmatism. *J. Volcanol. Geotherm. Res.*, **185**, 139–156.

Dohm, J. M., Baker, V. R., Boynton, W. V., *et al.* (2009b). GRS evidence and the possibility of ancient oceans on Mars: Special. *Planet. Space Sci.*, **57**, 664–684, doi:10.1016/j.pss.2008.10.008.

Dohm, J. M., Ferris, J. C., Baker, V. R., *et al.* (2011). Did a large Argyre lake source the Uzboi Vallis drainage system?: Post-Viking era geologic mapping investigation. *Lunar Planet. Sci. Conf.*, **42**, abs. 2255.

Dohm, J. M., Spagnuolo, M. G., Williams, J. P., *et al.* (2015). The Mars plate-tectonic-basement hypothesis. *Lunar Planet. Sci. Conf.*, **46**, abs. 1741.

Dundas, C. M. and Keszthelyi, L. P. (2014). Emplacement and erosive effects of lava in south Kasei Valles, *Mars. J. Volcanol. Geotherm. Res.*, **282**, 92–102.

Dundas, C. M., Diniega, S., Hansen, C. J., Byrne, S, and McEwen, A. S. (2012). Seasonal activity and morphological changes in Martian gullies. *Icarus*, **220**, 124–143.

Dundas C. M., Diniega, S., and McEwen, A. S. (2015). Long-term monitoring of Martian gully formation and evolution with MRO-HiRISE. *Icarus*, **251**, 244–263.

Dundas C. M., McEwen, A. S., Diniega, S., *et al.* (2017). The formation of gullies on Mars today. In Conway S. J., Carrivick, J. L., Carling, P. A., de Haas, T., and Harrison, T. N., eds., *Martian Gullies and their Earth Analogues*. Geological Society, London, Special Publications, 467, doi:10.1144/SP467.5.

Duxbury, T., Kirk, R. L., Archinal, B. A., and Neumann, G. A. (2002). Mars Geodesy/Cartography Working Group Recommendations on Mars Cartographic Constants and Coordinate Systems. In *Int. Soc. Photogramm. Remote Sensing*, **34**, GeoSpatial Theory, Processing and Applications, Ottawa, www.isprs.org/proceedings/XXXIV/part4/pdfpapers/521.pdf.

Ehlmann, B. L. and Edwards, C. S. (2014). Mineralogy of the Martian surface. *Ann. Rev. Earth Planet. Sci.*, **42**, 291–315.

Ehlmann, B. L., Mustard, J. F., and Murchie, S. L. (2010). Geologic setting of serpentine-bearing rocks on Mars. *Geophys. Res. Lett.*, **37**, L06201, doi:10.1029/2010GL042596.

El Maarry, M. R., Dohm, J. M., Marzo, G. A., *et al.* (2012). Searching for evidence of hydrothermal activity at Apollinaris Mons, Mars. *Icarus*, **217**, 297–314.

El Maarry, M. R., Dohm, J. M., Michael, G., Thomas, N., and Maruyama, S. (2013). Morphology and evolution of the ejecta of Hale crater in Argyre basin, Mars: Results from high resolution mapping. *Icarus*, **226**, 905–922.

Erkeling, G., Hiesinger, H., Reiss, D., Hielscher, F. J., and Ivanov, M. A. (2011). The stratigraphy of the Amenthes region, Mars: Time limits for the formation of fluvial, volcanic and tectonic landforms. *Icarus*, **215**, 128–152, doi:10.1016/j.icarus.2011.06.041.

European Space Agency (2012). The pit-chains of Mars: possible place for life? www.esa.int/Our_Activities/Space_Science/Mars_Express/The_pit-chains_of_Mars_a_possible_place_for_life.

Fairén A. G., Davila, A. F., Gago-Duport, L., *et al.* (2011). Cold glacial oceans would have inhibited phyllosilicate sedimentation on early Mars. *Nature Geosci.* **4**, 667–670.

Farmer, J. D. (2000). Hydrothermal systems: Doorways to early biosphere evolution. *GSA Today*, **10**, 1–9.

Farmer, J.D. and Landheim, R. (1995). Site 144: Diacria southeast. In Greeley, R., and Thomas, P. E., eds., *Mars Landing Site Catalog, second edition*. NASA Reference Publication 1238, http://cmex .ihmc.us/marstools/mars_cat/Mars_Cat.html.

Fassett, C. I. and Head, J.W., III (2006). Valleys on Hecates Tholus, Mars: origin by basal melting of summit snowpack. *Planet. Space Sci.*, **54**, 370–378, doi:10.1016/j.pss.2005.12.011.

Fassett, C. I. and Head, J. W., III (2008). Valley network-fed, open-basin lakes on Mars: Distribution and implications for Noachian surface and subsurface hydrology. *Icarus*, **198**, 37–56, doi:10.1016/j.icarus.2008.06.016.

Fastook, J. L., Head, J. W., Marchant, D. R., and Forget, F. (2008). Tropical mountain glaciers on Mars: Altitude-dependence of ice accumulation, accumulation conditions, formation times, glacier dynamics, and implications for planetary spin-axis/ orbital history. *Icarus*, **198**, 305–317, doi:10.1016/j.icarus.2008.08.008.

Fawdon, P., Skok, J. R., Balme, M. R., *et al.* (2015). The geological history of Nili Patera, Mars. *J. Geophys. Res.*, **120**, 951–977, doi:10.1002/2015JE004795.

Feldman, W. C., Prettyman, T. H., Maurice, S., *et al.* (2004). Global distribution of near-surface hydrogen on Mars. *J. Geophys. Res*, **109**, E09006, doi:10.1029/2003JE002160.

Forget, F., Haberle, R. M., Montmessin, F., Levrard, B., and Head, J. W. (2006). Formation of glaciers on Mars by atmospheric precipitation at high obliquity. *Science*, **311**, 368–371.

Fortezzo, C. M. and Skinner, J. A., Jr. (2012). Geologic evolution of the Runanga–Jörn basin, northeast Hellas, Mars. *Lunar Planet. Sci. Conf.*, **43**, 2681.

Fortezzo, C. M. and Tanaka, K. L. (2010). Mapping Planum Boreum unconformities using Context Camera mosaics. *Lunar Planet. Sci. Conf.*, **41**, abs. 2554.

Frey, H. (1979). Thaumasia: A fossilized early forming Tharsis uplift. *J. Geophys. Res*, **84**, 1009–1023, doi:10.1029/JB084iB03p01009.

Frey, H. V. (2006). Impact constraints on, and a chronology for, major events in early Mars history. *J. Geophys. Res*, **111**, E08S91, doi:10.1029/ 2005JE002449.

Frey, H. V., Roark, J. H., Shockey, K. M., Frey, E. L., and Sakimoto, S. E. H. (2002). Ancient lowlands on Mars. *Geophys. Res. Lett*, **29**, doi:10.1029/ 2001GL013832.

Fuller, E. R. and Head, J. W., III (2002). Amazonis Planitia: The role of geologically recent volcanism and sedimentation in the formation of the smoothest plains on Mars. *J. Geophys. Res*, **107**, doi:10.1029/ 2002JE001842.

Gardin, E., Allemand, P., Quantin, C., and Thollot, P. (2010). Defrosting, dark flow features, and dune activity on Mars: Example in Russell crater. *J. Geophys. Res*, **115**, E06016, doi:10.1029/ 2009JE003515.

Genova, A., Boossens, S., Lemoine, F. G., *et al.* (2016). Seasonal and static gravity field of Mars from MGS, Mars Odyssey and MRO radio science. *Icarus*, **272**, 228–245, doi:10.1016/j.icarus.2016.02.050.

Ghatan, G. J. and Head, J. W., III (2002). Candidate subglacial volcanoes in the south polar region of Mars: Morphology, morphometry, and eruption conditions. *J. Geophys. Res*, **107**, doi:10.1029/ 2001JE001519.

Gloth, T. D. and Christensen, P. R. (2005). Geologic and mineralogic mapping of Aram Chaos: Evidence for a water-rich history. *J. Geophys. Res*, **110**, E09006, doi:10.1029/2004JE002389.

Golder, K. B. and Gilmore, M. S. (2012). Evolution of chaos terrain in the Eridania basin, Mars. *Lunar Planet. Sci. Conf.*, **43**, abs. 2796.

Golder, K. B., and Gilmore, M. S. (2013). Eridania basin, Mars: Evolution of Electris terrain, chaos, and paleolake. *Lunar Planet. Sci. Conf.*, **44**, abs. 2995.

investigation. *Space Sci. Rev*, **170**, 5–56, doi:10.1007/s11214-012-9892-2.

Golombek, M. P., Cook, R. A., Economou, T., and 11 others (1997). Overview of the Mars Pathfinder Mission and assessment of landing site predictions. *Science*, **278**, 1743–1748.

Golombek, M. P., Anderson, F. S., and Zuber, M. T. (2001). Martian wrinkle ridge topography: Evidence for subsurface faults from MOLA. *J. Geophys. Res*, **106**, 23,811–23,821, doi:10.1029/ 2000JE001308.

Golombek, M., Grant, J., Kipp, D., *et al.* (2012). Selection of the Mars Science Laboratory landing site. *Space Sci. Rev*, **170**, 641–737, doi:10.1007/ s11214-012-9916-y.

Gomes, R., Levison, H. F., Tsiganis, K., and Morbidelli, A. (2005). Origin of the cataclysmic Late Heavy Bombardment period of the terrestrial planets. *Nature*, **435**, 466–469, doi:10.1038/nature03676.

Grant, J. A. and Parker, T. J. (2002). Drainage evolution of the Margaritifer Sinus region, Mars. *J. Geophys. Res*, **107**, 5066, doi:10.1029/2001JE001678.

Grant, J. A., Wilson, S. A., Fortezzo, C. M., and Clark, D. A. (2009). Geologic map of MTM-20012 and -25012 quadrangles, Margaritifer Terra region of Mars. *USGS Sci. Inv. Map 3041*, scale 1:500,000.

Greeley, R. and Crown, D. A. (1990). Volcanic geology of Tyrrhena Patera, Mars. *J. Geophys. Res*, **95**, 7133–7149.

Greeley, R. and Guest, J. E. (1987). Geologic map of the eastern equatorial region of Mars. *USGS Misc. Inv. Ser. Map I-1802–B*, scale 1:15,000,000.

Greeley, R., Bernard, F. H., McSween, H. Y., *et al.* (2005). Fluid lava flows in Gusev crater, Mars. *J. Geophys. Res*, **110**, E05008, doi:10.1029/ 2005JE002401.

Greeley, R., Williams, D. A., Fergason, R. L., *et al.*, and the HRSC Co-Investigator Team (2007). Amphitrites and Peneus: New insight into highlands paterae. *Lunar Planet. Sci. Conf.*, **38**, abs. 1373.

Gregg, T. K. P., Crown, D.A., and Greeley, R. (1998). Geologic map of MTM quadrangle -20252, Tyrrhena Patera region of Mars. *USGS Misc. Inv. Ser. Map I-2556*, scale 1:500,000.

Grimm, R. E., Harrison, K. P., and Stillman, D. E. (2014). Water budgets of Martian recurring slope lineae. *Icarus*, **233**, 316–327.

Grosfils, E. B. and Head, J. W. (1994). The global distribution of giant radiating dike swarms on Venus: Implications for the global stress state. *Geophys. Res. Letters*, **21**, 701–704.

Grotzinger, J., Beaty, D., Dromart, G., *et al.* (2011). Mars sedimentary geology: Key concepts and outstanding questions. *Astrobiology*, **11**, 77–87, doi:10.1089/ast.2010.0571.

Grotzinger, J. P., Crisp, J., Vasavada, A. R., *et al.* (2012). Mars Science Laboratory mission and science

Hartmann, W. K. and Raper, O. (1974). *The New Mars: The Discoveries of Mariner 9*. NASA Special Publication SP-337, Washington, DC, Government Printing Office.

Hauber, E., Grott, M., and Kronberg, P. (2010). Martian rifts: Structural geology and geophysics. *Earth Planet. Sci. Lett.*, **294**, 393–410, doi:10.1016/ j.epsl.2009.11.005.

Hayward, R. K., Mullins, K. F., Fenton, L. K., *et al.* (2007a). Mars Global Digital Dune Database and initial science results. *J. Geophys. Res*, **112**, E11007, doi:10.1029/2007JE002943.

Hayward, R. K., Mullins, K. F., Fenton, L. K., *et al.* (2007b). *Mars Global Digital Dune Database: MC2– MC29*. US Geological Survey Open-File Report 2007–1158, http://pubs.usgs.gov/of/2007/1158/.

Head, J. W., III and Pratt, S. (2001). Extensive Hesperian-aged south polar ice sheet on Mars: Evidence for massive melting and retreat, and lateral flow and ponding of melt-water. *J. Geophys. Res.*, **106**, 12,275–12,299.

Head, J. W., III, Mustard, J. F., Kreslavsky, M. A., Milliken, R. E., and Marchant, D. R. (2003). Recent ice ages on Mars. *Nature*, **426**, 797–802.

Head, J. W., III, Neukum, G., Jaumann, R., *et al.*, and the HRSC Co-Investigator Team (2005). Tropical to mid-latitude snow and ice accumulation, flow and glaciation on Mars. *Nature*, **434**, 346–351.

Head, J. W., III, Marchant, D. R., and Kreslavsky, M. A. (2008). Formation of gullies on Mars: Link to recent climate history and insolation microenvironments implicate surface water flow origin. *Proc. Natl. Acad. Sci.*, **105**, 13258–13263, doi:10.1073/pnas.0803760105.

Heldmann, J. L., Toon, O. B., Pollard, W. H., *et al.* (2005). Formation of Martian gullies by the action of liquid water flowing under current Martian environmental conditions. *J. Geophys. Res*, **110**, E05004, doi:10.1029/2004JE002261.

Hiesinger, H. and Head, J. W., III (2002). Topography and morphology of the Argyre basin, Mars: Implications for its geologic and hydrologic history. *Planet. Space Sci*, **50**, 939–981.

Hiesinger, H. and Head, J. W., III (2004). The Syrtis Major volcanic province, Mars: Synthesis from Mars Global Surveyor data. *J. Geophys. Res*, **109**, E01004, doi:10.1029/2003JE002143.

Hill, J. R., and Christensen, P. R. (2016). A quality constrained THEMIS daytime infrared global mosaic. *Lunar Planet. Sci. Conf.*, **47**, abs. 2326.

Hill, J. R., Edwards, C. S., and Christensen, P. R. (2014). Mapping the Martian surface with THEMIS global infrared mosaics. *Eighth Int. Mars Conf.*, abs. 1141.

Holt, J. W., Safaeinili, A., Plaut, J. J., *et al.* (2008). Radar sounding evidence for buried glaciers in the southern mid-latitudes of Mars. *Science*, **322**, 1235–1238, doi:10.1126/science.1164246.

Horgan, B. H. N. and Bell, J. F., III (2012). Seasonally active slipface avalanches in the north polar sand sea of Mars; evidence for a wind-related origin. *Geophys. Res. Letters*, **39**, L09201.

Howard, A. D. (1978). Origin of the stepped topography of the Martian poles. *Icarus*, **34**, 581–599.

Howard, A. D. (2000). The role of eolian processes in forming surface features of the Martian polar layered deposits. *Icarus*, **144**, 267–288.

Howard, A. D., Moore, J. M., and Irwin, R. P., III (2005). An intense terminal epoch of widespread fluvial activity on early Mars: 1. Valley network incision and associated deposits. *J. Geophys. Res.*, **110**, E12S14, doi:10.1029/2005JE002459.

HRSC (2004a). Dao Vallis (orbit 0528). *Press Release #067*, www.planet.geo.fu-berlin.de/en/projects/mars/hrsc067-DaoVallis.php.

HRSC (2004b). Reull Vallis (orbit 0451). *Press Release #138*, www.planet.geo.fu-berlin.de/en/projects/mars/hrsc138-ReullVallis.php.

HRSC (2005). Hour glass (orbit 0451). *Press Release #178*, www.planet.geo.fu-berlin.de/en/projects/mars/hrsc178-HourGlass.php.

HRSC (2006). Hour glass movies (orbit 0451). *Press Release #242*, www.planet.geo.fu-berlin.de/en/projects/mars/hrsc242-HourGlassMovies.php.

HRSC (2013). Upper Reull Vallis (orbit 10657). *Press Release #582*: www.planet.geo.fu-berlin.de/en/projects/mars/hrsc582-UpperReullVallis.php.

Hubbard, S. (2011) *Exploring Mars: Chronicle from a Decade of Discovery*. Tucson, AZ: University of Arizona Press.

Humayun, M., Nemchin, A., Zanda, B., *et al.* (2013). Origin and age of the earliest Martian crust from meteorite NWA7533. *Nature*, **242**, 513–516.

Hynek, B. M., Phillips, R. J., and Arvidson, R. E. (2003). Explosive volcanism in the Tharsis region: Global evidence in the Martian geologic record. *J. Geophys. Res.*, **111**, 5111, doi:10.1029/2003JE002062.

Hynek, B. M., Beach, M., and Hoke, M. R. T. (2010). Updated global map of Martian valley networks and implications for climate and hydrologic processes. *J. Geophys. Res.*, **115**, E09008, doi:10.1029/2009JE003548.

Irwin, R. P., III and J. A. Grant (2013). Geologic map of MTM -15027, -20027, -25027, and -25032 quadrangles, Margaritifer Terra region of Mars. *USGS Sci. Inv. Map SIM-3209*, scale 1:1,000,000.

Irwin, R. P., III, Maxwell, T. A., Howard, A. D., Craddock, R. A., and Leverington, D. W. (2002).

A large paleolake basin at the head of Ma'adim Vallis, Mars. *Science*, **296**, 2209–2212.

Irwin, R. P., III, Howard, A. D., and Maxwell, T. A. (2004). Geomorphology of Ma'adim Vallis, Mars, and associated paleolake basins. *J. Geophys. Res.*, **109**, E12009, doi:10.1029/2004JE002287.

Irwin, R. P., III, Howard, A. D., Craddock, R. A., and Moore, J. M. (2005). An intense terminal epoch of widespread fluvial activity on early Mars: 2. Increased runoff and paleolake development. *J. Geophys. Res.*, **110**, E12S15, doi:10.1029/2005JE002460.

Irwin, R. P., III, Tanaka, K. L., and Robbins, S. J. (2013). Distribution of Early, Middle, and Late Noachian cratered surfaces in the Martian highlands: Implications for resurfacing events and processes. *J. Geophys. Res.*, **118**, 1–14, doi:10.1002/jgre.20053.

Isherwood, R. J., Jozwiak, L. M., Jansen, J. C., and Andrews-Hanna, J. C. (2013). The volcanic history of Olympus Mons from paleo-topography and flexural modeling. *Earth Planet. Sci. Lett.*, **363**, 88–96.

Ivanov, M. A., Hiesinger, H., Erkeling, G., Hielscher, F. J., and Reiss, D. (2012). Major episodes of geologic history of Isidis Planitia on Mars. *Icarus*, **218**, 24–46, doi:10.1016/j.icarus.2011.11.029.

Jacobsen, R. E. and Burr, D. M. (2017). Dichotomies in the fluvial and alluvial fan deposits of the Aeolis Dorsa, Mars: Implications for weathered sediment and paleoclimate. *Geosphere*, **13**, 2154–2168, doi:10.1130/GES01330.1.

Jaumann, R., Neukum, G., Behnke, T., *et al.*, and the HRSC Co-Investigator Team (2007). The High-Resolution Stereo Camera (HRSC) experiment on Mars Express: Instrument aspects and experiment conduct from interplanetary cruise through the nominal mission. *Planet. Space Sci.*, **55**, 928–952.

Jaumann, R., Tirsch, D., Hauber, E., *et al.* (2015). Quantifying geologic processes on Mars: Results of the High Resolution Stereo Camera (HRSC) on Mars Express. *Planet. Space Sci.*, **112**, 53–97, doi:10.1016/j.pss.2014.11.029.

Johnsson, A., Reiss, D., Hauber, E, Hiesinger, H, and Zanetti, M. (2014). Evidence for very recent melt-water and debris flow activity in gullies in a young mid-latitude crater on Mars. *Icarus*, **235**, 37–54.

Jouanic, G., Gargani, J., Costard, F., *et al.* (2012). Morphological and mechanical characterization of gullies in a periglacial environment: The case of the Russell crater dune (Mars). *Planet. Space Sci.*, **71**, 38–54, doi:10.1016/j.pss.2012.07.005.

JPL (Jet Propulsion Laboratory) (2011). PIA15098: Chemical alteration by water, Mawrth Vallis. http://photojournal.jpl.nasa.gov/catalog/PIA15098.

Kadish, S. J., Head, J. W., Parsons, R. L., and Marchant, D. R. (2008). The Ascraeus Mons fan-shaped deposit: Volcano–ice interactions and the climatic implications of cold-based tropical mountain glaciation. *Icarus*, **197**, 84–109.

Karasozen, E., Andrews-Hanna, J. C., Dohm, J. M., and Anderson, R. C. (2012). The formation mechanism of the south ridge belt, Mars. *Lunar Planet. Sci. Conf.*, **43**, abs. 2592.

Kargel, J. S. and Strom, R. G. (1992). Ancient glaciation on Mars. *Geology*, **20**, 3–7.

Kargel, J. S., Baker, V. R., Begét, J. E., *et al.* (1995). Evidence of continental glaciation in the Martian northern plains. *J. Geophys. Res.*, **100**, 5351–5368, doi:10.1029/94JE02447.

Kerber, L. and Head, J. W. (2010). The age of the Medusae Fossae Formation: Evidence of Hesperian emplacement from crater morphology, stratigraphy, and ancient lava contacts. *Icarus*, **206**, 669–684.

Kerber, L., Head, J. W., Madeleine, J.-B., Forget, F., and Wilson, L. (2011). The dispersal of pyroclasts from Apollinaris Patera, Mars: Implications for the origin of the Medusae Fossae Formation. *Icarus*, **216**, 212–220, doi:10.1016/j.icarus.2011.07.035.

Kereszturi, A. (2012). Review of wet environment types on Mars with focus on duration and volumetric issues. *Astrobiology*, **12**, 586–600, doi:10.1089/ast.2011.0686.

Kereszturi, A., Möhlmann, D., Berczi, S., *et al.* (2011). Possible role of brines in the darkening and flow-like features on the Martian polar dunes based on HiRISE images. *Planet. Space Sci.*, **59**, 1413–27, doi:10.1016/j.pss.2011.05.012.

Kieffer, H. H., Jakosky, B. M., and Snyder, C. W. (1992a). The planet Mars: From antiquity to the present. In Kieffer, H. H., Jakosky, B. M., Snyder, C. W., and Matthews, M. S., eds., *Mars*. Tucson, AZ: University of Arizona Press, Space Science Series, pp. 1–33.

Kieffer, H. H., Jakosky, B. M., Snyder, C. W. and Matthews, M. S., eds. (1992b). *Mars*. Tucson, AZ: University of Arizona Press, Space Science Series, 1455 pp.

Kieffer, H. H., Christensen, P. R., and Titus, T. N. (2006). CO_2 jets formed by sublimation beneath translucent slab ice in Mars' seasonal south polar ice cap. *Nature*, **442**, 793–796, doi:10.1038/nature04945.

Kite, E. S. and Hindmarsh, R. C. A. (2007). Did ice streams shape the largest channels on Mars? *Geophys. Res. Lett.*, **34**, L19202, doi:10.1029/2007GL030530.

Kite, E. S., Lewis, K. W., Lamb, M. P., Newman, C. E., and Richardson, M. I. (2013). Growth and form of the mound in Gale crater, Mars: Slope wind

enhanced erosion and transport. *Geology*, **41**, 543–546, doi:10.1130/G33909.1.

Klein, H. P., Horowitz, N. H., and Biemann, K. (1992). The search for extant life on Mars. In Kieffer, H. H., Jakosky, B. M., Snyder, C. W., and Matthews, M. S., eds., *Mars*. Tucson, AZ: University of Arizona Press, Space Science Series, pp. 1221–1245.

Kleine, T., Münker, C., Mezger, K., and Palme, H. (2002). Rapid accretion and early core formation on asteroids and the terrestrial planets from Hf–W chronometry. *Nature*, **418**, 952–955.

Koeppen, W. C. and Hamilton, V. E. (2008). Global distribution, composition, and abundance of olivine on the surface of Mars from thermal infrared data. *J. Geophys. Res.*, **113**, E05001, doi:10.1029/2007JE002984.

Kopparapu, R. K., Ramirez, R., Kasting, J. F., and 7 others (2013). Habitable zones around main-sequence stars: New estimates. *Astroph. J.*, **265**, 131, doi:10.1088/0004-637X/765/2/131.

Kreslavsky, M. A. and Head, J. W., III (2000). Kilometer-scale roughness of Mars: Results from MOLA data analysis. *J. Geophys. Res.*, **105**, 26,695–26,711, doi:10.1029/2000JE001259.

Kreslavsky, M. A. and Head, J. W., III (2009). Slope streaks on Mars: A new "wet" mechanism. *Icarus*, **201**, 517–527, doi:10.1016/j.icarus.2009.01.026.

Laskar, J., Levrard, B., and Mustard, J. F. (2002). Orbital forcing of the Martian polar layered deposits. *Nature*, **419**, 375–377.

Laskar, J., Correia, A. C. M., Gastineau, M., *et al.* (2004). Long term evolution and chaotic diffusion of the insolation quantities of Mars. *Icarus*, **170**, 343–364, doi:10.1016/j.icarus.2004.04.005.

Leighton, R. B, Murray, B. C., Sharp, R. P., Allen, J. D., and Sloan, R. K. (1965). Mariner IV photography of Mars: Initial results. *Science*, **149**, 627–630.

Leighton, R. B, Horowitz, N. H., Murray, B. C., *et al.* (1969). Mariner 6 and 7 television pictures: Preliminary analysis. *Science*, **166**, 49–67.

Leonard, G. J. and Tanaka, K. L. (2001). Geologic map of the Hellas region of Mars. *USGS Misc. Inv. Ser. Map I-2694*, scale 1:5,000,000.

Leone, G. (2016). Alignments of volcanic features in the southern hemisphere of Mars produced by migrating mantle plumes. *J. Volcanol. Geotherm. Res.*, **309**, 78–95, doi:10.1016/j.jvolgeores .2015.10.028.

Leone, G., Tackley, P. J., Gerya, T. V., May, D. A., and Zhu, G. (2014). Three-dimensional simulations of the southern polar giant impact hypothesis for the origin of the Martian dichotomy. *Geophys. Res. Letters*, **41**, 8736–8743, doi:10.1002/2014GL062261.

Leverington, D. W. (2007). Was the Mangala Valles system incised by volcanic flows? *J. Geophys. Res.,* **112,** E11005, doi:10.1029/2007JE002896.

Leverington, D. W. (2011). A volcanic origin for the outflow channels of Mars: Key evidence and major implications. *Geomorphology,* **132,** 51–75, doi:10.1016/j.geomorph.2011.05.022.

Levinthal, E. C., Green, W. G., Cutts, J. A., *et al.* (1973). Mariner 9: Image processing and products. *Icarus,* **18,** 75–101.

Levy, J. S. and Head, J. W., III (2005). Evidence for remnants of ice-rich deposits: Mangala Valles outflow channel, Mars. *Terra Nova,* **17,** 503–509.

Li, H., Robinson, M. S., and Jurdy, D. M. (2006). Martian southern hemisphere debris aprons. *Lunar Planet. Sci. Conf.,* **37,** abs. 2390.

Lias, J. H., Dohm, J. M., and Tanaka, J. M. (1997). Geologic history of Lowell impact. *Lunar Planet. Sci. Conf.,* **28,** abs. 1650, 813–814.

Lias, J. H., Tanaka, K. L., and Hare, T. M. (1999). Geologic, tectonic, and fluvial histories of the Eridania region of Mars. *Lunar Planet. Sci. Conf.,* **30,** abs. 1074.

Loizeau, D., Mangold, N., Poulet, F., *et al.* (2010). Stratigraphy in the Mawrth Vallis region through OMEGA, HRSC color imagery and DTM. *Icarus,* **205,** 396–418, doi:10.1016/j.icarus.2009.04.018.

Lopes, R. M. C., Guest, J. E., Hiller, K. H., and Neukum, G. P. O. (1982). Further evidence for a mass movement origin of the Olympus Mons aureole. *J. Geophys. Res.,* **87,** 9917–9928, doi:10.1029/JB087iB12p09917.

Lucchitta, B. K. (1981). Mars and Earth: Comparison of cold-climate features. *Icarus,* **45,** 264–303.

Lucchitta, B. K. (1982). Ice sculpture in the Martian outflow channels. *J. Geophys. Res.,* **87,** 9951–9973.

Malin, M. C. (1979). Mars: Evidence of indurated deposits of fine materials. *NASA Conf. Pub.,* **2072,** 54.

Malin, M. C. and Edgett, K. S. (2000a). Evidence for recent groundwater seepage and surface runoff on Mars. *Science,* **288,** 2330–2335.

Malin, M. C. and Edgett, K. S. (2000b). Sedimentary rocks of early Mars. *Science,* **290,** 1927–37.

Malin, M. C. and Edgett, K. S. (2003). Evidence for persistent flow and aqueous sedimentation on early Mars. *Science,* **302,** 1931–1934. doi:10.1126/science.1090544.

Malin, M. C., Danielson, G. E., Ravine M. A., and Soulanille, T. A. (1991). Design and development of the Mars observer camera. *Int. J. Imag. Syst. Tech.,* **3,** 76–91.

Malin, M. C.., Edgett, K. S., Posiolova, L. V., McColley, S. M., and Noe Dobrea, E. Z. (2006). Present-day impact cratering rate and contemporary gully activity on Mars. *Science,* **314,** 1573–1577.

Malin, M. C., Bell, J. F., III, Cantor, B. A., *et al.* (2007). Context Camera Investigation on board the Mars Reconnaissance Orbiter. *J. Geophys. Res.,* **112,** E05S04, doi:10.1029/2006JE002808.

Malin, M. C., Edgett, K. S., Cantor, B. C., *et al.* (2010). An overview of the 1985–2006 Mars Orbiter Camera science investigation. *MARS,* **5,** 1–60, doi:10.1555/mars2010.0001.

Malin Space Science Systems (MSSS) (2003). Schiaparelli sedimentary rocks. *MGS MOC Release no. MOC2–403,* www.msss.com/mars_images/moc/2003/06/26/.

Mandt, K. E., de Silva, S. L., Zimbelman, J. R., and Crown, D. A. (2008). Origin of the Medusae Fossae Formation, Mars: Insights from a synoptic approach. *J. Geophys. Res.,* **113,** E12011, doi:10.1029/2008JE003076.

Manfredi, L. and Greeley, R. (2012). Origin of ridges seen in Tempe Terra, Mars. *Lunar Planet. Sci. Conf.,* **43,** abs. 2599.

Mangold, N. and Howard, A. D. (2013). Outflow channels with deltaic deposits in Ismenius Lacus, Mars. *Icarus,* **226,** 385–401.

Mangold, N., Costard, F., and Forget, F. (2003). Debris flows over sand dunes on Mars: Evidence for liquid water. *J. Geophys. Res.,* **108,** E4, 5027, doi:10.1029/2002JE001958.

Mangold, N., Mangeney, A., Migeon, V., *et al.* (2010). Sinuous gullies on Mars: Frequency, distribution, and implications for flow properties. *J. Geophys. Res.,* **115,** E11001, doi:10.1029/2009JE003540.

Marshak, S. (2012). *Earth: Portrait of a Planet,* 4th edition. New York, NY: Norton.

Martin. L. J., James, P. B., Dollfus, A., Iwasaki, K., and Beish, J. D. (1992). Telescope observations: Visual, photographic, polarimetric. In Kieffer, H. H., Jakosky, B. M., Snyder, C. W., and Matthews, M. S., eds. *Mars.* Tucson, AZ: University of Arizona Press, Space Science Series, pp. 34–70.

Masursky, H. (1973). An overview of geological results from Mariner 9. *J. Geophys. Res.,* **78,** 4009–4030.

Masursky, H., Batson, R., Borgeson, W., *et al.* (1970). Television experiment for Mariner Mars 1971. *Icarus,* **12,** 10–45.

McEwen, A. (2012). Lava lamp terrain on the floor of Hellas basin (HiRISE image caption). www.uahirise.org/ESP_025780_1415.

McEwen, A. S., Malin, M. C., Carr, M. H., and Hartmann, W. K. (1999). Voluminous volcanism on early Mars revealed in Valles Marineris. *Nature,* **397,** 584–586, doi:10.1038/17539.

McEwen, A. S., Eliason, E. M., Bergstrom, J. W., *et al.* (2007a). Mars Reconnaissance Orbiter's High Resolution Imaging Science Experiment (HiRISE). *J. Geophys. Res.,* **112,** E05, doi:10.1029/2005JE002605.

McEwen, A. S., Hansen, C. J., Delamere, W. A., *et al.* (2007b). A closer look at water-related activity on Mars. *Science,* **317,** 1706–1709.

McEwen, A. S., Keszthelyi, L. P., and Grant, J. A. (2012). Have there been large, recent (Mid-Late Amazonian) water floods on Mars? *Lunar Planet. Sci. Conf.,* **43,** abs. 1612.

McEwen, A. S., Dundas, C. M., Mattson, S. S., and 7 others (2013). Recurring slope lineae in equatorial regions of Mars. *Nature Geosci,* **7,** 53–58, doi:10.1038/ngeo2014.

McGovern, P. J. and Morgan, J. K. (2009). Volcanic spreading and lateral variations in the structure of Olympus Mons, Mars. *Geology,* **37,** 139–142, doi:10.1130/G25180A.1.

McGovern, P. J., Smith, J. R., Morgan, J. K., and Bulmer, M. H. (2004). The Olympus Mons aureole deposits: New evidence for a flank failure origin. *J. Geophys. Res.,* **109,** E08008, doi:10.1029/2004JE002258.

McKee, E. D. (1979), Introduction to a study of global sand seas. In McKee, E. D., ed., *A Study of Global Sand Seas,* US Geological Survey Professional Paper 1052, pp. 3–17.

McSween, H. Y., Jr. (2008). Martian meteorites as crustal samples. In Bell, J. F., III, ed, *The Martian Surface: Composition, Mineralogy, and Physical Properties.* Cambridge: Cambridge University Press, pp. 383–395.

Mellon, M. and Byrne, S. (2010). Glacier? (HiRISE image caption). See www.uahirise.org/ESP_018857_2225.

Mellon, M. T., Jakosky, B. M., Kieffer, H. H., and Christensen, P. R. (2000). High resolution thermal inertia mapping from the Mars Global Surveyor Thermal Emission Spectrometer. *Icarus,* **148,** 437–455.

Michalski, J. R. and Bleacher, J. E. (2013). Supervolcanoes within an ancient volcanic province in Arabia Terra, Mars. *Nature,* **502,** 47–52, doi:10.1038/nature12482.

Mishkin, A. (2003). *Sojourner: An Insider's View of the Mars Pathfinder mission.* New York, NY: Berkley Publishing.

Mitchell, D. P. (2004). Soviet Mars Images. http://www.mentallandscape.com/C_CatalogMars.htm.

Montgomery, D. R., Som, S. M., Jackson, M. P. A., *et al.* (2009). Continental-scale salt tectonics on Mars and the origin of Valles Marineris and associated outflow channels. *Geol. Soc. Amer. Bull.,* **121,** 117–133, doi:10.1130/B26307.1.

Moore, H. J. (2001). Geologic map of the Tempe-Mareotis region of Mars. *USGS Misc. Inv. Ser. Map I-1277,* scale 1:1,000,000.

Moore, J. M. and Edgett, K. S. (1993). Hellas Planitia, Mars: Site of net dust erosion and implications for the nature of basin floor deposits. *Geophys. Res. Lett.* **20,** 1599–1602.

Moore, J. M. and Howard, A. D. (2003). Ariadnes-Gorgonum knob fields of north-western Terra Serenum, Mars. *Lunar Planet. Sci. Conf.,* **34,** abs. 1402.

Moore, J. M. and Wilhelms, D. E. (2001). Hellas as possible site of ancient ice-covered lakes on Mars. *Icarus,* **154,** 258–276.

Moore, P. (1977). *Guide to Mars.* New York, NY: W. W. Norton.

Morris, E. C. (1982). Aureole deposits of the Martian volcano Olympus Mons. *J. Geophys. Res.,* **87,** 1164–1178, doi:10.1029/JB087iB02p01164.

Morton, O. (2002). *Mapping Mars: Science, Imagination, and the Birth of a World.* New York, NY: Picador.

Moscardelli, L. and Wood, L. (2011). Deep-water erosional remnants in eastern offshore Trinidad as terrestrial analogs for teardrop-shaped islands on Mars: Implication for outflow channel formation. *Geology,* **39,** 699–702.

Mouginis-Mark P. J. (2015). Geologic map of Tooting Crater, Amazonis Planitia region of Mars. *USGS Sci. Inv. Map SIM-3297,* scale 1:200,000.

Mouginis-Mark, P. J., Wilson, L, and Zuber, M. T. (1992). The physical volcanology of Mars. In Kieffer, H. H., Jakosky, B. M., Snyder, C. W., and Matthews, M. S., eds., *Mars.* Tucson, AZ: University of Arizona Press, Space Science Series, pp. 424–452.

Mouginis-Mark, P. J., Harris, A. J. L., and Rowland, S. K. (2007). Terrestrial analogs to the calderas of the Tharsis volcanoes on Mars. In Chapman, M., ed., *The Geology of Mars: Evidence from Earth-Based Analogs.* Cambridge: Cambridge University Press, pp. 71–94.

Murchie, S., Roach, L., Seelos, F., *et al.* (2009). Evidence for the origin of layered deposits in Candor Chasma, Mars, from mineral composition and hydrologic modeling. *J. Geophys. Res.,* **114,** E00D05, doi:10.1029/2009JE003343.

Murray, J. B. and Heggie, D. C. (2014). Character and origin of Phobos' grooves. *Planet. Space Sci.,* **102,** 119–143.

Murray, J. B. and the HRSC co-investigator team (2005). Evidence from the Mars Express High Resolution Stereo Camera for a frozen sea close to Mars' equator. *Nature,* **434,** 352–356, doi:10.1038/nature03379.

Musiol, S. and Neukum, G. (2012). Finite element models of lithospheric flexure and volcanic spreading at Olympus Mons, Mars. *Lunar Planet. Sci. Conf.*, **43**, abs. 1772.

NASA (1967). *Mariner Mars 1964: Final Project Report.* NASA Special Publication SP-139, Washington, DC: Government Printing Office.

NASA (1969). *Mariner-Mars 1969, A Preliminary Report.* NASA Special Publication SP-225, Washington, DC: Government Printing Office.

NASA (1974). *Mars as Viewed by Mariner 9.* NASA Special Publication SP-329, Washington DC: Government Printing Office.

NASA (1978). *Apollo Over the Moon: A View from Orbit.* NASA Special Publication SP-362, Washington DC: Government Printing Office.

NASA (1997). *Exploring the Moon: A Teacher's Guide with Activities.* NASA Publication EG-1997–10-116-HQ.

NASA (1998). *Planetary Geology: A Teacher's Guide with Activities in the Physical and Earth Sciences.* NASA publication EG-1998–03-109-HQ.

NASA (2017). Prolific Mars Orbiter completes 50,000 orbits. https://mars.nasa.gov/news/prolific-mars-orbiter-completes-50000-orbits/.

Nayak, M. and Asphaug, E. (2016). Sesquinary catenae on the Martian satellite Phobos from reaccretion of escaping ejecta. *Nature Commun.,* **7**, doi:10.1038/ncomms12591.

Neesemann, A., van Gasselt, S., Walter, S. (2014). Detailed geomorphologic–tectonic mapping of the Tempe Terra region, Mars under consideration of chronostratigraphic aspects. *Lunar Planet. Sci. Conf.*, **45**, abs. 2313.

Neukendorf, K. K. E., Mehl, J. P., Jr., and Jackson, J. A., eds. (2005). *Glossary of Geology*, 5th edition. Alexandria: American Geological Institute.

Neukum, G., Jaumann, R., Hoffmann, H., *et al.,* and HRSC team (2004). Recent and episodic volcanic and glacial activity on Mars revealed by the High Resolution Stereo Camera. *Nature,* **432**, 971–979, doi:10.1146/annurev.earth.33,092203,122637.

Neumann, G. A., Zuber, M. T., Wieczorek, M. A., *et al.* (2004). Crustal structure of Mars from gravity and topography. *J. Geophys. Res.,* **109**, E08002, doi:10.1029/2004JE002262.

Nimmo, F. and Tanaka, K. (2005). Early crustal evolution of Mars. *Ann. Rev. Earth Planet. Sci.,* **33**, 133–161, doi:10.1146/annurev.earth.33,092203,122637.

Nummedal, D. and Prior, D. B. (1981). Generation of Martian chaos and channels by debris flows. *Icarus,* **45**, 77–86.

Nussbaumer, J. (2008). The Granicus and Tinjar Valles channel system. *Lunar Planet. Sci. Conf.,* **39**, abs. 1724.

Ody, A., Poulet, F., Langevin, Y., *et al.* (2012). Global maps of anhydrous minerals at the surface of Mars from OMEGA/MEx. *J. Geophys. Res.,* **117**, E00J14, doi:10.1029/2012JE004117.

Oehler, D. Z. and Allen, C. C. (2012). Giant polygons and mounds in the lowlands of Mars: Signatures of an ancient ocean? *Astrobiology,* **12**, 601–615, doi:10.1089/ast.2011.0803.

Ojha, L., Wilhelm, M. B., Murchie, S. L., *et al.* (2015). Spectral evidence for hydrated salts in recurring slope lineae on Mars. *Nature Geosci,* **8**, 829–832, doi:10.1038/ngeo2546.

Parker, T. J., Clifford, S. M., and Banerdt, W. B. (2000). Argyre Planitia and the Mars global hydrologic cycle. *Lunar Planet. Sci. Conf,* **31**, abs. 2033.

Pasquon, K., Gargani, J., Massé, M., and Conway, S. J. (2016). Present-day formation and seasonal evolution of linear dune gullies on Mars. *Icarus,* **274**, 195–210, doi: 0.1016/j.icarus.2016.03.024.

Pelletier, J. D., Kolb, K. J., McEwen, A. S., and Kirk, R. L. (2008). Recent bright gully deposits on Mars; wet or dry flow? *Geology,* **36**, 211–14, doi:10.1130/G24346A.1.

Perminov, V. G. (1999). *The Difficult Road to Mars: A Brief History of Mars Exploration in the Soviet Union.* NASA Monographs in Aerospace History number 15, NP-1999–06-251-HQ.

Phillips, R. J., Davis, B. J., Tanaka, K. L., *et al.* (2011). Massive CO_2 ice deposits sequestered in the south polar layered deposits of Mars. *Science,* **332**, 838–841.

Pierce, T. L. and Crown, D. A. (2003). Morphologic and topographic analyses of debris aprons in the eastern Hellas region, Mars. *Icarus,* **163**, 46–65.

Planetary Data System (2009). PDS Standards Reference, version 3.8, http://pds.nasa.gov/tools/standards-reference.shtml.

Plaut, J. J., Safaeinili, A., Holt, J. W., *et al.* (2009). Radar evidence for ice in lobate debris aprons in the mid-northern latitudes of Mars. *Geophys. Res. Letters,* **36**, L02203, doi:10.1029/2008GL036379.

Plescia, J. B (1990). Recent flood lavas in the Elysium region of Mars. *Icarus,* **88**, 465–490, doi:10.1016/0019-1035(90)90095-Q.

Plescia, J. B. (2004). Morphometric properties of Martian volcanoes. *J. Geophys. Res.,* **109**, doi:10.1029/2002JE002031.

Plescia, J. B. (2013). Olympica Fossae Valles: Newly recognized fluvial-volcanic system. *Lunar Planet. Sci. Conf.,* **44**, abs. 2478.

Putzig, N. E., Phillips, R. J., Campbell, B. A., *et al.* (2009). Subsurface structure of Planum Boreum from Mars Reconnaissance Orbiter shallow radar soundings. *Icarus,* **204**, 443–457.

Putzig, N. E., Foss, F. J., II, Campbell, B. A., and Phillips, R. J. (2014). New views of Planum Boreum interior in a migrated 3-D volume of SHARAD data. *Eighth Int. Mars Conf.,* abs. 1336.

Quantin, C., Mangold, N., Hartmann, W. K., and Allemand, P. (2007). Possible long-term decline in impact rates: 1. Martian geological data. *Icarus,* **186**, 1–10, doi:10.1016/j.icarus.2006.07.008.

Reiss, D., Erkeling, G., Bauch, K. E., and Hiesinger, H. (2010). Evidence for present day gully activity on the Russell Crater dune field, Mars. *Geophys. Res. Lett,* **37**, L06203.

Robbins, S. J. and Hynek, B. M. (2012). A new global database of Mars impact craters ≥1 km: 1. Database creation, properties, and parameters. *J. Geophys. Res.,* **117**, E05004, doi:10.1029/2011JE003966.

Robbins, S. J., Di Achille, G., and Hynek, B. M. (2010). Dating the most recent episodes of volcanic activity from Mars' main volcanic calderae. *Lunar Planet. Sci. Conf.,* **41**, abs. 2252.

Robbins, S. J., Hynek, B. M., Lillis, R. J., and Bottke, W. F. (2013). Large impact crater histories of Mars: The effect of different model crater age techniques. *Icarus,* **225**, 173–184, doi:10.1016/j.icarus.2013.03.019.

Robinson, M. S., Mouginis-Mark, P. J., Zimbelman, J. R., *et al.* (1993). Chronology, eruption duration, and atmospheric contribution of the Martian volcano Apollinaris Patera. *Icarus,* **104**, 301–323, doi:10.1006/icar.1993.1103.

Rodriguez, J. A. P. and Tanaka, K. L. (2006). Sisyphi Montes and southwest Hellas Paterae: Possible impact, cryotectonic, volcanic, and mantle tectonic processes along Hellas basin rings. *Fourth Mars Polar Sci. Conf.,* abs. 8066.

Rodriguez, J. A. P., Tanaka, K. L., Yamamoto, A., *et al.* (2010). The sedimentology and dynamics of crater-affiliated wind streaks in western Arabia Terra, Mars and Patagonia, Argentina. *Geomorphology,* **121**, 30–54, doi:10.1016/j.geomorph.2009.07.020.

Rodriguez, J. A. P., Fairén, A. G., Tanaka, K. L., *et al.* (2016). Tsunami waves extensively resurfaced the shorelines of an early Martian ocean. *Scientific Reports,* **6**, 25106; doi:10.1038/srep25106.

Rogers, A. D. and Christensen, P. R. (2007). Surface mineralogy of Martian low-albedo regions from MGS-TES data: Implications for upper crustal evolution and surface alteration. *J. Geophys. Res.,* **112**, E01003, doi:10.1029/2006JE002727.

Rotto, S. and Tanaka, K. L. (1995). Geologic/geomorphologic map of the Chryse Planitia region of Mars. *USGS Misc. Inv. Ser. Map 1-2441,* scale 1:5,000,000.

Ruff, S. W. and Christensen, P. R. (2002). Bright and dark regions on Mars: Particle size and mineralogical characteristics based on Thermal Emission Spectrometer data. *J. Geophys. Res.,* **107**, 5127, doi:10.1029/2001JE001580.

Rummel, J. D., Beaty, D. W., Jones, M. A., *et al.* (2014). A new analysis of Mars "Special Regions": Findings of the Second MEPAG Special Regions Science Analysis Group (SR-SAG2). *Astrobiology,* **14**, 887–968, doi:10.1089/ast.2014.1227.

Russell, J. F., Snyder, C. W., and Keiffer, H. H. (1992). Appendix: Origin and use of Martian nomenclature. In Kieffer, H. H., Jakosky, B. M., Snyder, C. W., and Matthews, M. S., eds. *Mars.* Tucson, AZ: University of Arizona Press, Space Science Series, pp. 1305–1314.

Ryan, A. J. and Christensen, P. R. (2012). Coils and polygonal crust in the Athabasca Valles region, Mars, as evidence for a volcanic history. *Science,* **336**, 449–452.

Sagdeev, R. Z. and Zakharov, A. V. (1989). Brief history of the Phobos mission. *Nature,* **341**, 581–585.

Sakimoto, S. E. H., Gregg, T. K. P., Hughes, S. S., and Chadwick, J. (2003). Martian plains volcanism in Syria Planum and Tempe Mareotis as analogs to the eastern Snake River Plains, Idaho: Similarities and possible petrologic contributions to topography. *Lunar Planet. Sci. Conf,* **34**, abs. 1740.

Salvatore, M. and Christensen, P. (2014). Evidence for widespread aqueous sedimentation in the northern plains of Mars. *Geology,* **42**, 423–426, doi:10.1130/G35319.1.

Sautter, V., Toplis, M. J., Wiens, R. C., *et al.* (2015). *In situ* evidence for continental crust on early Mars. *Nature Geosci.,* **8**, 605–609, doi:10.1038/NGEO2474.

Scanlon, K. E. and Head, J. W. III (2014). Insights into the Late Noachian–Early Hesperian Martian climate change from fluvial features in the Dorsa Argentea Formation. *Eighth Int. Mars Conf.,* abs. 1357.

Schorghofer, N. and King, C. M. (2011). Sporadic formation of slope streaks on Mars. *Icarus,* **216**, 159–168.

Schorghofer, N., Aharonson, O., Gerstell, M. F., and Tatsumi, L. (2007). Three decades of slope streak activity on Mars. *Icarus,* **191**, 132–140.

Schultz, P. H. and Lutz, A. B. (1988). Polar wandering on Mars, *Icarus,* **73**, 91–141.

Schultz, P. H., Schultz, R. A., and Rogers, J. (1982). Structure and evolution of ancient impact basins on Mars. *J. Geophys. Res,* **87**, 9803–9820.

Schultz, R. A. (2003). Seismotectonics of the Amenthes Rupes thrust fault population, Mars. *Geophys. Res. Letters,* **30**, 1303. doi:10.1029/2002GL016475.

Schultz, R. A. and Tanaka, K. L. (1994). Lithospheric-scale buckling and thrust structures on Mars: The Coprates rise and south Tharsis ridge belt. *J. Geophys. Res*, **99**, 8371–8385.

Schultz, R. A., Moore, J. M., Grosfils, E. B., Tanaka, K.L., and Mège, D. (2007). The Canyonlands model for planetary grabens: Revised physical basis and implications. In Chapman, M. G. (ed.), *The Geology of Mars: Evidence from Earth-based Analogs*. Cambridge, Cambridge University Press, pp. 371–399.

Schulze-Makuch, D., Dohm, J. M., Fan, C., *et al.* (2007). Exploration of hydrothermal targets on Mars. *Icarus*, **189**, 308–324, doi:10.1016/j. icarus.2007.02.007.

Scott, D. H. and Carr, M. H. (1978). Geologic map of Mars. *USGS Misc. Inv. Ser. Map I-1083*, scale 1:25,000,000.

Scott, D. H. and Chapman, M. G. (1991). Geologic map of science study area 6, Memmonia region of Mars (MTM-10172). *USGS Misc. Inv. Ser. Map I-2084*, scale 1:500,000.

Scott, D. H. and Tanaka, K. L. (1982). Ignimbrites of the Amazonis Planitia region of Mars. *J. Geophys. Res*, **87**, 1179–1190.

Scott, D. H. and Tanaka, K. L. (1986). Geologic map of the western equatorial region of Mars. *USGS Misc. Inv. Ser. Map I-1802–A*, scale 1:15,000,000.

Scott, D. H. and Zimbelman, J. R. (1995). Geologic map of Arsia Mons volcano, Mars. *USGS Misc. Inv. Ser. Map I-2480*, scale 1:1,000,000.

Scott, D. H., Dohm, J. M., and Applebee, D. (1993). Geologic map of science study area 8, Apollinaris Patera region of Mars. *USGS Misc. Inv. Ser. Map I-2351*, scale 1:502,000.

Scott, D. H., Dohm, J. M., and Zimbelman, J. R. (1998). Geologic map of Pavonis Mons volcano, Mars. *USGS Misc. Inv. Ser. Map I-2561*, scale 1:1,000,000.

Seidelmann, P. K., Abalakin, V. K., Bursa, M., *et al.* (2002). Report of the IAU/IAG Working Group on Cartographic Coordinates and Rotational Elements of the Planets, and Satellites: 2000. *Celest. Mechan. Dynam. Astron.*, **82**, 83–110.

Shean, D. E., Head, J. W., III, and Marchant, D. R. (2005). Origin and evolution of a cold-based tropical mountain glacier on Mars: The Pavonis Mons fan-shaped deposit. *J. Geophys. Res.* **110**, E05001, doi:10.1029/2004JE002360.

Sheehan, W. (1996). *The Planet Mars: A History of Observation and Discovery*. Tucson, AZ: University of Arizona Press.

Skinner, J. A., Jr. and Tanaka, K. L. (2007). Evidence for and implications of sedimentary diapirism and mud volcanism in the southern Utopia highland–

lowland boundary plain, Mars. *Icarus*, **186**, 41–59, doi:10.1016/j.icarus.2006.08.013.

Skinner, J. A., Jr. and Fortezzo, C. M. (2012). Efficiency of scale in photogeologic mapping using the Runanga-Jörn basin, Mars and the Verde basin, Arizona: Project introduction and technical approach. In Skinner, J. A., Jr., Tanaka, K. L., and Kelley, M. S., eds., *Abstracts of the Annual Meeting of Planetary Geologic Mappers, Flagstaff, AZ, June 18–20*, NASA Conference Paper, pp. 29–30.

Skinner, J. A., Jr., Tanaka, K. L., Hare, T. M., *et al.* (2004). Mass-wasting of the circum-Utopia highland/lowland boundary: Processes and controls. *Workshop on Hemispheres Apart: The Origin and Modification of the Martian Crustal Dichotomy*, abs. 4031.

Skinner, J. A., Jr., Tanaka, K. L., and Platz, T. (2012). Widespread loess-like deposit in the Martian northern lowlands identifies Middle Amazonian climate change. *Geology*, **40**, 1127–1130, doi:10.1130/ G33513.1.

Skok, J. R., Mustard, J. F., Ehlmann, B. L., Milliken, R. E., and Murchie, S. L. (2010). Silica deposits in the Nili Patera caldera on the Syrtis Major volcanic complex on Mars. *Nature Geosci*, 3, 838–841, doi:10.1038/ngeo990.

Sleep, N. H. (1994). Martian plate tectonics. *J. Geophys. Res*, **99**, 5639–5655.

Slipher, E. C. (1962). *The Photographic Story of Mars*. Cambridge, MA: Sky Publishing Corp.

Smith, D. E., Zuber, M. T., Solomon, S. C., *et al.* (1999). The global topography of Mars and implications for surface evolution. *Science*, **284**, 1495–1503.

Smith, D. E., Zuber, M. T., Frey, H. V., *et al.* (2001). Mars Orbiter Laser Altimeter: Experiment summary after the first year of global mapping of Mars. *J. Geophys. Res.*, **106**, 23689–23722, doi:10.1029/ 2000JE001364.

Smith, I. B. and Holt, J. W. (2010). Onset and migration of spiral troughs on Mars revealed by orbital radar. *Nature*, **465**, 450–453, doi:10.1038/ nature09049.

Smith, P. H., Tamppari, L. K., Arvidson, R. E., *et al.* (2009). H_2O at the Phoenix landing site. *Science*, **325**, 58–61, doi:10.1126/science.1172339.

Snyder, C. W. and Moroz, V. I. (1992). Spacecraft exploration of Mars. In Kieffer, H. H., Jakosky, B. M., Snyder, C. W., and Matthews, M. S., eds. *Mars*. Tucson, AZ: University of Arizona Press, Space Science Series, pp. 71–119.

Snyder, J. P. (1987). Map projections: A working manual. *US Geol. Surv. Prof. Paper* 1395.

Soare, R. J., Conway, S. J., Dohm, J. M., and El-Maary, M. R. (2014). Possible hydraulic (open-system)

pingos in and around the Argyre impact basin, Mars. *Lunar Planet. Sci. Conf*, **45**, abs. 1121.

Souness, C. J. and Hubbard, B. (2012). Crevasse-like openings as indicators of flow in Martian glacier-like forms. *Lunar Planet. Sci. Conf*, **43**, abs. 1070.

Souness, C. J., Hubbard, B., Quincey, D. J., and Milliken, R. (2011). Geographical controls on the distribution of glacier-like forms in Mars' mid-latitudes: Observations from a survey of MRO CTX camera data. *Lunar Planet. Sci. Conf*, **42**, abs. 1021.

Souness, C. J., Hubbard, B., Milliken, R., and Quincey, D. J. (2012). An inventory and population-scale analysis of Martian glacier-like forms. *Icarus*, **217**, 243–255.

Spitzer, C. R., ed. (1980). *Viking Orbiter Views of Mars*. NASA Special Publication SP-441, Washington, DC: Government Printing Office.

Spudis, P. (1993). *The Geology of Multi-Ring Impact Basins: The Moon and Outer Planets*. Cambridge: Cambridge University Press.

Squyres, S. W. (1989). Urey Prize Lecture: Water on Mars. *Icarus*, **79**, 229–288.

Squyres, S. W. (2005). *Roving Mars: Spirit, Opportunity, and the Exploration of the Red Planet*. New York, NY: Hyperion.

Squyres, S.W., Arvidson, R. E., Bell, J. F., III, *et al.* (2004). The Spirit Rover's Athena science investigation at Gusev crater, Mars. *Science*, **305**, 794–799.

Stillman, D. E., Michaels, T. I., Grimm, R. E., and Harrison, K. P. (2014). New observations of Martian southern mid-latitude recurring slope lineae (RSL) imply formation by freshwater subsurface flows. *Icarus*, **233**, 328–341.

Stooke, P. J. (2012). *The International Atlas of Mars Exploration: Volume 1, 1953 to 2003: The first five decades*. Cambridge: Cambridge University Press.

Stooke, P. J. (2016). *The International Atlas of Mars Exploration: Volume 2, 2004 to 2014: From Spirit to Curiosity*. Cambridge: Cambridge University Press.

Sullivan, R., Thomas, P., Veverka, J., Malin, M., and Edgett, K. S. (2001). Mass movement slope streaks imaged by the Mars Orbiter Camera. *J. Geophys. Res.*, **106**, 23607–23633, doi:10.1029/ 2000JE001296.

Tanaka, K. L. (1985). Ice-lubricated gravity spreading of the Olympus Mons aureole deposits. *Icarus*, **62**, 191–206.

Tanaka, K. L. (1990). Tectonic history of the Alba Patera–Ceraunius Fossae region of Mars. *Lunar Planet. Sci. Conf*, **20**, 515–523.

Tanaka, K. L. (2000). Dust and ice deposition in the Martian geologic record. *Icarus*, **144**, 254–266, doi:10.1006/icar.1999.6297.

Tanaka, K. L. and Chapman, M. G. (1992). Kasei Valles, Mars: Interpretation of canyon materials and flood sources. *Proc. Lunar Planet. Sci. Conf*, **22**, 73–83.

Tanaka, K. L. and Davis, P. A. (1988). Tectonic history of the Syria Planum province of Mars. *J. Geophys. Res.*, **93**, 14893–14917.

Tanaka, K. L. and Fortezzo, C. M. (2012). Geologic map of the north polar region of Mars. *USGS Sci. Invest. Map SIM-3177*, scale 1:2,000,000.

Tanaka, K. L. and Kolb, E. J. (2001). Geologic history of the polar regions of Mars based on Mars Global Surveyor data: 1. Noachian and Hesperian Periods. *Icarus*, **154**, 3–21, doi:10.1006/icar.2001.6675.

Tanaka, K. L. and Leonard, G. J. (1995). Geology and landscape evolution of the Hellas region of Mars. *J. Geophys. Res*, **100**, 5407–5432.

Tanaka, K. L. and Schultz, R. A. (1993). Large, ancient, compressional structures on Mars. *Lunar Planet. Sci. Conf.*, **24**, abs. 1702.

Tanaka, K. L. and Scott, D. H. (1987). Geologic map of the polar regions of Mars. *USGS Misc. Inv. Ser. Map I-802–C*, scale 1:15,000,000.

Tanaka, K. L., Golombek, M. P., and Banerdt, W. B. (1991). Reconciliation of stress and structural histories of the Tharsis region of Mars. *J. Geophys. Res*, **96**, 15,617–15,633.

Tanaka, K. L., Scott, D. H., and Greeley, R. (1992). Global stratigraphy. In Kieffer, H. H., Jakosky, B. M., Snyder, C. W., and Matthews, M. S., eds. *Mars*. Tucson, AZ: University of Arizona Press, Space Science Series, pp. 345–382.

Tanaka, K. L., Dohm, J. M., and Watters, T. R. (1996). Possible coronae structures in the Tharsis region of Mars. *Lunar Planet. Sci. Conf*, **27**, abs. 1658.

Tanaka, K. L., Dohm, J. M., Lias, J. H., and Hare, T. M. (1998). Erosional valleys in the Thaumasia region of Mars: Hydrothermal and seismic origins. *J. Geophys. Res.*, **103**, 31,407–31,419, doi:10.1029/ 98JE01599.

Tanaka, K. L., Skinner, J. A., Jr., and Hare, T. M. (2005). Geologic map of the northern plains of Mars. *USGS Sci. Invest. Map SIM-2888*, scale 1:15,000,000.

Tanaka, K. L., Rodriguez, J. A. P., Skinner, J. A., Jr., *et al.* (2008). North polar region of Mars: Advances in stratigraphy, structure, and erosional modification. *Icarus*, **196**, 318–358.

Tanaka, K. L., Rodriguez, J. A. P., Fortezzo, C. M., Hayward, R. K., and Skinner, J. A., Jr. (2010). Nature of Hesperian resurfacing in the Scandia–north polar region of Mars. *Lunar Planet. Sci. Conf.*, **41**, abs. 2323.

Tanaka, K. L., Fortezzo, C. M., Hayward, R. K., Rodriguez, J. A. P., and Skinner, J. A., Jr. (2011).

History of plains resurfacing in the Scandia region of Mars. *Planet. Space Sci.*, **59**, 1128–1142.

Tanaka, K. L., Skinner, J. A., Jr., Dohm, J. M., *et al.* (2014). Geologic map of Mars. *USGS Sci. Inv. Map SIM-3292*, scale 1:20,000,000.

Tauxe, L. (2010). *Essentials of Paleomagnetism.* Cambridge: Cambridge University Press.

Thelin, G. P. and Pike, R. J. (1991). Landforms of the conterminous United States: A digital shaded-relief portrayal. *USGS Misc. Inv. Ser. Map I-2206*, scale 1:3,500,000.

Thomas, P., Veverka, J., Bell, J., Lunine, J., and Cruikshank, D. (1992). Satellites of Mars: Geologic history. In Kieffer, H. H., Jakosky, B. M., Snyder, C. W., and Matthews, M. S., eds., *Mars.* Tucson, AZ: University of Arizona Press, Space Science Series, pp. 1257–1282.

Thompson, A. and Taylor, B. N. (2008). *The National Institute of Standards and Technology (NIST) Guide for the Use of the International System of Units.* NIST Special Publication 811, www.nist.gov/pml/pubs/sp811/index.cfm.

Treiman, A. H. (2003). Geologic settings of Martian gullies: Implications for their origins. *J. Geophys. Res.*, **108**, 8031, doi:10.1029/2002JE001900.

van der Kolk, D. A., Tribbett, K. L., Grosfils, E. B., *et al.* (2001). Orcus Patera, Mars: Impact crater or volcanic caldera? *Lunar Planet. Sci. Conf.*, **32**, abs. 1085.

van Gasselt, S., Hauber, E., Rossi, A.-P., Dumke, A., and Neukum, G. (2010). Geomorphology of the Tempe Terra lobate debris aprons. *Lunar Planet. Sci. Conf.*, **41**, abs. 2324.

Vaucher, J., Baratoux, D., Mangold, N., *et al.* (2009). The volcanic history of central Elysium Planitia: Implications for Martian magmatism. *Icarus*, **204**, 418–442, doi:10.1016/j.icarus.2009.06.032.

Voelker, M., Platz, T., Tanaka, K. L., *et al.* (2012). Geological mapping of Havel Vallis, Xanthe Terra, Mars: Stratigraphy and reconstruction of valley formation. *Lunar Planet. Sci. Conf.*, **43**, abs. 2738.

Voelker, M., Platz, T., Tanaka, K. L., *et al.* (2013). Hyperconcentrated flow deposits and valley formation of Havel Vallis, Xanthe Terra, Mars. *Lunar Planet. Sci. Conf.*, **44**, abs. 2886.

Watters, T. R. (1993). Compressional tectonism on Mars. *J. Geophys. Res.*, **98**, 17,049–17,060.

Watters, T. R. (2003). Thrust faults along the dichotomy boundary in the eastern hemisphere of Mars. *J. Geophys. Res.*, **108**, 5054, doi:10.1029/2002JE001934.

Watters, T. R. and Janes, D. M. (1995). Coronae on Venus and Mars: Implications for similar structures on Earth. *Geology*, **23**, 200–204.

Watts, A. B. (2001). *Isostasy and Flexure of the Lithosphere.* Cambridge: Cambridge University Press.

Wendt, L., Gasselt, S. V., and Neukum, G. (2008). Possible salt tectonics in Ariadnes Colles? *Third Eur. Planet. Sci. Congress*, abs. EPSC2008-A-00345.

Wendt, L., Bishop, J. L., and Neukum, G. (2012). Knob fields in the Terra Cimmeria/Terra Sirenum region of Mars: Stratigraphy, mineralogy, and morphology. *Lunar Planet. Sci. Conf.*, **43**, abs. 2024.

Wichman, R. and Schultz, P. (1989), Sequence and mechanisms of deformation around the Hellas and Isidis impact basins on Mars. *J. Geophys. Res.*, **94**, 17,333–17,357, doi:10.1029/JB094iB12p17333.

Williams, R. M. E. and Malin, M. C. (2004), Evidence for late stage fluvial activity in Kasei Valles, Mars. *J. Geophys. Res.*, **109**, E06001, doi:10.1029/2003JE002178.

Wilson, S. A., Howard, A. D., Moore, J. M., and Grant, J. A. (2007). Geomorphic and stratigraphic analysis of Crater Terby and layered deposits north of Hellas basin, Mars. *J. Geophys. Res.*, **112**, E08009. doi:10.1029/2006JE002830.

Witbeck, N. E., Tanaka, K. L., and Scott, D. H. (1991). Geologic map of the Valles Marineris region of Mars. *USGS Misc. Inv. Ser. Map I-2010*, scale 1:2,000,000.

Wood, L. J. (2006). Quantitative geomorphology of the Mars Eberswalde delta. *Geol. Soc. Amer. Bull.*, **118**, 557–566, doi:10.1130/B25822.1.

Wray, J. and Squyers, S (2010). Layers exposed in crater near Mawrth Vallis (image caption), http://hirise.lpl.arizona.edu/PSP_004052_2045.

Wray, J. J., Ehlmann, B. L., Squyres, S. W., Mustard, J. F., and Kirk, R. L. (2008). Compositional stratigraphy of clay-bearing layered deposits at Mawrth Vallis, Mars. *Geophys. Res. Letters*, **35**, doi:10.1029/2008GL034385.

Wray, J. J., Murchie, S. L., Ehlmann, B. L., *et al.* (2011) Evidence for regional deeply buried carbonate-bearing rocks on Mars. *Lunar Planet. Sci. Conf.* **42**, abs. 2635.

Yin, A. (2012). An episodic slab–rollback model for the origin of the Tharsis rise on Mars: Implications for initiation of local plate subduction and final unification of a kinematically linked global plate-tectonic network on Earth. *Lithosphere*, **4**, 553–593.

Zhong, S. and Zuber, M. T. (2001). Degree-1 mantle convection and the crustal dichotomy on Mars. *Earth Planet. Sci. Lett.*, **189**, 75–84.

Zimbelman, J. R., Craddock, R. A., and Greeley, R. (1994). Geologic map of the MTM-15147 quadrangle, Mangala Valles region of Mars. *USGS Misc. Inv. Ser. Map I-2402*, scale 1:500,000.

Zubrin, R. (2011). *The Case for Mars: The Plan to Settle the Red Planet and Why We Must* (revised edition). New York, NY: Free Press.

Index

Note: page numbers in italic indicates figures or tables